GROUP TECHNOLOGY AND CELLULAR MANUFACTURING

A State-of-the-Art Synthesis of Research and Practice

GROUP TECHNOLOGY AND CELLULAR MANUFACTURING

A State-of-the-Art Synthesis of Research and Practice

Editors

Nallan C. Suresh
School of Management,
State University of New York
Buffalo, NY, USA

and

John M. Kay
School of Industrial and Manufacturing Sciences
Cranfield University,
Cranfield, Bedfordshire, UK

Kluwer Academic Publishers
Boston / Dordrecht / London

Distributors for North America:
Kluwer Academic Publishers
101 Philip Drive
Assinippi Park
Norwell, Massachusetts 02061 USA

Distributors for all other countries:
Kluwer Academic Publishers Group
Distribution Centre
Post Office Box 322
3300 AH Dordrecht, THE NETHERLANDS

Library of Congress Cataloging-in-Publication Data

A C.I.P. Catalogue record for this book is available from
the Library of Congress.

Copyright © 1998 by Kluwer Academic Publishers

All rights reserved. No part of this publication may be reproduced, stored in a retrieval system or transmitted in any form or by any means, mechanical, photo-copying, recording, or otherwise, without the prior written permission of the publisher, Kluwer Academic Publishers, 101 Philip Drive, Assinippi Park, Norwell, Massachusetts 02061.

Printed on acid-free paper.

Printed in the United States of America

CONTENTS

Preface ix

Dedication xiii

Editorial Advisory Board xv

A. INTRODUCTION

A. Group Technology and Cellular Manufacturing: Updated Perspectives: *Nallan C. Suresh and John M. Kay* 1

B. PRODUCT DESIGN

B1. Group Technology, Concurrent Engineering and Design for Manufacture and Assembly: *Winston A. Knight* 15

B2. Classification & Coding Systems: Industry Implementation and Non-Traditional Applications: *Charles T. Mosier* 37

B3. Part Family Identification: The Role of Engineering Data Bases: *Richard E. Billo and Bopaya Bidanda* 58

B4. Part Family Formation and Classification Using Machine Learning Techniques: *Young B. Moon* 77

C. PROCESS PLANNING

C1. GT and CAPP: Towards An Integration of Variant and Generative Approaches: *Dusan N. Sormaz* 93

C2. GT and CAPP: Observations on Past, Present and Future: *James Nolen and Carey Lyman-Cordes* 112

D. DESIGN OF CELLULAR LAYOUTS: PART-MACHINE-LABOR GROUPING

D1. Part-Machine-Labor Grouping: The Problem and Solution Methods: *Scott M. Shafer* 131

D2. Design of Manufacturing Cells: PFA Applications in Dutch Industry: *Jannes Slomp* 153

D3. Artificial Neural Networks and Fuzzy Models: New Tools for Part-Machine Grouping: *V. Venugopal* 169

D4. Cell Formation Using Genetic Algorithms: *Jeffrey A. Joines, Russell E. King, C. Thomas Culbreth* 185

D5. A Bi-chromosome GA for Minimizing Intercell and Intracell Moves: *Chun H. Cheng, Wai. H.Lee and John Miltenburg* 205

E. DESIGN OF CELLULAR LAYOUTS: USE OF ANALYTICAL AND SIMULATION MODELS

E1. Design / Analysis of Manufacturing Systems: A Business Process Approach: *N. Viswanadham, Y. Narahari and N.R. S. Raghavan* 221
E2. Cellular Manufacturing Feasibility at Ingersoll Cutting Tool Company: *Danny J. Johnson and Urban Wemmerlöv* 239
E3. A Decision Support System For Designing Assembly Cells in Apparel Industry: *Mohamad Kalta, Tim Lowe and David Tyler* 255
E4. Evaluation of Functional and Cellular Layouts Through Simulation and Analytical Models: *Nallan C. Suresh* 273

F. OPERATIONS PLANNING AND CONTROL

F1. Production Planning and Control Systems for Cellular Manufacturing: *Jan Riezebos, Girish Shambu and Nallan Suresh* 289
F2. Period Batch Control Systems for Cellular Manufacturing: *Daniel C. Steele* 309
F3. Scheduling Rules for Cellular Manufacturing: *Farzad Mahmoodi and Charles T. Mosier* 321
F4. Operation and Control of Cellular Systems at Avon Lomalinda, Puerto Rico: *Gürsel A. Süer* 339
F5. Classification and Analysis of Operating Policies for Manufacturing Cells: *Ronald G. Askin and Anand Iyer* 362

G. CELLS AND FLEXIBLE AUTOMATION

G. Cells & Flexible Automation: A History and Synergistic Application: *Kathryn E. Stecke and Rodney P. Parker* 381

H. HUMAN FACTORS IN CELLULAR MANUFACTURING

H1. Human Resource Management and Cellular Manufacturing: *Richard Badham, Ian P. McLoughlin and David Buchanan* 401
H2. Teams and Cellular Manufacturing: The Role of the Team Leader: *Peter D. Carr and Gwyn Groves* 423

I. INTEGRATED IMPLEMENTATION EXPERIENCES

I1. Group Technology-Based Improvements in Danish Industry: *Jens O. Riis and Hans Mikkelson* 441
I2. Benefits from PFA in Two Make-To-Order Manufacturing Firms in Finland: *Sauli Karvonen, Jan Holmström & Eero Eloranta* 458

I3. Cellular Manufacturing at Zanussi-Electrolux Plant, Susegana, Italy: *Roberto Panizzolo* 475
I4. Microsoft Ireland: Realigning Plant, Sales, Key Suppliers By Customer Family: *Richard J. Schonberger* 491
I5. Group Technology at John Deere: Production Planning and Control Issues: *Michael S. Spencer* 501
I6. Design / Reengineeing of Production Systems: Yugoslavian (IISE) Approaches: *D. Zelenović, I. Cosić and R. Maksimović* 517

Appendix. Bibliography: The Works of John L. Burbidge 537

Index 545

PREFACE

As early as 1924, Flanders clearly articulated many of the problems encountered in batch production, and proposed a solution that would now be referred to as cellular manufacturing (Nimmons 1996). But it was essentially the 1960s before these concepts were formalized by Mitrofanov's Scientific Principles of Group Technology (1966) and numerous publications by John (Jack) Leonard Burbidge.

It is now three decades since these early pioneers of group technology (GT) first described their ideas. Group technology ideas have come a long way since then. Despite a relative diminution of interest during 1970s, there has been a revival, particularly in USA, with the advent of Japanese, just-in-time (JIT) manufacturing philosophies, flexible manufacturing technologies, computer-integrated manufacturing and several production planning and control philosophies. Group technology has come to be regarded as a prerequisite, and an integrative philosophy for computer integrated manufacturing (CIM) and as a core element of several manufacturing paradigms.

GT/CM has been a widely researched area during the last 15 years, judging by the number of articles published in various research journals. Much progress has been made in all branches of this broad-based philosophy, thanks to the efforts of numerous researchers and practitioners. There has been a proliferation of techniques for part-machine grouping, engineering data bases, expert system-based design methods and computer aided process planning techniques, new pattern recognition methods for identifying part families, new analytical and simulation tools for design and evaluation of cells, new types of cells incorporating robotics and flexible automation, team-based approaches for organizing the workforce, etc.

Compiling specially-commissioned, state-of-the-art articles from leading researchers and implementers, this book essentially attempts to address the needs of, and bring together, three groups of individuals: academic researchers, industry practitioners and students.

This book is also motivated by a desire to pay tribute to Professor Jack Burbidge, who has made seminal contributions to this knowledge domain. A bibliography of his works may be found in Appendix A of this book. Professor Burbidge unarguably belongs to the earliest group of production management experts who championed the adoption of part-family-based manufacturing systems.

From the standpoint of academics, several observations may be made. First, it may be unequivocally stated that GT/CM is an area of topical and lively interest among researchers in production/operations management, industrial engineering and manufacturing systems engineering. This is evident from the number of papers devoted to GT/CM topics in journals such as *International Journal of Production Research (IJPR), Journal of Operations Management (JOM), European Journal of Operational Research (EJOR), Decision Sciences and IIE Transactions*. Several of these journals have brought out special issues devoted to this topic in recent years.

From a researcher's standpoint, the desirability of a book that provides an up-to-date perspective, incorporating the advances made in various sub-topics of GT/CM during the last 15 years, is very clear. A clear need also exists for synthesizing the latest industry practices, and industry inputs to guide research towards greater real-world relevance. It would clearly be beneficial for researchers to gain a sense of practical applications, so as not to lose sight of industry reality. This would, of course, be in line with the sentiments of Jack Burbidge, who always insisted that one must try out new ideas in real factories, and with real data.

Research results in the GT/CM area have tended to be scattered in a wide range of journals: Industrial engineers seem to prefer journals like *IIE Transactions*; production/operations management researchers seem to prefer *IJPR, JOM, Management Science, Decision Sciences*; and production engineers publish their findings in *CIRP Annals, ICPR Proceedings* and other avenues. It would be useful to bring the essence of recent research findings into one venue; it might encourage researchers belonging to different disciplines to access additional sources that they might otherwise ignore.

There have also been some national divides in this field: researchers from USA have tended to ignore research that has emanated from Europe (in a different terminology), for instance, and have occasionally "reinvented the wheel" (which runs counter to GT philosophy itself). Thus, it would be beneficial, as this book strives to do, to bring researchers and practitioners, from several countries, together.

From the standpoint of industry practitioners, it may again be stated that GT/CM concepts are of topical interest and relevance, in the current context of efforts to reengineer manufacturing and service processes through many different approaches. Numerous firms in industry (manufacturing & service industries) are currently in the process of streamlining their operations via just-in-time (JIT) production systems, total quality management (TQM), computer integrated manufacturing (CIM), business process reengineering (BPR), concurrent engineering, total systems engineering, etc. As made clear in this book, the ideas behind cellular manufacturing are strongly related to, or form core elements of, these new developments which are aimed at reducing developmental and manufacturing lead times, costs, and improving quality and delivery performance.

From a practitioner's point of view, it would be useful to provide descriptions of state-of-the-art applications in leading-edge firms, that would be worthy of emulation (or benchmarking, in contemporary parlance). This book also attempts to present a summary of research results in a user-friendly manner, so that practitioners may take advantage of research results more fully.

The utility of such a text book for students of engineering and management should also be evident.

Thus, the basic objective of the book has been to compile a set of specially-commissioned, state-of-the-art papers by leading researchers and implementers, drawn from many countries, and in various topic areas of GT/CM. Essentially three types of articles have been compiled: theory (research) articles, application articles, and combined theory & application articles.

The theory articles, authored by leading researchers, have sought to provide an in-depth summary of the state-of-the-art in research, indicating high pay-off areas for future research, to be of benefit to research community world-wide. They also attempt to provide taxonomies, updated bibliographies, and a summary of prior results, to benefit researchers, engineering and business students, and industry practitioners. There has also been an attempt to focus on research of real-world relevance, in keeping with the sentiments of Professor Burbidge.

The application articles, authored by leading implementers from several nations, have attempted to provide a state-of-the-art summary of applications, in a wide variety of industry and national contexts. They were requested to provide insights on special problems encountered in practice, to comment on the applicability of various research techniques, and to include anecdotal information that may supplement the formal knowledge base in the field. Application articles have also, where possible, attempted to document implementation sequence, economic justification, costs and benefits, and the system's impact on company performance.

The editors have many individuals to thank for the successful completion of this project: members of the Editorial Advisory Board; all the Authors for submitting state-of-the-art, high-quality manuscripts; Mr. Gary Folven and Ms. Carolyn Ford at Kluwer; Deans Howard Foster, Rick Winter and John Thomas, and Professors John Boot, Carl Pegels, Stanley Zionts and Winston Lin of State University of New York, for their support and encouragement; and, Ms. Valerie Limpert and Ms. Trish Wisner at SUNY, Buffalo for their able support.

The editors express special thanks to Cranfield secretaries Lynne Allgood and Linda Nicholls who not only helped with this book, but for many years helped Jack produce his numerous publications; and, Samantha Johnson who maintains the collection of all Jack's publications safe and sound in the Cranfield University Library.

The editors also express their appreciation to Mr. Sung June Park, doctoral candidate at SUNY, Buffalo and Dr. Girish Shambu of Canisius College, Buffalo, for valuable assistance with editorial work. And last, but not the least, the editors thank their wives, Radhika and Gwyn, and family and friends for their support and patience during the development of this book.

Nallan C. Suresh, SUNY, Buffalo, NY, USA
John M. Kay, Cranfield University, Bedfordshire, UK.

DEDICATION

John Leonard Burbidge (Jack to all who knew him) was born on the 15th January 1915 in Canada. He spent his early years in New York State, USA but was sent to school in England. He started to study Mechanical Engineering at Cambridge but his family members were hit hard by the depression in the USA where they still lived. Undeterred, Jack sought employment and in 1934, after a short period as a door-to-door salesman, he became apprenticed to the Bristol Aeroplane Company. We owe that company a great thank-you for introducing him to the complexities of production. He was to spend the next 60 years telling all of us how we could do it better. As part of the war effort, he moved to the Ministry of Supply and was involved with the wartime manufacture of Spitfires, perhaps the world's first Just-In-Time project.

After the war he returned to industry, serving variously as Chief Planner, Sales Manager, Works Director and Managing Director in such companies as the Bristol Aeroplane Company, R A Lister, David Brown. Such wide practical experience laid the foundation for his next career in 1962 when he became a Production Expert with the International Labour Organisation, serving for five years on assignments in Poland, Cyprus and Egypt before becoming Professor, in 1967, at the International Training Centre run by the ILO in Turin. He remained in Turin until his retirement from the ILO in 1976.

His so called retirement was short and he quickly became a Visiting Professor at Cranfield. There he continued to teach new generations of production (he refused to use the word manufacturing!) engineers ensuring that his fundamental principles of material flow will persist. He was known to the students as "The Professor" despite Cranfield having many personnel with that title. His questioning of students was legendary, especially any who dared to suggest that MRP might be useful. He gained the respect of all his students and fellow academics. Even when in his last year at Cranfield, aged 79, he taught and worked with his students with great dedication and energy and delivered lectures in Israel, Ireland, Poland and Yugoslavia.

It was however Jack's drive to publish his ideas that led to him becoming known as the Father of Group Technology. He wrote 15 books and published over 150 papers. This is an amazing volume of work especially when you realise that his first publication came out when he was 43 years old! He attended conferences all over the world and was inevitably surrounded by admirers wishing to discuss his ideas. His papers were invariably accepted (even if twice the maximum length!).

When he delivered a paper, it was with the passion of a preacher. He was always trying to convert the unbelievers to his Group Technology religion. His faith was very strong - he knew that GT worked.

His philosophy was to simplify the material flow - which is easier to say than to achieve - but he developed his Production Flow Analysis over many years of trial and error working with industry. Watching him work with an enormous part-machine matrix on a large sheet of paper was an education. (His IT decisions were when to sharpen his pencil.) He could spot groupings in an instant and his depth of knowledge of production processes made dealing with exceptions much easier. Although the current computer based grouping algorithms are much faster, they can easily by-pass the "practical" solution. He always insisted on working with the engineers from a company and not just producing a solution in isolation. He claimed that PFA would find the grouping of parts and machines unlike coding and classification methods that only group the parts. Once a group is correctly formed (and Jack would never allow any cross-flow!), it is like a mini-factory. It has all the necessary facilities and skills to complete a range of parts. These groups are the building blocks that allowed Just-In-Time principles to be developed.

When he first started publishing his message fell on deaf ears in the UK (with the exception of Ferranti, Edinburgh) but was listened to with great interest by Japanese manufacturers. They realised the potential and used Jack's published work to adopt the ideas and develop them into the concepts we now know as JIT. Their success has led to a revival of interest in Europe and North America. The 'groups' were now known as 'cells' but only in one of his last publications did Jack accept the use of this word. "Nobody wanted to work in cells - they were for criminals". He did surprisingly allow cells to be defined as synonymous with groups in his IFIP book on "Glossary of terms used in production control". Whether we call them groups or cells, they are the foundation of modern manufacturing and Jack's ideas, writings, lectures and presentations have been a major factor in their development.

He was honoured by the Universities of Novi Sad, Yugoslavia and Strathclyde, UK, was elected as an Honorary Fellow of the Institute of Electrical Engineers and was awarded the Order of the British Empire by Queen Elizabeth II at Buckingham Palace in 1991 for Services to Industry.

He gives credit in his publications to Mitrofanov, Gigli, Patrigani, Sidders and Edwards for his inspiration but his mission in life to educate and inspire the manufacturing world with his ideals have left a legacy that we will remember and cherish for generations to come.

He was an excellent colleague at Cranfield and he is missed by his friends all over the world. We are honoured to dedicate this book to Jack Burbidge, the "Professor" and "Father of Group Technology".

John M. Kay, Cranfield University, Bedfordshire, UK
Nallan C. Suresh, SUNY, Buffalo, USA

Editorial Advisory Board

Professor Leo Alting
Technical University of Denmark
Lyngby, Denmark

Professor Ronald G. Askin
University of Arizona,
Tucson, AZ, USA

Professor J. T. Black
Auburn University
Montgomery, AL, USA

Professor Jim Browne
University College Galway
Galway, Ireland

Professor Allan S. Carrie
University of Strathclyde,
Glasgow, Scotland, UK

Dr. Chun Hung Cheng
Chinese University of Hong Kong
Shatin N.T., Hong Kong

Professor Ezey Dar-El
Technion, Israel Inst. of Technology
Haifa, Israel

Professor Eero Eloranta
Helsinki University of Technology
Espoo, Finland

Professor F. Robert Jacobs
Indiana University
Bloomington, IN , USA

Professor Winston A. Knight
University of Rhode Island
Kingston, RI, USA

Professor J. Miltenburg
McMaster University
Hamilton, Ontario, Canada

Professor Yasuhiro Monden
University of Tsukuba
Tsukuba, Ibaraki, Japan

Mr. James Nolen
James Nolen & Company
Coventry, RI, USA

Professor Jens O. Riis
Aalborg University
Aalborg, Denmark

Dr. Richard J. Schonberger
Schonberger Associates, Inc.
Seattle, WA, USA

Professor Kathryn E. Stecke
The University of Michigan
Ann Arbor, MI, USA

Professor Mario T. Tabucanon
Asian Institute of Technology
Pathumthani, Thailand

Professor Urban Wemmerlöv
University of Wisconsin
Madison, WI, USA

Professor Jacob Wijngaard
University of Groningen
Groningen, Netherlands

Professor Dragutin M. Zelenović
University of Novi Sad
Novi Sad, Yugoslavia.

GROUP TECHNOLOGY AND CELLULAR MANUFACTURING

A State-of-the-Art Synthesis of Research and Practice

A

Group Technology & Cellular Manufacturing: Updated Perspectives

N. C. Suresh and J. M. Kay

1. INTRODUCTION

Since the human race first decided to specialize, make goods, and trade goods for other goods, there has been competition in terms of not only the technology used but also the organizational aspects of production. Although group technology (GT) sounds like it should be the former, it is essentially a system of production organization that allows firms to compete by minimizing work-in-progress, minimizing lead times, and minimizing costs while still producing a wide range of products. By combining different machines, equipment and personnel into a group or cell, total responsibility for making a set of parts can be delegated to the group. The traditional method has been to arrange a factory such that all similar machines and corresponding skilled personnel are located together. This has resulted in complex job routings inside the facility.

Nimmons (1996) stated that, *Flanders (1924) recognized that work organization based on groups of similar machines was disadvantaged by "the constant movement of work from department to department with its consequent slowing up of the work flow, division of responsibility and difficulty of control." The alternative he suggested, was to arrange facilities by product such that any individual piece stays in a single department until it is completely finished. He explained that, "All long waits ... are eliminated, and with them the expensive items of storage space and idle capital for inactive stock. The ideal aimed at has been that of a small, fast flowing stream of work instead of a large, sluggish one." Flanders also described simplifications to production control, inventory control, and cost accounting procedures that are made possible by changing to product organization.*

This seems to be the first published mention of what we now refer to as group technology (GT) and/or cellular manufacturing (CM). These concepts have continued to evolve and grow over the years. Early pioneers of these concepts included Mitrofanov (1959), Burbidge (1963), Petrov (1966), Ivanov (1968), Durie (1970), Edwards (1971), Grayson (1971), and Gallagher and Knight (1973) and others. After a relative diminution of interest in GT during 1970s, there has been a revival, with the advent of Japanese manufacturing philosophies, flexible manufacturing technologies, computer integrated manufacturing (CIM) and new production planning and control philosophies. In the context of CIM, group technology began to be regarded as a prerequisite, simplification methodology, as well as an integrative philosophy. The cellular manufacturing aspect of GT was also recognized as a core element of just-in-time (JIT) production systems.

GT/CM has been a widely researched area during the last two decades and numerous implementation case studies have been reported. Much progress has been made in all branches of this broad-based philosophy, thanks to the efforts of numerous researchers and practitioners. There has been a proliferation of techniques for part-machine grouping, expert systems for design and computer aided process planning, new pattern recognition methods for identifying part families, analytical and simulation modeling for cell design, new types of cells incorporating robotics and flexible automation, team-based approaches for organizing work force and other developments.

This introductory chapter is organized as follows. §2 summarizes the traditional framework of GT/CM and discusses various elements within this framework. §3 provides an updated perspective, summarizing recent developments in various areas and their relationships to GT/CM. This enhanced perspective also serves to provide a structure and organization to ensuing chapters of this book.

2. GT/CM: THE TRADITIONAL FRAMEWORK

The traditional body of knowledge pertaining to GT/CM may be grouped under the following categories (Suresh and Meredith 1985):

- Part family identification
- GT in engineering design
- GT in process planning
- GT in shop floor: Cellular manufacturing
- Production planning and control systems for CM
- Impact on other functions: quality control, costing, personnel, etc.

Figure 1 is a schematic diagram showing the linkages among the above areas.

2.1. Part Family Identification

Identification of part families, and the *systematic exploitation of similarities* in all phases of the operations cycle constitutes a basic objective of GT/CM. This may be of significant value in intermittent production situations where economies of scale are limited. Identification of part families, a prerequisite for GT/CM, enables savings in design, process planning, tooling, in production processes, and also in downstream areas of warehousing and shipping.

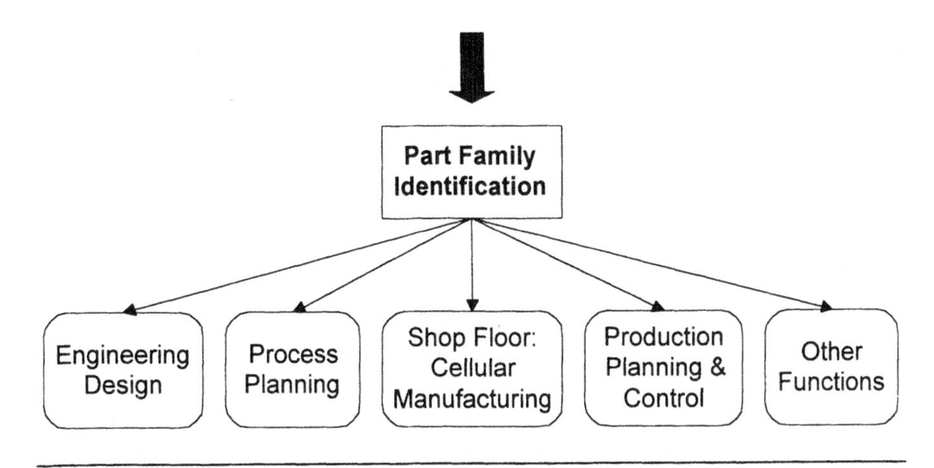

FIGURE 1. GT/CM: The traditional framework (Suresh and Meredith 1985).

To identify part families, broadly three approaches have been identified over the years: 1) use of a classification and coding system; 2) use of approaches that directly analyze process plans to identify part and machine families; and, 3) visual and intuitive approaches.

The implementation of classification and coding systems was recommended as a prerequisite by early GT advocates. GT was often equated to classification and coding systems. Numerous classification and coding systems have been developed over the years, e.g., Brisch (UK), Opitz (Germany), MICLASS (Netherlands), and CODE (USA). These coding systems were often classified into *monocodes* (hierarchical), or *polycodes* (faceted), or a combination of the two. Coding systems were also classified as those that emphasize design aspects, or systems based on process features. But codes such as MICLASS/MULTICLASS have attempted to incorporate design, manufacturing and other aspects, and are quite long, running to more than 30 digits (Houtzeel 1975).

Classification and coding systems involve considerable time and effort, and by and large, have not been very popular. The investment has been partially recovered at times in many companies when adding a CAD system, which requires the data to be keyed into the database in any case. Frequent errors during codification and resulting recodification have also been major problems, and semi-automated approaches have been widely used.

A second major method of identifying part families is based on direct analysis of process plans and identifying part and machine families, without recourse to classification and coding. This has been advocated by many GT pioneers including Burbidge. Production flow analysis (PFA) method developed by Burbidge (1963), for instance, was intended to identify part and machine families, without the use of a coding system. The payoffs from GT/CM are clearly faster with this approach, though they are based on the assumption that existing process plans are consistent, sufficiently optimal and appropriate for CM. In recent years, there has been a large number of methods developed to analyze process plans and develop block diagonal matrices from the part-machine incidence matrix.

A large number of firms have also adopted the third, intuitive approach, often visually identifying part families and creating machine cells for manufacturing these part families.

2.2. GT In Engineering Design

The creation of part data base, accessed through GT codes, results *in little re-invention of the wheel, and eventually, design rationalization and variety reduction.* When a new part has to be designed, the data base is accessed to check whether such a part or a similar part has been designed before. This represents in effect the institutional memory and serves to counter the adverse effects of turnover among technical personnel.

This reliance on a part data base accessed through GT codes, also results in *reduced engineering effort, reduced engineering lead time, increased productivity, more reliable and maintainable designs*, besides the benefits of standardization and variety reduction. The proliferation of part drawings and process plans is countered with GT, as shown in Figure 2.

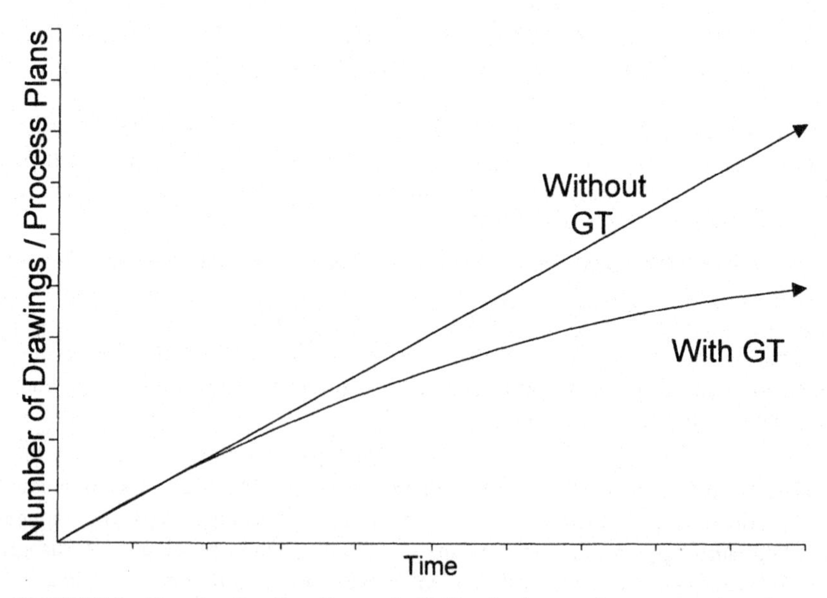

FIGURE 2. Standardization through GT in design and process planning.

2.3. GT In Process Planning

The impact of GT in the process planning area is essentially similar to that in engineering design. Group technology principles have been utilized in computer aided process planning (CAPP) under both *generative method* and *variant method.*

In the generative method, the process planning logic is stored in the system as a knowledge base, together with the various algorithms that define various technical decisions. This is similar to a decision support or artificial intelligence system. Taking into account the production processes available within the firm, every new drawing is analyzed from its fundamentals to arrive at a process plan. Obviously, this is a difficult system to realize, compared to the more popular, variant method. The variant approach involves, using the GT code, retrieving the process plan of a similar part designed earlier, and making minimal modifications to it.

GT-based CAPP systems lead to many benefits, including lower process planning lead times, greater accuracy and consistency, routing standardization, and considerable savings in tooling investments (Nolen 1989). The same CAPP principles are used in tool design and tooling standardization.

It was anticipated that GT may serve as an effective integrator of design engineering and process planning, through the use of a common data base accessed with GT codes, as shown in Figure 3. However, as we see below, this integration of design and process planning has occurred in a different manner in recent years, through the *concurrent engineering* approach.

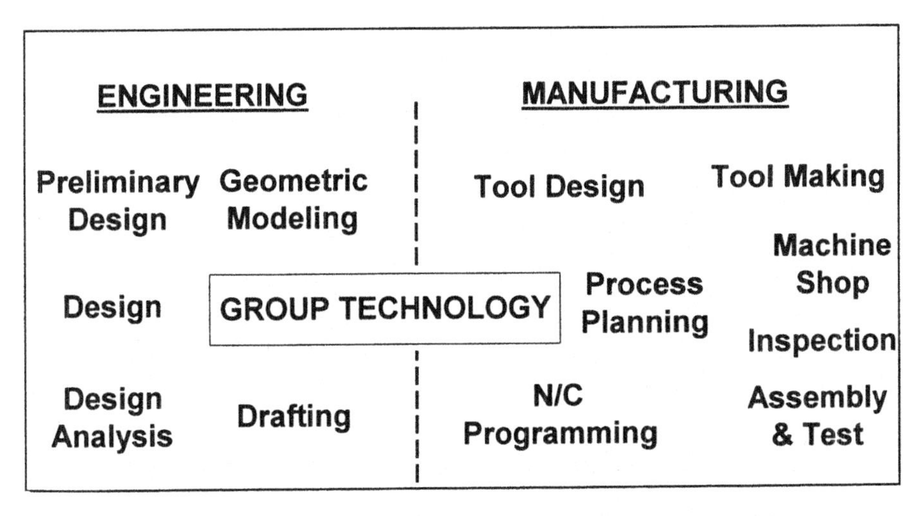

FIGURE 3. GT as an integrative element (Taraman 1980).

2.4. GT In Shop Floor: Cellular Manufacturing

GT is implemented in the shop floor primarily via *group layouts*. In traditional job shops, the layout of the machines is on the basis of process (functional) specialization, locating similar machines, and labor with similar skills, together. But the group layout is based on part-family specialization.

The well-known benefits from cellular manufacturing include:

- Reduction in manufacturing lead time and work-in-process (WIP) inventory; also greater predictability of lead times, unlike in job shops; less floor space requirements due to less WIP
- Reduction in setup times due to part-family similarities, which reduces lead times further, enables small-lot production, releases capacity, etc.
- Simplified work flow, leading to reduction in material handling, since all or most operations are confined to the cell
- Better accountability and control, since there is no operation-wise division of responsibility as in the functional layout
- Possibilities of overlapped operations due to proximity of different functions; this reduces lead times further, and results in earlier detection of defects that normally surface later, in a subsequent work center, in the functional layout.

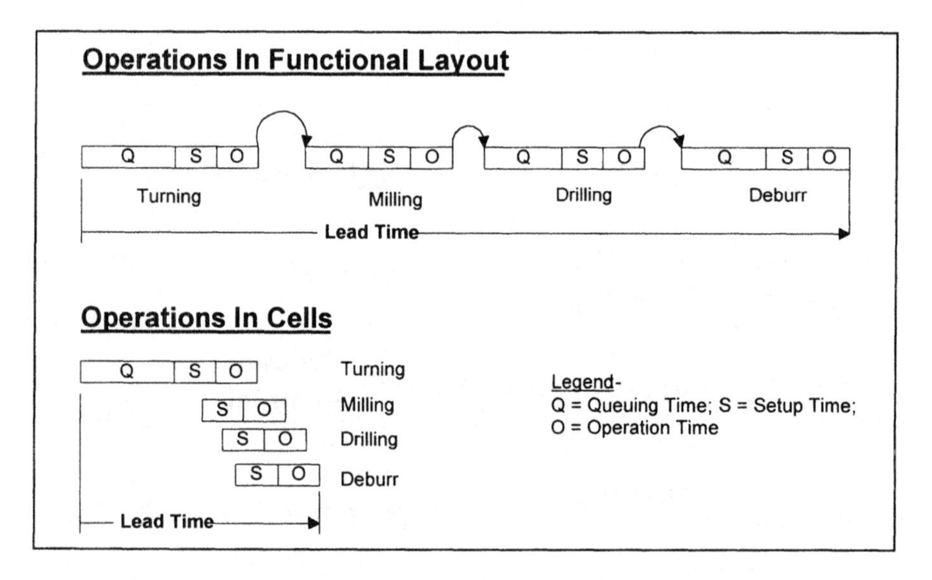

FIGURE 4. Operations in functional and cellular layouts.

The contrasting nature of operations in functional and group layouts is seen in Figure 4. Three types of layout variations have occurred in practice: 1) group layouts, 2) GT flow lines, or 3) machining centers using composite parts. The difference between a GT flow line and a group layout is that in the flow line, the routing of all parts within the cell are similar, whereas in the group layout the routings may involve different paths in the cell.

2.5. Production Planning And Control Systems For CM

Given the short cycle operations found in group layouts, one of the first effects noted was the inapplicability of classical methods such as reorder points ("stock control"). This led to the development of *flow control* and period batch control (PBC) methods (Burbidge 1995). But materials requirements planning (MRP) systems have been more widely adopted, and increasingly combined with JIT methods in CM. Lately there has been a resurgence of interest in PBC methods. GT pioneers, notably Burbidge, have also been critical of the use of economic order quantity (EOQ) model, and independent component lot-sizing methods in general.

2.6. Impact On Other Functions

GT/CM have also had an impact on quality control, total quality management (TQM) movement, JIT logistics, cost accounting systems, and human resource management. In the case of human resources, CM has clearly pointed the way towards breadth of skill, cross-functional skill requirements, etc., which constitute a radical departure from the era of functional specialization and expertise.

3. UPDATED PERSPECTIVES

In this section we consider some new developments that have taken place in the context of GT/CM, as well as developments that are conceptually similar and/or related to various elements of GT/CM. Foremost among these developments, perhaps, is the adoption of a *business process perspective* in recent years.

3.1. Business Processes: Undermining Of Functional Organizations

A major initiative is currently under way in industry to fundamentally reorganize firms based on the *business process reengineering* philosophy (Hammer 1990; Hammer and Champy 1993; Davenport 1993).

Traditionally, organizations have been viewed as a sequential arrangement of functions such as design, process planning, R&D, marketing, finance, etc. But contemporary view of organizations is in terms of a collection of value-delivering *business processes*.

Business process reengineering seeks to reorganize a commercial undertaking in a competitive, and radical way, based on a clean-slate approach. The purpose of a business process is to offer each (internal or external) customer the right product/service with a high degree of performance, measured against cost, longevity, service and quality. Some examples of business processes include order-processing systems, new product development, supply chain process, etc. New product development, as a business process is shown in Figure 5.

Business processes cut through traditional functional organizations, requiring cross-functional assemblage of human and other resources to meet certain customer needs. A business process is a specific ordering of work activities across time and place with a beginning and an end, and clearly identified inputs and outputs: a structure of action (Davenport 1993). A business process perspective is a horizontal view, with inputs at the beginning and customers at the end. It de-emphasizes functional views of the business, and focuses instead on how the organization delivers value to the customer. .

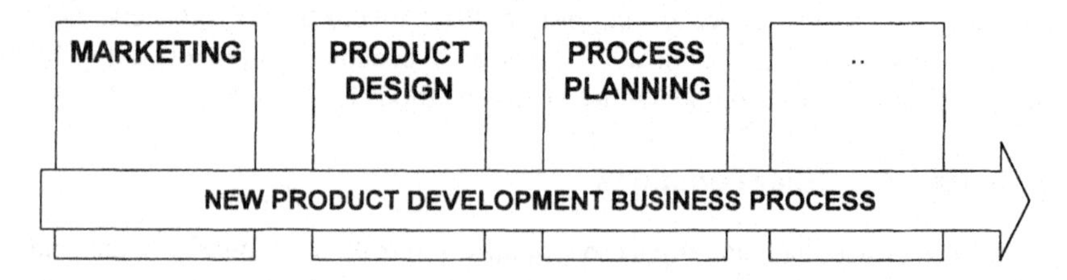

FIGURE 5. New product development as a business process.

Functional divisions have led to several familiar problems. Each function tends to act as a "silo" and hands its output "over the wall" to the next function. They are characterized by departmental barriers and parochialism, and generally slow progress of work through the system. Much of the job lead time consists of non-value-adding times devoted to move, wait, information collection, etc.

Group technology and cellular manufacturing have traditionally emphasized replacement of the functional organization within the shop floor. But, it is now becoming clear that, a concerted attack on the functional organization, throughout the firm, is under way. This has taken many forms: streamlining order entry/processing systems, supply chain management process, new product development activity, etc.

3.2. Developments In Engineering Design

Traditionally, design activities have been carried out in a sequential fashion, in a functional mode. In recent years, this has given way to the *concurrent engineering approach.* Concurrent or simultaneous engineering involves a cross-functional, group activity in which many activities normally downstream from preliminary design phase are given greater consideration early in the design cycle (Boothroyd, Dewhurst and Knight 1994) (Figure 6). This has had a major impact industry in recent years, serving to reduce development lead times, improving quality through early detection of defects, etc.

One readily notes a parallel between concurrent engineering and cellular manufacturing (compare Figure 4 and Figure 6). In traditional job shops, the operations are carried out in a sequential manner: drilling operations in the drilling work center, milling operations in the milling work center, etc. In cellular manufacturing, this *department-after-department processing* is replaced by concurrent operations. Each cell, dedicated to one or more part families, comprises a cross-functional assemblage of machines and labor. Lead times reduce, quality/machining problems that would normally surface much later, in a subsequent department, are made evident and addressable much earlier, etc.

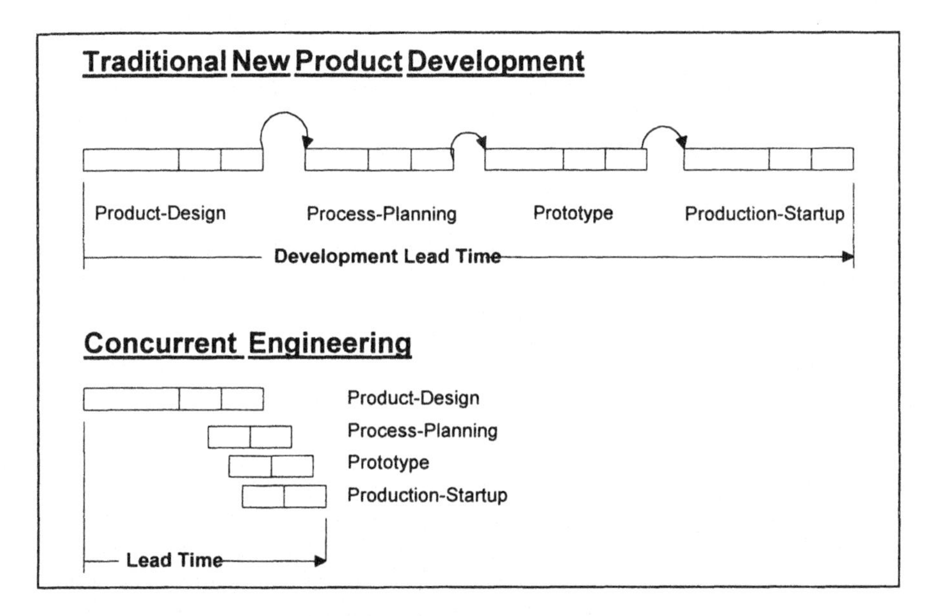

FIGURE 6. Concurrent engineering.

Concurrent engineering and design for manufacturability and assembly (DFMA) have been the most-widely adopted solution in recent years for streamlining the business processes relating to development of new products. As described by Knight in chapter B1, concurrent engineering, besides reducing development lead time, has also led to more comprehensive approaches to design, serving to address manufacture, assembly, service and environmental considerations, and much earlier in the design cycle.

Other major developments in the context of design engineering and process planning have been:

1. The use of classification and coding systems for non-traditional applications, such as robotic end-of-arm-tooling; these developments are described by Mosier in chapter B2.

2. In recent years, there has been a considerable progress in database management systems, notably relational data bases. These, and other developments such as object-oriented modeling, have altered the role and need for classification and coding, and, in some ways, obviated the need for tedious classification and coding. These developments are described by Billo and Bidanda in chapter B3.
3. Important developments in pattern recognition technologies, primarily by way of machine learning and artificial neural networks, are enabling identification of part families through extraction of feature data from CAD data bases. These methods are described by Moon in chapter B4.

3.3. Other Developments And Organization of Topic Areas

Other developments in GT/CM have been, by and large, within the confines of the structure presented in Figure 1. The various chapters in this book are organized under these topic areas, as shown in Table 1.

3.3.1. GT In Process Planning.

GT was seen to be a useful guiding philosophy for CAPP systems under both variant and generative methods, as discussed above in §2.3. However, there has been a clear divergence in this area, with most CAPP systems not rooted in part-family orientation of GT. The benefits and implementation modes of GT-based CAPP, in a concurrent engineering framework are discussed by Sormaz in chapter C1. The developments and implementation experiences are also described in chapter C2 by Nolen and Lyman-Cordes, whose work was cited earlier in §2.3.

3.3.2. Design Of Cells: Part-Machine-Labor Grouping.

The problem of grouping machines and parts using an incidence matrix and converting them into "a block diagonal form" was addressed by early GT researchers, especially Burbidge, as part of the PFA procedure. There has been a proliferation of methods during the last two decades to achieve this. Shafer presents an overview and taxonomy of these methods in chapter D1. Slomp presents implementation experiences with PFA in several Dutch firms in chapter D2. Chapter D3, by Venugopal, describes artificial neural networks for the same problem, given the promise of neural networks for handling large part-machine data sets. Chapter D4, by King, Culbreth and Joines, and chapter D5, by Cheng, Lee and Miltenburg describe new, biologically-inspired methods based on genetic algorithms, for part-machine grouping.

3.3.3. Use Of Analytical And Simulation Models.

There have also been notable advances in the areas of analytical and simulation modeling, both in theory and practice. Viswanadham, Narahari and Raghavan discuss, in chapter E1, the use of queuing networks and simulation models, within the new, business process perspective. This is followed by an application article illustrating the use of such tools in an industry context by Johnson and Wemmerlöv in chapter E2. Chapter E3, by Kalta, Lowe and Tyler discusses the use of a decision support system (DSS) that integrates various models and applies them for design of assembly cells in the apparel industry in UK. The use of simulation models for comparing functional and cellular layouts has given rise to a controversy in recent years, based on findings that contradict some traditionally-held notions about cellular operations. These are discussed in chapter E4 by Suresh.

3.3.4. Operations Planning And Control.

An overview of production planning and control frameworks for CM is presented by Riezebos, Shambu and Suresh in Chapter F1. The topic of period batch control (PBC), and recent research in this area receive a focused attention in chapter F2 by Steele. Chapter F3, by Mahmoodi and Mosier, summarizes research results in the area of part-family-oriented scheduling in CM. This has been a growing area of research in recent years. Chapter F4, by Süer, presents operations planning and control problems faced in an industry context in Puerto Rico. Chapter F5, by Askin and Iyer, presents a classification of operating policies for cells, along with simulation-based results.

3.3.5. Cells And Flexible Automation.

The advent of computer numerical control (CNC) technology has given rise to cells based on flexible automation. These include flexible manufacturing systems (FMSs) and robotic multi-machine systems. Cellular manufacturing, utilizing flexible automation, makes possible several flexibility-related advantages, and synergistic benefits. New problems also arise; these and other issues are discussed in chapter G, by Stecke and Parker.

3.3.6. Human Factors In Cellular Manufacturing.

There have been major developments in the area of human resource management, in the context of CM. Team-based operations, empowerment, new roles for team leaders, breadth of skill, as opposed to depth of skill and specialization, etc., are discussed in chapters H1 and H2. Chapter H1, by Badham, McLaughlin and Buchanan, presents new, multi-dimensional views of HRM in CM contexts. Chapter H2, by Carr and Groves is an application article that deals with, among other issues, the new role of team leaders in CM context.

3.3.7. Implementation Experiences.

Finally, integrative implementation experiences are summarized by leading implementers in various parts of the world: Danish GT experiences by Riis in chapter I1; Finnish experiences by Karvonen, Holmström and Eloranta in chapter I2; Electrolux AB, Italy by Panizzolo, in chapter I3; Microsoft Ireland, by Schonberger, in chapter I4; GT experinces at John Deere, USA by Spencer in chapter I5; and, Yugoslavian (IISE) approaches by Zelenović, Cosić and Maksimović, in chapter I6.

TABLE 1. Organization of Topic Areas.

Topic Area	Chapter
B. Engineering Design	
GT, Concurrent engineering and DFMA	B1
Classification and coding systems & non-traditional applications	B2
Part family identification: Role of engineering data bases	B3
Part family identification: New pattern recognition methods	B4
C. Process Planning: GT-based CAPP	C1, C2
D. Design of Cells: Part-Machine-Labor Grouping	
Part-machine-labor grouping methods	D1
PFA applications	D2
Use of artificial neural networks	D3
Use of genetic algorithms	D4, D5
E. Use of Analytical/Simulation Models	
Modeling within a business process approach	E1
An industry application	E2
A DSS for design of assembly cells	E3
Comparison of functional and cellular layouts	E4
F. Operations Planning & Control for CM	
Production planning and control in CM	F1
Period Batch Control	F2
Part family scheduling rules	F3
Operation control: a case study	F4
Cell loading and control policies	F5
G. Cells & Flexible Automation	G
H. Human Factors in CM	H1, H2
I. Integrative Implementation Experiences	I1..I6

4. CONCLUSIONS

Thus, an outstanding group of researchers and implementers, in various parts of the world, have come forward to share their findings and experiences in GT/CM area, and we hope the material compiled is of topical interest, relevance and utility, to researchers, practitioners and students.

REFERENCES

Boothroyd, G., Dewhurst, P. and Knight, W.A., 1994, *Product Design for Manufacture and Assembly*, Marcel Dekker, New York.

Burbidge, J.L., 1963, Production flow analysis, *Production Engineer*, 42, 12, 742-752.

Burbidge, J.L., 1970, Production Flow Analysis, in *Proceedings of Group Technology International Seminar*, International Center for Advanced Technical and Vocational Training, Turin, 89.

Burbidge, J.L., 1975, *The Introduction of Group Technology*, Heinemann, London.

Burbidge, J.L., 1979, *Group Technology in the Engineering Industry*, Mechanical Engineering publications, London.

Burbidge, J.L., 1989, *Production Flow Analysis*, Clarendon Press, Oxford.

Burbidge, J.L., 1995, *Period Batch Control*, Clarendon Press, Oxford.

Davenport, T. H., 1993, *Process Innovation*, Harvard Business School Press, Cambridge, MA.

Durie, F.R.E., 1970, A survey of group technology and its potential for user application in the U.K., *Production Engineer*, 49, 51.

Edwards, G.A.B., 1971, *Readings in Group Technology*, The Machinery Publishing Company, London.

Flanders, R.E., 1924, Design, manufacture and production control of a standard machine, *Transactions of the American Society of Mechanical Engineers*, 46, 691-738.

Gallagher, C.C. and Knight, W.A., 1973, *Group Technology*, Butterworths, London.

Gallagher, C.C. and Knight, W.A., 1985, *Group Technology Production Methods in Manufacture*, Ellis Horwood Limited, Chichester, U.K.

Grayson, T.J., 1971, An International Review of Group Technology, Society of Manufacturing Engineers Conference, Dearborn, Michigan, 2.

Grayson, T.J., 1972, *Group Technology - A Comprehensive Bibliography*, United Kingdom Atomic Energy Authority, Group Technology Centre.

Ham, I., Hitomi, K. and Yoshida, T., 1985, *Group Technology: Applications to Production Management*, Kluwer Nihjoff Publishing, Boston.

Hammer, M. and Champy, J., 1993, *Reengineering the Corporation: A Manifesto for Business Revolution*, Harper Collins, New York.

Hammer, M., 1990, Reengineering work: Don't automate, obliterate, *Harvard Business Review*, July-August, 104.

Houtzeel, A., 1975, MICLASS- a Classification System Based on Group Technology, *SME Technical Paper*, MS75-721.

Hyer, N. and Wemmerlöv, U., Group technology in the US manufacturing industry: a survey of current practices, *International Journal of Production Research*, 27, 8, 1287-1304.

Ivanov, E.K., 1968, *Group Production Organization and Technology*, Business Publications.

Mitrofanov, S.P., 1959, (Russian Text); 1966, *The Scientific Principles of Group Technology*, National Lending Library Translation, UK.

Nimmons, T.A.K., 1996, Improving the Process of Designing Cellular Manufacturing Systems, Unpublished PhD Thesis, Cranfield University, UK, p.18.

Nolen, J., 1989, *Computer-Automated Process Planning for World-Class Manufacturing*, Marcel Dekker, New York.

Opitz, H., 1970, *A Classification to Describe Workpieces*, Pergamon Press, Oxford, UK.

Petrov, V.A., 1966, *Flowline Group Production Planning*, National Lending Company, Yorkshire.

Taraman, K., (Ed.), 1980, CAD/CAM, CASA/Society of Manufacturing Engineers, Dearborn, MI.

Tatikonda, M.V. and Wemmerlöv, U., 1992, Adoption and implementation of group technology classification and coding systems: insights from seven case studies, *International Journal of Production Research*, 30, 9, 2087-2110.

Schonberger, R.J., 1983, Plant layout becomes product-oriented with cellular, just-in-time production concepts, *Industrial Engineering*, November, 66-71.

Singh, N. and Rajamani, D., 1996, *Cellular Manufacturing Systems: Design, Planning and Control*, Chapman & Hall.

Suresh, N.C. and Meredith, J. R., 1985, Achieving factory automation through group technology principles, *Journal of Operations Management*, 5, 2, 151-167.

Wemmerlöv, U. and Hyer, N. L., Cellular manufacturing in the US industry: A survey of users, *International Journal of Production Research*, 27, 9, 1511-1530.

Wemmerlöv, U. and Hyer, N.L., 1987, Research issues in cellular manufacturing, *International Journal of Production Research*, 25, 3, 413-431.

AUTHORS' BIOGRAPHY

Nallan C. Suresh is Associate Professor of Manufacturing & Operations Management at School of Management, State University of New York, Buffalo, USA (website: http://www.acsu.buffalo.edu/~ncsuresh). Dr. Suresh has published numerous papers in journals such as *Management Science, Decision Sciences, IIE Transactions, Journal of Operations Management, International Journal of Production Research, European Journal of Operational Research, Journal of Manufacturing Systems, Computers and Industrial Engineering, Production and Inventory Management, Journal of Intelligent Manufacturing, Operations Management Review and SME Technical Papers*, besides numerous conference proceedings articles. He has served in industry as a systems analyst specializing in manufacturing information systems and has consulted for many manufacturing organizations. He is an Associate Editor for *International Journal of Flexible Manufacturing Systems* and Editorial Advisory Board member for *Journal of Operations Management*. He is a recipient of Alfred Bodine/Society of Manufacturing Engineers Award for Studies in Machine Tool Economics, Joseph T.J. Stewart Faculty Scholarship and other research awards.

John M Kay is Professor of Manufacturing Systems Engineering and Head of the Department of Manufacturing Systems at Cranfield University, U.K. He served as Dean of Manufacturing Technology and Production Management at Cranfield from 1993 to 1996. His research interests are in simulation, GT/CM, the use of information systems in manufacturing, and manufacturing system design. He has published over 60 papers in journals and conference proceedings and worked with over 50 companies. He has had the pleasure of working with Professor Burbidge for 10 years at Cranfield and was nominally his manager, but everyone knew who was really in charge.

B1

Group Technology, Concurrent Engineering and Design for Manufacture and Assembly

W. A. Knight

1. INTRODUCTION

Group technology (GT) is an approach to organizing and rationalizing various aspects of design and manufacturing, and it is applicable to any engineering industry where small-batch, variety production is used (Gallagher and Knight 1975; Gallagher and Knight 1986; Burbidge 1975; Ham, Hitomi and Yoshida 1985). Group Technology is a manufacturing philosophy with far reaching implications. The basic concept is relatively simple: identify and bring together items that are related by similar attributes, and then take advantage of similarities to develop simplified and rationalized procedures in all stages of design and manufacture. The term similar attributes may mean similar design features, similar production requirements, similar inspection requirements, etc. The principles of GT can be applied in all facets of a company: in design, through standardization and variety reduction, in cost and time estimating and work planning; as the basis for computer-aided process planning (CAPP) systems; in production, as the underlying principle of cellular manufacturing systems, etc. Group Technology forms the basis for the development of flexible automation and computer integrated manufacturing.

Since GT involves identification of items with similar attributes, the topic has always been closely associated with classification processes, as a structured way to bring similar items together, usually by the design features of shape, size, material, and function, or, alternatively, by similar production requirements, as in Production Flow Analysis (PFA), for example, developed by Burbidge (1970) and others.

Product design and development has been influenced dramatically by a number of important trends in recent years. In particular, the development of computer-aided design and manufacturing systems, has changed the whole approach to product design, from largely manual, drafting board-based procedures to a highly integrated computer-based environment (Singh 1996). There has also been a gradual change from product design and development as a series of sequential activities (product concept, preliminary design, detailed design, then manufacture) to a concurrent operation utilizing cross-functional teams. This approach has become known as concurrent or simultaneous engineering (Hartley 1992).

As part of these developments, a need has arisen for systematic product analysis tools to enable design teams to investigate and quantify in a predictive way the consequences of different design concepts and decisions. Of particular relevance, are tools such as Quality Function Deployment (QFD) (Sullivan 1986), Failure Mode and Effect Analysis (FMEA) (Automotive Industry Action Group, 1993) and the whole area of what has been called Design for X (DFX) (Huang 1996), including, Design for Assembly and Manufacture (DFMA) (Boothroyd, Dewhurst and Knight 1994), Design for Service (DFS) (Dewhurst and Abbatiello 1996) and, more recently, Design for Environment (DFE) (Rapoza, Harjula, Knight and Boothroyd 1996).

Concurrent engineering tools are based soundly on the principles of GT and cellular manufacturing in many ways, and this chapter provides an updated perspective of design and development, and relationships to group technology.

2. FOUNDATIONS OF GROUP TECHNOLOGY IN PRODUCT DESIGN

The history of GT is quite long, with a few examples of what would now be called group or cell layouts prior to 1925 (Flanders 1924). However, a flurry of interest occurred in the 1960s and '70s, initially in Europe and somewhat later in the United States. Wide interest was generated by publication in Russia of the book by Mitrofanov, "The Scientific Principles of Group Technology", indicating widespread use of the principles of GT and cellular manufacture in the Soviet Union (Mitrofanov 1966). It was reported that in 1965 at least 800 factories in the Soviet Union were using GT (Grayson 1971). In the United Kingdom, a number of industrial firms implemented GT and cellular layouts (Durie 1970; Gallagher and Knight 1975). With Government support, a Group Technology Centre was established in 1968, together with several state funded research programs, the results of which were reported by Burbidge (1979). A large amount of literature existed at that time, as is indicated by the comprehensive bibliography on GT published in 1972 (Grayson 1972), which contained over 800 references. Obviously this body of literature has been added to many fold in the intervening 25 years. Some interest in the United States followed this flurry of activity in Europe and since that time cellular layouts have become reasonably widely used, together with considerable developments in CAPP systems based on GT principles (Nolen 1989).

The association between GT and classification and coding has always been prominent. Much discussion took place during the 1960s as to the most appropriate classification scheme to use, with the protagonists of various systems being outspoken at times, perhaps losing sight of the fact that such systems are merely tools to facilitate GT applications. A large number of classification coding schemes have been developed, some associated with specific manufacturing processes, such as forging, casting, etc. (Gallagher and Knight 1975; Gallagher and Knight 1986).

It is interesting to note that two classification schemes which significantly influenced GT development were initially developed for other purposes. The first of these is the Brisch system, which was initially conceived as a classification coding system for design retrieval and variety reduction (Brisch 1954). Soon after World War II, Brisch and his colleague Gombinski founded a consulting firm specializing in classification and coding systems. They had recognized the problems in industry caused by excessive variety, largely resulting from the inability of designers to identify and retrieve similar existing items, due to the inadequacies of available numbering and identification schemes. They developed a hierarchical classification scheme, which was extended to cover all items within a company, from personnel, to machines, to components and sub-assemblies (Gombinski 1964). This system was eventually installed in a large number of companies in Europe and subsequently in the US, through a branch of the company call Brisch, Bern and Partners, which still operates today (Hyde 1981). These developments spawned some competing firms offering similar classification schemes, for example, Lovelace and Lawrence (Tangerman 1966; Lovelace 1975). With the growing interest in GT, Brisch and Partners were conveniently positioned to further develop their classification approach as one of the tools for GT and cellular system development (Gombinski 1967).

It might be imagined that since the problem of variety reduction was addressed as early as the late 1940s, and, with the advent of computer based data management, this should not be a problem any more for industry. However, several recent texts devoted to variety control in industry indicate that this is still a major problem (Suzue and Kohdate 1988; Galsworth 1994). Another example, perhaps, for 'what goes around comes around'.

The second important classification system in GT development is from University of Aachen Machine Tool Laboratory, usually referred to as the Opitz system (Opitz 1970). This was developed initially to collect statistical data on workpieces manufactured in German industry with the aim of specifying machine tool design features to suit the distribution of part types and attributes being processed (Opitz, Rohs and Stutte 1960). Again, soon after completion of the workpiece statistics investigation, interest in GT in Europe grew and the Opitz classification became one of the tools for implementation (Opitz, Eversheim and Wiendahl 1969). The code was extended and developed further for application in variety reduction and standardization, cell design, etc. The basic structure of classification was extended to cover all items in a company and installed in the large German conglomerate DEMAG (Brankamp and Dirzus 1970). Out of this activity at Aachen came some of the early developments in variant CAPP systems and part family CAD systems (Eversheim and Miese 1975). A very similar classification structure was later proposed by TNO in the Netherlands (Houtzeel 1975). This system later became known as MICLASS and formed the basis for a set of computer based procedures for design retrieval, variant CAPP and cell design. These systems, marketed by a consulting group called Organization for Industrial Research (OIR) (Houtzeel 1979), came to be used in a number of firms in USA and Europe.

3. CONCURRENT ENGINEERING AND DESIGN FOR MANUFACTURE

Many design text books outline the activities for new product design and development in the sequence shown in Figure 1 (Pugh 1990). During preliminary design, decisions about the general configuration or structure of the product and sub-systems are made, including initial choices of materials and processes for component parts. In detailed design, depending on the type of product, numerous technical design analyses may be required, such as stress analysis, fluid flow calculations, etc. The overall design and development cycle can be very lengthy for complex products, which adversely affects the important, time-to-market for new products.

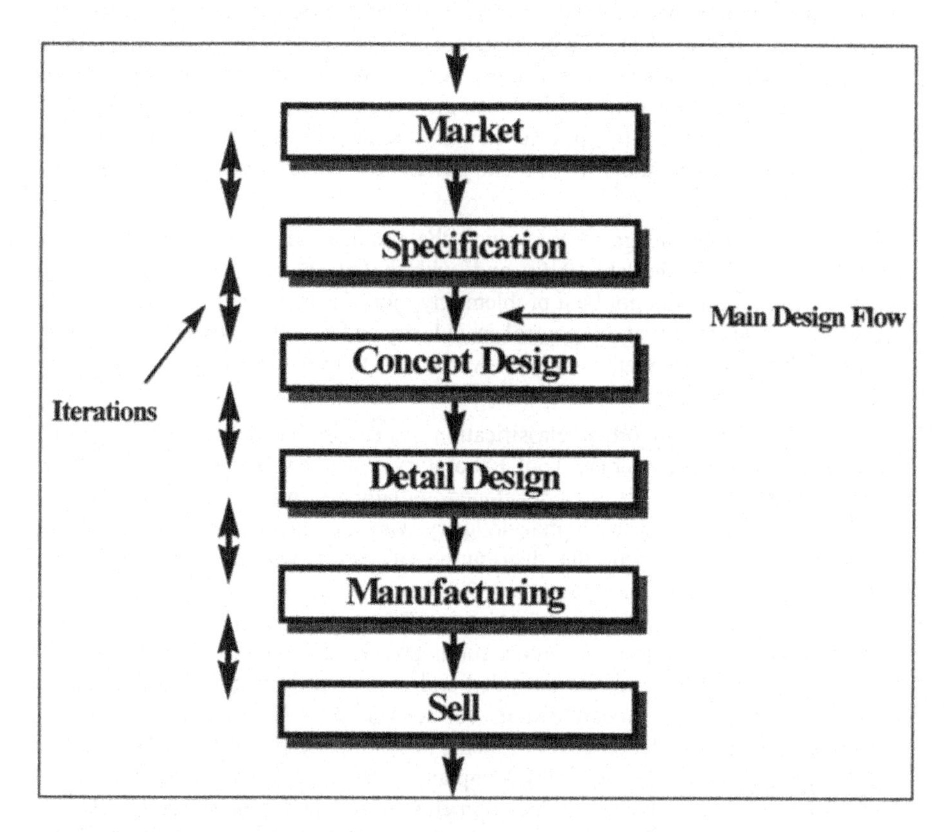

FIGURE 1. Design and development of new products (from Pugh 1990).

Traditionally the activities in the design cycle have been carried out sequentially, with the design information flowing in succession from phase to phase. It is now recognized that this approach, with lengthy design cycles, results in a reduced ability to respond effectively to inevitable design changes that occur in the product life cycle. Thus, in recent years, there has been a move from this sequential approach to a concurrent engineering approach, in which the various activities, normally downstream from preliminary design phase, are given greater consideration early in the design cycle (Hartley 1992).

There have been many definitions of concurrent engineering, but the following definition sums up the essential features: "Concurrent engineering is a systematic approach to the integrated concurrent design of products and their related processes, including manufacture and support. This approach is intended to cause the developers, from the outset, to consider all elements of the product life cycle from conception to disposal, including quality, cost, schedule and user requirements" (Pennell and Winner 1989). Many variations of concurrent engineering are prevalent in practice, but some common features can be identified. The most distinguishing feature is reliance on multidisciplinary (cross-functional teams). This results in numerous benefits, including:

- Reduced time-to-market by moving peak periods of problem solving from start of manufacture to the concept phases of design.
- More flexible skills, and making team members aware of the concerns of complementing functions.
- Collective ownership of design, thereby reducing resistance to change commonly experienced in sequential design/ development process.
- More competitive products due to closer communication, enabling effective responses to unexpected changes in product requirements.

A second aspect of concurrent engineering is an integrated approach to product design, and early consideration of all aspects of the product life cycle. This requires integrated product data management systems and utilization of systematic analysis procedures to facilitate early design decisions (Pugh 1990; Ulrich and Eppinger 1995). These analysis procedures include techniques such as QFD, FMEA and DFMA. The purpose of these procedures is to force design teams to systematically analyze alternative design concepts in such a way that their thinking is guided in a certain direction and the consequences of different design decisions can be quantified in a predictive way. The analysis process becomes a catalyst for new ideas and product improvements. The emphasis, overall, is on a somewhat increased effort during initial design phases, to shorten total design cycle time, particularly by reducing the number of design changes later in the cycle (Figure 2).

It is now well accepted that early design decisions influence a large proportion of subsequent manufacturing costs, and other life cycle costs and efficiencies. Figure 3 illustrates the typical cost build-up as a product goes through preliminary design, detail design and into manufacture. It is well recognized that preliminary design is a relatively low-cost activity, but 70-80% of manufacturing costs are committed at this stage. During detail design, much more design effort is involved, but the influence on final manufacturing costs is much less. Figure 3 also shows that the later the design changes are made, to improve manufacturability, service, etc., the more costly it becomes, with the cost of changes rising in an exponential-like manner. Therefore, design analysis procedures must enable design teams to predict the consequences of design decisions, and help make the most appropriate choices to reduce manufacturing costs. Procedures such as DFMA are squarely aimed at this issue (Boothroyd et al. 1994).

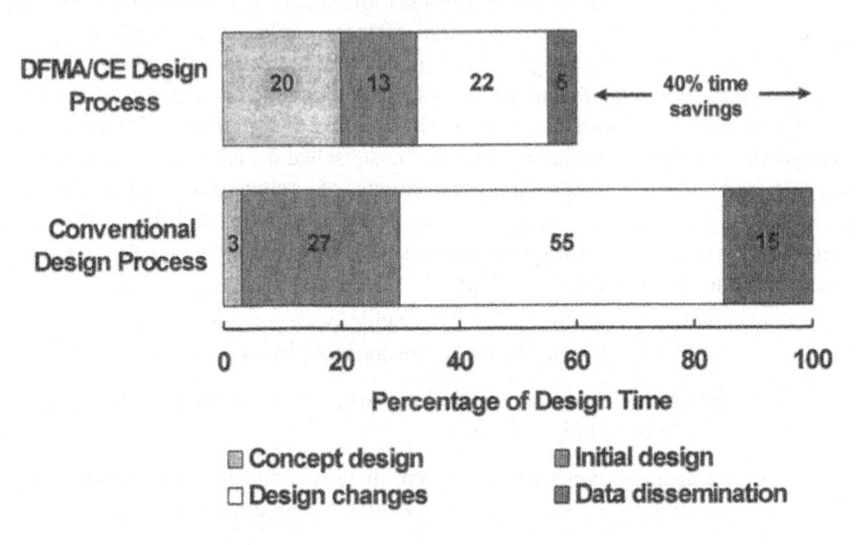

FIGURE 2. Effect on product design cycles.

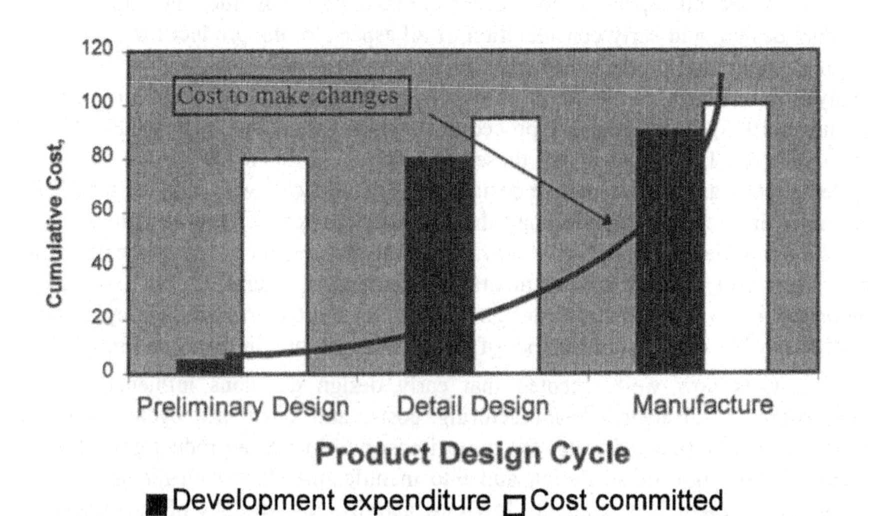

FIGURE 3. Cost build-up during the stages of product design.

Similarly, it follows that early design decisions also have a large impact on service costs, end-of-life product management and environment. Techniques such as

DFS (Dewhurst and Abbatiello 1996) and DFE (Rapoza et al. 1996) are aimed at influencing design decisions which affect these aspects.

The various product analysis procedures used as part of concurrent engineering are based on the underlying principles of Group Technology. Besides this, one also notes a basic parallel with cellular manufacturing (as described in the previous chapter by Suresh and Kay: see Figures 4 & 6). In traditional job-shops, the operations are carried out in a sequential manner: drilling operations in the drilling work center, milling operations in the milling work center, etc. In cellular manufacturing, this *department-after-department functioning* is replaced by concurrent operations, and each cell, dedicated to one or more groups of parts, comprises a *cross-functional assemblage of machines and labor*. Lead times reduce, quality/machining problems that would normally surface much later, in a subsequent work center, are made evident, and addressable much earlier. Similarly, concurrent engineering, besides reducing development lead times, has also led to more comprehensive approaches to design, addressing manufacture, assembly, service and environmental considerations much earlier in product design cycle.

4. DESIGN FOR MANUFACTURE AND ASSEMBLY (DFMA)

Serious developments in product analysis procedures for DFMA began in mid 1970s. At the heart of these procedures are group technology classification systems which cover assembly and manufacturing attributes of items. The most widely used systems for DFMA analysis are those developed by Boothroyd and co-workers (Boothroyd et al. 1994), but other assembly analysis systems have also been applied (Miyakawa, Ohashi and Iwata 1990; Miles 1989). Development of DFA analysis procedures built on earlier work on analysis of feeding and orienting of parts for automatic assembly (Boothroyd, Poli and Murch 1978). This work was presented in the form of a handbook containing a series of design charts for feeding and orienting devices, accessed by a classification and coding scheme. From this, NSF and industry funding enabled development of DFA procedures for manual, automatic and robot assembly, although it has been the system for manual assembly analysis which has received the most widespread use in industry (Boothroyd et al. 1994). The basic aim is to predict assembly costs in the early stages of design, through time standard databases accessed by means of a classification of part features (Boothroyd and Dewhurst 1985). These procedures have become widely used in industry and numerous case studies of product improvements exist (Boothroyd et al. 1994).

Figure 4 summarizes the steps in DFMA, consisting of the two main stages, DFA and DFM. DFA analysis is carried out first, aimed at *product simplification* and *part count reduction*, together with detailed analysis of the product for ease of assembly. An essential feature of DFA is the analysis of the proposed product by multidisciplinary teams. Product simplification and part count reduction are achieved by applying simple criteria for the existence of separate parts, including:

- During operation, does the part move relative to other parts already assembled?
- Must the part be of a different material, or be isolated from other parts already assembled? Only fundamental reasons, e.g., material properties are acceptable.
- Must the part be separate from other parts assembled because necessary assembly or disassembly of other parts would otherwise be impossible?

If none of these criteria is fulfilled by an item, then it is theoretically unnecessary. This analysis leads to the establishment of a theoretical minimum part count for the product. It does not mean that the product can practically or economically be built of this number of items, but it represents a level of simplification to be aimed for. DFA is not a design system: the innovation must come from the design team. However team members are challenged and stimulated to consider simplified product structures with less parts.

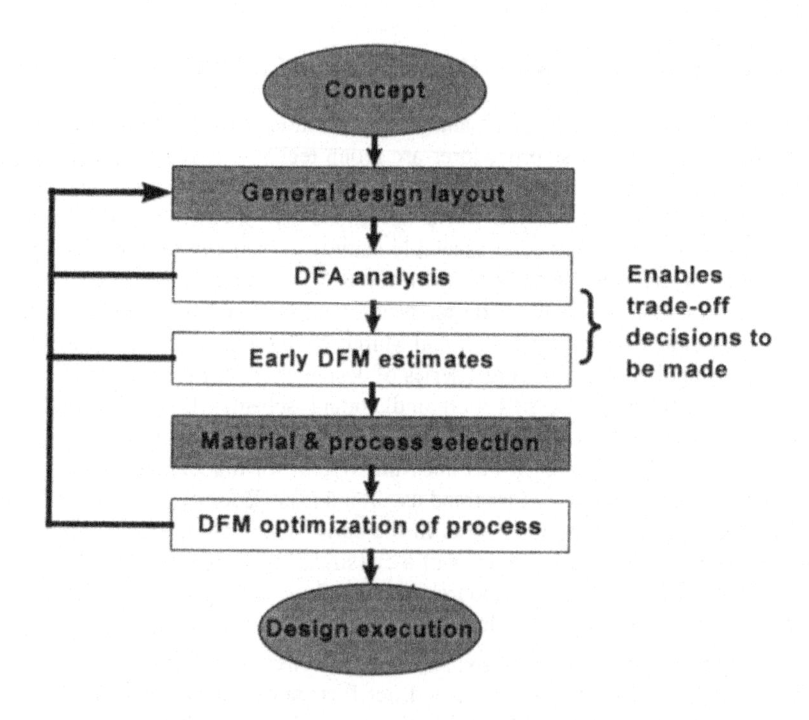

FIGURE 4. Key components of DFMA.

Detailed DFA analysis considers the geometry and other features which influence assembly difficulties, and hence assembly times. To estimate assembly time, each part is examined for two considerations: how the part is to be acquired /

fetched, oriented, and made ready for insertion, and how it is inserted and/or fastened. The difficulty of these operations is rated, and, from this rating, standard times are determined for all operations necessary to assemble each part. DFA time standard is a classification of design features which affect part assembly. The data has proved to be quite accurate for overall times, generally within 6 percent of the actual times. Besides enabling total assembly time to be estimated, items which contribute the most to this time are also identified, which again leads design teams to consider ways of improving ease of assembly. From the assembly time and minimum part count, a design efficiency or index can be determined, which can be used as a metric to gage product improvements (Boothroyd and Dewhurst 1985).

Design for Manufacture (DFM) analysis is focussed on the early assessment of component manufacturing costs and selection of best process/material based on realistic cost estimates. During this process the relationship between part features and manufacturing costs for a given process is indicated. The objective is to influence the early decisions on choice of material and process for a part, before detailed design is undertaken. This is because, once this decision is made, the items must be designed in detail to meet the processing requirements of the chosen process. The overall aim of DFM is to enable design teams to make cost trade-off considerations while these decisions are being made.

The development of early cost estimating procedures is somewhat similar to the development of computer-aided process planning (CAPP) systems. The process must be studied in detail to determine factors which influence processing sequences and to determine the main cost drivers, thus establishing how these are influenced by part features. This often leads to classification schemes as one of the main inputs to the cost models for determining items costs. Such systems, based on classification schemes, include those for forging and other forming processes (Knight and Poli 1981; Gokler, Dean and Knight 1982); procedures for estimating tooling costs for cold forged parts (Bariani, Berti and D'Angelo 1993), etc. Detailed early cost estimating procedures have been developed for a range of processes including machining, injection molding, sheet metal forming, die casting and powder metal processing. These systems have become quite widely used in industry as an integral part of DFMA process (Boothroyd et al. 1994).

4.1. DFMA Case Study

The main requirements of a motor-drive assembly are that it must be designed to sense and control its position on two steel guide rails. The motor must be fully enclosed for protection, and have a removable cover for access so that the position sensor can be adjusted. The principal requirements are a rigid base that slides up and down the guide rails, and supports the motor and sensor. The motor and sensor have wires that connect them to a power supply and a control unit, respectively (Boothroyd et al. 1994).

A proposed solution to this specification is shown in Figure 5. The base is provided with two bushes for friction and wear characteristics. The motor is secured to the base with two screws, and a hole in the base accepts the sensor, which is held in place with a set screw. To provide the required covers, an end plate is secured by two screws, to two standoffs, which are, in turn, screwed into the base. This end plate is fitted with a plastic bush through which the connecting wires pass. Finally, a box-shaped cover slides over the whole assembly from below the base, and is held in place by four cover screws, two passing into the base, and two into the end cover.

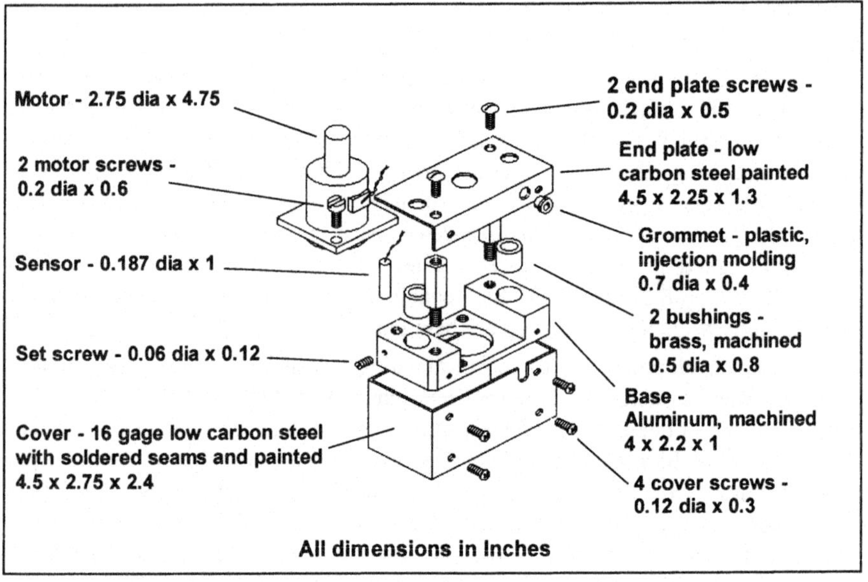

FIGURE 5. Initial design proposal for motor drive assembly.

Table 1 shows a DFA analysis of this product with a total assembly time of 160s. This analysis shows that the product consists of 19 items and the assembly times are obtained from time-standard data base based on a classification of assembly attributes. Application of the criteria for separate parts by the design team shows that only four items are considered theoretically necessary for the product's basic function The theoretical minimum number of parts is four and, if these parts were easy to assemble, they would take about 3s each to assemble on average. Thus, the theoretical minimum (or ideal) assembly time is 12s, a figure which can be compared with the estimated time of 160s, giving an assembly efficiency (or DFA index) of 12/160, or 7.5 percent.

In the process of carrying out the analysis the design team will be led to consider alternative design configurations. For example, the elimination of parts not meeting the minimum part-count criteria, and which cannot be justified on practical grounds, results in the design shown in Figure 6. Here, the bushes are combined with the base, and the standoffs, end plate, cover, plastic bush and six associated screws are replaced by one snap-on plastic cover. The eliminated items entailed an assembly time of 97.4 s. The new cover takes only 4s to assemble, and it avoids the need for a reorientation. In addition, screws with pilot points are used and the base is redesigned so that the motor is self-aligning. Table 1 also presents the results of DFA analysis of the redesigned assembly. The new assembly time is only 46s, and design efficiency has increased to 26%. Assuming an assembly labor rate of $30 per hour, the savings in assembly cost amount to $0.95.

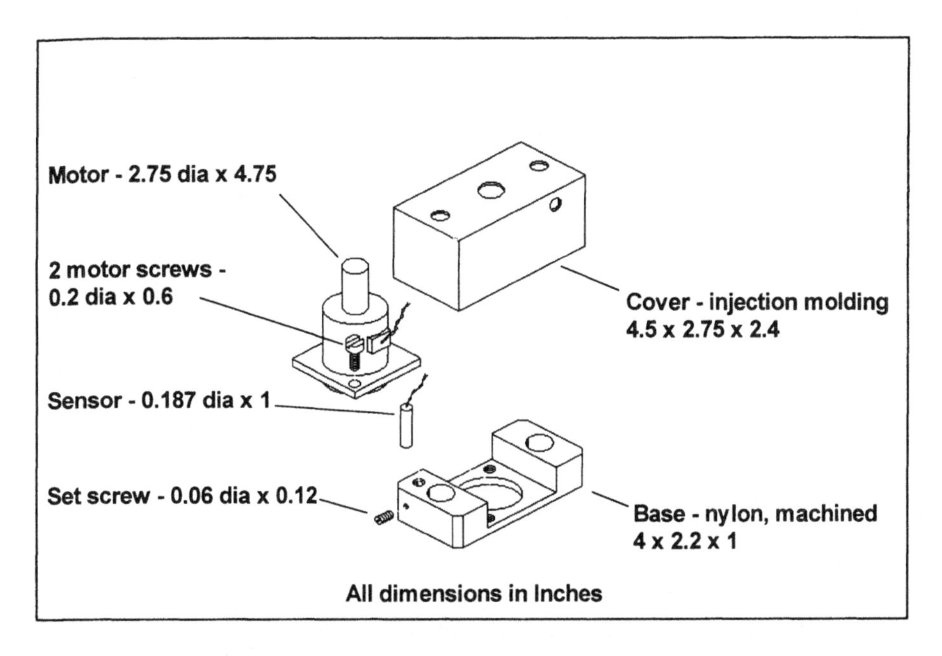

FIGURE 6. Proposed redesign of motor drive assembly.

It is necessary to also consider the parts costs for each design and the results of this are shown in Table 2. For items produced by machining, injection molding and sheet metal working costs have been estimated using DFM early costing procedures. For the remaining items, catalog prices have been used. It can be seen that the savings in part costs considerably outweigh the savings in assembly costs. This is generally found to be the case for design simplifications of this type.

It should be noted that a significant hidden cost saving, relating to savings in overhead, has not been included in this comparison. With the original design, nineteen items must be processed, each with its own documentation, inventory records, purchasing and ordering procedures, inspection methods, etc. With the redesign, only seven items need to be dealt with, which will lead to a significant reduction in overheads.

TABLE 1. DFA Analysis for Proposed Design and Redesign.

	Proposed				Redesign			
Name	Repeat	Min. Items	Assy Time	Assy Cost	Repeat	Min. Items	Assy Time	Assy Cost
Base	1	1	3.45	0.03	1	1	3.45	0.03
Bushing	2	0	15.26	0.13				
Motor	1	1	9.00	0.08	1	1	9.00	0.08
Motor Screw	2	0	17.30	0.14	2	0	17.3	0.14
Stand-off	2	0	25.10	0.21				
Sensor	1	1	11.60	0.10	1	1	11.6	0.10
Set Screw	1	0	14.65	0.12	1	0	14.65	0.12
End Plate	1	1	7.15	0.06				
End Pl. Screw	2	0	17.90	0.15				
Grommet	1	0	8.45	0.07				
Push/Pull Wire	2	-	18.79	0.16	2	-	18.79	0.16
Reorientation	1	-	4.50	0.04				
Cover	1	0	9.45	0.08	1	1	3.75	0.03
Cover Screw	4	0	32.90	0.27				
	22	4	195.50	1.63	9	4	78.54	0.65
DFA INDEX				5.99				14.92

TABLE 2. Part Costs For Motor Drive Assembly.

Items In Proposed Design	Cost ($)	Items in Redesigned Assembly	Cost ($)
Base (aluminum)	15.29	Base (nylon)	13.04
Bush (2)	3.06	Motor Screw (2)	0.20
Motor Screw(2)	0.20	Set Screw	0.10
Set Screw	0.10	Plastic Cover (includes	
Stand-off (2)	9.74	tooling cost of $8k)	8.66
End Plate	2.26		
End Pl. Screw (2)	0.20		
Plastic Bush	0.10		
Cover	3.73		
Cover Screw (4)	0.40		
TOTAL	35.08	TOTAL	22.00

5. DESIGN FOR SERVICE (DFS)

The serviceability of products has a major influence on customer satisfaction, warranty repair costs for manufacturers, and often, the useful life of products, since products are often discarded when service costs become higher than purchasing a replacement product. Product analysis tools for DFS enable design teams to evaluate the efficiency of service tasks during design and compare the serviceability of alternative design configurations (Dewhurst and Abatiello 1996). This is done through the effective simulation of each service task. The output of a DFS analysis is estimates of service times and costs, determined from time standard databases of service times, accessed by a classification of the disassembly and assembly attributes of the items removed during a service task. The analysis is most readily carried out from a DFA analysis, as the items removed for the service task can be selected from the product structure and then reassembly developed automatically from initial disassembly sequences. Two other aspects considered are, a service efficiency index and an importance ranking of service tasks/repairs.

5.1. Service Task Efficiency

A time-based service task efficiency index can be formulated, similar to the DFA efficiency index (Abbatiello 1994; Dewhurst and Abbatiello 1996). Three criteria, to establish the theoretical minimum number of parts which must be removed or unsecured to access a service item, or carry out a service procedure, have been proposed. A part, subassembly or operation is justified for disassembly if:

- The part/subassembly removed is the service item or contains the item and will be replaced as a component; or the operation performed is the service operation.
- The part/subassembly must be removed, or operation performed to isolate the service item, or subassembly containing the service item, from other parts and subassemblies with which it has functional connections, which typically transfer fluids, electricity, electrical signals, motion or force. An example may be a wire harness connection, drive belt or a fan blade connected to a motor.
- The part/subassembly removed is a cover part, or the operation performed reorients a cover part, which must completely obstruct access to the service item and related parts to seal or protect it from the environment. A cover part may also protect the user from potential injury.

Multiple identical operations that have the same purpose must be accounted for separately. For example, a set of connectors released to isolate a service item must be considered independently. If several of the connectors could be combined, into one larger connector, then only one of the set may be counted as necessary in the disassembly sequence.

The ideal service time is the time required to perform a service task which has no associated difficulties. This is the minimum time to remove, set aside and reassemble an item from the DFS database. The time-based efficiency of a service task can be then expressed as:

$$\eta_{time} = \frac{t_{min} \times N_m}{T_s} \tag{1}$$

where t_{min} = the minimum time to carry out a service operation; N_m = theoretical minimum number of items that can be justified for removal in the service task, according to the three criteria, and T_s = estimated time to perform service task (sec). To calculate the time-based service efficiency for a product that requires several service tasks several service tasks, a combined index can be calculated as:

$$\eta_{total} = \frac{\eta_1 \times f_1 + \eta_2 \times f_2 + ... + \eta_n \times f_n}{f_1 + f_1 + ... + f_n} \tag{2}$$

where η_1, η_2,..., η_n = Efficiency values for task 1, 2, ..., n respectively; and,

f_1, f_2, ..., f_n = Failure frequencies for items 1, 2, ..., n respectively.

5.2. Service/Repair Task Importance Analysis

Important factors which should be considered in the design of any product are the frequency (or probability) of different failures, and the consequences of failures. If an item is expected to fail frequently, then the required service procedure should clearly be designed for easy execution. Furthermore, if a failure may result in grave consequences then easy preventive maintenance procedures are necessary even if the chance of the failure is small.

To this end, a Frequency Ranking and Consequence Ranking for service tasks can be developed (Dewhurst and Abbatiello 1996). The failure frequency rank (F_r) is determined from a scheme shown in Table 3. The seriousness of the item's failure is captured in terms of a consequence rank (C_r).

TABLE 3. Failure Frequency and Consequence Ranking.

Fr	Likely Failure Rate
10	one or few times per week
9	one or few times per month
..	
1	very unlikely - FF < 0.0001 per year
Cr	**Criteria for Consequence Ranking**
10	Catastrophic failure, with no warning, high personal risk
9	Loss of operating capability, damage and personal risk
..	
2	Slight effect on performance
1	No effect

The importance rank denotes the importance of designing for quick and easy service of a failure source. It is calculated as: $I_r = F_r \times C_r$, which yields a number between 1 and 100. The higher the importance rank, the easier it should be to perform the service task. In general, regardless of importance rank, special attention should be given to items when frequency of failure and/or consequences of failure are high.

The DFS methodology has been successfully applied to a wide range of products, including domestic appliances and computer peripherals (Lee 1995).

6. PRODUCT ANALYSIS FOR DISASSEMBLY AND ENVIRONMENT

It is estimated that, in USA, approximately 11 million vehicles were scrapped in 1990 and the rate of disposal continues to increase (Anon 1990); over 350 million appliances of various types were discarded in 1992 (Parkes 1993); approximately 150 million personal computers could potentially be discarded into landfills each year by the year 2005 (Wittenburg 1992). Landfill disposal is becoming increasingly unacceptable due to the growing shortage of space, and potential detrimental effects on the environment. New sites are difficult to open since site selection and operation must meet more rigid standards and regulations. In addition, continued loss of valuable and reusable materials to landfills has become unacceptable. These concerns, along with public opinion, have led to more stringent legislation on product disposal and recycling (Anon 1994; Bylinsky 1995; Cairncross 1992). Consequently, there is a growing pressure to develop effective take-back and recycling of manufactured products.

Materials have significant recycled value when they are divided into clean, separate types, and viable recycling depends greatly on the efficiency with which materials can be separated. The separation and recovery of items and materials from products can be approached in two main ways, which can be used together. These are careful disassembly, and bulk recycling, which involves shredding of mixes of items and using separation techniques on the stream of particles produced. In the long term, however, recycling can be made more effective by *design of products for greater ease of disassembly and recycling.* This requires development of product analysis tools to enable design teams to evaluate ease of disassembly and recycling, in the early stages of design, similar to DFMA and DFS.

Development of analysis procedures for disassembly has been receiving attention lately (Johnson and Wang 1995; Glen and Kroll 1996). These procedures concentrate mainly on the economics of disassembly, and not on environmental impact. Recently, tools that enable consideration of a balance between the financial and environmental aspects have been developed (Girard and Boothroyd 1995; Rapoza et al. 1996).

Product analysis procedures for disassembly are aimed at simulation of disassembly at end-of-life disposal and quantifying the resulting costs, benefits and environmental impact. Disassembly costs are determined from time-standard databases developed for end-of-life disassembly processes, which are accessed by classifying disassembly attributes of items. Each item in the assembly is allocated to an appropriate end-of-life destination (recycle, reuse, regular or special disposal and so on) based on its material content, and this information, together with the item weight, enables the possible costs or profits to be determined from a materials database. The user also provides information on disassembly precedence of each item (items that must be removed immediately prior to the component). This information enables the disassembly sequence to remove valuable items as early as possible to be determined.

An initial disassembly sequence can be entered directly by the user or is generated automatically from a DFA analysis, by reversing the initial assembly list. Based on the resulting disassembly sequence, two main analyses are performed: 1) Financial return assessment of disassembly and disposal, including remanufacturing and recycling, and, 2) Environmental impact assessment, resulting from initial product manufacture and disposal, including remanufacturing and recycling.

6.1. Financial Analysis Of Disassembly

The financial assessment, for the disassembly of a 386 PC, is shown by the upper line in Figure 7 as the return or cost as disassembly progresses. This is determined as the difference between cost incurred to disassemble each item and recovered value, plotted against disassembly time. Disassembly times and costs are obtained from time standard databases accessed by a classification of disassembly attributes of each item removed. A point on this curve represents the profit or net cost if disassembly is stopped at this stage.

Thus the first point represents the costs of product disposal by landfill or incineration (user selected) without disassembly, which includes the take-back costs of the product (collection, inventory, etc.) and the product weight multiplied by the appropriate disposal cost per unit weight. Included in the contributions to each point on the graph are: 1) take-back cost of the whole product; 2) cost of disassembly up to this point (cumulative disassembly time * labor rate); 3) cumulative value(from recycling or reuse) of items disassembled to this point; 4) cost of disposal of remainder of the product yet to be disassembled (rest fraction), assuming that the rest fraction is disposed of by special waste treatment (landfill or incineration) as long as any item requiring such treatment remains in the rest fraction; Special disposal is generally more expensive than regular disposal.

Generally, at the start of disassembly, small items such as fasteners are removed, for which the cost of disassembly is greater than value of material recovered, and the graph goes down. Some items have a significant positive effect on financial return e.g., items that have high recycle values, are reused or are toxic (the rest fraction of the product becomes less costly to dispose of once these items are removed). These items are referred to as critical items. For the curve shown in Figure 7, it was assumed that some items, such as memory SIMMs, disk drives and power supplies, have significant resale value when removed. It should be noted that the careful disassembly of small products, such as PCs, is usually not profitable unless some parts or sub assemblies had significant reuse/resale value. Otherwise the material content does not compensate for the labor cost of disassembly. But bulk recycling of all or part of the product may be appropriate.

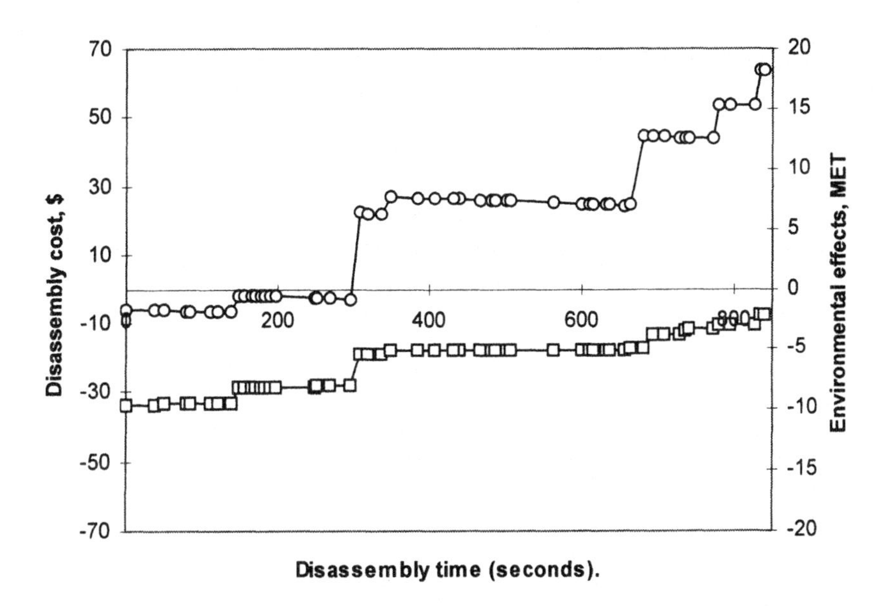

FIGURE 7. Initial disassembly analysis of a 386 PC.

6.2. Environmental Impact Assessment

To develop the environmental line, a method suggested by the TNO Product Center has been used. It takes into account the effects of materials, energy and toxicity (MET) effects of the product on the environment (Kalisvaart and Remmerswaal 1994). Application of these procedures to a range of large consumer products has been described elsewhere (Rapoza et al. 1996). Other single-figure indicators, such as EcoIndicator 95, could also be used (Netherlands NOH 1995). The TNO method has been chosen because the results are more readily understood and interpreted by designers than the complex data often developed from other, full life-cycle-analysis (LCA) procedures.

The environmental impact line (lower line in Figure 7) shows the net MET points at any stage of disassembly. A specific point on the curve represents the net environmental impact of the product if disassembly is stopped at this stage, on the basis of several assumptions. The contributions to each point on the curve are due to: 1) MET points for initial material and processes for manufacturing the whole product; 2) Cumulative effect of reprocessing for recycling and disposal for all items disassembled to this stage; 3) MET points for disposal of the rest-fraction of the product; 4) Less MET points recovered for items disassembled so far which are recycled or reused (remanufactured); and, 5) MET points associated with take-back of the whole product (transport, etc.).

Note that all MET points are negative, since they measure effects on the environment through emissions, use of scarce materials, etc. Re-manufacturing or recycling of items reduces the negative effects of product initial manufacture and end-of-life disposal. The first point on the curve represents the environmental impact from initial manufacture of the whole product plus environmental impact of disposal of the whole product, by regular or special waste treatment. As disassembly proceeds, the curve moves up if items are recycled or remanufactured, and some of the MET points are effectively 'recovered.' The curve also moves up or down if items are removed which result in the rest-fraction being changed from special waste to regular waste disposal. This normally happens when the last item that has to be treated as special waste is removed from the assembly. At this point, the curve may in fact move down, because of increased environmental impact for all materials processed as regular waste relative to special waste, but at lower cost.

6.3. Optimization Of Disassembly Sequences

In complex products, there may be numerous possible disassembly sequences, limited by disassembly precedences assigned to each item. For example many items in subassemblies can be removed without removing the subassemblies themselves from the product first or the subassembly can be removed before disassembling further to reach the required item. Determination of the most appropriate disassembly sequence for a product is therefore of interest.

Procedures have been developed (Girard and Boothroyd 1995; Rapoza et al. 1996), which will reorder the critical items (those which produce significant steps in the financial curves) as early as possible in the disassembly sequences, but limited by the precedence constraints input with the initial disassembly sequences. The order in which items are moved forward in the disassembly sequence is determined from the greatest yield (rate of improvement). This same procedure can be applied to either the financial or environmental assessment curves.

6.4. Example Results For Disassembly Analysis

Consider the case of disassembly of the 386 PC. As a first step in disassembly analysis, a DFA analysis was carried out, using DFA software, which showed that the product consisted of 99 parts with 3 subassemblies and had a total of 102 items assembled, plus 18 additional operations. The assembly time was estimated to be 1071s. The PC weighed 7.32 kg. The initial disassembly sequence was arbitrary, being a reversal of the DFA assembly lists. Some editing of disassembly steps was necessary, so that parts made of the same or of compatible materials that would not have to be separated for the purposes of reuse were treated as one item. The financial line and environmental line for this edited sequence were shown in Figure 7. A profit of around $64 can be seen to be realized for the product, but complete disassembly of the computer would be required. The point of least environmental impact also corresponds to complete disassembly, which is often the case.

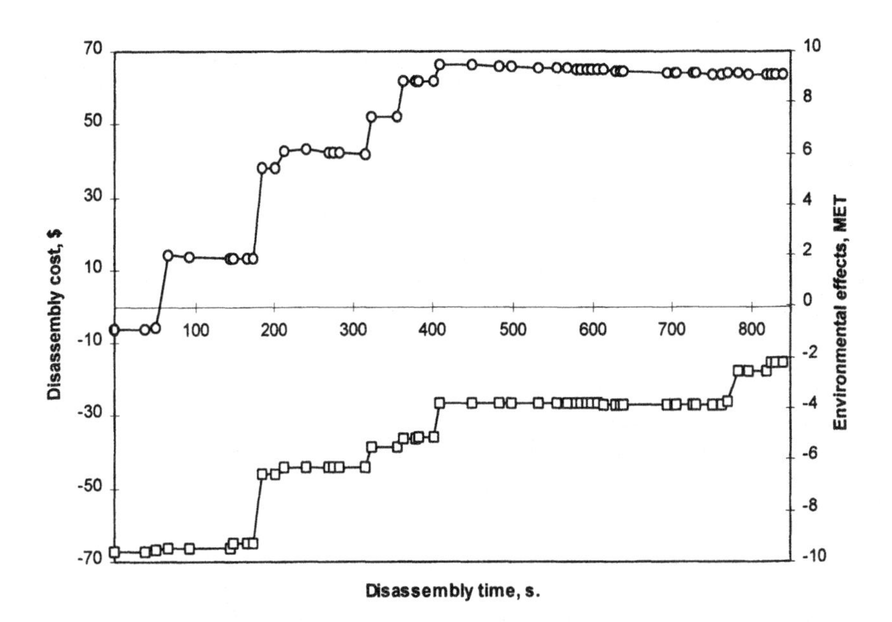

FIGURE 8. Financially optimized disassembly sequence for a 386 PC.

The analysis allows different disassembly sequences to be investigated in two ways: 1) by allowing the user to move selected items forward in the disassembly sequence alone and, 2) by implementing the optimization strategy outlined in Section 6.3. Figure 8 shows the financial and environmental lines for the financially optimized disassembly sequence. The maximum financial return of $65.50 now occurs with the removal of item number 24, the hard drive, at a disassembly time of 410s. From the financial viewpoint disassembly should cease at this point, but for least effective environmental impact almost complete disassembly of the computer is required.

7. CONCLUSIONS

The concurrent engineering approach has radically transformed design activities and it has become widely accepted in industry. Similar to cellular manufacturing, concurrent engineering constitutes a major departure from the traditional, department-by-department sequence of operations. The net result has been a significant reduction in time-to-market, improved product quality and other benefits. The various product analysis tools developed, for use in the early design stages, when important decisions which affect the manufacturability, serviceability and environment impact of the product are made, have been found to be very beneficial. All these tools are based on the foundations of GT and classification of various product attributes to organize and access data.

REFERENCES

Anon, 1990, *Wards Automotive Annual.*

Abbatiello, N., 1994, *Development of a Design for Service Methodology*, Unpublished Thesis, University of Rhode Island.

Anon, 1994, Durable goods manufacturers explore links between recycling and sustainable materials, *Business and the Environment*, November, 2-4.

Automotive Industry Action Group, 1993, *Potential Failure Mode and Effect Analysis (FMEA)- Reference Manual*, AIAC, Dearborn, Michigan.

Bariani, P., Berti, G. and D'Angelo, L., 1993, Tool cost estimating at the early stages of cold forging process design, *Annals of CIRP*, 42, 279-282.

Boothroyd, G. and Dewhurst, P., 1985, *Product Design for Assembly*, Boothroyd Dewhurst, Inc., Wakefield, RI.

Boothroyd, G., Dewhurst, P. and Knight, W.A., 1994, *Product Design for Manufacture and Assembly*, Marcel Dekker, New York.

Boothroyd, G., Poli, C. and Murch, L.E., 1978, *Handbook of Feeding and Orienting Techniques*, University of Massachusetts, Amherst.

Brankamp, K. and Dirzus, E., 1970, Zeilsetzungen und Aufbau eines einheitlichen Sachnummernsystems fur einen Industriekonzern, *Industrie Anzeiger*, 92, 2501.

Brisch, E.G., 1954, Maximum Ex Minimo, *Institution of Production Engineers Journal*, June, 344-351.

Burbidge, J.L., 1970, Production flow analysis, *Proceedings of Group Technology International Seminar*, International Center for Advanced Technical and Vocational Training, Turin, 89.

Burbidge, J.L., 1975, *The Introduction of Group Technology*, Heinemann Press, London.

Burbidge, J.L., 1979, *Group Technology in the Engineering Industry*, Mechanical Engineering publications, London.

Bylinsky, G., 1995, Manufacturing for Reuse, *Fortune*, Feb., 102.

Cairncross, F., 1992, How Europe's companies reposition to recycle, *Harvard Business Review*, March, 34-45.

Dewhurst, P. and Abatiello, N., Design for service, in *Design for X: Concurrent Engineering Imperatives*, (Ed. G. Q. Huang), Chapman and Hall, London, 298-717.

Durie, F.R.E., 1970, A survey of group technology and its potential for user application in the UK., *The Production Engineer*, 49, 51.

Eversheim, W. and Miese, M., 1975, Group technology developments and modes of application, *Proceedings of 15th International Machine Tool Design and Research Conference*, Pergamon Press, 7.

Flanders, R.E., 1924, Design, manufacture and production control of a standard machine, *Transactions of The American Society of Mechanical Engineers*, 46, 691.

Gallagher, C.C. and Knight, W.A., 1975, *Group Technology*, Butterworths Press, London.

Gallagher, C.C. and Knight, W.A., 1985, *Group Technology Production Methods in Manufacture*, Ellis Horwood Limited, Chichester, U.K..

Galsworth, G.D., 1994, *Smart-Simple Design*, Oliver Wright Publications, Vermont.

Girard, A. and Boothroyd, G., 1995, Design for disassembly, *Proceedings of the International Forum on Product Design for Manufacture and Assembly*, Newport, RI, June 12-13.

Glen, P. and Kroll, E., 1996, Diane: disassembly analysis software for product recycling, *Proceedings of National Design Show Conference*, March, Chicago, ASME, NY.

Gokler, M.I., Dean, T.A. and Knight, W.A., 1982, Computer-aided sequence design for hot upset forgings, *Proceedings of 20th International Conference on Machine-Tool Design and Research*, Pergamon Press, 457.

Gombinski, J., 1964, Classification and coding, *Engineering Materials and Design*, September, 600-605.

Gombinski, J., 1967, Group technology - an introduction, *The Production Engineer*, September, 564-570.

Grayson, T.J., 1971, An international review of group technology, Society of Manufacturing Engineers Conference, Dearborn, Michigan, 2.

Grayson, T.J., 1972, *Group Technology - A Comprehensive Bibliography*, United Kingdom Atomic Energy Authority, Group Technology Centre.

Ham, I., Hitomi, K. and Yoshida, T., 1985, *Group Technology: Applications to Production Management*, Kluwer Nihjoff Publishing, Boston.

Hartley, J.R., 1992, *Concurrent Engineering: Shortening Lead Times, Raising Quality and Lowering Costs*, Productivity Press, Cambridge.

Houtzeel, A., 1975, MICLASS- a classification system based on group technology, *SME Technical Paper*, MS75-721.

Houtzeel, A., 1979, The many faces of GT, *American Machinist*, 123, 115.

Huang, G.Q.(Ed.), 1996, *Design for X: Concurrent Engineering Imperatives*, Chapman and Hall, London.

Hyde, W.F., 1981, *Improving Productivity by Classification, Coding and Data Standardization*, Marcel Dekker, New York.

Johnson, M.R. and Wang, M.H., 1995, Planning product disassembly for material recovery, *International Journal of Production Research*, 43, 3119.

Kalisvaart, S. and Remmerswaal, J., 1994, The MET-Points Method: a new single figure environmental performance indicator, *Proceedings of Integrating Impact Assessment into LCA*, SETAC, Brussels, Oct.

Knight, W.A. and Poli, C., 1981, Product design for economical use of forging, *Annals of CIRP*, 30, 337.

Lee, A., 1995, *Design for Service and Damage Repair*, unpublished Thesis, University of Rhode Island.

Lovelace, G.E., 1975, Part family classification and group technology, prerequisites for computer-aided manufacture, *Proceedings of CAM-I Coding and Classification Workshop*, Arlington, Texas, 151.

Miles, B.L., 1989, Design for Assembly- a key element within design for manufacture, *Proceedings of Institution of Mechanical Engineers., Part D: Journal of Automobile Engineering*, 203, 29-38.

Mitrofanov, S.P., 1966, *The Scientific Principles of Group Technology*, National Lending Library Translation, UK.

Miyakawa, S., Ohashi, T. and Iwata, M., 1990, The Hitachi new assemblability evaluation method, *Transactions of the North American Manufacturing Research Institution (NAMRI)*, SME, Dearborn, Michigan.

Netherlands NOH, 1995, *Eco-indicator 95 - Manual for Designers*, Report No. 9524.

Nolen, J., 1989, *Computer-Automated Process Planning for World-Class Manufacturing*, Marcel Dekker, New York.

Opitz, H., 1970, *A Classification to Describe Workpieces*, Pergamon Press, Oxford, UK.

Opitz, H., Eversheim, W. and Weindahl, H-P., 1969, Workpiece classification and its industrial application, *International Journal of Machine Tool Design and Research*, 9, 39.

Opitz, H., Rohs, H. and Stute, G., 1960, Statistical investigations on the utilization of machine-tools on one-off and mass production, *Aachen Technical University Research Report*, No. 831.

Parkes, K., 1993, Being green does'nt hurt, *Manufacturing Systems*, October.

Pugh, S., 1990, *Total Design*, Addison Wesley, New York.

Pennel, J.P. and Winner, R.I.., 1989, Concurrent engineering: practices and prospects, *Proceedings: IEEE Global Telecommunications Conference and Exhibition (GLOBCOM 89)*, Part I, IEEE, IEEE Service Center, Piscataway, NJ, 647-655.

Rapoza, B., Harjula, T., Knight, W.A. and Boothroyd G., 1996, Product design for disassembly and environment, *Annals of CIRP*, Vol..45, 109.

Singh, N., 1996, *Systems Approach to Computer Integrated Design and Manufacture*, John Wiley, New York.

Sullivan, L.P., 1986, Quality Function Deployment, *Quality Progress*, June, 39-50.

Suzue, T. and Kohdate, A., 1988, *Variety Reduction Program: A Production Strategy for Product Diversification*, Productivity Press, Cambridge, MA.

Tangerman, E.J., 1966, It only takes minutes to retrieve a design, *Production Engineering*, 37, 4, 83-86.

Ulrich, K.T. and Eppinger, S.D., 1995, *Product Design and Development*, McGraw Hill, New York.

Wittenburg, G., 1992, Life after death for consumer products, *Assembly Automation*, 12, 2, 21.

AUTHOR'S BIOGRAPHY

Dr. Winston Knight is Professor and Chairman of Industrial & Manufacturing Engineering Department, University of Rhode Island, and Vice-President of Boothroyd Dewhurst, Inc. He moved to University of Rhode Island in 1986. Prior to this he was Lecturer in Mechanical Engineering at University of Birmingham, England (1972-80), University Lecturer in Engineering Science at University of Oxford and Fellow of St. Peter's College, Oxford (1980-85) and Professor of Engineering at University of Bath, England (1985-86). During these periods he has conducted research on many aspects of manufacturing engineering, including machine-tool vibrations and noise, GT, CAD/CAM, with particular reference to metal forming dies and DFM. With colleagues, Geoffrey Boothroyd and Peter Dewhurst, a major industry-sponsored research program was established at URI focused on DFM. This has resulted in the development of software tools and procedures which are currently having a major impact in industry in the development of more competitive products. Dr. Knight received his Ph.D. from University of Birmingham, UK. He is a Corresponding Member of CIRP (since 1983), a senior member of SME/NAMRI and member of Institution of Mechanical Engineers, London. He is the author of over one hundred papers on various aspects of manufacturing engineering, CAD/CAM and design for manufacture, and he is the co-author of six text books.

B2

Classification and Coding Systems: Industry Implementation & Non-Traditional Applications

C.T. Mosier

1. INTRODUCTION

The purpose of this chapter is to explore group technology (GT) identification and coding concepts studied and applied in a *non-traditional* manner. Accordingly, the discussion is limited to innovations in techniques and/or technologies associated with coding systems, and innovations in the focus of retrieval, with respect to the information used for encoding and retrieval, or the population of entities being retrieved.

Asking practitioners or researchers of GT for a global definition often yields very limited viewpoints. To some, GT is the realignment of manufacturing facilities into cellular layouts; to others, it is the controlled retrieval of part designs from a design database. One is reminded of the *six blind men of Indostan* and their attempt to describe an elephant. Their "local" views (an elephant is very much like a wall, a spear, a snake, a tree, a fan, or a rope, etc.) are not unlike the local views of GT, and the nature of consequent academic disagreements, as articulated in the final two stanzas of Saxe's version of the story:

"And so these men of Indostan
Disputed loud and long,
Each in his own opinion
Exceeding stiff and strong,
Though each was partly in the right,
And all were in the wrong!

Moral:
So, oft in theologic wars,
The disputants, I ween,
Rail on in utter ignorance
Of what each other mean,
And prate about an Elephant
Not one of them has seen!"

Thankfully, one finds definitions of GT in simple terms, yet successfully addressing the operational nature and breadth of the application. For example, Shunk (1985) defines GT as: "*... a disciplined approach to* ***identify*** *things such as parts, processes, equipment, tools, people, or customer needs by their attributes looking for similarities between and among the things;* ***grouping*** *the things into families according to similarities; and, finally, increasing the efficiency and effectiveness of* ***managing*** *the things by taking advantage of the similarities.*"

Shunk's definition identifies three key elements intrinsic to GT: **identification**, **grouping**, and **managing**. Operationally, GT tends to involve all three of these elements. Further, the notions of identification and grouping are not always considered independently. Groups of ***things*** are most often determined by the consideration of attributes identified as being critical to the characterization of candidate ***group***s - unfortunately, a rather circular state of affairs.

Further, the categories of ***things*** are frequently not considered independently. For example, the well known part-machine incidence matrix used by many as the basis of analysis for determining families of parts and cells of machines (King 1980) is an example of things where the focus is on the *interaction* of parts and machines in the context of the manufacturing arena.

The most basic notion of GT is that if we are made aware of patterns in ***things*** that can lead to improvements in our interactions with these ***things***, we will, in fact, ***manage*** our interactions with these ***things*** in a different manner to make improvements. This leads to some rather basic and intuitive implications:

1. Seeking patterns (and pattern recognition methods) is a worthwhile endeavor.

2. Design engineering and manufacturing arenas, with their myriad of systemic complexities, is fertile ground for seeking patterns.

3. Within this fertile ground, the things worthiest of investigation are the things being manufactured (parts and assemblies) and the things utilized for manufacturing, (machinery, people and other resources).

There has been a great deal of activity (both research and application) in the arena of design retrieval. The productivity gains to be made by the process of part coding and design retrieval are significant and relatively immediate. The focus of design retrieval is on the ***management*** aspect of Shunk's definition. Simply, it is to develop a systemic way of *"finding out what was done in the past."*

This leads our discussion to the potential of the management aspect of GT not as well studied in the context of design retrieval, that is, the notion of evaluating the *thing*s under consideration. The fact is that the focus on a family of parts and a cell of machines should enable those involved with them to "know them better." In a cell, the broad expertise of the workers involved with the productive processes, including the physical transformation processes and the overall management of the processes should improve. In the design retrieval process, more than geometric information may be retrieved. The evolution of processes designed for retrieving part information will provide the basis for organization of the rest of the discussion in this chapter.

2. THE EVOLUTION OF DESIGN RETRIEVAL PROCESSES

First, our notion of the term "evolution" needs clarification. This evolution process in the context of design retrieval is not necessarily temporal. Oftentimes it is financially prudent to balance the potential benefits to be gained by exotic and powerful design retrieval approaches against their potentially high costs, particularly when those benefits have not been generally proven. A common strategy is to initially take a small "step" into the domain of design retrieval to provide evidence of benefits and justification of costs to higher management, and to test the *discipline* necessary to manage and operate a design retrieval mechanism.

The form of the design retrieval evolution has two principal facets. The first is the *mechanism for retrieval*, which may be categorized as:

- smart part numbers,
- simplistic code development,
- features-based code development, and
- database-oriented procedures

This ordering is generally aligned with the power of the computer technologies required.

Smart part numbering is the simplest of our procedures, where part numbers provide the user with varying degrees of information about the attributes of the part; most often with a focus on the part location in the bill-of-materials. In reality, smart part numbering was a precursor to GT part coding and identification, and in more recent applications it is a way to avoid the cost of implementing a coding system. Research quite often involves the development of some innovative numerical techniques (e.g., Kini et al. 1991).

Simplistic code development most often involves the use of "home grown" coding systems for developing GT codes for parts for a specific purpose or in a localized environment. This approach is often the first step in the education (to the concepts and benefits of part retrieval) of a particular manufacturing location. This approach typically involves the development of a coding system on systems ranging greatly in sophistication; ranging from systems built in BASIC (Mosier and Janaro 1990) running on primitive microcomputers, to systems written in C (Mosier, Ruben and Talluru 1993). The primary activities required are:

1. Determination of the critical attributes that reasonably characterize the parts (Tatikonda and Wemmerlöv 1992), and,
2. Partitioning the parts database in a manner consistent with frequency distribution of parts within each category, or consistent with a pre-existing or reality-based categorization.

This partitioning of the levels of each digit can require a number of iterations. For example, in a past study (Mosier and Janaro 1990), the *radius* of a rotational part was considered to be a critical attribute. As shown in Table 1, codes 0-9 were identified on the basis of frequency of occurrence of parts in each radius category. Later, this partition was mapped to a series of designated mold blanks used for casting the parts. However, this new mapping dictated a revision of GT codes, which is also shown in Table 1.

Features-based coding procedures cover a broad set of approaches for linking code development process to the computer-aided design (CAD) database. This category of procedures brings to the table the need to logically deal with the difference between the two primary activities associated with the code development process (as mentioned previously) and the use of the coding system. Features-based coding involves the development of the part code *directly* from the CAD database. The sophistication varies widely, but with respect to the power of the inference mechanism underlying the structure. Initially, design identification was the primary objective (Shah and Bhatnagar 1989); more recent studies involve the *direct* (from the CAD geometrical database) evaluation of assembly designs as part of design-for-assembly (DFA) analyses (Li and Huang 1992).

Features-based coding systems are complex and expensive - expensive enough so that they are not appropriate for all firms. Clearly, the firm must be using a single advanced CAD system for all design tasks, containing essentially a single parts database. This in itself is somewhat rare. Many of these systems are still in research and development stages. It is fairly clear that without an exceptional level of expertise in-house, these systems are not "home-grown." Further, it will probably be some time in the future before a mature product makes the feature-based approach widespread in industry.

TABLE 1. Frequency-Based Partition and Modified Partition (Mosier & Janaro 1990).

Frequency-Based Partition			Modified Partition		
Initial GT Digit	For Radius Ranges	Blanks Identified	Revised GT Digit	Radius	Blanks
0	$r < 8$	A_0	0	$r < 12$	A_0
1	$8 \leq r < 12$	A_0	1	$12 \leq r < 14$	A_1
2	$12 \leq r < 14$	A_1	2	$14 \leq r < 16$	A_2
3	$14 \leq r < 18$	A_2, A_3	3	$16 \leq r < 18$	A_3
4	$18 \leq r < 20$	A_4	4	$18 \leq r < 20$	A_4
5	$20 \leq r < 22$	A_5	5	$20 \leq r < 22$	A_5
6	$22 \leq r < 30$	A_6	6	$22 \leq r < 30$	A_6
7	$30 \leq r < 32$	A_7	7	$30 \leq r < 32$	A_7
8	$32 \leq r < 40$	A_{S1}	8	$32 \leq r < 40$	A_{S1}
9	$40 \leq r$	A_{S2}	9	$40 \leq r$	A_{S2}

Database-oriented procedures range from code development processes using information from design, engineering, or manufacturing databases, to processes which bypass notion code development entirely and focus upon retrieval using the mechanisms inherent to modern database technologies. The dynamics of these procedures induce a variety of applications. Given the ease of use of modern database languages, creating the logic and code to develop "querying" system is not beyond design engineering personnel with training in computer software. Unfortunately, the required links to the available databases require that the *optimal* development of these querying systems be accomplished in conjunction with the actual database design. Ad-hoc systems "pasted" onto existing design, engineering, or manufacturing databases are not likely to dominate industry. Rather, major software vendors (most probably with expertise in CAD and database technologies) or computer service providers will dominate the marketplace. Table 2 presents a partial list of studies within each category.

The second facet in design retrieval evolution is the focus and form of the information being retrieved:

- basic part design retrieval,
- focused identification and retrieval,
- engineering and manufacturing information retrieval,
- functional design evaluation,
- manufacturing evaluation, and
- assembly evaluation.

TABLE 2. Representative Studies in Each Category of Coding.

Smart Numbers	Lyle (1988), Udoka (1991), Kini et al. (1991), Kini (1994)
Simplistic Code	Mosier and Janaro (1990), Mosier et al. (1993), Mosier and Eichler (1994)
Features-Based	Shah and Bhatnagar (1989), Sreevalsan and Shah (1992), De Fazio et al. (1993), Shah and Rogers (1993)
Database Procedures	Urban et al. (1993), Bhadra and Fischer (1988)

Basic part design retrieval falls under the traditional category. It is a focus with well-proven benefits. **Focused identification and retrieval** involves focus on a subclass of part or manufactured item, or the logical expansion of the GT retrieval concept to other items where entity coding and retrieval will hold value either from insights gained by the process of code development or the retrieval of information. The purpose and benefits of such systems are necessarily localized within the firm. For example, the coding system of robotic end-of-arm-tooling (EOAT) developed in Mosier, Ruben, and Talluru (1993) was most usable to the personnel directly involved with EOAT design. In situations where interest in coding is driven locally within the firm, the long term, broad-based discipline necessary to develop and maintain the entity retrieval database may be lacking.

Alternatively, there are two good reasons to expend the time and effort required to develop systems with a limited focus. First it is a constructive way to educate design personnel of the benefits of GT design retrieval. It is easier to illustrate the usefulness of a system to a more limited audience who can see the immediate impact on *their* productivity. The other benefit comes into play if the particular environment involves a costly design activity (either the item is expensive, or the design activities require specialized (thus expensive) expertise.

Functional design evaluation is basic to the initial notion of design retrieval, i.e., the retrieval should be of a part design which functioned as planned. It is assumed that those designs that did not function as planned would not be included in the database. The notion of functional design evaluation extends beyond the simple "yes-or-no" implied by inclusion in the database. The functionality of a given part design is evaluation information that should be contained in the database. Retrieving a design should provide the designer with an evaluation of the retrieved design with *relative* meaning. To illustrate, a well-known sport utility vehicle model recently had difficulty with their fuel tanks. In fact, an expensive recall was required to remedy the problem. A design engineer in this firm charged with the task of designing a new fuel tank to go into a similar model could retrieve a number of old fuel tank designs. All records associated with the design's functionality, including engineering analyses and records of operational performance in the field should be made available to the designer.

It should be noted that information concerning the functionality of a given part has become more readily available using the processes intrinsic to CAD systems. As these systems become more sophisticated the profile of needs in this area should change radically. Still, the value of design retrieval providing functional (or even feasible) information to the designer will have value. A favorite story around the semiconductor fabrication industry has to do with the conflict between a chip design engineer and a manufacturing engineer working on a development line:

> *After numerous meetings and arguments, the designer flatly stated that manufacturing engineers were "bums" who weren't doing their jobs. "You always say it's impossible to fabricate new chip designs." The response was that the "line" of gold specified by the design engineer in a new chip had a width of 80% of the diameter of a gold atom. "Do you know who manufactures machines for dividing up gold atoms?"*

New CAD systems for designing semiconductors have often have built-in design rule systems which would prohibit such design errors, but the prevention of more esoteric errors, not traditionally included in design rule sets, would induce a more efficient design process with fewer error-induced redesign iterations.

The extension of the logic of evaluation to shop floor involves retrieval of a particular design's **manufacturing evaluation**. This extension is of particular benefit because it can be directly associated with real (often well-tracked) performance data. For example, a new design can be compared with similar (on the basis of the code) to existing designs both for the purpose of aiding the development of unit manufacturing cost (UMC) estimates as well as referencing these estimates to actual UMCs of past designs. Mosier and Mahmoodi (1995) discuss, in detail, notions of reference to past design manufacturing costs. If the estimated UMC of a newly designed part falls in region C of Figure 1, the designer should investigate why the cost is high; if the estimate is in region A, and the accuracy and completeness of the estimate, and of the part design itself, should be examined.

In the context of manufacturing the most common evaluation approach involves **assembly evaluation**. The approaches and benefits of design-for-assembly (DFA) are well known. What is new in the domain of GT part coding and identification is the use of the code itself as the mode of evaluation. This can be simple, as in the study by Mosier and Eichler (1994) where the evaluation of the assembly was intrinsic in the code. The population of entities encoded was of assemblies. The higher each digit's value, the more complex or difficult this assembly attribute was. If an assembly design had not previously been manufactured in-house, no designs in the database would have similar code values. This study used this feature to provide some degree of DFA oriented evaluation, *with respect to in-house assembly capabilities*. The justification for the project included perceived "distance" between the corporate design and manufacturing communities.

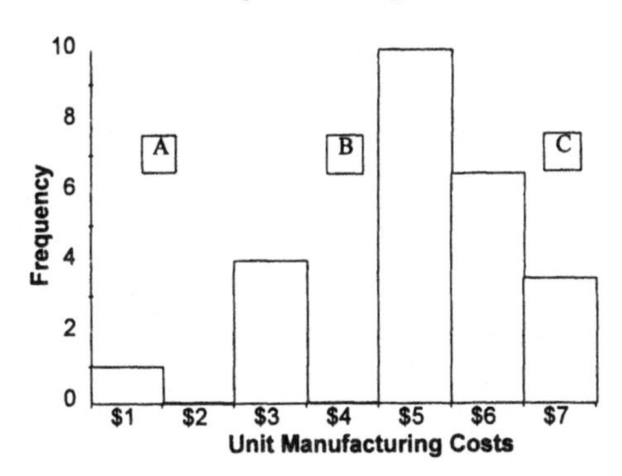

FIGURE 1. Unit cost of a part family (Mosier & Mahmoodi 1995).

To illustrate, the client firm, a domestic manufacturer had a history of efficiently assembling "user replaceable components" (URCs), using highly automated, robotic assembly lines. A very profitable situation. A highly skilled engineer working on a new product made a judgment call that the URC should be fastened together using ultrasonic welding rather than screw fasteners. The decision was based on good engineering knowledge, but little knowledge firm's the existing manufacturing capabilities. Ultimately, the retooling required to accommodate the new design was prohibitively expense. So the new URC design was assembled by hand - not a very profitable situation.

Table 3 presents a partial list of studies within each *focus-and-form-of-the-information-being-retrieved* category.

The rest of this chapter will focus on three studies that illustrate code development efforts, including research studies and industry applications, of a relatively non-standard nature. The studies do not necessarily represent a comprehensive survey, but they illustrate some interesting and insightful approaches and applications of GT coding.

TABLE 3. Studies Categorized by Focus and Form of Information Being Retrieved.

Information Focus	**Industry or Focus**	**Citation**
Basic Design Retrieval	MACHINE TOOL	Opitz 1970
Focused Retrieval	FIXTURES: ROBOT GRIPPERS: SIMULATION MODELS: RECYCLING: CIRCUIT BOARDS ELECTRONIC COMPONENTS	Sun and Chen (1995, 1996) Mosier et al. (1993) Ozdemirel et al. (1993) Hentschel et al. (1995) Styslinger & Melkanoff (1989) Bao (1989)
Engineering and Manufacturing Data Retrieval	PAPER MANUF. EQUIP.	Mosier and Janaro (1990)
Functional Evaluation	GENERAL	Mosier and Mahmoodi (1995)
Manufacturing Evaluation	GENERAL	Mosier and Mahmoodi (1995)
Assembly Evaluation		Liu and Fischer (1995) Mosier and Eichler (1994)

At this juncture, it makes sense to return to the basics of GT identification and coding. The primary purpose of the exercise is to retrieve information. The primary difficulty is how to identify the information you wish. An insightful example (blatantly stolen from a presentation by Inyong Ham) is provided by the initial work experience of a young, intelligent Korean mechanical engineer:

> *Our hero (actually Ham) was employed by Cushman, where his first assignment was to design a new disk brake for one of their vehicles. After a stressful week mulling about in his cubicle our hero finally built up the courage to speak to his supervisor. In his supervisor's office he finally confessed that although he excelled in his engineering courses, and he had a great deal of industrial experience throughout his academic career, he was unfamiliar with disk brakes. In fact, he had never seen a disk brake. "Was it possible for him to see an existing design to use as a basis for his design assignment?" The supervisor simply grinned, pulled a blueprint off a shelf behind his desk, and handed it to our hero.*

2.1. Retrieving Modular Fixture Design

In a recent paper by Sun and Chen (1996), the focus of code development was on modular fixtures. The authors proposed the use of their index (their terminology) system as the basis for a case-based reasoning (CBR) system. The logic proposed for a CBR system was *"When a CBR system faces a new problem, the most similar previous case is retrieved and modified to satisfy the new situation."* Their coding (or index) system was hierarchical, contains 7 digits, described in the following.

WORKPIECE DESCRIPTION:

- **Digit 1:** Workpiece class: This digit had seven categories first describing whether or not the part is rotational, non-rotational, or requires variable rotation. Further, if the part is rotational, three categories of *length to diameter ratio* were designated - $L/D \leq 0.5$ [*flat disk*], $0.5 < L/D < 3$, and $L/D \geq 3$ [*long cylinder*]. If the part is non-rotational, the ratio the three dimensions were similarly assigned to one of three categories.

- **Digit 2:** Workpiece external shape: Being hierarchical, this digit had specification dependent upon the value of digit 1. For example, rotational parts, were categorized as: (1) smooth (uniform diameter along the entire length), (2) cone shaped (diameter stepped to one end), or (3) having a variety of diameters along the its length. Depending upon the value of digit 1, the values in digit range from "1 to 3" to "1 to 5."

- **Digit 3:** Workpiece internal shape: This digit was not hierarchical, ranging from 1 to 6, i.e., (1) no internal hole, (2) small single hole (< 30 mm), (3) bigger single hole, (> 30 mm), (4) more than one hole with parallel axes, (5) more than one hole with non-parallel axes, or (6) non-circular holes.

RELATIONSHIP BETWEEN WORKPIECE AND FIXTURE:

- **Digit 4:** Workpiece Locating Direction: This digit described the orientation of the workpiece with respect a *base plate* incorporated into every modular fixture. If the part is rotational, there were three options, i.e., the main rotation axis is: (1) perpendicular to the base plate, (2) parallel to the base plate, or (3) inclined to the base plate. If the part is non-rotational, two available options were the surface of the *first locating datum* is (1) parallel to the base plate, or (2) not parallel to the base plate.

FIXTURE DESCRIPTION:

- **Digit 5:** Locating Device, the primary datum: The location of the primary reference point used by the CNC machine to initialize location. There were 8 options: (1) the base plate, (2) three or more points, (3) protruding flange surface, (4) step surfaces, (5) V-block, (6) pin, (7) hole, or (8) drilling support.

- **Digit 6:** Locating Device - secondary datum: Similar as before but with 9 options: (1) the base plate, (2) two or more points, (3) contour surface, (4) protruding flange surfaces, (5) hole perpendicular to the primary datum, (6) V-block, (7) arc-block, (8) drilling support, or (9) free.

- **Digit 7:** Locating Device - tertiary datum: (1) one point, (2) two or more points, (3) protruding flange surfaces, (4) hole perpendicular to the primary datum, or (5) free.

Interestingly, clamping devices were excluded from their coding system. A number of general insights were offered toward the development of a set of principles to guided in the building coding systems, including the following:

1. The system must agree with the *"thinking way"* of the human designer.
2. A fixture design is represented by an index code, i.e., the general shape of the code infers the function of the fixture. This implies that *"the order of the digits ... is also the order of priority of the features."*
3. The code should be as simple as possible.
4. The use of the CBR methodology requires that the indexing system be capable of being computerized.

2.2. End-of-arm-tooling (EOAT) Design Retrieval

In a paper by Mosier, Ruben, and Talluru (1993), the focus of the code development project was on EOAT (End-of-arm-tooling), i.e., the mechanisms required for grasping the part while performing the insertion and assembly processes on a dedicated robotic assembly line.

The primary justification of the project was to reduce the high cost of the design and fabrication of specialized EOAT. Further, since the nature of the business involved many products with extremely short life-cycles, they found that they were often designing new EOAT for each new product, while old EOAT ended up idle in storage. Thus, the coding system had the potential of saving design and engineering costs, and, if old EOAT could be reused, potentially saving the fabrication costs.

An anticipated secondary benefit could come directly from the product design engineering process. The thought was that if the "code generator" focused on the part being inserted or assembled, and, thus, if part design engineers using the system was made aware of existing designs of EOAT designs that were very "close" to being able to handle the required insertion, then rationale decisions could be made concerning part design modifications to make the use of existing EOAT. Thus the objectives of this coding project were:

- to induce efficiencies in the design and fabrication of EOAT,
- to develop a process where information about existing EOAT would be made available to designers, potentially resulting in part design modification to accommodate the use of existing EOAT, and, to generally reduce the information gap between manufacturing and design so prevalent in today's industry.

The structure of the code was of a numerical monocode. It was an eleven-digit code where each digit could range from 0 to 9 (or less). The first digit indicated the class of mechanism to be coded, and initially only included EOAT. The plan was that the structure would be expanded in the future to the population of entities broadly described as "portable tooling." The information in other digits were:

Digit 2: Part Geometry
Digit 3: Part Contact Configuration
Digit 4: Gripper-part adhesion method
Digit 5: Lateral movement required for insertion
Digit 6: Rotation required for insertion
Digits 7-9: Yaw, pitch, and roll required for insertion
Digit 10: Actuator power
Digit 11: Number of contact points

Figures 2 and 3 illustrate the options available for coding digits 2 and 3. These figures indicate the focus on the part being insertion, and, to some degree, the format of the code development software. A user was presented with a series of graphical screens. Selecting the appropriate option in each digit screen in sequence developed the code for an EOAT.

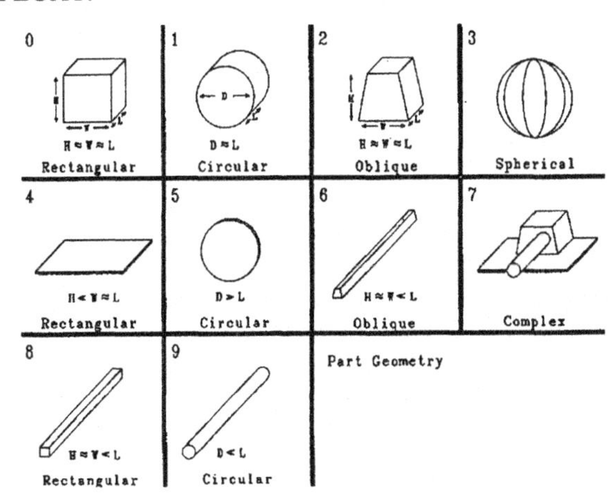

FIGURE 2. Coding part geometry (Mosier et al. 1993).

These figures also indicate the tactical focus of the coding system. It was intended that designer engineers using the system would be able to gain some level of insight concerning the complexity of the EOAT required to complete the assembly task, and thus, infer some (albeit imperfect) degree of evaluation of the design-for-assembly (DFA) characteristics of the part. For example, the client firm judged that the *Part Contact Configuration* options available in Figure 3 were presented in order of *increasing* EOAT complexity. The accuracy of this judgment is arguable, but the intent was that the system provides some degree of information concerning the part's "ease of assembly" to the design engineer.

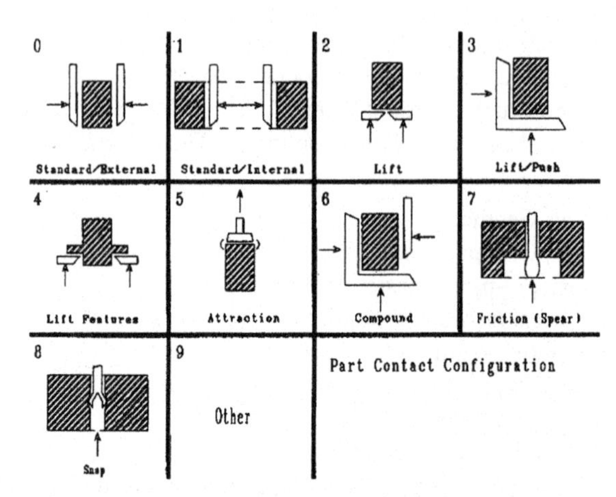

FIGURE 3. Coding part contact configuration (Mosier et al. 1993).

Additionally, this characteristic of the system (termed *ordered* by the authors) and a characteristic they termed *comparable* were intended to make the system more amenable to subsequent numerical analysis - enabling the potential of future analyses of the EOAT database for meaningful groupings. The immediate implication of these system characteristics was that the magnitude of measures of *distance* between two EOAT designs had comparative meaning. For example, using three parts with code values as listed below:

X	1	2	3	6	6	0	2	3	1	1
Y	1	4	3	1	0	0	0	0	0	1
Z	1	5	3	1	1	0	0	0	0	1

summing the code values for a given part, i.e., X has a sum of 27, for Y, 10, and Z, 12, users may infer that part X requires the most complex EOAT to accomplish the required assembly. Further, if the analyst attempts to measure how different the required EOAT are, say by using the Euclidean distance metric, i.e.,

$$d(A,B) = \sqrt{\sum_{i=1}^{11}\left(v_{A,i} - v_{B,i}\right)^2} \tag{1}$$

where $v_{A,i}$ is the code value for digit i of the code for part A, which in this case yields $d(X,Y) \approx 9.11$ and $d(Y,Z) \approx 1.414$, it is valid to say that parts Y and Z are much closer to each other than are parts X and Y.

The hardware platform for the system was the INTERGRAPH MicroStation, to enable future connection directly to the planned corporate wide CAD database. The user interface and a "flat" database structure used for querying for EOAT designs with matching or "close" codes was written in C. The options for conducting searches, or querying, the data base were not sophisticated. Users could direct the system to search on matches, of complete or partial (with the wildcard "*" specification) code values that were entered into the system. Future plans were to migrate the system to the microcomputer environment; specifically to laptops, to facilitate the use of the system in meetings held by design and manufacturing personnel during the new product design cycle.

Again, a number of general principles of design coding systems were proposed:

1. **System Completeness:** The coding system must include all entities in a specified population. The notion of a Pareto-like focus on the 80% with the highest analytic value is not appropriate.

2. **Specification Consistency:** The coding system must behave like a mathematical function, where the code development process will consistently yield the same code values for a given or similar design.

3. **Accurate Discrimination:** The code development process must yield retrievals of design families of reasonably similar size (number of designs retrieved). For example, dimensional categorizations must be based upon a reasonable analysis of the dimensional characteristic's frequency distribution.

2.3. Evaluative Identification Code

Features-Based.

Liu and Fischer (1995) report the development of a feature-based information model, using PDES (*Product Data Exchange*)/STEP (*Standard for the Exchange of Product model data*) information entities as a basis. Figure 4 presents the assembly used to illustrate the operation of their coding system. A unique feature of their system is the incorporation of a coding (and implicit evaluation) mechanism for required assembly processes. Parts P1, P3, and part P4 are assembled using two "agents" (bolts) at joints J1 and J4, respectively. Further, part P2 joins (using what are termed "operational" joints) P1 and P3 at joints J2 and J3, respectively.

Figure 5 presents the salient dimensions of P1, and Figure 6 presents an abbreviated STEP analysis of P1 (the left most part in Figure 4). Figures 7 and 8 present similar details for P2. In their coding structure, each part *or joint* is assigned a code (alphanumeric). The structure of their coding system is partitioned as:

- Physical Characteristics: Rotational, Size, and Weight.
- Geometric Characteristics: α Symmetry, β Symmetry, and Number of Assembly Ports.
- Handling Characteristics: Stability, Sticky, Tangling, Nesting, Overlapping, and Flexibility.

Key and unique elements of the system are:

- The alignment of the code and, intrinsically, the code development process established part analysis procedures (i.e., STEP).
- The inclusion of parts *and joints* into the coding system (see Figures 9 and 10 for their part and joint dictionaries).

- A code structure facilitating the extraction of DFA evaluation information using Liu and Fischer's (1994) multi-attributes utility theory assembly evaluation method (MAUTAEM).

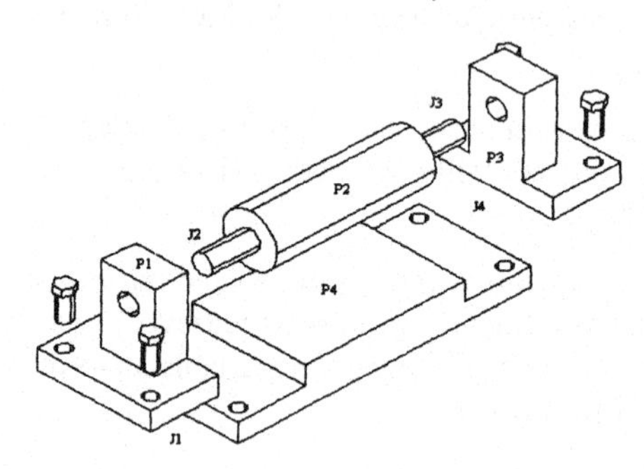

FIGURE 4. Example assembly (Liu and Fischer 1995).

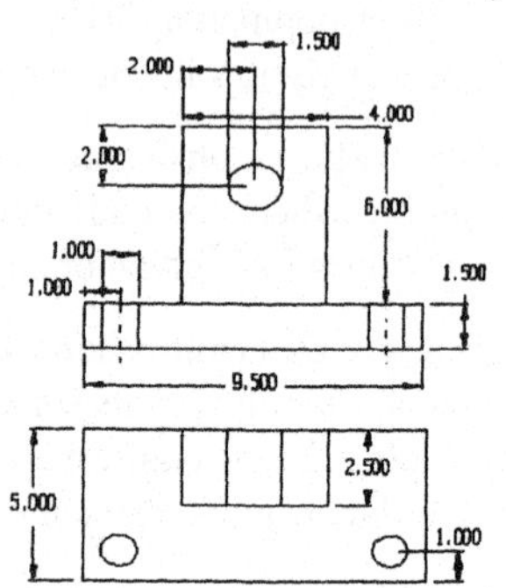

FIGURE 5. Dimensions of part P1 (Liu and Fischer 1995).

FIGURE 6. Product definition of P1 (Liu and Fischer 1995).

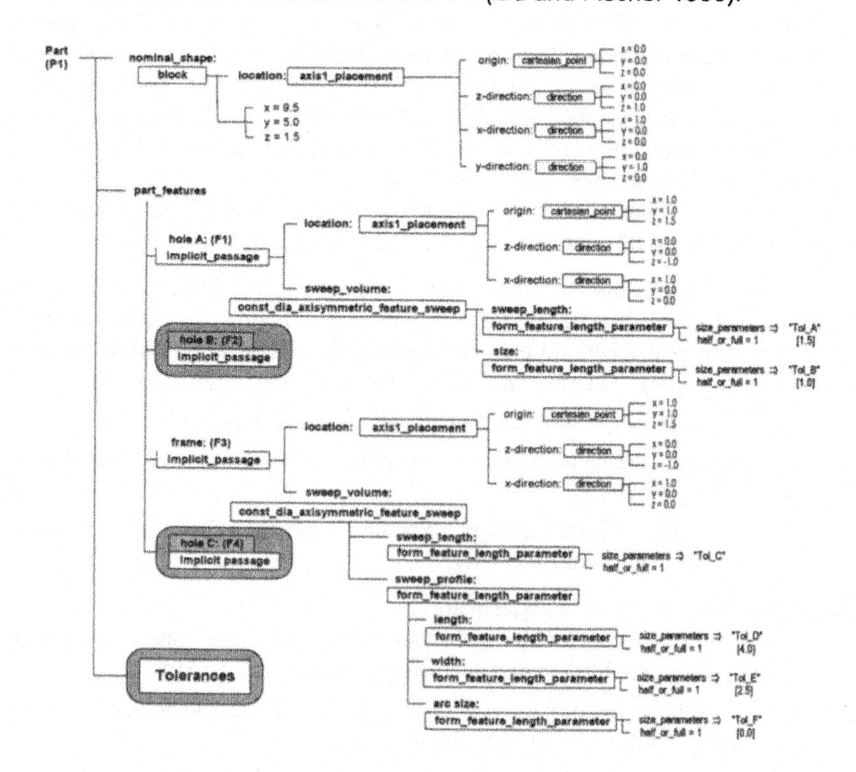

FIGURE 7. Dimensions of P2.

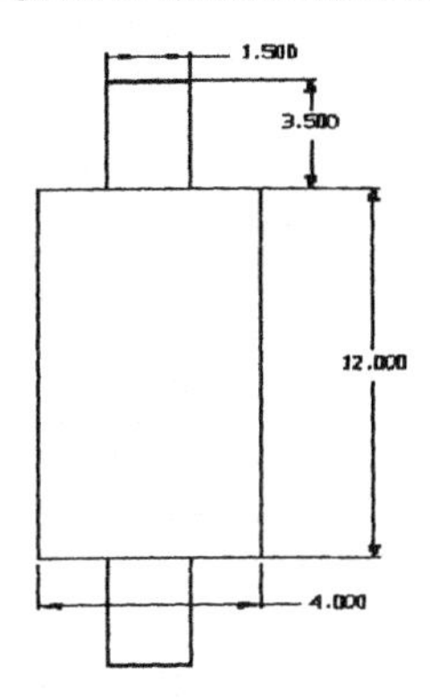

FIGURE 8. Product definition of P2 (Liu & Fischer 1995).

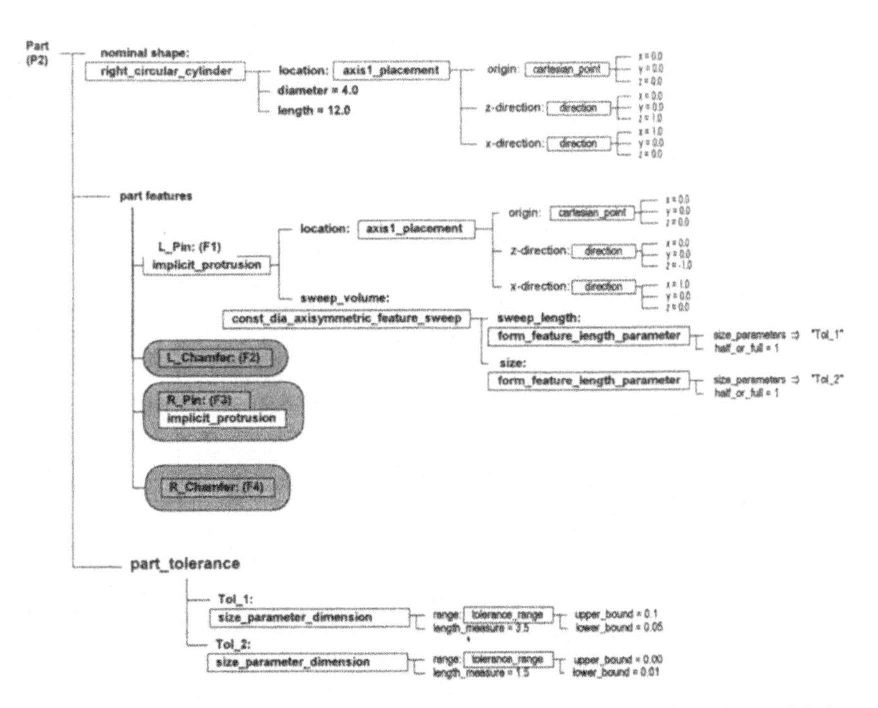

Examples of their part code structures using part P1 (see Figure 4) and part P2 (see Figure 4) are presented in Figure 11, and examples of their **joint** code structures using joints J1 (see Figure 4) and joint J2 (again, see Figure 4) are presented in Figure 12.

Part Characteristic	Assembly Factor	Definition	Part Code Index And Number of Digits
Part ID		Identification of a part.	1 [3 digits]
Physical	Rotational	Turned parts or milled parts?	21 [1 digit]
	Size	Measure of size, essentially a combination of length, width, and height	22 [1 digit]
	Weight	Measure of mass of an assembly component	23 [1 digit]
Geometry	α Symmetry	Measure of rotational symmetry of a part about an axis perpendicular to the axis of insertion (Boothroyd and Dewhurst 1987)	31 [1 digit]
	β Symmetry	Measure of rotational symmetry of a part about its axis of insertion (Boothroyd and Dewhurst 1987)	32 [1 digit]
	Number of Assembly Ports	Number of joints of a part that connect with other parts.	33 [2 digits]
Handling	Stability	Measure of low center of gravity, and features that prevent jamming during feeding and assembly.	41 [1 digit]
	Sticky	Measure of size and smooth surfaces tending to stick together.	42 [1 digit]
	Tangling	Measure of open loops easy to become interconnected with other parts.	43 [1 digit]
	Nesting	Measure of unwanted self-alignment features	44 [1 digit]
	Overlapping	Measure of level of shingling of parts one atop the other	45 [1 digit]
	Flexible	Measure of thin and long parts with low stiffness increasing difficulty of assembly	46 [1 digit]

FIGURE 9. Code dictionary of parts (Liu and Fischer 1995).

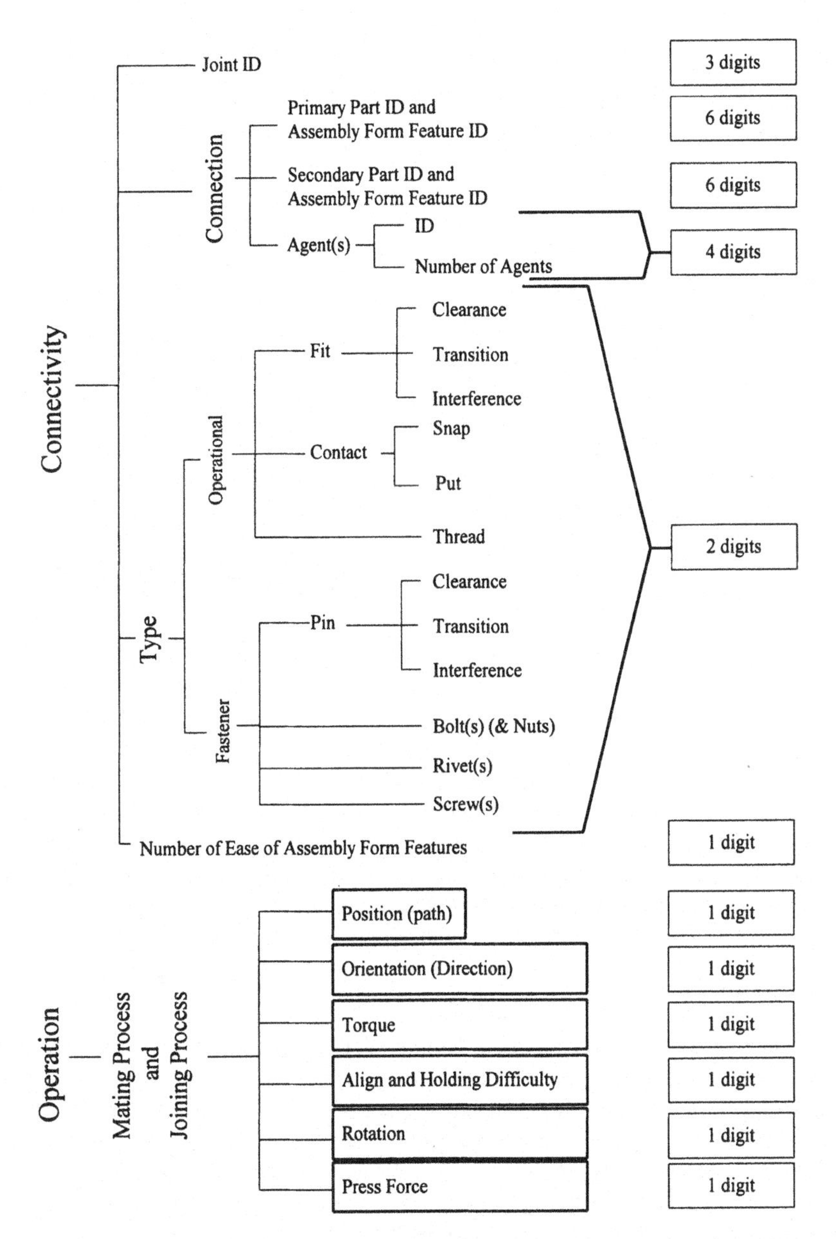

FIGURE 10. Code dictionary of joints (Liu and Fischer 1995).

Liu and Fischer's (1995) report is of a system where automated links do not yet exist. The development of part and joint codes and the mechanism for evaluating the assembly designs are still accomplished through user interaction.

FIGURE 11. Part codes for P1 and P2 (Liu and Fischer 1995).

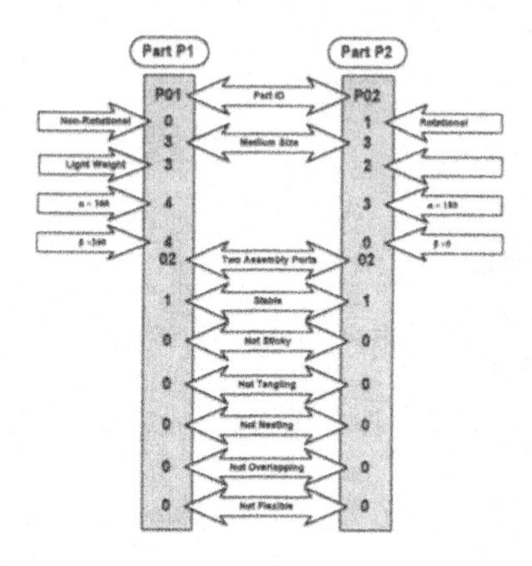

FIGURE 12. Joint codes for J1 and J2 (Liu and Fischer 1995).

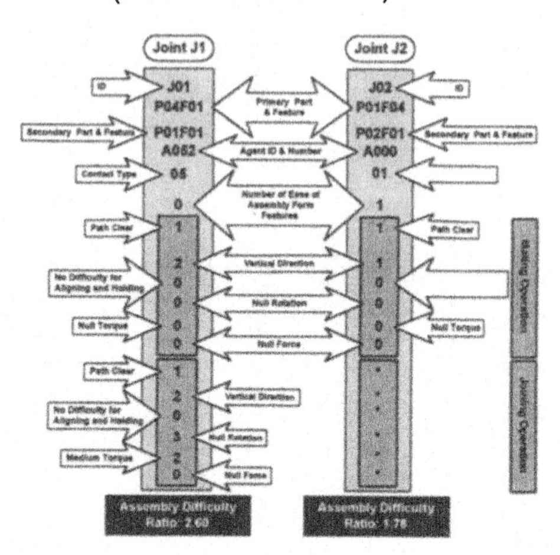

The mechanisms for moving from CAD database through a feature recognition system into coding structures and ultimately into assembly evaluation tools are still experimental. Experimental systems often involve intensive human interaction and targeted parts and assemblies intrinsically compatible with the analysis procedures. This may be characteristic of systems for which ultimate product maturity and widespread usage is not technically feasible. Alternatively, it is said that Sutherland's doctoral thesis designing the first CAD system used contrived examples to illustrate its function. Certainly, CAD technologies have achieved a maturity level sufficient for widespread use.

3. CONCLUSIONS

The conclusions for non-standard coding and identification systems should be intrinsically similar to those for standard systems. The domain of application is unlimited. In the future, given a need for part, assembly, or mechanism, the system would be able to retrieve (and automatically modify) a design, possibly identify and locate a candidate which already exists physically in a warehouse somewhere. A current view (LeBlanc 1995) of the operation of such a system could be couched in the reality of modern manufacturing design process.

Consider a meeting of design and manufacturing engineers and those who "envision" products (possibly marketers; possibly others). In this meeting, a part design is brought into question. Boothroyd and Dewhurst (1987) have spurred thinking towards DFA. The specific design discipline(s) define "rule" structures associated with that part's ultimate functionality. Other design disciplines are concerned with the interaction of that part with other portions of the parent product that may be impacted by decisions made here. Manufacturing types (and optimally all others in the meeting) are ultimately concerned with the bottom line.

Questions concerning form, shape, and materials requirements associated with the particular part fly across the table. An artificially intelligent system able to respond to queries concerning the part is desired. Questions range from: "What other materials are available for this part, and what are our functionality and cost trade-offs if we change materials? to "Is the part necessary at all for ultimate product functioning?"

This suggests that the ultimate coding and identification system form will not be a coding system at all. Rather, such a system will be able to observe nature and make decisions based upon that nature. Ultimately, it must involve some sort of machine "inference," making extrapolations from facts to things that do not exist.

The computer technologies where most of the work is yet undone, as we evolve toward such AI systems involve, (a) data gathering, including the computer abilities in the area of database "reading," environmental observation, and generally seeking information, (b) the development of decision making logic (e.g., inference engines), and, possibly the most important, (c) the development of pattern recognition techniques and the characterization of their operational characteristics.

The author has identified many non-standard situations where such systems would be of value. For brevity, we consider only the following situation:

The Takeoff: A large (4-engine) airliner is climbing away from a major metropolitan airport. The attachment of the right-most engine fails and the engine separates from the airliner and begins to fall. Two computerized systems immediately take over. Small systems onboard the engine realize that it is disconnected and in free fall. These systems explosively disintegrate the engine to minimize ground damage (recall the radio controlled destruction of the solid rocket boosters for the same purpose accomplished by ground control in the shuttle tragedy). These onboard systems shut off the "utilities" routed to the damage region. They also adjust the thrust of the other engines and adjust the air control configurations to prevent immediate disaster. Lastly, they prescribe all required user actions necessary for successful landing of the damaged airliner.

These examples involve problem environments that are more critical than those normally found in manufacturing, but the intrinsic logic is the same: the notion of a computer system converting a "need" into the best alternative action, best component, or best *design* is all pervasive.

REFERENCES

Bao, H.P., 1989, Group technology classification and coding for electronic components: Development and applications, *Productivity and Quality Improvements in Electronics Assembly*, (Eds. J. A. Edosomwan and A. Ballakur), McGraw-Hill, New York.

Bhadra, A. and Fischer, G.W., 1988, A new GT classification approach: A data base with graphical dimensions, *Manufacturing Review*, 1, 1, 44-49.

Billo, R.E., Rucker, R. and Shunk, D.L., 1987, Enhancing group technology modeling with database abstractions, *Journal of Manufacturing Systems*, 7, 2, 95-106.

Boothroyd, G. and Dewhurst, P., 1987, *Product Design for Assembly Handbook*, Boothroyd and Dewhurst, Inc., Wakefield, RI.

Chen, Y.T. and Young, R.E., 1988, PACIES: A Part Code Identification Expert System, *IIE Transactions*, 20, 2, 132-136.

De Fazio, T.L., Edsall, A.C., Gustavson, R.E., Hernandez, J., Hutchins, P.M., Leung, H.-W., Luby, S.C., Metzinger, R.W., Nevins, J.L., Tung, K. and Whitney, D.E., 1993, A prototype of feature based design for assembly, *Journal of Mechanical Design*, 115, 4, 723-734.

Henderson, M.R. and Musti, S., 1988, Automated group technology part coding from three dimensional CAD database, *ASME Transactions Engineering for Industry*, Aug.

Hentschel, C., Seliger, G. and Zussman, E., 1995, Grouping of used products of cellular recycling systems, *Annals of CIRP*, 44, 1, 11.

Jung, J-Y. and Ahluwalla, R.S., 1991, FORCOD: A coding and classification system for formed parts, *Journal of Manufacturing Systems*, 10, 3, 223-232.

King, J.R., 1980, Machine-component grouping in production flow analysis, *International Journal of Production Research*, 18, 2, 213-232.

Kini, R.B., 1994, A data tree structure for a hierarchical structure processing, *Computers and Industrial Engineering*, 26, 3, 551-563.

Kini, R.B., Taube, L. and Mosier, C.T., 1991, Part identification and group technology: A new approach, *Journal of Manufacturing Systems*, 10, 2, 134-145.

LeBlanc, E., 1995, Xerox Corporation, personal communication.

Li, R.-K. and Huang, C.-L., 1992, Assembly code generation from a feature-based geometric model, *International Journal of Production Research*, 30, 3, 627-646.

Liu, T.-H. and Fischer, G.W., 1995, An assembly code classification and coding scheme based on a step mechanical product model, *Manufacturing Review*, 8, 1, 33-46.

Liu, T.-H. and Fischer, G.W., 1994, Assembly evaluation method for pdes/step-based mechanical system, *Journal of Design and Manufacturing*, 4, 1-19.

Lyle, L.L., 1988, What part numbering method is right for you? *Production Inventory Management Review & APICS News*, 8, 11, 35-37.

Marion, D., Rubinovich, J. and Ham, I., 1986, Developing a group technology coding and classification scheme, *Industrial Engineering*, 18, 7, 90-97.

Mosier, C.T. and Eichler, E.F.,1994, A prototype group technology design-for-assembly database and analysis system, *Journal of Design and Manufacturing*, 4, 41-47.

Mosier, C.T. and Janaro, R.E., 1990, Toward a universal classification and coding system for assemblies, *Journal of Operations Management*, 9, 1, 44-64.

Mosier, C.T. and Mahmoodi, F., 1995, Design quality: The untapped potential of group technology, *Manufacturing Research and Technology 24: Planning, Design, and Analysis of Cellular Manufacturing Systems*, Elsevier Science B.V., Amsterdam, 63-71.

Mosier, C.T., Ruben, R.A. and Talluru, L., 1993, Group technology classification and coding design considerations and Extension to Robotic End-of-Arm Tooling (EOAT), *Robotics and Computer-Integrated Manufacturing*, 10, 5, 377-389.

Opitz, H., 1970, *A Classification System to Describe Workpieces*, Pergamon Press, Oxford, UK.

Ozdemirel, N.E., Mackulak, G.T. and Cochran, J.K., 1993, A group technology classification and coding scheme for discrete manufacturing simulation, *International Journal of Production Research*, 31, 3, 579-601.

Saxe, J.G., 1936, The blind men and the elephant, *Best Loved Poems of The American People*, Originally in *Udana*, a Canonical Hindu Scripture.

Shah, J.J. and Bhatnagar, A.S., 1989, Group technology classification from feature-based geometric models, *Manufacturing Review*, 2, 3, 204-213

Shah, J.J. and Rogers, M.T., 1993, Assembly modeling as an extension of feature-based design, *Research in Engineering Design*, 5, 218-237.

Shunk, D., 1985, Group technology provides organized approaches to realizing benefits of CIMS, *Industrial Engineering*, 17, 416-423.

Sreevalsan, P.C. and Shah, J.J., 1992, Unification of form feature definition methods, *Intelligent Computer Aided Design*, North Holland - Elsevier Science, 83-99.

Srinivasan, M. and Moon, Y.B., 1995, Goal driven part classification for multiple group technology applications: a machine learning approach, *Flexible Automation and Intelligent Manufacturing*, Begell House, New York, 357-371.

Styslinger, T.P. and Melkanoff, M.A., 1989, Group technology for electronics assembly, *Productivity and Quality Improvements in Electronics Assembly*, (Eds. J. A. Edosomwan and A. Ballakur), McGraw-Hill, New York.

Sun, S.H. and Chen J.L., 1995, A modular fixture design system based on case-based reasoning, *International Journal of Advanced Manufacturing Technology*, 10, 6, 389-395.

Sun, S.H. and Chen, J.L., 1996, An index system for modular fixture design: applied to case study reasoning, *International Journal of Production Research*, 34, 12, 3487-3497.

Tatikonda, M.V. and Wemmerlöv, U., 1996, Adoption and implementation of group technology classification and coding systems: insights from seven case studies, *International Journal of Production Research*, 30, 9, 2087-2110.

Udoka, S.J., 1991, Automated data capture techniques: a prerequisite for effective integrated manufacturing systems, *Computers and Industrial Engineering*, 21, 1/4, 217-221.

Urban, S.D., Shah, J.J. and Rogers, M.T., 1993, Engineering data management: achieving integration through database technology, *Computing and Control Engineering Journal*, 4, 3, 119-126.

AUTHOR'S BIOGRAPHY

Charles T. Mosier is Professor of Management at the School of Business, Clarkson University, Potsdam, NY. USA. His degrees are from Clarkson (BS, Industrial Distribution), Potsdam College (MA, Mathematics), and University of North Carolina at Chapel Hill (Ph.D., Business Administration). He has published numerous GT-related papers in leading production journals, including: *Management Science, Interfaces, Decision Sciences, International Journal of Production Research, Journal of Operations Management, International Journal of Computer Integrated Manufacturing, International Journal of Production Planning and Control, Omega, Production and Operations Management, Journal of Design and Manufacturing, Journal of Manufacturing Systems,* and in numerous conference proceedings. He has been involved in many research and consulting projects with firms such as AT&T, IBM, Xerox, Gleason, and Kraft. His research interests are in design and scheduling of cellular and flexible manufacturing systems, shop order release and due date assignment, systems modeling & simulation, design of efficient configurations for automated guided vehicle systems and, currently, in design of virtual cellular manufacturing environments.

B3

Part Family Identification: The Role of Engineering Data Bases

R. E. Billo and B. Bidanda

1. INTRODUCTION

Group Technology (GT) is a manufacturing philosophy designed to simplify and standardize production operations by taking advantage of similarity. Manufacturing performance parameters such as cycle time, quality and costs can be improved by capitalizing on similarities that can be found in such things as workpiece design, assembly methods, purchasing of materials, or tooling designs (Apple 1977).

Two formal methods have principally been applied to group parts into similar *families*. These include cell formation techniques such as production flow analysis and cluster analysis (Chan and Milner 1984; King 1980; McAuley 1980) where routings are systematically compared; and classification and coding (C&C) where interactive software is used to aid the user in categorizing parts under study according to such design and manufacturing attributes as function, size, and shape (Vakharia 1986). Due to the increased popularity of cellular manufacturing, cell formation techniques have recently seen increased attention while C&C systems appear to have fallen into disuse.

However, cell formation techniques are not suited to applications other than manufacturing cell design. Tasks such as materials purchasing, workpiece design, and tooling design require an approach that can assess similarity based on a multitude of attributes unlike current cell formation techniques that rely on a single attribute for grouping. For these types of applications, C&C systems best serve as the mechanism for grouping parts. C&C allows a user to classify and retrieve information based on a variety of attributes considered important to the task at hand.

In addition to the increased popularity of cellular manufacturing, we believe two additional reasons for decreased usage of C&C approaches have been the high cost of C&C software and a general lack of understanding of organizing principles used for design of robust C&C models (Billo, Rucker and Shunk 1987, 1988; Billo and Bidanda 1995). To be most effective, a C&C system should be customized for the application at hand. However, commercial C&C software packages with embedded workpiece or tooling models are most often used. In an attempt to encompass as diverse a range of applications as possible, these packages contain exorbitant amounts of general information included in the model, and not enough specific information to make them useful for the particular task at hand. As a result, many of these packages are expensive to procure and often are unnecessarily complex to program and use.

Rather than procure such external software packages, we believe that a much better C&C information system can be developed with no more effort than what goes into the development of any other type of management report, assuming modern database tools and software engineering practices are followed. This chapter describes how typical problems associated with the design of C&C systems can be overcome by adopting a more formal approach for representing the underlying C&C model and then implementing the model in a modern database management system (DBMS). Such database design principles allow implementation of C&C data models in low-cost, modern, PC-based DBMS's.

Specifically, organizing principles available in object oriented modeling (OOM) can be readily applied to C&C model design. Analogies will be developed between OOM principles and existing informal methods used for developing C&C models. Additional OOM principles will be described to illustrate how C&C models can be enhanced and integrated with other data to minimize redundancy of data entry. Finally, a case study illustrating the application of the modeling principles to the design of a purchased parts classification and coding system will be presented.

2. OBJECT-ORIENTED MODELING CONCEPTS

Object-Oriented modeling provides a mechanism for representing information by using models organized around real-world objects and concepts. The basic construct is the *object*, which combines both the data structure and its behavior into a single entity. An object may be defined as a concept, abstraction, or thing with crisp boundaries and meaning for the problem at hand (Rumbaugh, Blaha, Premerlani, Eddy and Lorensen 1991) For GT, individual part numbers, material identifications or designs would typically represent the objects of the system.

An *object class* describes a group of objects with similar properties, common behavior, and common relationships to other objects. For a procured part to be used in an assembly process, *Part* may represent an object class. The objects stored in this class may include information for each part used in assembly process.

Objects and object classes typically contain *attributes* that are data values held by objects in a class. For a Part class, attributes may include such fields as the part number, part name, the manufacturer identification number, and the part number assigned by the manufacturer of the part.

Object-Oriented models also contain *operations*. An operation is a function or transformation that may be applied to, or by, objects in a class. For the group technology problem, transactions such as classification, coding, and retrieval of part families typically represent the operations of the object.

Along with objects, object classes, and operations, OOM relies on *abstractions* to model real-world constructs. An abstraction of a system is a collection of details which can be conveniently named as a whole (Billo and Bidanda 1995). For example, when attempting to store the dimensions of a part in a C&C system, we may define a class called *Dimensions* to abstract a collection of attributes of a part including its length, width, and height. When we refer to the collection of attributes, we do not name them individually, but we abstract the collection by referring to the term Dimensions. Several types of abstractions can be used to formally design and enhance C&C models. The abstractions to be discussed include Classification, Aggregation, Association and Generalization.

Additional concepts that can be applied to formalize the design of classification models include such principles as *derived attributes*, *predicate defined subclasses,* and *recursive classes*. Each of these will be described as they relate to the C&C model design problem.

3. GT CLASSIFICATION PRINCIPLES

Depending on the requirements of the application, C&C models may be based on a variety of characteristics. Traditionally, these have included such properties as part geometry, material, function, initial form, and the like. Once these characteristics are chosen, a C&C model can be designed. There are three major modes of construction. Models that are hierarchical in nature proceed from general to specific with respect to each characteristic. For example, Parts may be Rotational or Non-Rotational. In turn, Rotational Parts may be Centric, Concentric or Gear-like. This type of coding scheme where the values of each digit is a function of the value of the previous digit is called a *monocode.* Other classification models attempt to include several related features, not necessarily hierarchical in nature. These representations attempt to incorporate diverse and possibly independent information such as workpiece geometry, material, and initial form. Here, the value of information contained within each digit is absolute and independent. Such models are represented with *polycodes.* Still other models include both hierarchical and nonhierarchical characteristics. *Hybrid codes* represent the shorthand for these models (Groover and Zimmers 1984).

This differentiation of types of coding systems used in representing C&C systems can be categorized into a variety of decision trees (Allen 1984). Such decision trees form the basis of the popular DCLASS™ classification and coding software package. These decision tree structures include a) *E-trees* for mutually exclusive branching, b) *N-trees* for non-exclusive branching, c) *C-trees* for combinations of exclusive and non-exclusive branching, d) *X-trees* for evaluating mathematical expressions, and e) *D-trees* for describing sequenced decision rules. E-trees correspond to monocode classification schemes, while N-trees correspond to polycode systems. C-trees correspond to hybrid code systems. X-trees and D-trees are unique features of the DCLASS™ system and do not correspond to the more traditional GT classification and coding schemes. However, all of the GT design principles mentioned here can be modeled with the OOM principles described below.

4. MODELING C&C SYSTEMS WITH OOM CONCEPTS

The following sections describe the OOM organizing principles typically used to model the five decision trees described above as C&C models.

4.1. Generalization With Disjoint Subclasses

Hierarchical classification models with mutually exclusive data are termed E-trees. They divide a large collection of items into mutually exclusive families. With E-trees, only one branch may be selected at each decision node. Figure 1 illustrates this concept . In this example, the Shape of a part can be either Rotational or Non-Rotational. In turn, a Rotational part can be either Centric, Concentric, or Gear-like. A part can be classified as only one of these types at each decision point.

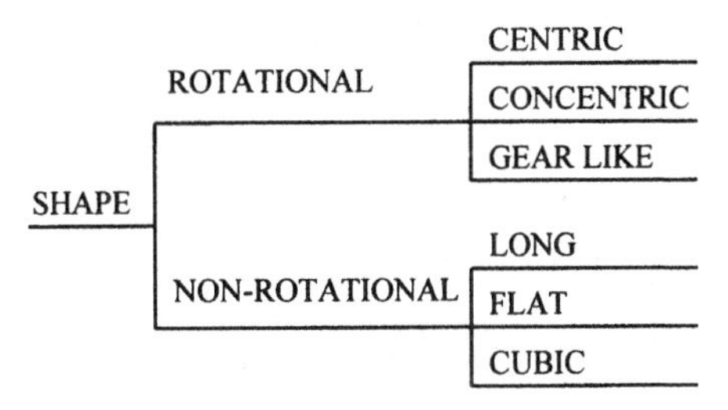

FIGURE 1. E- tree.

In OOM, mutually exclusive classification can be modeled using the *generalization* abstraction with disjoint subclasses. Generalization is the relationship between a class and one or more refined versions of it (Rumbaugh *et al.* 1991). The class being refined is called the superclass and each refined version is called a subclass. For example, from Figure 2 it can be seen that Shape is the superclass of the two subclasses Rotational and Non-Rotational. In turn, Rotational acts as the superclass of the three subclasses centric, concentric, and gear-like. Using Enhanced Entity Relationship Modeling (EER), the notation for generalization with disjoint subclasses to represent mutually exclusive classification is a circle with the letter *d* connecting a superclass to its subclasses (Elmasri and Navathe 1994). Double lines connecting a class to disjoint operator designates *total participation* of class to subclass i.e., all objects of class will be classified in one of the subclasses.

4.2. Generalization With Overlapping Subclasses

Some subclasses overlap. Allen (1984) describes this type of branch as Non-Exclusive Classification, or N-trees, in which multiple paths may be traversed. Consider the example in Figure 3a in which the user may select a combination of features, i.e., the option of traversing one or more paths to classify the part. In OOM, the type of situation depicted in Figure 3a is best modeled as generalization with overlapping subclasses. Figure 3b illustrates the equivalent OOM representation of the N-tree. In this case, the subclasses Chamfer, Groove, Hole, and Thread overlap because some parts may have one or some combination of these features. A circle with the letter *O* indicates overlapping subclasses.

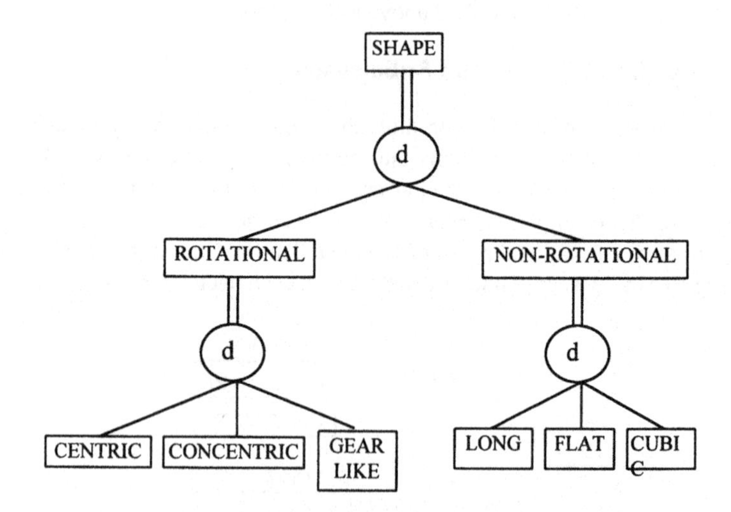

FIGURE 2. Generalization with disjoint subclasses.

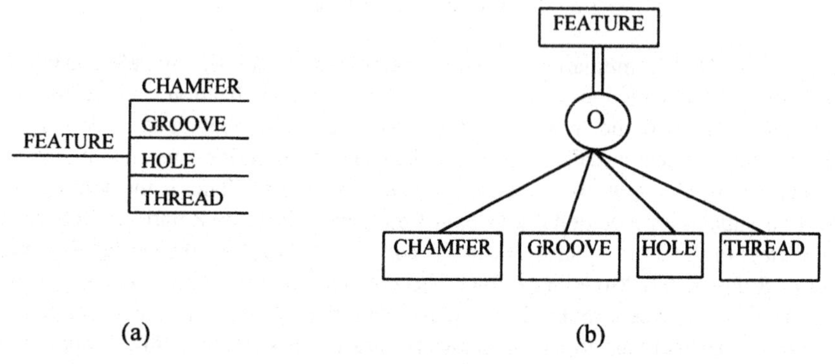

FIGURE 3. (a) N-tree; (b) generalization with overlapping subclasses.

4.3. Classification And Derived Attributes

X-trees are used for instantiating attributes and evaluating mathematical expressions. Figure 4a illustrates this process for classifying the size of a rotational part. In this example, the user enters the Length and Diameter of the part, and the expression LD = Length / Diameter is then evaluated.

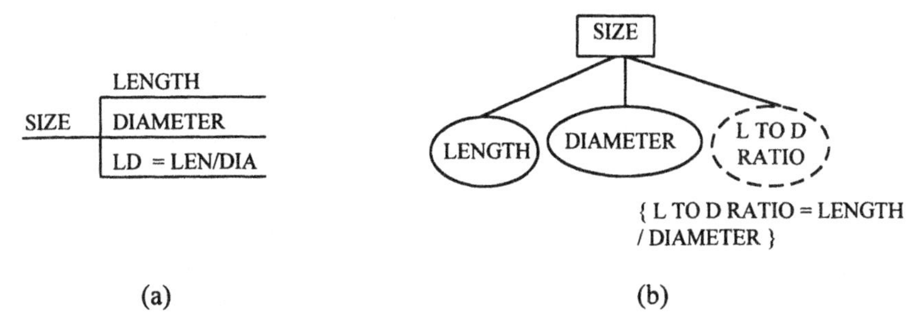

FIGURE 4. (a) X-tree; (b) classification with derived attribute.

In OOM, this process can be modeled using the abstraction *Classification* and making use of *derived attributes*. Classification means that objects that share the same attributes are grouped into a class (Rumbaugh *et al* 1991). For example, when classifying a workpiece, one typically must capture the part's size. Therefore, the class Size may be defined. For rotational parts, the developer may decide that the size attributes of interest will be the Length and Diameter of the rotational part. Classification may be used to properly model the Size class with the primitive attributes Length and Diameter. Figure 4b shows this model.

Mathematical expressions can be evaluated and stored through derived attributes. A derived attribute is defined as a function of one or more attributes (Rumbaugh *et al.* 1991; Elmasri and Navathe 1994). The notation for a derived attribute using EER modeling notation is a dashed circle around the attribute. The expression that determines the derived value is denoted as a *constraint* which is dictated in a declarative manner within braces below the derived attribute.

4.4. Predicate Defined Subclasses And Recursive Classes

Allen (1984) discusses the need for decision rules in C&C models and describes two types of D-trees for this purpose. The Type 1 D-tree permits the developer to incorporate decision rules into the tree structure for which a single consequence results. For example, consider the decision tree illustrated in Figure 5a. In this example, the type of tooling chosen is dependent on the length to diameter ratio of the rotational part under consideration. If $LD < 3$, then a chuck is selected. If $3 < LD < 8$, then a center is used, and if $LD > 8$ then a steadyrest is used.

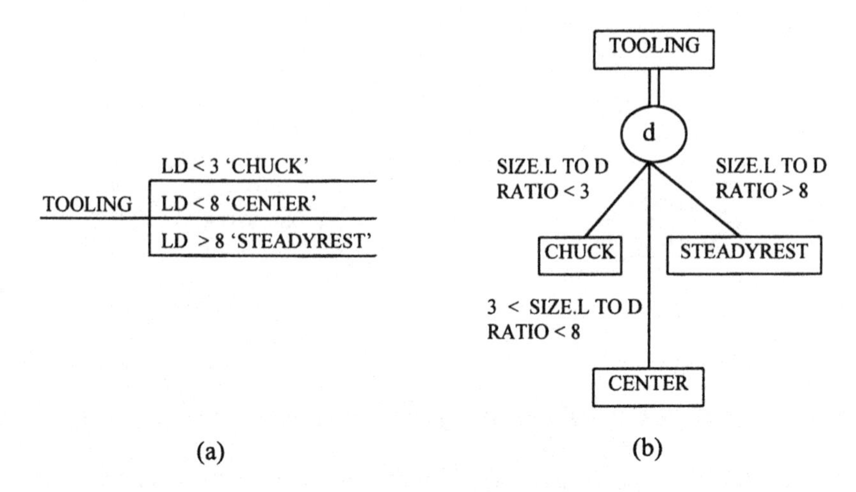

FIGURE 5. (a) Type 1 D-tree; (b) predicate defined subclasses.

Usage of *Predicate defined subclasses* would be the technique to emulate similar decision rules in OOM, since they serve to place constraints on the attributes of a superclass. In this case, subclass membership is defined by rule, and all objects whose values satisfy the rule belong to the subclass. For example, Figure 5b shows how the length to diameter ratio from the Size class is formulated into a rule and denoted using EER modeling notation. The Tooling class consists of three subclasses: Chuck, Center, and Steadyrest. A constraint is placed on each of the subclasses. If an object is to be instantiated in the subclass Chuck, then the Length To Diameter Ratio from the Size class illustrated earlier in Figure 4b must be less than 3 units. Classes Center, and Steadyrest have similar constraints restricting membership into their respective subclasses.

A Type 2 D-tree contains rules in which a consequence contains a sequence of actions. Consider the example in Figure 6 in which the sequenced set of operations are displayed to create a countersink hole. From the Holes branch, if the countersunk hole is selected, then the operations Drill Diameter 1, Drill Diameter 2, and Deburr are automatically selected.

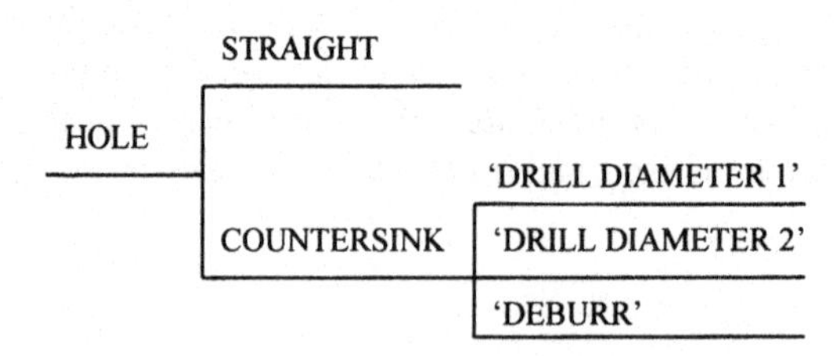

FIGURE 6. Type 2 D-tree.

To emulate this sequence of actions, a *recursive class* is applied. A recursive class shows ordered lists of objects by having attributes in a class to point to other attributes of the same class thereby linking ordered objects. Figure 7 illustrates a recursive class for the process plan of the countersink hole. In this model, the recursive class Process is created and associated with the Countersink class. The Process class will allow the operations to be listed as an ordered list of actions.

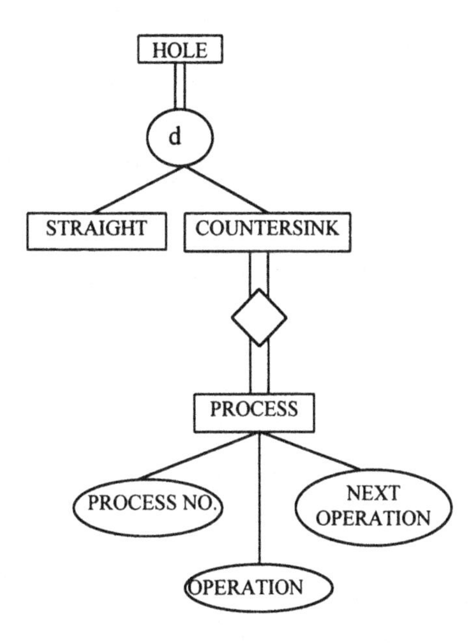

FIGURE 7. Recursive class.

Table 1 illustrates how the Process recursive class operates if implemented in a relational DBMS. In the first tuple of this table, the first operation (Drill Diameter 1) listed under the Operation attribute points to the second operation (Drill Diameter 2) listed under the Next Operation attribute. In turn, the DBMS software is designed such that it has capabilities to query remaining tuples to find under the Operation attribute the tuple listing the same attribute values. In this case, Drill Diameter 2 can be found in the second tuple. In turn, Drill Diameter 2 is associated with Deburr, and the process repeats itself.

TABLE 1. Recursion In A Relational Database.

PROCESS

PROCESS NO.	OPERATION	NEXT OPERATION
101	DRILL DIAMETER 1 ⟶	DRILL DIAMETER 2
101	DRILL DIAMETER 2 ⟷	DEBURR

4.5. Aggregation

Allen (1984) suggests the use of C-trees to combine E-trees, N-trees, X-trees, and D-trees. This action allows the designer to have a completed classification model that incorporates mutually exclusive branching, nonexclusive branching, instantiation of data into attributes, evaluation of mathematical expressions, and execution of simple decision rules. Figure 8 shows a C-tree that combines several of the models illustrated in previous figures.

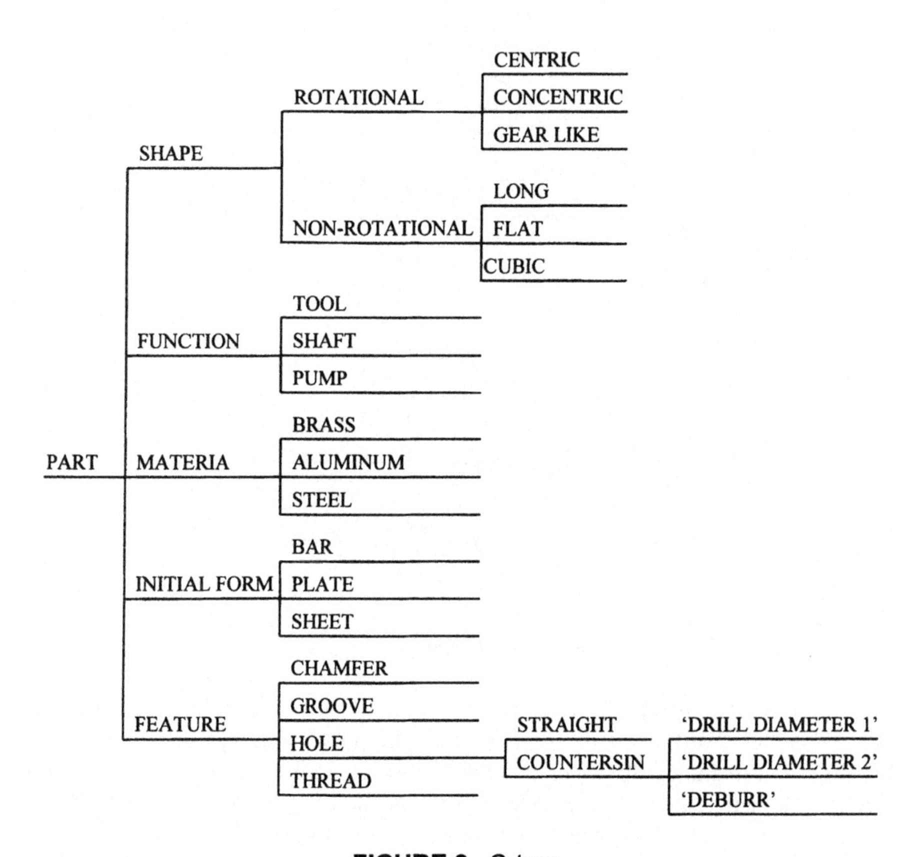

FIGURE 8. C-tree.

In OOM, C-trees can be represented through the *Aggregation* abstraction. Aggregation is the *part-whole* relationship in which objects representing the components of an assembly are associated with an object representing the entire assembly. An example is the bill-of-materials tree where an assembly consists of a collection of discrete components. Aggregation is denoted by connecting classes with a diamond and a slash. The slash denotes the assembly end of the relationship. Figure 9 illustrates aggregation. Here, Part is an assembly class that contains component classes Shape, Function, Material, Initial Form, and Features.

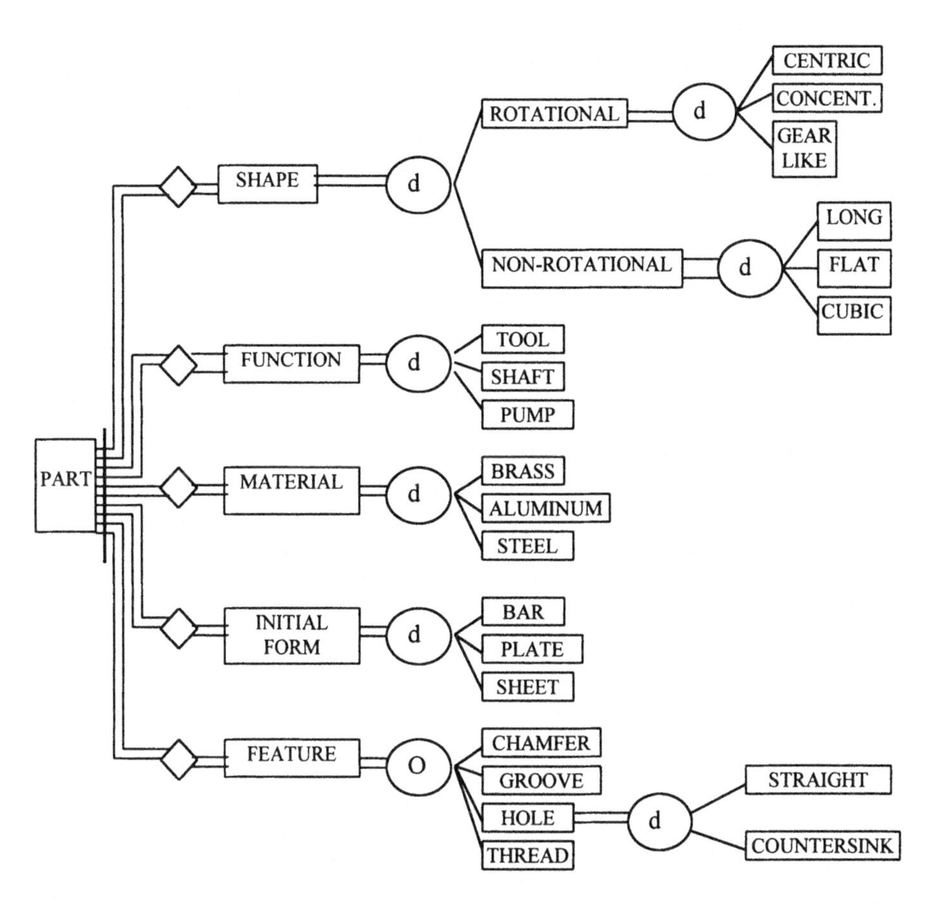

FIGURE 9. Aggregation.

5. ENHANCING GT WITH ASSOCIATION, ROLE NAME, AGGREGATION

OOM principles can further formalize the design of C&C models by minimizing redundancy and providing a mechanism to enhance the integration of the C&C system with other databases. Additional organizing principles often found to be helpful included *Association* and *Role Names.* In addition, *Aggregation,* as described above, helps in making the C&C model more efficient by eliminating redundant object classes.

5.1. Association

An association is used to connect objects from several independent classes (Elmasri and Navathe, 1994). Associations can be described as a looser form of aggregation in that aggregation is best expressed as the part-whole relationship while association is expressed as a linking of two independent object classes. In the English language, associations often appear as verbs. For example, "An operator is *assigned to* a machine", would represent the linking of the class Operator with the class Machine through the *assigned* association.

Associations serve an important role in relating the C&C model with other classes that may not be part of the model, but are part of a much larger database. For example, in designing and implementing a C&C system for grouping parts to be purchased, it is often useful to associate the classification model with classes that are part of a larger purchase order database. Consider the model illustrated in Figure 10. In this example, Part represents a class with many subclasses representing the start of a GT classification tree. The association *describes* shows the relationship between the separate class Purchase Order Line Item and the GT class Part. This association can be read as "Purchase Order Line Item *describes* Part". Normally, purchase order information is not part of a C&C system. This association allows the developer to easily provide a link between a particular part with its GT descriptive information and the purchase order that will serve to purchase that part. When this particular C&C system was developed for the customer, the two systems (Purchase Order module and C&C module) were integrated and completely transparent to the customer allowing the customer to complete a purchase order by using the C&C system to electronically select and insert specific part line item information into the Purchase Order.

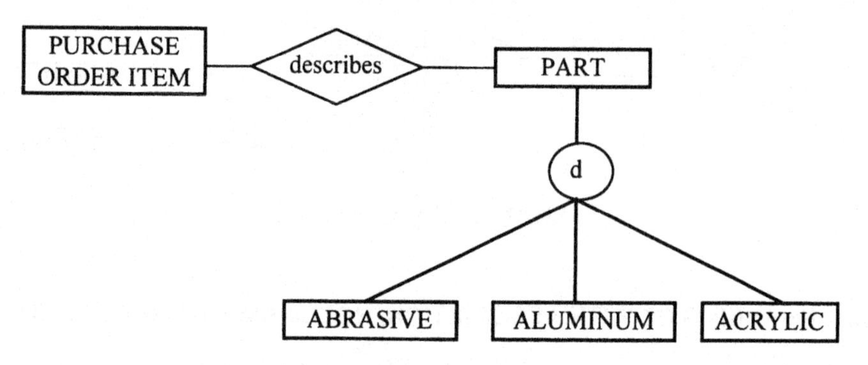

FIGURE 10. Association.

5.2. Role Names

A role name uniquely identifies one end of an association or an aggregation. Role names are a helpful tool in reducing unnecessary redundancy in C&C models. For example, consider the decision tree and the top view of the compacting die of Figure 11a. The die consists of two distinguishing geometric features: the outside shape and the inside shape. From the decision tree in Figure 11b, it can be seen that there are several choices for both the outside and inside shapes of the die: rectangular, elliptical, or circular. To accommodate these choices, the decision tree must repeat the geometric choices. If repeated often enough in the C&C model, the data model and resulting database can become cumbersome and unwieldy.

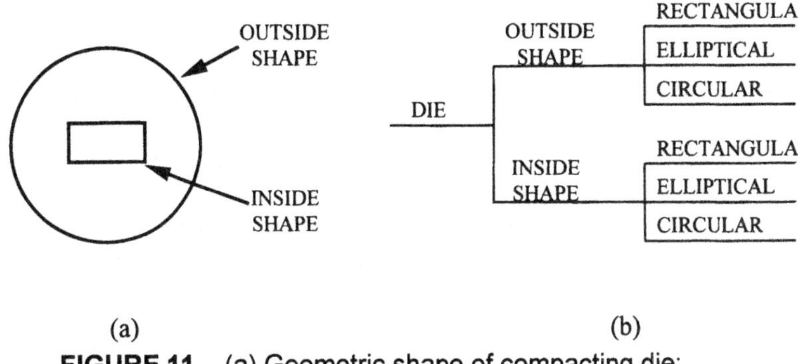

FIGURE 11. (a) Geometric shape of compacting die; (b) decision tree representation of die.

Figure 12 shows how this problem is resolved with role names. Here, the die is modeled as an aggregate class with a component class called Shape Type. In turn, Shape Type acts as a superclass for subclasses Rectangular, Elliptical and Circular. Shape Type actually plays two roles as a component of Die. In one role, it acts as the outside shape; in the other role it acts as the inside shape; therefore, the role names outside shape and inside shape are placed at one end of the aggregation nearest to the class in which they represent the role.

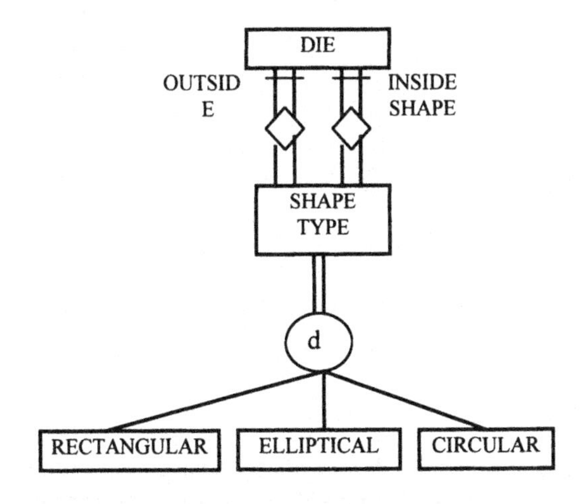

FIGURE 12. Use of role names to represent die.

5.3. Reducing Redundancy With Aggregation

A designer can take advantage of aggregation to eliminate repetitive classes in the model. Consider the decision tree illustrated in Figure 13a. In this example, Aluminum, Copper and Steel all have branching mechanisms to represent a choice of the Initial Form of the material. In the model, Initial Form along with its choices Bar, Plate, Rod, Sheet or Tube are represented three times in the model.

This redundancy can be eliminated through the usage of aggregation in which multiple aggregate objects share common component objects. This is illustrated in Figure 13b. In this example, Initial Form serves as a component class to all three aggregate classes Aluminum, Copper and Steel, and therefore, there is no need to repeat the component class for each aggregate class. This efficiency is, in turn passed on to the subclasses of Initial Form in that they no longer need to be repeated in the model nor in the implementation of the model in the database.

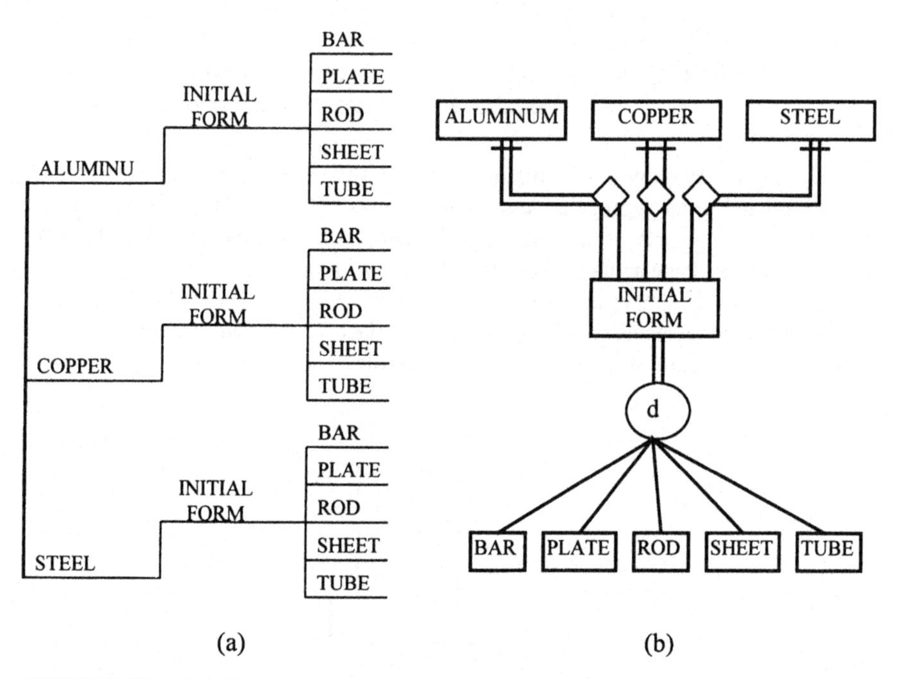

FIGURE 13. (a) Redundancy in a decision tree; (b) redundancy reduction.

6. CASE STUDY

The OOM modeling principles described above have been applied to a large variety of engineering and manufacturing applications over a 10 year period. The authors have developed C&C systems for the classification of composite materials, machined parts, fixture designs, and purchased parts.

One of the most useful applications of C&C systems is in retrieving information for parts to be purchased in support of manufacturing operations. In the current application, a mid-size manufacturer of spectroscopy equipment was spending a significant amount of time (30 hours per week) in the manual identification and order placement of raw materials and purchased parts needed on the shop floor. Over 4,000 parts were used in the manufacturing operation and included raw materials such as steel, copper and aluminum, as well as electrical components, and consumable supplies used in the manufacturing process.

Problems associated with the manual procedures for identification and procurement of parts included the following: a) large amounts of time reviewing hundreds of catalogs in an attempt to find specific needed parts; b) large amounts of time reviewing old purchase orders to locate specific part numbers and vendors; c) frequent errors resulting from ordering the wrong parts or parts with wrong features (e.g., wrong alloy, thickness, shape, electrical characteristics); and d) large backlog often building up in the ordering process during peak production periods, causing frequent stockouts and late product deliveries.

The intent of the project was to develop a classification and retrieval system that would significantly reduce the time and errors associated with the purchasing process. In addition, the system was required to be developed in a general purpose relational database operating in a PC environment that would be transparently integrated with other key modules of the factory including an inventory control module, purchasing module, routing module, and a bill of materials module. The C&C system was to help staff reduce the time to complete purchasing task by placing important part feature information in electronic format so that existing families of parts could be easily retrieved and electronically placed into purchase order.

Figure 14 shows a portion of the Purchased Parts C&C model that was subsequently developed. Over 225 classes comprised the actual classification and coding model developed to classify the purchased parts. The OOM concepts directly applied to the design of the Purchased Parts C&C model. A large part of the model utilized the Generalization abstraction. Much redundancy in the model was reduced by part classes able to be associated with common features made available through the Aggregation abstraction.

The culmination of these development tasks is illustrated in Figures 15, 16 and 17. Figure 15 shows the Part Number Classification form and illustrates the classification process of a new part into the database. Here, the user enters a part number for a new part to be classified and stored in the database. As the user selects identifying information, appropriate subforms are shown on the master form allowing for further detailed selections. As selections are made, this information is concatenated and shown the user at the bottom of the form. Once the user is satisfied with the classification of the part, she presses the button on the form to add it to the database. This module then instantiates the appropriate tables of the database so that the detailed information can be queried and retrieved at a later time.

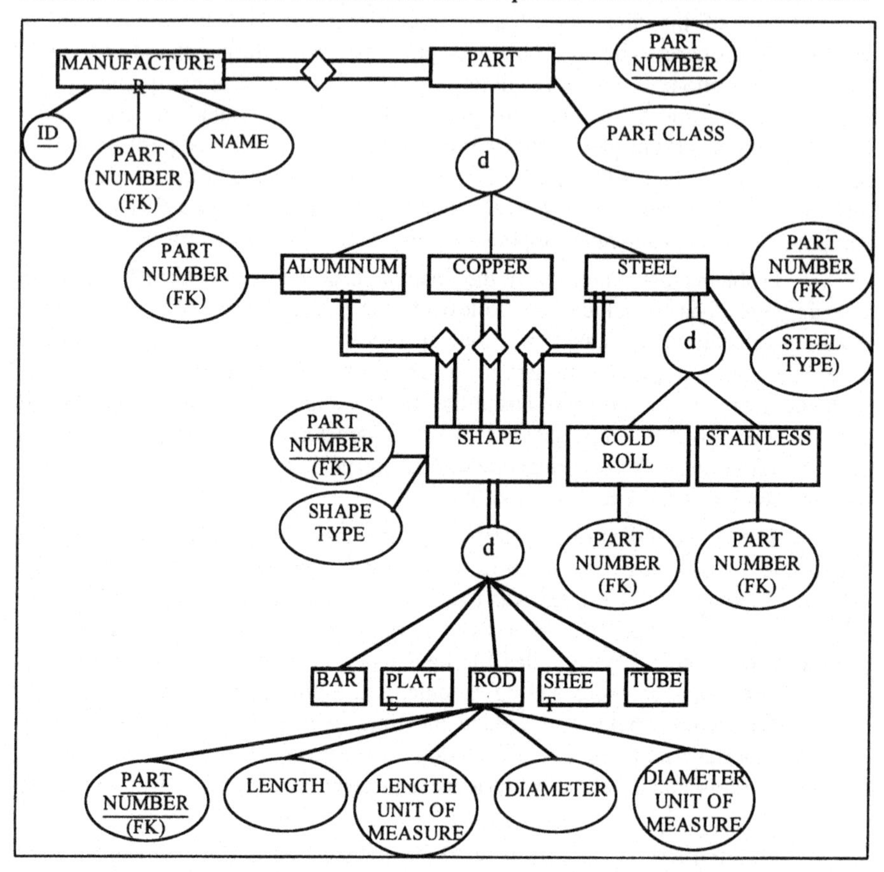

FIGURE 14. Portion of a purchased part classification model.

Figure 16 shows a typical retrieval session through the *Create Report* form. Like the classification session, subforms of detailed part feature selection options are displayed to the user as she makes early part identification decision. Once selection of requested part family information is completed, the software is designed to systematically generate a SQL query on the appropriate tables containing the part numbers and their descriptions that meet the query request.

Figure 17 shows the *Class Description Maintenance* form. This module allows the user to enter a new classification model and to modify an existing model. This feature allows the software to be robust enough to incorporate any properly designed classification model.

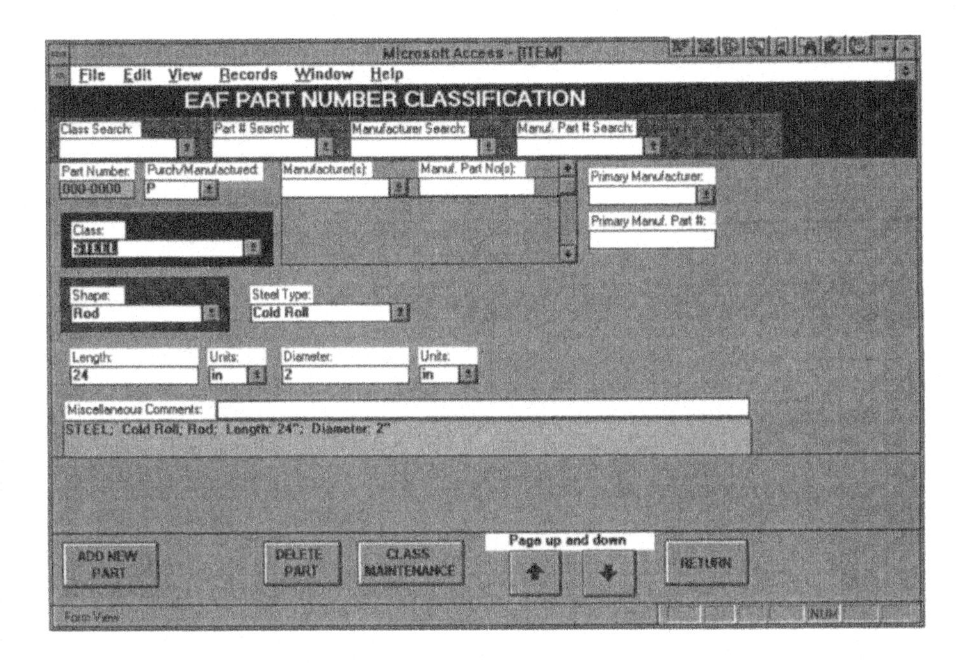

FIGURE 15. *Part Number Classification* form illustrating the classification of a part into the C&C database.

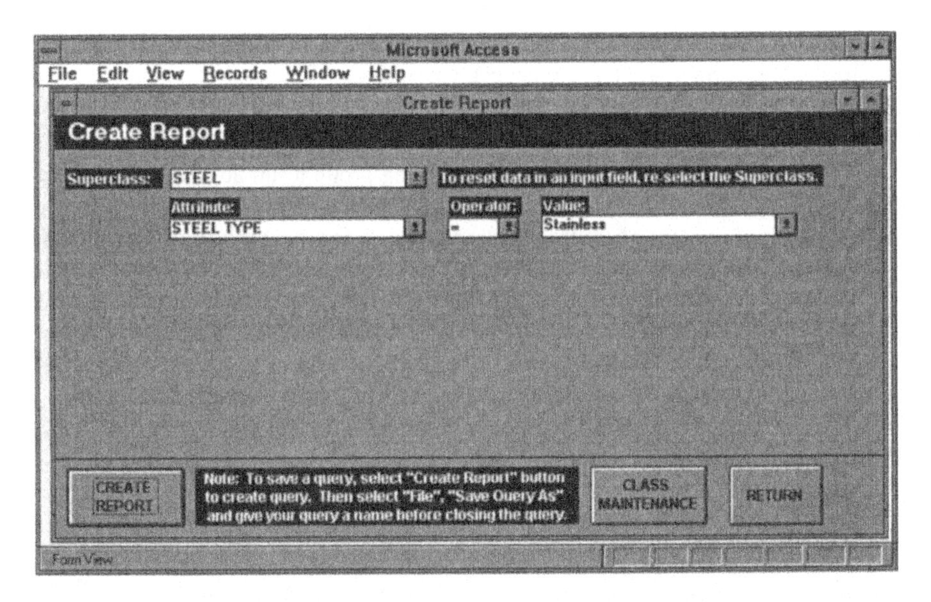

FIGURE 16. *Create A Report* form to retrieve similar parts from database.

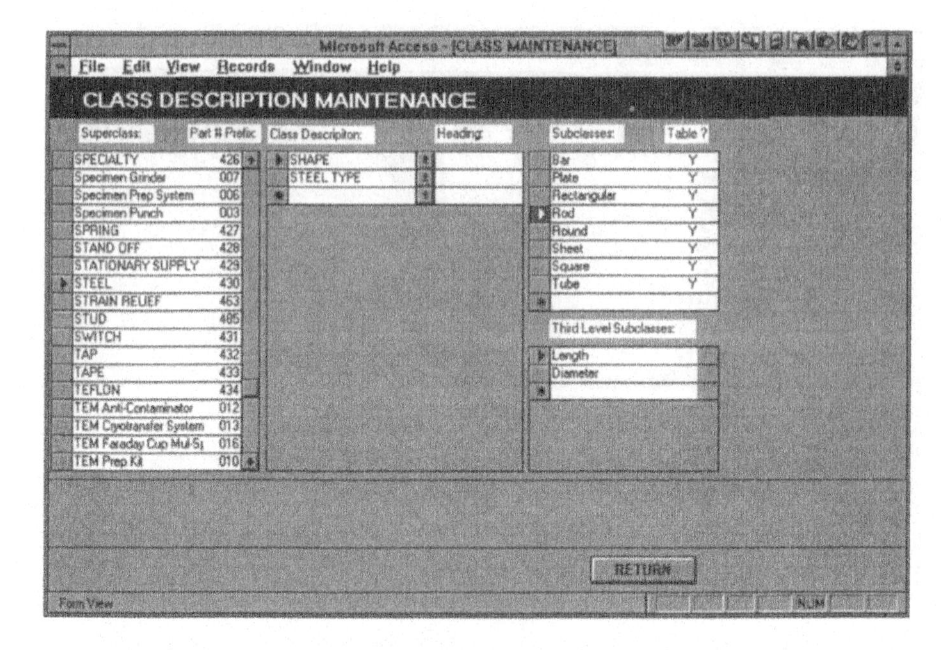

FIGURE 17. *Class Description Maintenance* form to implement and maintain the classification model.

The Purchased Parts C&C software system was implemented in the Microsoft ACCESS 2 database management system. The system operated on a Pentium 100 MHz personal computer. The system is entirely integrated with the other required manufacturing modules.

Improvements achieved by implementing group technology C&C software have been significant in the short time it has been in place. The manufacturing company can now place all of their purchase orders for the week in approximately two hours. Errors in purchase orders have been eliminated. Stockouts due to such errors as well as elimination of the purchasing backlog have also been eliminated. In addition, benefits have been added due to the other modules integrated with the C&C software. Staff no longer need do a manual check of inventory status each week which was often fraught with errors as well as time consuming. In addition, inventory levels have been reduced due to accurate inventory status which reduced the need for large safety stocks.

7. CONCLUSIONS

By viewing classification and coding systems from a formal object modeling perspective, these systems can be both simple to design and implement. Cost savings are achieved in several significant ways including reductions in labor costs, material costs, avoidance of expensive software packages, and transparent integration with other databases.

In working with all of the major database schemes (e.g., hierarchical, network, relational, frames), as well as decision tree systems, the authors have found OOM principles to be the most thorough in representing structures most commonly needed in C&C software, leading to shorter development time and lower development costs.

Although low-cost development time and implementation cost is a key feature of some PC-based decision tree software packages, these systems do not offer transparent integration with other databases. This problem is due to their lack of implementation in modern relational and object-oriented database management systems which are firmly established on the representation principles described in this paper (Smith and Smith 1977a,b). Given the OOM representation principles, the C&C development effort can be seen as no different than any other small database development effort that can be implemented with many existing tools available to a company.

The OOM representation principles are extendible to any C&C paradigm beyond decision trees. As was described at the beginning of this paper, the C&C basic building blocks of the monocode, polycode and hybrid code structuring schemes were in correspondence to the decision tree principles described in this paper. In turn, it is obvious that the abstractions generalization, classification, and aggregation are analogous concepts to these basic coding systems.

The benefits of the approach adopted in this paper are that two currently isolated bodies of knowledge, information systems and classification and coding can meet on common semantic ground via shared principles and tools. We hope this work will contribute to the movement toward integration of both groups and tasks based on common understandings and tools.

Dedication. *This work is dedicated to the loving memory of Debra M. Wolfe, January 7, 1954 - August 1, 1997.*

REFERENCES

Allen, D. K., 1984, Computer-aided process planning: Software tools, in *Integrated & Intelligent Manufacturing*, (Eds. C. R. Lui and T. C. Chang), ASME Winter Annual Meeting.

Apple, J. M., 1977, *Plant Layout and Material Handling*, John Wiley and Sons, New York.

Billo, R. E. and Bidanda, B., 1995, Representing group technology classification and coding techniques with object oriented modeling principles, *IIE Transactions*, 27, 542-554.

Billo, R. E., Rucker, R., and Shunk, D. L., 1988, Enhancing group technology with database abstractions, *Journal of Manufacturing Systems*, 7, 2, 95-106.

Billo, R. E., Rucker, R., and Shunk, D. L., 1987, Integration of a group technology classification and coding system with an engineering database, *Journal of Manufacturing Systems*, 6, 1, 37-45.

Chan, H. M., and Milner, D. A., 1984, Direct clustering algorithm for group formation in cellular manufacture, *Journal of Manufacturing Systems*, 1, 1, 65-75.

Elmasri, R. and Navathe, S., 1994, *Fundamentals of Database Systems*, Benjamin/Cummins.

Groover, M. P. and Zimmers, E. W., 1984, *CAD/CAM: Computer-Aided Design and Manufacturing*, Prentice-Hall, Englewood Cliffs, NJ.

King, J. R., 1980, Machine-component grouping in production flow analysis, *International Journal of Production Research*, 18, 2.

McAuley, J., 1980, Machine grouping for efficient production, *Production Engineer*, February, 53 - 57.

Rumbaugh., J., Blaha, M., Premerlani, W., Eddy, F., and Lorensen, W., 1991, *Object-Oriented Modeling and Design*, Prentice Hall.

Smith, J. M. and Smith, D. C. P., 1977, Database abstractions: Aggregation and generalization, *Communications of the ACM*, 20, 6.

Vakharia, A. J., 1986, Methods of cell formation in group technology: A framework for evaluation, *Journal of Operations Management*, 6, 3, 260.

AUTHORS' BIOGRAPHY

Richard E. Billo is Associate Professor of Industrial Engineering at the University of Pittsburgh and the Director of the University of Pittsburgh's Automatic Data Collection Laboratory. Dr. Billo performs research in Computer Integrated Manufacturing Systems, Automatic Identification, Cellular Manufacturing, Classification & Coding System Design, and Material Tracking System Design. Previously, he managed the Production Systems Analysis Group at Battelle Pacific Northwest Laboratory. In addition, Dr. Billo has been a consultant to many mid-size and large national and international corporations. He received his MS and Ph.D. degrees in Industrial Engineering from the Arizona State University. He is a senior member of the Institute of Industrial Engineers, the Society of Manufacturing Engineers, and AIM^{USA} -- the industry trade organization of Automatic Identification Manufacturers. Dr. Billo has been recognized by his peers as an awardee of the University of Pittsburgh Board of Visitors Outstanding Faculty and the Whiteford Faculty Fellowship.

Bopaya Bidanda is Associate Professor of Industrial Engineering and Whiteford Faculty Fellow at The University of Pittsburgh. In addition to his doctoral degree from the Pennsylvania State University, he has worked in the area of aerospace manufacturing. His research focus is in the area of Computer Integrated Manufacturing Systems, Rapid Prototyping, Robotic Applications, Shared Manufacturing, and Manufacturing Management. He has published in a variety of international journals and has edited/authored two books. He is a senior member of the Institute of Industrial Engineers and the Society of Manufacturing Engineers. Dr. Bidanda is Past President of the Pittsburgh Chapter of IIE and is currently responsible for organization of the 1998 Industrial Engineers Conference.

B4

Part Family Formation and Classification Using Machine Learning Techniques

Y. B. Moon

1. INTRODUCTION

An ideal tool for successful part family identification, as part of group technology (GT) efforts, should address several important technical issues. First, the tool must have an adequate representation scheme to cover a wide variety of part features, which are necessary for different applications. Second, the tool must be flexible enough to capture a changing array of information because of increasing demand placed by variations in the types of products or new applications. Third, it is desirable to have a tool which can provide a meaningful description of the part groups so those humans can easily interpret and evaluate the groups with respect to different applications.

Two fundamental engineering problems associated with such an ideal GT tool are the initial part family formation and the subsequent classification of parts. It is not unusual that hundreds or thousands of parts are considered in a typical GT implementation. Therefore, it is very difficult and laborious to determine critical part features, then form part families by examining the entire spectrum of parts. Furthermore, newly designed parts are constantly introduced to the manufacturing floor without any reference to similar parts that have been manufactured before. It is difficult to identify similar parts or part families because of the number of parts that must be considered for resemblance checking. These two, yet closely related problems have been addressed by various approaches and techniques such as visual inspection, production flow analysis, classification and coding systems, rank order clustering method, mathematical programming, syntactic pattern recognition, expert systems, etc. (see chapter D1 for a taxonomy of this literature).

The laborious and often inconsistent recognition process of part features (sometimes called coding) would be eliminated if part features were derived directly from already-existing computer-aided design (CAD) database (Shah and Bhatnagar 1989; Moon and Roy 1991; Kaparthi and Suresh 1991). Current solid modelers can easily represent the nominal objects in a computer, but its database does not contain the complete information required for the GT applications. To be useful for GT, the feature information needs to be extracted from the model database or decoded from the object database so that the parts are expressed in terms of form features.

Even though form features (or symbolic representation) are adopted as the base representation scheme for the parts, most of the traditional tools for the family formation and part classification cannot take advantage of the richer set of information presented in the form features. This is because traditional tools use alphanumeric format as their base data representation and such an alphanumeric format is simpler to handle than symbolic representation. In this article, we present a few techniques, which enable form features to be used as the base representation scheme for the GT implementation. These techniques were initially derived from the domain of Machine Learning, a sub-field of Artificial Intelligence (Shavlik and Dietterich 1990).

The next section discusses various representation schemes, which have been used for GT applications. §3 presents the definition of form features and feature extraction system. In §4, multiple part family formations using an unsupervised sub-symbolic machine learning technique are presented. §5 deals with the problem of consistency maintenance using sub-symbolic machine learning techniques. §6 presents a technique to use the same set of information for different goals. This chapter concludes with §7.

2. PART REPRESENTATION SCHEMES FOR FAMILY FORMATION

An appropriate representation scheme is essential for an effective GT tool. An ideal representation scheme should be able to encompass a wide variety of part features since different applications require different information to be processed and analyzed. Several representation schemes have been proposed and used for various GT applications, including 0-1 representation, alphanumeric representation, binary representation with drawings, and symbolic representation.

2.1. 0-1 Representation

In this representation scheme, an existence of a certain part feature for a given part is denoted as 1 or 0. One popular example is the machine-part incident matrix used for Group Analysis of the Production Flow Analysis (Burbidge 1970). This representation scheme is simple, therefore readily available as long as process plans are already existing. However, this representation carries a minimal set of information, thus limits its applicability for other applications than the initial part-machine grouping.

2.2. Alphanumeric Representation

This form of representation is used mostly in classification and coding systems. The process involved in this representation is mapping from N-dimensional space to 1-dimensional space, where the 1-dimensional space is pre-determined in terms of alphabet and numerals. Compared with the previous representation method, this one can provide a much wider range of information including tolerances, material, dimensions, etc., therefore, is more versatile in terms of its applicability.

However, since the system should be determined in advance, the number of part features to be represented is fixed and limited. Also the number of possible values for a part feature is usually determined a priori. In other words, the designer of the system ultimately determines the scope of information, which describes each part. For these reasons, update of an existing system using the alphanumeric representation is very difficult even though variations in the types of products or new applications are inevitable during a course of GT applications. It is very costly to develop such a classification and coding system and maintain it properly over the years. It is typical for a company to take several months to years to finish the mapping procedure for all the parts they are producing. Furthermore, the meanings of alphanumeric representation are not apparent to humans.

2.3. Binary Representation With Drawings

A flexible binary representation scheme is also proposed to represent part information as an improvement over traditional fixed-number coding systems. A vector of binary digits describes each part and each digit corresponds to a single feature. Each feature is accompanied with a simple drawing for clearer reference. This scheme is especially appropriate for the applications, which are expanding in its number of features to be considered. A better consistency may be sustained with the aid of drawings. Beyond this advantage, this representation scheme shares the same drawbacks as those of 0-1 representation.

2.4. Symbolic Representation

The symbolic representation scheme can generate conceptual descriptions of part families, thus provides human with immediate understanding. Also, it has the potential for integrating the parts grouping task directly with feature-based CAD systems and the flexibility to provide for part representation. Such an approach can capture the similarity of parts at different levels of detail, thereby representing the part information necessary for different applications.

Another benefit of employing the symbolic representation scheme is for multiple applications using a same set of information. Since each application has its own unique goal, each application is different in the kind and amount of information to be analyzed and interpreted and in the method of representation and analysis. For instance, the goal of design retrieval and process planning is to minimize duplication of effort by using information about existing parts to help design, manufacture, and refine details of new parts. In contrast, the goal of production planning is to maximize manufacturing productivity by implementing cellular manufacturing and optimizing production schedules. The symbolic representation provides a unique advantage for such multiple applications.

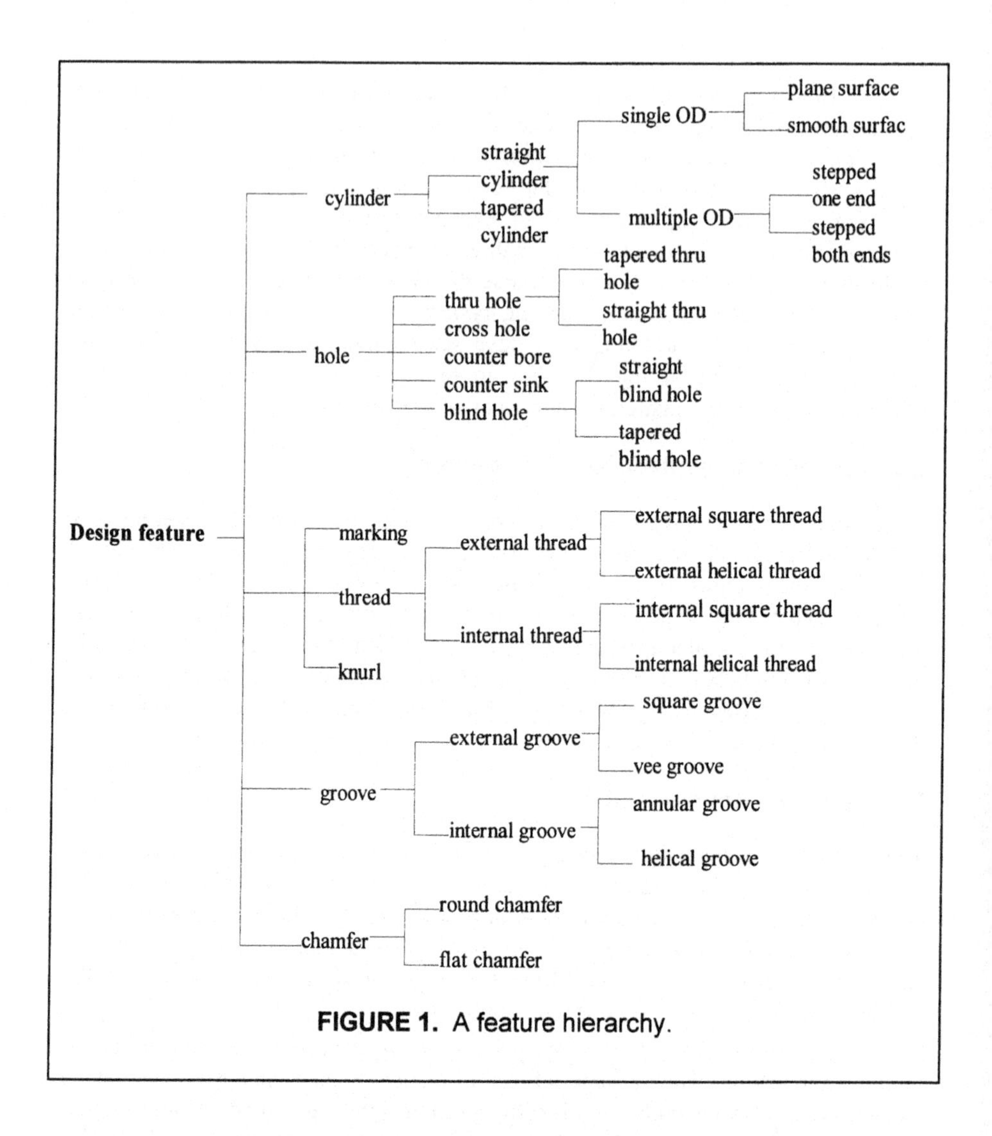

FIGURE 1. A feature hierarchy.

3. AUTOMATIC FEATURE RECOGNITION FROM SOLID MODELS

Part features are a logical candidate for symbolic representation. Part features are used to provide additional evaluation of part families and form the basis for assignment of parts to machines through the machining sequence embedded in the features. A way of extracting part features directly from solid models is described in this section.

3.1. Part Features

The term 'feature' is defined as a general term applied to a physical portion of the object. It may be a basic entity, or it may include more than one surface such as holes or slots. Two options are available to define the feature primitive: (a) as surface primitives and (b) as form primitives, defining a specific geometric configuration formed on the surface, edge or corner of an object. Each form feature must be tangible and should have a single characteristic. A form feature may be composed of several other features such as a tapped hole composed of a hole, a thread and a chamfer. As there are frequently several occurrences of a component in the object, features are required to be instantiated more than once.

Form features are predefined geometric shapes. CAM-I's primary form feature has been adopted in this study. Each form feature is specified by:

- A feature specification name such as slot, bevels or counterbore.
- A feature type (primitive or complex).
- An orientation (orthogonal or non-orthogonal).
- A feature identification number and its status.
- Feature parameters.

The feature parameters are dimensional variables whose instantiated values establish the sizes and relationships of various geometric elements comprising a predefined set. These features are defined relative to geometric entities such as vertices, edge, or faces according to their geometric and form constraints within a specific datum reference frame.

3.2. Feature-based Design System

The feature-based design system consists of two modules – a feature library and a parser – built upon an existing CAD system. Parts are represented by a collection of features described using geometric and technical attributes. Each part is represented at two levels, the part level and the feature level. At the part level, information about the entire part such as name, material condition, overall shape (e.g. rotational or non-rotational) and special processes (chrome-plating, anodizing, heat treatment, etc.) are provided. At the feature level, individual information concerning each feature such as name, type, dimension, locations relative to a predefined coordinate system, surface finish, etc. are enumerated.

A generic feature library is employed next. A portion of an example feature library is shown in Figure 1. The user incorporates each feature into the part by selecting the appropriate feature from the hierarchy. For example, if the part contains a tapered through-hole, the user first selects the option hole from the first level in the hierarchy, after which the different options available at the second level are presented for selection. After the user reaches the leaf node (i.e. selects the feature), dimension, location, orientation, etc. can be defined. In this way, the user specifies each feature until the entire part is designed. The CAD system uses a CSG representation scheme and each part feature can be added to or subtracted from the initial form in order to design the part. Once the part is designed, a program file is automatically created by the CAD system.

A parsing program implemented in C reads the program file and extracts the relevant information useful for GT. The information is then formatted so that the GT tool to generate part classifications can use it. Before the part information can serve as input to the GT tool, additional information from the user about a particular application of the GT such as feature type, dimension, locations relative to a predefined coordinate system, surface finish etc. may be sought out.

4. FORMATION OF MULTIPLE FAMILIES

Traditionally GT family formation has been restricted to part families and machine groups. Two primary tools that have been used for the GT family formation are classification and coding system (CCS) and production flow analysis (PFA). CCS is more adequate for accomplishing the tasks of the retrieval of similar parts and the assignment of new parts, while PFA is more natural for accomplishing the tasks of the part family formation and the machine cell formation.

Main limitations of CCS and PFA lie in the types of relationships utilized in each approach. CCS exploits the relationship between part and part feature while PFA relies on the relationships between part and machine. These problems can be addressed by simultaneously using both types of relationships: the relationships between parts and part features and the relationships between part features and machines. The idea is to use part feature as a medium to connect parts and machines and to use memory association for part family and machine cell formation.

Part family and machine cell formations are viewed as a memory association process. There are two types of memory association: auto-associative memory (or auto-association) and hetero-associative memory (or hetero-association). Auto-associative memory is an association process that can distinguish an object that is partially or incompletely described. Therefore, auto-associative memory can relate an object with itself or the same type of objects. For example, one part reminds of another similar part that was seen before. On the contrary, hetero-associative memory is a process that can associate an object with other different types of objects. For instance, a part reminds us of a set of machines that can make the part. Auto-associative memory is used for part family formation while hetero-associative memory is adopted for forming machine cells.

The approach is based on the interactive activation and competition (IAC) neural network (Rumelhart et al. 1986). This class of neural network is a fixed-weight competitive network with one layer structure for the tasks of associative memory. In the IAC networks, processing units are divided into several pools. There are three types of pools: attribute pools, object pools and an instance pool. The relations between objects and attributes are mapped into network connections between pools. The weight values of those connections are determined a priori and not changed in the execution stage (Carpenter and Grossberg 1987).

The Procedure.

Three neural network pools, part pool, machine pool and instance pool, are constructed for GT applications (Moon 1990). The relations between parts and machines are mapped into a part similarity matrix and a machine similarity matrix. The part similarity matrix is used to determine the connection weights within the part pool while the machine similarity matrix is used to determine the connection weights within the machine pool. The inter-pool connection weights are all set to 1. After the neural network is constructed, it performs part family and machine cell formation simultaneously.

Two equations are used for neural computation during the network operation: the net-input equation and activation-updating equation (Rumelhart et al. 1986). The net-input equation (equation 1) calculates the net input to a processing unit by summing up all the inputs from other units or from outside the network. Equation 2 or 3 is used to calculate the output activation of each neuron.

$$net_i = alpha*\_{j} (+W_{ij})(output_j) + gamma*\_{j} (-W_{ij})(output_j) + estr*extinput_i \quad (1)$$

where

net_i: net input of unit i.

W_{ij}: the connection weight value from unit j to unit i,
the negative value represents an inhibitory connection,
the positive value represents an excitatory connection.

$output_j$: the current output of unit j in the network.

$extinput_i$: the external input to unit i from outside of the network.

estr: the parameter scales the strength of external input to unit i.

alpha: the parameter scales the strength of the excitatory input to unit i from other units.

gamma: the parameter scales the strength of the inhibitory input to unit i from other units in the network.

$$\text{If } net_i > 0 \ \&\ \_act_i = 0.0,\ act_i = [(max)(net_i) + (rest)(decay)] / (net_i + decay); \quad (2)$$

$$\text{otherwise, } act_i = [(min)(net_i) - (decay)(rest)] / (net_i - decay). \quad (3)$$

where,

$\_act_i$: the resulting change in the activation of the unit.

act_i: the activation of the unit i.

max: the maximum activation parameter that is normally set to 1.0.

min: the minimum activation parameter that is sometimes equal to (-1.0).

rest: the resting activation level, min <= rest <= 0.0.

If $extinput_i = 0.0$, then act_i tends to go to the resting level decay: the decay rate, which is the strength of the tendency that act_i returns to the resting level.

There are seven network parameters contained in the above equations. These parameter values are determined by a trial and error considering two important effects of IAC models: "the rich get richer" and resonance. The first one means that each neuron tries to suppress other neurons which are in the same pool while the second means that each neuron tries to excite those neurons belonging to different pools. Even though there is no guarantee that these are a set of best values, the above values are found to be satisfactory in numerous experiments. Variances in these values did not generate significantly different results in those experiments. This set of parameter values is presented for an initial guide for those who will implement the approach, but may have to be adjusted.

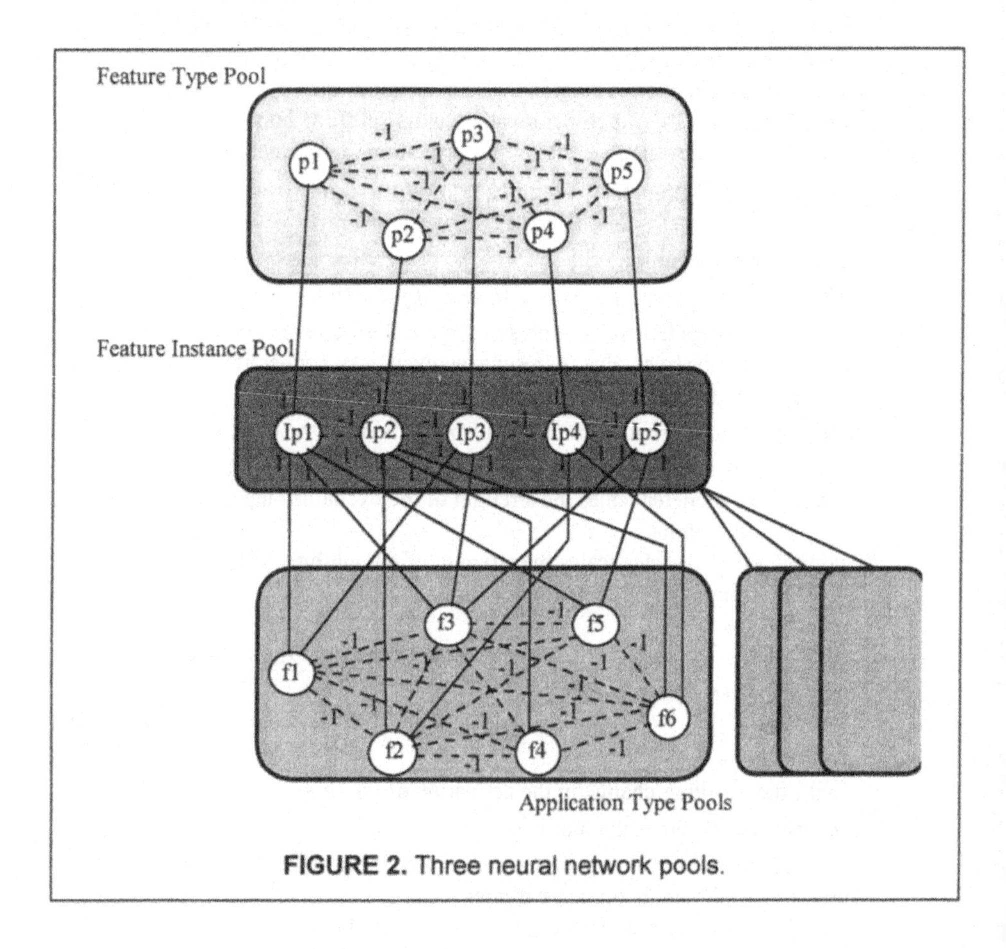

FIGURE 2. Three neural network pools.

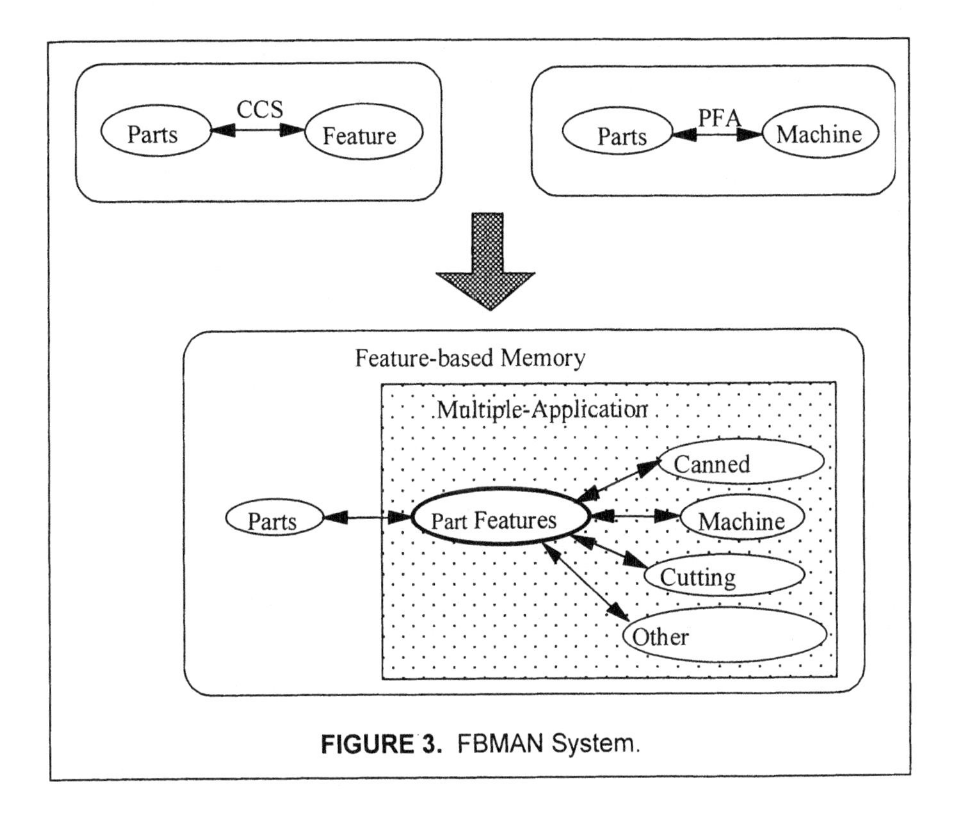

FIGURE 3. FBMAN System.

Extension.
The above described system can be easily extended for multiple-application set formation and was name as FBMAN (Feature-Based Memory Association Networks) system (Kao and Moon 1997). Multiple-application sets include not only part families and machine cells but also cutting tool sets, CNC canned cycle sets, etc.

The FBMAN system to GT can form multiple application sets for several applications concurrently. It is quite difficult for human engineers to associate various types of applications by examining part features only. However, the FBMAN system can not only form multiple application sets, but also form those groups at the same time.

5. MAINTAINING CONSISTENCY

The two engineering problems, that is, part family formation and part classification are closely related, and usually the same approach is used for both problems in most cases. For example, if parts are categorized into part families by visual inspection, the closest part family to a newly introduced part is identified by the same method –

visual inspection. Alternatively, a part classification and coding system may be used for both family formation and classification. If a production flow analysis is adopted for the formation process, the same type of information from a routing sheet and the same logic of the PFA are used for the classification process.

Regardless of the approach adopted for the formation and classification, a critical problem is how to maintain consistency. The consistency problem can be addressed most effectively if the formation and classification is a single procedure rather than two separate procedures. Therefore, how to create new families during the classification process becomes important in dealing with the consistency maintenance problem (Kao and Moon 1991).

5.1. Backpropagation Learning Rule

A multi-layer feedforward neural network with the backpropagation learning algorithm (Rumelhart et al. 1986) is adopted for this task. This network is able to associate input patterns of part features with output part families. A typical structure of multilayer feedforward network contains one input layer, one output layer and some hidden layers. As the learning procedure begins, the input-target pattern pairs from training sets are presented to the network. The input units receive the input patterns directly and the target patterns are associated with output units.

Procedure

Notation:

Fn = the nth family
I = the iteration cycle number
M = the maximum digit number in the target pattern code
NFn = the number of new family members in Fn
NP = the number of parts in the part pool, P
P = {p1, p2, ..., pi}, the part set
S = {s1, s2, ..., sj}, the seed set (used as the training set)
Ti = the activation value of a digit in a target pattern code, Ti = 1 or 0
Xi = the activation value of a digit in an output pattern code, |Xi| < 1
TT = the threshold value

Step 0. I = 1
NS = 0
NP = NP
S = { }
P = {p1, p2, ..., pNP}

Step 1. Determine the value of N.
N>1 and N < 6 or 7
(The maximum value of N depends on the number of apparently distinctive parts.)

Step 2. Choose N seed parts from part pool.

Step 3. Assign predetermined family ID numbers to each seed part.
These input-target pattern pairs from the training set, S.
Add these seed parts into the seed pool.
$NS = NS + N$, a new set S
Eliminate the training parts from the part pool.
$NP = NP - N$,
$P = P - \{pi \mid pi \quad S\}$

Step 4. Present the training set to the BP network for training it.

Step 5. Present the remaining parts in the part pool for the assigning phase.
For apart,
If $|Ti - Xi| < TT$, $I = 1,2, \ldots m$
Then the part belongs to the corresponding family;
Otherwise this part is not assigned to any family yet.
After the assigning phase, each family seed is associated with its family members.

Step 6. Eliminate family members chosen in this cycle from the part pool.
$NP = NP - NFn$, $n = 1, \ldots, NS$
The Ith cycle is finished.

Step 7. If $NP > 0$,
Then $I = I + 1$ and
Go to step 2,
Otherwise stop.

The described approach deals with the consistency problem by utilizing a common feedforward neural network for both part family formation and part classification. The only human involvement is when several seed parts are selected from a pool of parts. Once the distinctive parts are identified, the neural network learns the generalized characteristics of each part in terms of part features. When a new part is introduced, the neural network indicates the closest family in terms of part features. The formation and classification processes are treated in a unified framework.

Since a constructed neural network is continuously used while more new parts are being introduced, the number of output units should be large enough so that new part families are being created. If there are n output units, then 2n distinctive part families are possibly considered. Initially, only the selected number of output units are utilized by assigning values of 0 to the connection weights leading to the unnecessary output units. As new part families are needed, random numbers are assigned to the connection weights leading to the corresponding output units before the described procedure starts.

5.2. Adaptive Resonance Theory

The difficulty of the automatic generation of part families is that of maintaining a consistent standard of classification while allowing the modification of the standard when the creation of new families is necessary. The nature of the problem is similar to a stability-plasticity dilemma in neural network modeling: how the neural network learns unseen input patterns without losing the material previously learned.

Adaptive resonance theory (ART) is an approach to the problem of a stability-plasticity dilemma. The ART network accepts input vectors and classifies them into one of the existing categories, according to the degree of similarity between new input vectors and stored family prototypes. If there is no resemblance between them then a new category is created in order to store these 'unseen' input vectors. If a new input pattern is similar to one of the stored prototypes then this new input will be grouped into that category and the network will start the learning process to adjust the related weight values such that the stored prototype becomes more similar to the input vector.

The classification decision is made by an important parameter, the vigilance threshold. This determines the maximum degree of difference between two vectors to be classified in the same category. It is equivalent to a distance threshold or a cluster diameter in clustering theory. The higher it is, the lower is the permissible invariance within a cluster, and the greater the number of clusters that will be created. On the other hand, if a larger number of members in the same category are permitted then the vigilance threshold should be set to a lower value. Therefore, the number of members in a family and the total number of families can be controlled.

If the current input vector does not match any of the stored prototype patterns, then, instead of modifying or erasing these existing stored patterns an additional classification category will be formed by the network. In this way, the stability-plasticity dilemma is addressed. More detailed discussions about the ART architecture can be found in (Carpenter and Grossberg 1987).

The ART model deals with both part family formation and part classification concurrently. The most important advantage of using the ART model is the capability of maintaining consistency throughout the GT implementation and operation. This consistency problem can be addressed most effectively if the formation and classification is a single procedure rather than two separate procedures as in the described approach (Moon and Kao 1993).

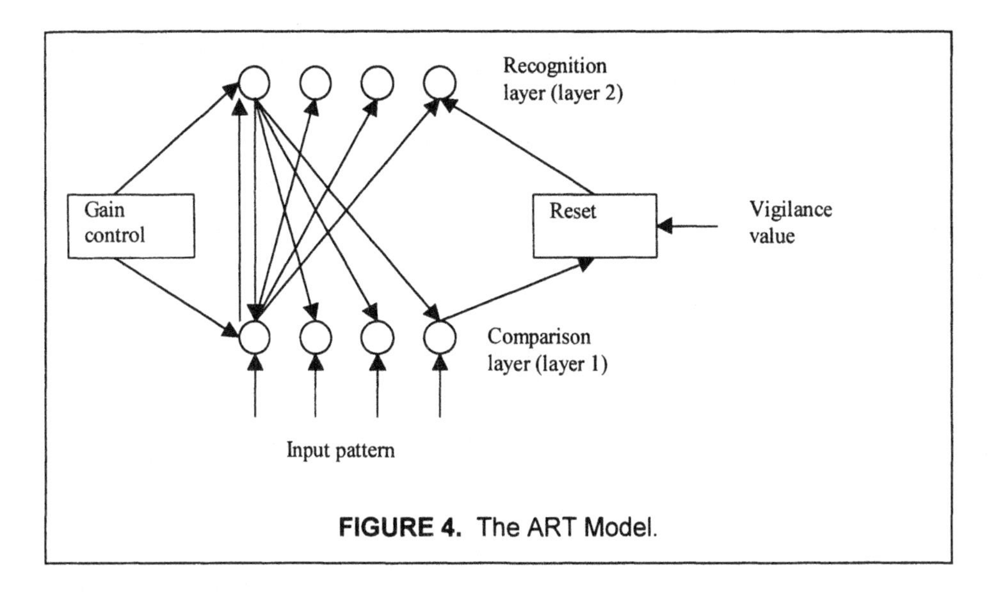

FIGURE 4. The ART Model.

6. MULTIPLE GOALS

In addition to the method of representation, the method of organizing and analyzing part information differs from one application to another. To learn useful classifications for different goals of various applications, the clustering process should be influenced by the goal of the application. One way to influence part groupings is to select attributes useful for the application. A more effective approach is to incorporate the goals of the application in the clustering process. Conceptual clustering techniques employ an evaluation function during the clustering process to evaluate alternative classifications. Compared to distance functions used in most numerical clustering approaches, these evaluation functions measure the quality of clusters based on characteristics such as prediction capability, balance among classes, concepts that describe the object classes, and the match between concepts and the classes they cover.

Conceptual clustering is a class of inductive learning systems which forms classification schemes over an initially unclassified set of data and groups objects into classes that can be represented by simple conceptual descriptions. These systems do not require a 'teacher' to pre-classify objects, but use an evaluation function to discover classes with 'good' conceptual descriptions. Despite differences in representation, quality judgment, and mode of learning (incremental or agglomerate), all conceptual clustering systems evaluate class quality by looking to a summary of concept description of the class. A fundamental difference between conceptual clustering and numerical clustering approaches is based on the methods used to measure similarity between objects. In methods of numerical taxonomy, the similarity between two objects is the value of a numeric function applied to the description of the two objects.

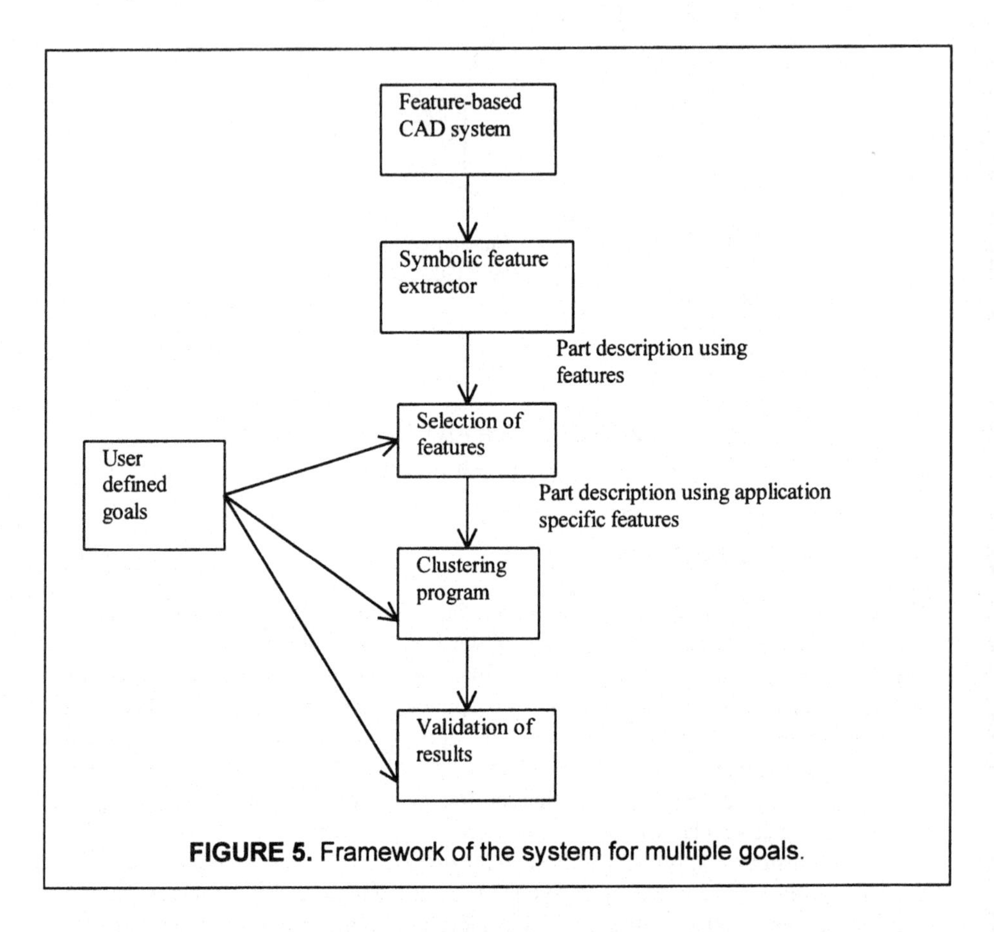

FIGURE 5. Framework of the system for multiple goals.

GT applications also lend themselves to such measures. For example, applications such as design retrieval, generative process planning and cost estimations use existing parts data to infer useful information about new parts. For such applications, classifications should be determined and evaluated based on their ability to accurately predict missing information of a part. In contrast, in the design of cells, the goals may include preferred number of cells, an even distribution of parts among different cells, minimizing load imbalances among the cells, minimizing machine duplication in different cells, etc. The evaluation function must be modeled in terms of the goal of the application.

The goal of a particular GT application should also influence the choice of the inductive learning system. A design retrieval task does not need the explicit formation of a specific number of part families. It is more likely to benefit when the parts are organized in a hierarchy and are presented at different levels of detail. A hierarchical conceptual clustering approach is more likely to be useful for this application. On the other hand, cellular manufacturing requires that explicit part families be developed so that a set of machines can be dedicated to one or more families of parts. Optimizing techniques or clumping techniques of conceptual clustering can perform the tasks of generating meaningful part families for such an application. Conceptual descriptions of part families are useful here for engineers to evaluate alternative classifications with respect to manufacturing facilities available.

GT applications differ along four lines: the information to be represented, the method of representation, the technique for generating part groups and the evaluation of the part groups. The goals of a GT application must be considered at each stage for the results to be meaningful for that application. A feature based CAD system is employed to design parts in terms of high level features. By adopting a flexible symbolic and structured representation scheme, it addresses the disadvantages of traditional coding systems and numerical representation methods. Conceptual clustering is utilized for the task of generating part groups and it provides a much better tool for GT in comparison to numerical clustering techniques, by its ability to deal with symbolic representation and by using measures which are more appropriate in the GT domain. The described approach also demonstrates the need for evaluating the generated part groups explicitly for the application that it is intended for (Srinivasan and Moon 1997).

7. CONCLUSION

Group technology has served manufacturing effectively for the past several decades and is universally recognized as the building block for new ideas such as lean manufacturing, concurrent engineering, design for assembly, agile manufacturing, etc. With the advancement in related fields such as CAD and Machine Learning, the concept now develops far-reaching implications. This article presented a number of recent research efforts to utilize the automatic feature extraction from CAD database and to adopt machine learning techniques for part family formation and classification. The problems of multiple family formation, consistency maintenance, and multiple goals were addressed by symbolic as well as sub-symbolic machine learning techniques based on symbolic representation of part features.

REFERENCES

Burbidge, J.L., 1970, Production flow analysis, in *Proceedings of Group Technology International Seminar*, International Center for Advanced Technical and Vocational Training, Turin, 89.

Carpenter, G. and Grossberg, S., 1987, A massively parallel architecture for a self-organizing neural pattern recognition machine, *Computer Vision, Graphics and Image Understanding*, 37, 54-115.

Han, C. and Ham, I., 1986, Multiobjective cluster analysis for part family formation, *Journal of Manufacturing Systems*, 5, 4, 223-230.

Kao, Y. and Moon, Y.B., 1991, A unified group technology implementation using the backpropagation learning rule of neural networks, *Computers and Industrial Engineering*, 20, 4, 425-437.

Kao, Y. and Moon, Y.B., 1997, Feature-based memory association for group technology, *International Journal of Production Research*, (in press).

Kaparthi, S. and Suresh, N.C., 1991, A neural network system for shape-based classification and coding of rotational parts, *International Journal of Production Research*, 29, 9, 1771-1784.

Moon, Y.B., 1990, An interactive activation and competition model for machine-part family formation in group technology, in *Proceedings of the International Joint Conference on Neural Networks*, Washington, D.C., 667-670.

Moon, Y.B. and Kao, Y., 1993, Automatic generation of group technology families during the part classification process, *International Journal of Advanced Manufacturing Technology*, 8, 160-166.

Moon, Y.B. and Roy, U., 1991, Learning group technology part families from solid models by parallel distributed processing, *International Journal of Advanced Manufacturing Technology*, 7, 109-118.

Rumelhart, D.E., Hinton, G.E., and Williams, R.J., 1986, Learning internal representations by error propagation, in *Parallel Distributed Processing: Explorations in the Microstructure of Cognition)*, (Eds. D.E. Rumelhart and J.L. McClelland), 1, 8, 318-362, MIT Press, New York.

Shah, J.J. and Bhatnagar, A.S., 1989, Group technology classification from feature-based geometric models, *Manufacturing Review*, 2, 3, 204-213.

Shavlik and Dietterich, 1990, *Readings in Machine Learning*, Morgan Kaufmann Publishers, Inc.

Srinivasan, M. and Moon, Y.B., 1997, Framework for a goal-driven approach to group technology applications using conceptual clustering, *International Journal of Production Research*, 35, 3, 847-866.

Wu, H.L., Venugopal, R. and Barash, M.M., 1986, Design of a cellular manufacturing system: a syntactic pattern recognition approach, *Journal of Manufacturing Systems*, 5, 2, 81-87.

AUTHOR'S BIOGRAPHY

Dr. Young B. Moon is an Associate Professor in the Department of Mechanical, Aerospace and Manufacturing Engineering at Syracuse University, Syracuse NY, USA. He holds a BS degree from Seoul National University, MS degree from Stanford University, and Ph.D. degree from Purdue University. His current research interests include design of manufacturing cells and machine learning applications to manufacturing systems. He has held visiting positions at NIST, Hewlett-Packard Labs, MIT and Windsor Manufacturing Company.

C1

GT and CAPP: Towards An Integration of Variant and Generative Approaches

D. N. Sormaz

1. INTRODUCTION

The goal of reducing process planning efforts for similar parts has been one of the objectives behind group technology (GT) right from inception. Later, however, the method extended its scope and shifted away from process planning, into machine cell formation, part family grouping, cellular manufacturing, etc. Process planning practice, and research, also moved away from group technology. Research in generative process planning shifted from considering families of similar parts into work based on design of individual parts. Application of artificial intelligence methods and CAD/CAM integration also contributed to this split.

In this chapter we discuss relationships between GT and process planning, and attempt to find a missing link in current research that will serve to connect them again. The chapter explains variant and generative process planning, identifies benefits of applying group technology in knowledge-based process planning and provides some research directions that will enable the link between two of them.

The chapter is organized as follows. In §2 we provide a model that integrates the views that process planning is a key ingredient for advances in manufacturing integration. Variant process planning procedure and systems are described in §3, while §4 is devoted to generative process planning. In §5 we overview application of GT principles in process planning and in §6 we discuss relations between process planning and cell design procedures. §7 points to possible research directions that will include GT into process planning systems, while §8 provides the conclusions.

2. PROCESS PLANNING WITHIN CONCURRENT ENGINEERING

Product development and manufacture involve several production management activities with a series of individual tasks that are to be completed in order to design and manufacture a product of a required quality. These tasks are usually carried out in a linear sequence, but very often feedback is necessary from a subsequent task to the previous one. Majority of these feedback loops are requests to modify the previous task's solution to generate a better solution in the subsequent one. This interlinking is what has become known as concurrent or simultaneous engineering.

In this section we will provide a model of these activities and tasks and identify how these tasks connect high-level activities. First we provide a model in which the product development cycle may be seen from asking several high level questions and providing answers to them. Then we develop a model that explains the role of process planning in product development. Finally, we identify areas within the model in which concurrent engineering and group technology are most applicable.

Product development cycle may be seen as a set of answers to a series of simple questions: What? How? Where? When? to produce (Ham and Lu 1988). Answers to these questions are given by the following activities: Product design (What? - product), Process planning (How? - by these processes), Resource management (Where? - on these machines), and Scheduling (When? - during this time period). However, there are tasks in these activities that can not be easily classified into any of these activities (e.g., stock selection is part of product design, but the selected stock significantly reduces process planning options i.e. by selecting cast iron as stock, later machining processes involve only needed finishing of small number of part faces). Process planning, as an activity that determines the methods by which to manufacture the products and necessary resources plays a very important role in product development. It may be seen as an activity that connects design, manufacturing, resources, and scheduling.

Starting from the above assumption and analyzing set of tasks of process planning and other activities it is possible to develop the model that shows interactions between process planning and them. The model of these interactions is shown in FIGURE 1. As seen from the figure, each activity consists of a set of tasks that are to be done in product development. For example, creation of the solid model of the part is done in design, while manufacturing resource planning is done within resource management. However, there are numerous tasks that require interaction between two or more activities. They are shown within overlapping circles of activities and represent integration links. For example, setup planning is part of process planning, but also

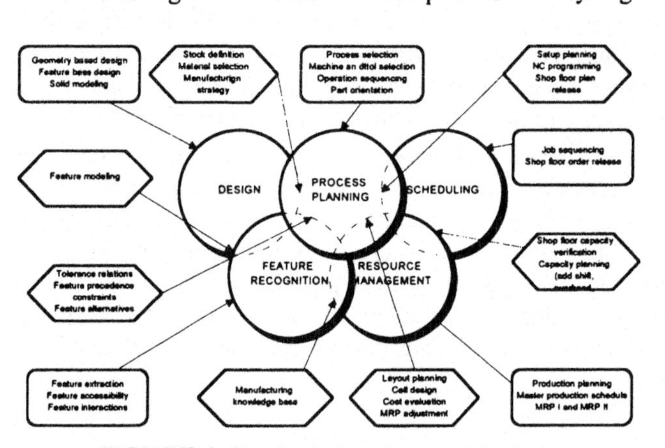

FIGURE 1. Product development tasks.

needs information about scheduling for efficient setups, or feature modeling belongs between design, feature recognition, and process planning.

Understanding of above explained interactions is important in order to completely utilize engineering knowledge and expertise. Each of these activities needs specialists in its domain, while intersections need group work and they are suitable for applying GT principles. The most important intersections from process planning perspective are: between design and process planning related to part family formation, between process planning and resource management related to manufacturing cell design, and between process planning and scheduling related to production control of cells.

3. VARIANT PROCESS PLANNING

In this section we describe variant process planning method. We describe the details of variant process planning with discussion of the procedure, and support the discussion with description of few variant systems.

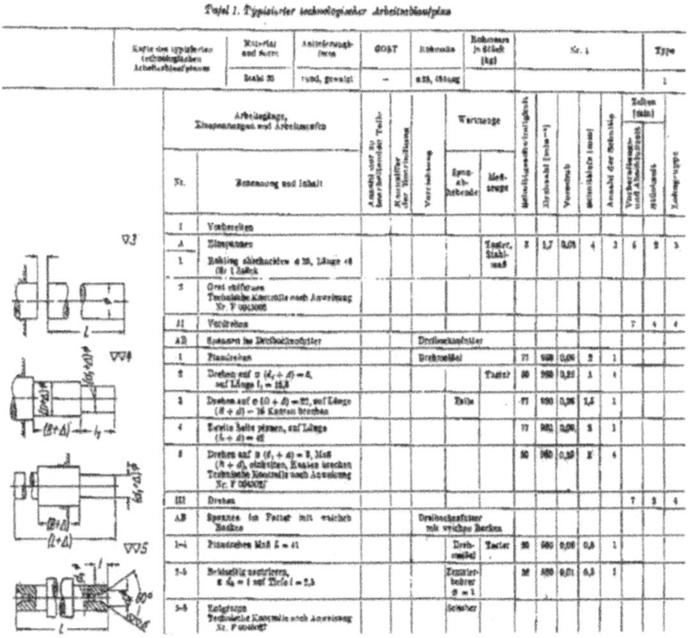

Tafel 1. Typisierter technologischer Arbeitsablaufplan

Karte des typisierten technologischen Arbeitsablaufplanes	Material und Sorte	Anlieferungsform	GOST	Rohmaße	Rohmasse je Stück (kg)	Nr. 1	Type
	Stahl 35	rund, gewalzt	–	⌀35, 48 lang			1

Nr.	Benennung und Inhalt	Anzahl der zu bearbeitenden Teile	Kennziffer der Vorrichtung	Vorrichtung	Werkzeuge: Spanabhebende	Werkzeuge: Meßzeuge	Schnittgeschwindigkeit	Drehzahl	Vorschub	Schnittiefe (mm)	Anzahl der Schnitte	Zeiten (min): Vorbereitungs- und Abschlußzeit	Zeiten (min): Stückzeit	Lohngruppe
I	Vorbereiten													
A	Einspannen					Taster, Stahlmaß	3	1,7	0,05	4	3	6	9	3
1	Rohling abstechen ⌀35, Länge 48 (für 1 Stück)													
2	Grat entfernen. Technische Kontrolle nach Anweisung Nr. F [illegible]													
II	Vordrehen											7	4	4
AB	Spannen im Dreibackenfutter			Dreibackenfutter										
1	Plandrehen			Drehmeißel			77	[illegible]	0,08	2	1			
2	Drehen auf ⌀ $(d_1 + \delta)$ = [illegible], auf Länge l_1 = [illegible]					Taster	[illegible]	[illegible]	0,25	1	4			
3	Drehen auf ⌀ $(D + \delta)$ = 22, auf Länge $(l + \delta)$ = 16 Kanten brechen				Feile		77	990	0,26	1,5	1			
4	Zweite Seite planen, auf Länge $(l_3 + \delta)$ = 45						77	982	0,06	3	1			
5	Drehen auf ⌀ $(d_1 + \delta)$ = 8, Maß $(R + \delta)$ einhalten, Kanten brechen. Technische Kontrolle nach Anweisung Nr. F [illegible]						[illegible]	960	0,32	[illegible]	4			
III	Drehen											7	3	4
AB	Spannen im Futter mit weichen Backen			Dreibackenfutter mit weichen Backen										
1-4	Plandrehen Maß L = 41				Drehmeißel	Taster	[illegible]	960	0,08	0,8	1			
3-5	Beidseitig zentrieren, ⌀ d_4 = 1 auf Tiefe l = 2,5				Zentrierbohrer ⌀ = 1		[illegible]	[illegible]	0,01	0,5	1			
3-6	Entgraten. Technische Kontrolle nach Anweisung Nr. F [illegible]				Schaber									

FIGURE 2. Process route for the complex part (Mitrofanov 1960).

3.1. Variant Process Planning Procedure

The roots of variant process planning systems go back to the work of Sokolovsky and Mitrofanov (1960) in which they generated the families of similar shafts and for each family created process routes that were used in planning individual parts from the family. In their work, the manufacturing engineers created the manual process routes (as shown in FIGURE 2 for each family with generic (parametric) dimensions. These route sheets (or process plans) were used to generate process plans for any new part from the family (i.e., new shaft). Process plan generation for individual parts consisted of selecting only necessary steps from the family route sheet and filling in necessary dimensions. This methodology was explored further and used in research and in practice in other countries.

Variant process planning procedure consists of two phases: preparation phase, and production phase. In preparation phase two tasks are performed: generation of the GT code for each component and generation of the process route for each component or for complex part. Both items are stored into the database.

Generation of the GT code is based on a previously created classification and coding system. Several researchers devoted significant amount of work to create classification systems that would be generally applicable in discrete part manufacturing. Few of well-known systems are: Opitz code, MICLASS, PRAHA, NIITMAS. Generally, all of them divide parts according to basic shape, dimensions, detailed shape features, material, and tolerances. An example of top level division in Opitz code is shown in FIGURE 3. For each of these properties subdivision is performed according to expected instances in parts. An example of the assigned GT codes (using customized classification system similar to Opitz code) for a sample part is shown in FIGURE 4.

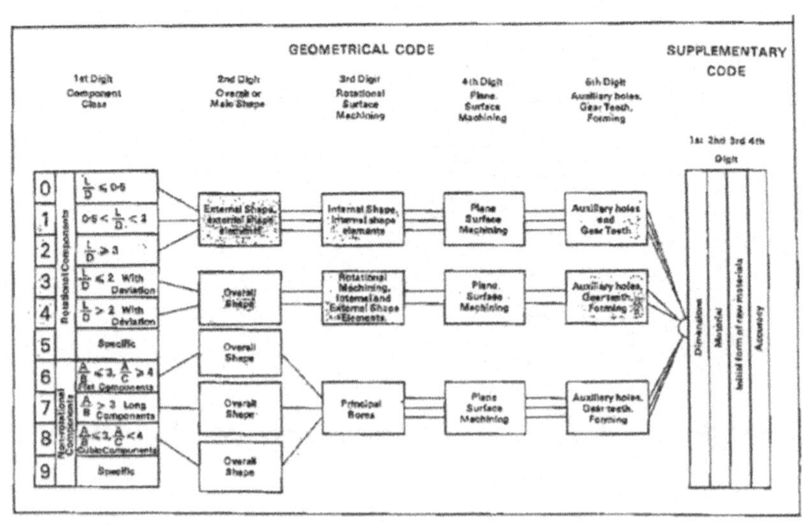

FIGURE 3. Top level classification of parts in Opitz system.

The next task in preparation phase is generation of process plans or routings for all components. This task is usually performed by copying the data from paper-based process routes and plans into the computer database. The flat text file of the process plan is generated. In some systems process plans may be slightly modified to adopt a unified approach while in some systems process plans are simplified to include only routing information. Usual part of these procedure is an inventory of existing machines and generation of the database of machines and equipment.

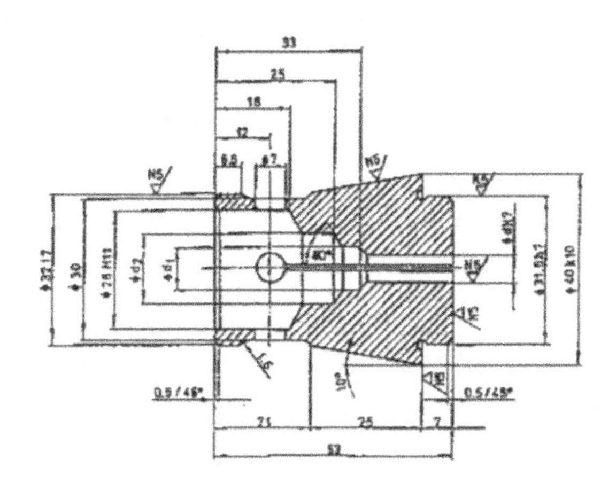

[1]	[2]	[3]	[4]	[5]	[6]	[7]	[8]	[9]	[10]	[11]	[12]	[13]	[14]
1	4	0	3	3	3	1	3	3	2	0	2	7	6

FIGURE 4. The part classification.

In the production phase a process plan for new parts is generated by using database created in the preparatory phase. Process planning procedure (see FIGURE 5) starts by analyzing new design (usually drawing) and developing a GT code for the new design. The GT code for new parts is created by using Classification and Coding system and performing an analysis of the part drawing and matching the part geometry with prescribed shapes from the C&C system. This task was usually done manually, but later some systems for automatic GT classification appeared. These systems used a user response to the queries about geometry and generated the code in an interactive session.

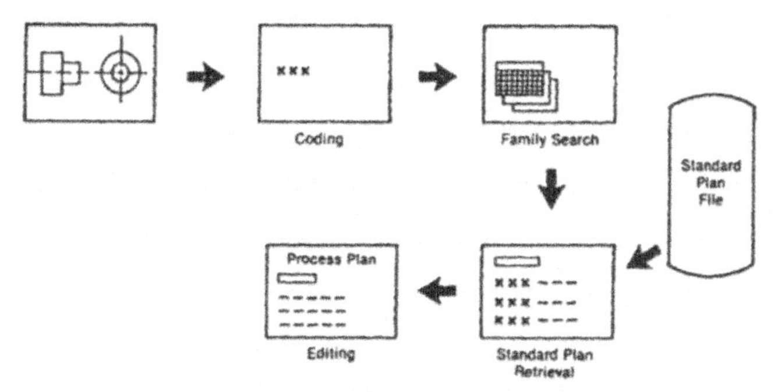

FIGURE 5. Production phase of variant systems (Chang et al. 1991).

Once the new part has received its code the product database is searched in order to find all parts in the database with the same GT code. These parts are, having the same GT code most likely candidates to have the same or similar manufacturing requirements. The process plans for such parts are retrieved into the editing session and a human process planner would edit such a plan to adjust it for the part. Usually adjustments were to enter correct values for the part dimensions, or to exclude one of machining operations, or to set particular cutting parameters. The modified plan thus becomes a process plan for the new part and as such is stored back into the database. This procedure may be repeated for each new part in the production phase of the process planning system.

The benefits of these systems were that the manufacturing and process planning knowledge of experienced human planners was captured in the form of process plans. There was no need to repeat planning of similar parts in different periods. Process planing procedure was performed in a uniform way. The main problem with this approach was that rapid and significant changes in the design of new products and parts would not allow for existence of similar process plan and therefore planning for such part had to start from the beginning, from the design.

3.2. Variant Process Planning Systems

Large number of variant process planning systems is described in the literature [see (Chang et al. 1991) and (Zhang and Alting 1994) for comprehensive surveys]. In this section we will mention only few best known as described in these surveys.

One of earliest process planning systems is CAM-I CAPP (which gave the generic name CAPP to all other process planning research). It was developed by CAM-I as a database management system written in FORTRAN. The system provided for alphanumeric GT code with up to 36 digits that was tailored for the user application. The system consisted from database, search and retrieval logic, and interactive editing capability for retrieved parts and plans. The system was widely accepted for the use in hundreds of companies.

Another well-known system is MIPLAN developed by the Organization for Industrial Research. The system was implemented as a data retrieval system that used the part code (MICLASS coding system was embedded for part classification), part number, and user defined similarity matrix for search and interactive plan editing. The system was later extended into MULTI-II to include classification system, process planning, time standards, and tool tracking.

Several other systems used similar approach in maintaining indexed database, searching for similar parts and interactively editing process plans. One such system, APOPS-08 (Zelenović et al. 1987) was embedded into GT-cell generation procedure. The system uses the classification and coding system for generating machine cells for similar parts. Once the cells are created process plans (in terms of process routes) for members of the part family are created by copying routes from the family complex part and editing if necessary. These routes are used as a feedback for cell formation procedure in terms of machine loading information and necessary adjustments are performed.

Few systems tried to combine elements of generative process planning with GT. One of these systems is described in (Joshi et al. 1994). The system consists of three modules: GTWORK for part coding, GTQUERY for search and retrieval, and PLANWORK for process planning. The system is implemented in relational database on PC. It provides for user defined classification system and separate query module to protect data integrity and security. Process planning module implemented decision rules and tables in the form of user definable decision graph.

4. GENERATIVE PROCESS PLANNING

Generative process planning is an approach that received research attention in early 80s by infusion of artificial intelligence methods into manufacturing planning. It starts with a product design specification and generates the process plan based on a previously acquired knowledge about manufacturing processes, tolerances, machines, tools and 'manufacturing practice'. Truly generative process planning system is a turnkey system that creates process plans based on built-in decision logic without human intervention (Chang et al. 1991). The heart of such a procedure is built-in manufacturing knowledge. However, in practice such an important task can not be transferred completely to the computer and relaxed definition of generative process planning includes systems that have various degree of implemented decision logic.

In this section we describe details of generative process planing procedure, identify necessary components of generative CAPP systems, and describe two stage in generative process planning development.

4.1. Generative Process Planning Procedure

The generative process planning procedure is based on knowledge about manufacturing. As we said above it starts with the part design (which should include complete design specification, i.e., geometry, dimensions, tolerances, etc.), then utilizes the existing manufacturing knowledge in order to infer necessary manufacturing processes, and finally provides full description for these processes (e.g., operator instructions, NC code, etc.) and generates the optimal sequence of processes for manufacturing. The comprehensive discussion on generative process planning is given in (Chang 1990). In this section we will describe the basic procedure and in the following sections we will describe specific features of early generative planning systems and discuss current research in the area.

The basic components of knowledge based process planning systems are: design representation, knowledge base, and inference mechanism. The process planning system should reason about and provide decisions for machining features and constraints between them, for selecting machining processes and equipment for them, and for clustering and sequencing of these processes into the final process plan. Below is a description of these components and tasks.

The design representation for process planning should be complete and should include all details of product specification. Such a representation includes the part geometry, dimensions, tolerances, part material, necessary material and surface finishes and all other notes and specs that define the quality product. Traditionally, CAD systems provide only portions of the design spec, most frequently the solid model of nominal geometry and set of unrelated notes, dimensions and tolerances. For efficient process planning it is necessary that the design representation includes machining features (portions of the part geometry that can be individually machined). Usually more than one part face belongs to such features. This task is performed by feature recognition procedure, which analyzes the part solid model and generates the set of machining features. Details of this research fall beyond the scope of this paper and may be found in (Requicha 1996) or (Mantyla et al. 1996). The feature model created by feature recognition has to be completed in order to provide missing links between nominal geometry, tolerances, and other design specifications. This task is precondition for automatic process planning and falls somewhere between feature recognition and process planning. In existing research systems it is usually performed by user input.

Knowledge base is a component of a generative process planning system that actually defines its scope and efficiency. The knowledge base is usually organized in one of two ways: rules or frames. Rule base is actually direct implementation of decision logic and rules as human planners use them. Examples of these rules may be found in numerous machining handbooks. Following are some examples from (Chang *et al.* 1991): If surface roughness is between 63 and 32 μin, then the required operation for hole making is reaming; If flat surface roughness is 8 μin, then surface finish can be obtained by grinding, polishing or lapping, but rarely by milling; If the diameter of the hole is less than 0.5" and greater than 0.01" and diametric tolerance is greater than 0.01", then use drilling, if tolerance is less than 0.01", then use drilling and reaming; and so on. These rules may be implemented directly into rule-based system such as OPS5. Many rule based process planning systems implement rule pattern matching module of more or less general form. An example of rule implemented in (Berenji and Khoshnevis 1986) is shown in FIGURE 6. This rule specifies that for any feature (variable *feature.1*) of the type of NOTCH, if its quality (i.e., tolerance, variable *q1*) is greater than 60 (ten-thousands of an inch) MILLING process should be used to machine it and with only one ROUGH-FINISH cut.

```
RECOMMEND.MILLING INFERENCE
 FTYPE feature.1 NOTCH
 QUALITY feature.1 q1
 GREATERP q1 60
 ->
 recommend-process feature.1 MILLING
 RECOMEND-CUT feature.1 ROUGH-FINISH
```

FIGURE 6. The process selection rule.

An extended approach to knowledge representation in process planning and manufacturing is the use of frames introduced into AI world by (Minsky 1975). A frame is a set of attributes and values of these attributes that describe some entity from the real world (Rich and Knight 1991). In process planning every machining feature, manufacturing process, machine or tool may be seen as a frame upon identifying relevant attributes and their values. For example, the frame that represents twist drilling process is shown in FIGURE 7. The reader may notice a large variety of attributes necessary to describe twist drilling and, in general any machining process.

```
{{ [illegible] :context #ROOT-CONTEXT
    IS-A: CORE-MAKING
    PREFERRED-TO: GUN-DRILLING SPADE-DRILLING END-DRILLING
    MAY-USE-TOOL: TWIST-DRILLING-TOOL
    CUTPARMS: CP-TWIST-DRILLING
    INPUT-LENGTH: 1/8
    MAKE-PROCESS-TIME: MAKE-DRILLING-TIME
    SWEPTSOLID: MAKE-DRILLING-SOLID
    SMALLEST-TOOL-DIAMETER: 0.0625
    LARGEST-TOOL-DIAMETER: 2.0
    NEGATIVE-TOL: TWIST-NEG-TOL
    POSITIVE-TOL: TWIST-POS-TOL
    STRAIGHTNESS: TWIST-STRAIGHT
    ROUNDNESS: 0.004
    PARALLELISM: TWIST-PARALLEL
    PERPENDICULARITY: TWIST-PARALLEL
    DEPTH-LIMIT: 12
    TRUE-POSITION: 0.008
    SURFACE-FINISH: 100
      }}
```

FIGURE 7. The twist drilling frame.

The frame approach has been extended into two directions: frame taxonomy and semantic network in several CAPP systems. Frame taxonomy or hierarchy as in (Nau 1987) utilized ideas from object-oriented programming to develop a hierarchy of concepts with subclass/superclass relations and inheritance of attributes. An example of hierarchy of frames for the domain of milling and hole making operations is shown in FIGURE 8. The fact that the one frame relates to other frames (e.g., hole frame relates to twist drilling and end milling but not to side milling) lead to the development of semantic network of interconnected frames and identification of various relations between them. An example of such semantic network built into 3I-PP system (Sormaz and Khoshnevis 1997) is shown in FIGURE 9.

Automatic reasoning in generative process planning is achieved by implementing an inference mechanism. Rule-based systems use rule-matching cycle that consists in three steps: pattern matching, conflict resolution, and rule execution. Pattern matching algorithm compares each rule conditions (which may contain variables for matching) with elements in the database (or working memory) called working memory elements (WME). If for all conditions a match between rule and WME the rule is satisfied and may be fired or executed. The second step, conflict resolution, considers all the rules that may be fired and selects one according to established conflict

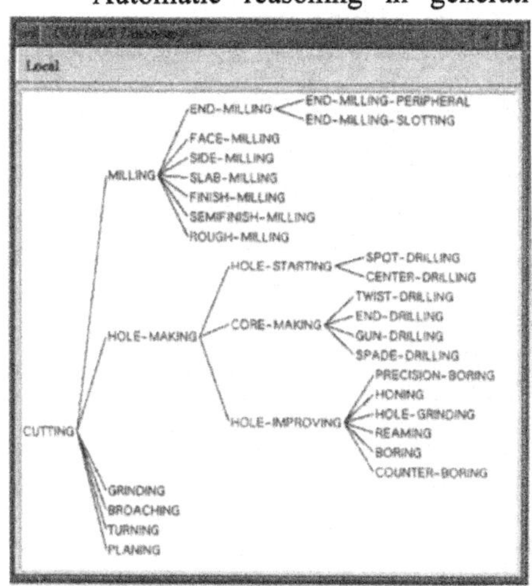

FIGURE 8. Taxonomy of manufacturing processes.

resolution strategy. The selected rule form this step is fired or executed in the last step. Execution of the rule usually changes some attributes in WME's, creates new WME or deletes existing ones. As a result of these changes rule satisfactions may change, therefore new cycle starts again. Actual implementations of this cycle usually avoids checking of all rules in each cycle because changes in WME are of local characters and impact only a few rules. An efficient implementation may be based on the Rete algorithm (Forgy 1983).

An alternative way for execution of process planning system is used when frame based knowledge is utilized. A hybrid procedures that utilize message passing and method of various objects are used. Manufacturing rules are implemented as components of object methods in this approach. Usual process planning procedure may be performed through the following procedure. An object of type feature is sent a message to select its machining processes. The object executes its method that initiates messages to appropriate processes (e.g., a hole creates drilling and end milling objects, while a linear slot may create end milling and side milling objects). Each process instance executes its method that verifies that the process is capable of generating the feature with required tolerances and creates instances of appropriate machine and tool objects (e.g., drill press and mill for twist drilling, while only mill for end milling). These instances in turn execute their own methods that specify required dimensions and parameters. When processing of all features is completed, a message is sent to the part object to sequence features and processes into the final process plan. As a result of this procedure the instances of all objects are created as it is shown in the central portion of FIGURE 9.

The above described inference mechanism is usually applied in the tasks of selecting manufacturing process, machine, and tool for individual features. These tasks require significant amount of manufacturing knowledge and understanding of manufacturing process and physics. By moving from the one domain of manufacturing and process planning (say, machining) to another (for example forging, or sheet metal processing, or assembly) it is necessary to acquire the knowledge about a new domain in order to generate

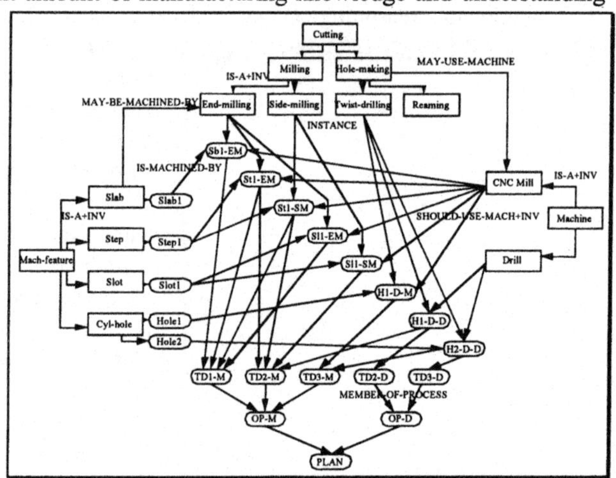

FIGURE 9. Semantic network of process planning entities.

good process plans.

Process sequencing is the part of process planning that is usually performed upon selecting processes for individual features. For this task a method based on optimization is very often used. An approach that utilizes dynamic programming is part of RTCAPP system, while various forms of state space search and appropriate algorithms were used in other systems.

4.2. Early Generative CAPP Systems

One of the first research results in generative process planning systems was work by Wysk at Purdue University (Wysk 1977). In this work a procedure based on decision tables was implemented in FORTRAN for process planning of machined parts. The system captured the significant portion of process planning decision logic with consideration of process capabilities. This work was followed by work in implementing various rule based procedures into process planning.

Several process planning systems were built using various flavors of symbolic processing and rule-based programming. One of the earliest rule based process planning systems is GARI (Descotte and Latombe 1984) which performs planning of the sequence of machining processes for mechanical parts. GARI uses as its input a symbolic representation of the part, features, and relations between features in the form of LISP lists. Most other rule based systems used this scheme (LISP list plus rules). HiMapp (Berenji and Khoshnevis 1985) uses a revised form of a planner called Deviser in its core while Machinist (Hayes and Wright 1987) was built as a set of OPS5 rules for reasoning about interactions between features and how they influence fixturing of prismatic parts.

RTCAPP (Park and Khoshnevis 1993) extended the previous work to include a manufacturing knowledge base which consists of frame representation (implemented as LISP lists) of knowledge that connects processes, machines, and tools. SIPS (Nau 1987) is based on a hierarchical abstraction technique called hierarchical knowledge clustering, where the knowledge is organized in a taxonomic hierarchy of objects. AMPS (Automated Machining Planning System) (Chang 1990) is a process planning expert system that supports a QTC (Quick Turnaround Cell), an automated manufacturing cell for prismatic parts of one-of-kind type. AMPS uses both frames (in KEE) for declarative knowledge representation and rules for procedural knowledge representation. The combination of object-oriented approach (for feature classification) and rule base (for process selection) was used in XCUT (Hummel and Brooks 1988).

Most of these early systems were based on symbolic processing of feature information only. They were very successful in implementing process planning decision logic, but did not perform tasks of geometric verification, process sequencing in sufficient details nor did they address issues related to CAD/CAM integration. These topics were included in later systems in 1990s described in the following section.

4.3. Current Research In Generative Process Planning

Majority of work last five to seven years is related to inclusion of CAD data and geometric design into process planning procedure and to integration of process planning with design and scheduling (NIST 1996).

The first significant work that addressed issue of integration between CAD and process planning was the system FirstCut, which was followed by NextCut (Brown and Cutkosky 1990). The system provided integrated framework in which design was created using manufacturing features, process planning was performed on such design and manufacturability feedback was sent back to designer. The system utilized knowledge base approach in process planning and provided graphical user interface to the user. This work was later extended to include broader integration of product development cycle.

A system that combined group technology procedures with automated process planning is described in (Joshi 1994). The system utilizes features of both variant and generative process planning. The work on IMACS system is described in (Gupta et al. 1994). The emphasis of this system is on performing geometric computations as part of process planning. The system generates machining features, analyzes alternative feature interpretations, and generates appropriate process plans.

The work on 3I-PP system (Sormaz 1994) extended the process planning research in two directions. The first direction is a comprehensive knowledge representation and knowledge base for process planning based on semantic networks and object-oriented approach (Sormaz and Khoshnevis 1997). A taxomony of objects, object attributes and relations between objects as means for representing process plans has been developed. The second direction is toward analyzing alternative process plans from shop floor resources and scheduling point of view. The authors have developed a process plan network that provides for alternative processes, machines and tools to be selected at a later stage. The system also provides a space search based algorithm for process clustering and sequencing (Sormaz and Khoshnevis 1996).

Implementation of generative process planning procedure into commercial products was achieved in two currently available systems: CIMPLEX Manufacturing Analyst (Cimplex 1994) and PART (PART 1994). Both of these systems perform feature recognition from CAD data file. For case that automatic recognition is not completely successful they provide user guided feature identification through GUI. Upon complete feature recognition and/or identification they generate process plan based on implemented knowledge base. They advance generated process plans toward tool path generation, graphical machining animation and generation of NC code for automated machining. Both of these systems also depend largely on previously built database of machines and tools. CIMPLEX utilizes an object-oriented database system Objectivity, while PART is based on SQL based relational database technology. Both systems are user extendible in order to allow for acquisition of company specific manufacturing practices and knowledge.

Integration of various design and manufacturing software modules has been addressed in several papers. The main issue addressed is the representation of process plan data in the form suitable for data exchange. A graphical representation of process plan has been explained in (Catron and Ray 1991). The authors described the language that represents process plans in the form of a graph. The graph includes nodes for hierarchical representation of processes and activities and addresses the issue of parallel, sequential and other constraints between nodes of the represented process plans. Another research project was reported in (Lee et al. 1994) with the goal of integrating of process planning and shop-floor control. The authors provide the process plan representation to be used in shop-floor control that carries hierarchical representation of process plans and alternative process plans. The work toward the standard format for the process plan exchange has been carried out at NIST (National Institute for Standards and Technology). Two current projects are: 1) generation of process plan interfaces as the part of ISO STEP standard (ISO 1994), and 2) research on process specification language (NIST 1997).

5. APPLICATION OF GT PRINCIPLES IN PROCESS PLANNING

Process planning and generation of process routes as described in an earlier section, is a very early example of applying group approach to process planning. Not only time and effort in planning are saved but also time and cost in manufacturing are saved when applying group routing due to the fact that optimal or near optimal (from cost and quality point of view) process plans are generated for the part family.

The application of group technology to process planning revealed the fact that machine setting time is not a constant factor, but can be easily reduced (Burbidge 1975). Another example pointed out in (Burbidge 1975) is the selection of new equipment, also one of the tasks in process planning. By its very nature, the new machine is designated to more than one product. Therefore, it is necessary and obligatory to look into requirements (shape, material, quality, etc.) for all of these products before making decision on which machine to purchase.

Issue of standardization in CAD/CAM and process planning has received significant attention in research community, government and industry (NIST 1996). This issue is related to broad definition of group technology and refers to efforts for generating a common language for process plan (and product) representation and methods for exchange of this information between various applications. The goal of this work is to reduce time and effort in overall product development. However, for successful integration through standard information exchange much more research in process planning is necessary.

Another issue related to application of group technology in process planning is also motivated by reduction of time and efforts. After the process plan is generated it is stored into database and sent to other applications. However, in real applications there is always possibility of changes in design that may require changes in process plan. Most of current process planning systems will require that process plan is generated again from design data, even for minimal change as hole diameter correction. Truly intelligent process planning system would be able to analyze the change in design and modify process plan only for the change keeping the rest of process plan unmodified.

6. PROCESS PLANNING AND MANUFACTURING CELL DESIGN

In this section, we briefly discuss the need for simultaneous consideration of process plans and manufacturing cells in an integrated procedure. First we briefly explain machine cell formation procedure Provide several scenarios of such procedure and explain research direction for the integration.

The fact that process planning data (in one or the other form) are necessary for generation of machine cells is acknowledged in all machine cell formation algorithms. The necessity to obtain so called part-machine incidence matrix, as a starting point in clustering algorithms is also well known. However, the need to revise process plans after part families are assigned to machine cells is very rarely discussed in the literature.

Most of cell formation algorithms usually assume that the process plans have been generated. They accept the part-machine incidence matrix as 0-1 matrix or in some other form (e.g., operation times, load, etc.). The one of few machine cell formation methods that addresses process plans is Production Flow Analysis (PFA) (Burbidge 1989). PFA provides for changes in process plans as a way to eliminate exceptions (parts that do not fit into primary material flow). Namely, if the result of PFA are cells with intercell flow it is necessary to modify process routes for parts that cause intercell flow in order to make cells completely independent.

The author feels that integral consideration of process planning and machine cell design will provide significant benefits. Research into methods and procedures for this integration is needed to obtain the full benefit of applying group technology in this area. Possible directions are to generate few alternative process plans and incorporate their consideration in machine cell formation as an additional variable. An alternative way would be to include cell formation into product development cycle and analyze machine cell impact on both product design and manufacturing. Another method would be to consider dynamic machine cells (cells for which part family may change), in which case it is necessary to perform process planning after such a change.

7. FUTURE DIRECTIONS

The author feels that future research in the area of generative process planning will move from consideration of individual products (parts) to analyze the group (or family) of parts. There are several reasons for this. First, the CAD/CAM integration in the current stage might be extended to include other manufacturing planning tasks (e.g., layout design, capacity planning, scheduling, etc.). All these tasks perform a significant shift from design and planning of individual parts toward planning for the group of parts. Second, CAD development in recent years moved toward parametric design (as in Parametric Technology's ProEngineer). Products are being designed in parametric and variable form in order to give the designer freedom to generate the functional products. Thus, design is performed not only for a single product but instead for a family of similar products. This is well suited for application of group technology. In the section, we briefly describe several recent research efforts in process planning and group technology that provide examples of integration of two areas. After that, we explain several research approaches that will provide for the extended use of combined methods.

7.1. Group Technology Using CAD Systems

GT application in product design was covered earlier in this volume. Here we point out the relation between design and process planning and the use of GT in that link. We also explain feature recognition and feature modeling from a GT point of view.

There has been significant research effort in the area of feature recognition and feature-based design in order to provide the bridge between geometric-oriented design and manufacturing-oriented process planning (Requicha 1996; Mantyla et al. 1996). The recognition of manufacturing features on its own represents the classification of the part shape elements into suitable manufacturing processes through assigned features. Thus, manufacturing features defined as machinable objects represent the first step towards integrating design and process planning.

Representations of designs in the form of manufacturing feature models that capture manufacturing information are still under development. Feature recognition systems e.g., in (Han 1996) and (Regli 1994), convert geometric design into features, but they do not generate complete feature models. Methods and techniques for generation of a feature model that will include complete design are still under development. Such model should include nominal geometry, tolerances, other requirements and notes as well as constraints imposed on available machining operations and order among them. Methods based on various forms of graphs (or networks), for example AND-OR graphs are under investigation.

7.2. Knowledge Base For Generative Process Planning

Knowledge base is a very important part of generative computer aided process planning systems. The capture of manufacturing process information in a systematic way may also be seen as an application of group technology in broad sense. The representation of machining processes as objects (as explained in section 4. Generative Process Planning) with attributes and methods for process selection, toll path generation, etc. in algorithmic way allows its uniform application for all parts, and therefore is indeed group technology in its broad sense.

The creation of manufacturing knowledge base is usually performed in research systems in a somewhat limited domain in order to provide for testing of the systems. However, for real application of automated process planning systems it is necessary to build an extended knowledge base that will include all machining process within a company. Such knowledge base also has to be custom build to include the existing expertise within the company. PART system mentioned earlier contains general knowledge base and user interface for customization of the knowledge base.

7.3. Case-Based Process Planning

The existence of various variant process planning systems in the use by industry provides for integration of variant and generative approaches in process planning. Such integration may be achieved by applying case-based reasoning technique from artificial intelligence field. There has not been much work in order to use the existing process planning databases nor to use case based reasoning.

The possible research direction in integration of variant and generative process planning may be as follows. Part families may be created from CAD database by applying the selected similarity criterion between parts based on geometry, tolerances and other data. Process plan for the representative of the family may be created using generative process planing system. The result may be the process plan network with alternative process plans. The network may be generated using reasoning explained in section 4.1. Generative Process Planning Procedure. Process plans for other parts may be generated by modifying the process plan network according to actual similarity and differences between an individual part and the representative. The benefit of this method is that time-consuming process planning procedure is not repeated, only incremental change in the process plan is performed.

7.4. Dynamic Process Planning With Alternative Plans

Process plans are used for manufacturing planning tasks, such as capacity planning, layout design and scheduling. On the other hand, the status of machine shop floor and number of available machines are used during process planning procedure. This interaction leads to the need to consider process plans in a dynamical way. Creation of flexible process plans is desirable. There are several ways to achieve such flexibility: simultaneous process planning and scheduling, incremental process planning, and alternative process planning.

Simultaneous process planning and scheduling is reported in a few` studies. Incremental process planing has been studied in the context of the link with CAD system. The method is applied for the generation of process plan in increments that correspond to changes in the design. Each modification in the design corresponds to incremental change in the process plan (e.g., addition of feature in the design causes only addition to process plan, change in part dimensions require only changes in tool dimensions, etc.). Alternative process planning has been addressed in (Sormaz and Khoshnevis 1997). The authors generated a network of alternative processes (with machine and tool selection) for each feature. The process plan network is created and saved as a representation for alternative process plans. The selection of the particular process plan (or route in the network) may be postponed if needed. If process plans are used in capacity planning, the constraints on number of available machines may influence the selection of the best plan from the network. If some other criteria are used (e.g., due dates in scheduling) a different process plan may be selected from the same network.

8. CONCLUSIONS

This chapter presented a brief overview of research in automated process planning. Two distinct approaches: variant and generative have been explained, application of group technology discussed, and possible research directions described. Details of both variant and generative process planing methods were provided and distinctions between two approaches were pointed out. Variant process planning has been developed into several commercially available products. To the contrary, generative process planning is still under development with first commercial products appearing in last several years.

The model of integrated manufacturing introduced in the chapter explains the significant role of process planning for successful integration. The need to use the process planning data in most manufacturing planning tasks provides an opportunity to extend GT application in process planning. The chapter ended with a selection of several research directions which would relink GT and process planning and, thus, provide for significant saving of time and effort in manufacturing planning tasks.

REFERENCES

Bedworth, D. D., Henderson, M. R., and Wolfe, P. M., 1991, *Computer-Integrated Design and Manufacturing*, McGraw-Hill, New York.

Berenji, H. R. and Khoshnevis, B., 1986, Use of artificial intelligence in automated process planning, *Computers in Mechanical Engineering*, 5, 2, 47 -55.

Brown, D. R. and Cutkosky, M. R., 1990, Next-Cut: A computational framework for concurrent engineering, *Proceedings of Second International Symposium on Concurrent Engineering*, Morgantown, WV, February.

Brownston, L., Farrell, R., Kant, E., and Martin, N., 1985, *Programming Expert Systems in OPS5, An Introduction to Rule-Based Programming*, Addison-Wesley, Inc., Menlo Park.

Burbidge, J. L., 1975, *Introduction of Group Technology*, John Wiley & Sons, New York.

Burbidge, J. L., 1989, *Production Flow Analysis for Planning Group Technology*, Oxford Press.

Catron, B. A. and Ray, S. R., 1991, ALPS: A Language for Process Specification, *International Journal of Computer Integrated Manufacturing*, 4, 2, 105-113.

Chang, T.-C., *Expert Process Planning for Manufacturing*, 1990, Addison-Wesley, Inc., Menlo Park, CA.

Chang, T.-C., Wysk, R. A., and Wang, H-P., 1991, *Computer-Aided Manufacturing*, Prentice Hall.

CIMPLEX User Manual, 1994, Cimplex Inc., San Jose, CA.

Chen, Q. and Khoshnevis, B., 1993, Scheduling with flexible process plans. *Production Planning & Control*, 4, 4, 333 -343.

Descotte, Y. and Latombe, J.-C., 1984, GARI: An expert system for process planning, in *Solid Modeling by Computers, From Theory to Applications*, (Eds. M. S. Pickett and J. W. Boyse), 329-346, Plenum Press.

Forgy, C. L., 1982, Rete: A fast algorithm for the many pattern/many object pattern match problem, *Artificial Intelligence*, 19, 1, 17-37.

Gupta, S. K., Nau, D., Regli, W., and Zhang, G., 1994, A methodology for systematic generation and evaluation of alternative process plans, in *Advances in Feature Based Manufacturing, Manufacturing Research and Technology*, (Eds. J. J. Shah, M. Mantyla, and D. S. Nau), 20, 161-184, Elsevier, Amsterdam.

Ham, I. and Lu, S. C.-Y., 1988, Computer-aided process planning: The present and the future, *Annals of CIRP*, 37, 2, 591-601.

Han, J.-H., 1996, *3D Geometric Reasoning Algorithms for Feature Recognition,* Ph.D. dissertation, University of Southern California.

Hayes, C. and Wright, P., 1987, Automating process planning: Using feature interactions to guide search, *Journal of Manufacturing Systems*, 8, 1, 1 -15.

Hummel, K. E., and Brooks, S. L., 1988, Using hierarchically structured problem-solving knowledge in a rule-based process planning system, in *Expert Systems and Intelligent Manufacturing*, (Ed. M. D. Oliff), 120 -137, Elsevier, Amsterdam.

ISO, 1996, *Industrial automation systems and integration – Product data representation and exchange, Part 213: Application protocol: Numerical control process plans for machined parts,* ISO.

Joshi, S. B., Hoberecht, W. C., Lee, J., Wysk, R. A., and Barrick, D. C., 1994, Design, development and implementation of an integrated group technology and computer aided process planning system, *IIE Transactions*, 26, 4, 2-18.

Khoshnevis, B., Park, J. Y., and Sormaz, D. N., 1994, A cost based system for concurrent part and process design, *The Engineering Economist*, 40, 1, 101 -124.

Lee, S., Wysk, R. A., and Smith, J.S., 1994, Process planning interface for a shop floor control architecture for computer integrated manufacturing, *International Journal of Production Research*, 33,9, 2415-2435.

Mantyla, M., Nau, D., and Shah, J., 1996, Challenges in feature-based manufacturing research, *Communications of the ACM,* 39, 2, 71-76.

Minsky M., 1995, A framework for representing knowledge, In *The Psychology of Computer Vision*, (Ed.) P. Winston, McGraw-Hill, New York.

Mitrofanov, S. P., 1960, *Wissenshaftliche Grundlagen der Gruppentechnologie*, Veb Verlag, Berlin.

Nau, D. S., 1987, Automated process planning using hierarchical abstraction, *Texas Instruments call for papers on AI for Industrial Automation*, July.

NIST Process Planning Workshop, 1996, Gaithersburg, June.

NIST Process Specification Language Project Web Site 1997.

Opitz, H., 1970, *A Classification System to Describe Workpieces, Part I*, Pergamon Press.

Park J. Y. and Khoshnevis, B., 1993, A real time computer aided process planning system as a support tool for economic product design, *Journal of Manufacturing Systems*, 12, 6, 181-193.

PART Reference Manual, 1994, ICEM Technologies, Ardan Hills, MN.

Regli, W. C., 1995, *Geometric Algorithms for Recognition of Features from Solid Models,* Ph.D. dissertation, University of Maryland.

Requicha, A. A. G., 1996, Geometric reasoning for intelligent manufacturing, *Communications of the ACM,* 39, 2, 71-76.

Rich, E. and Knight, K., 1991, *Artificial Intelligence*, McGraw-Hill, New York.

Sormaz, D. N., 1994, *Knowledge-based Integrative Process Planning System using Feature Reasoning and Cost-based Optimization*, Ph.D. thesis, University of Southern California.

Sormaz, D. N. and Khoshnevis, B., 1996, Process sequencing and process clustering in process planning using state space search, *Journal of Intelligent Manufacturing,* 7, 3, 189-200.

Sormaz, D. N. and Khoshnevis, B., 1997, Process planning knowledge representation using an object-oriented data model, *International Journal of Computer Integrated Manufacturing,* 10, (1-4), 92-104.

Sormaz, D. N. and Khoshnevis, B., 1997, Generation of alternative process plans in the integrated manufacturing system (working paper).

Wysk R. A., 1977, Ph.D. dissertation, Purdue University.

Zelenović, D. M., Cosić, I. P., Sormaz, D. N., and Sisarica, Z. Dj., 1987, An approach to the design of more effective production systems, *International Journal of Production Research,* 25, 1, 3-15.

Zhang, H.-C. and Alting, L., 1994, *Computerized Manufacturing Process Planning Systems*, Chapman & Hall.

AUTHOR'S BIOGRAPHY

Dusan N. Sormaz is Assistant Professor of Industrial and Manufacturing Systems Engineering at Ohio University. He received his BS. and M.Sc. degrees in industrial engineering from University of Novi Sad, and his M.Sc. in computer science and Ph.D. degree in industrial and systems engineering from University of Southern California. He worked one year as a post-doctoral research associate in manufacturing at University of Southern California before joining Ohio University. His principal research interests are in process planning, CAD/CAM integration, concurrent engineering, group technology and application of knowledge based systems in integrated manufacturing. Dr. Sormaz is senior member of SME, senior member of IIE, and member of IEEE. Dr. Sormaz has worked with Professor Burbidge on many GT research and implementation projects in Yugoslavia for several years. He has also had the privilege of attending a course at Cranfield University in 1987 in Production Control, taught by "The Professor".

C2

GT and Computer Aided Process Planning: Observations on Past, Present & Future

J. Nolen and C. Lyman-Cordes

1. INTRODUCTION

Group technology (GT) seeks to exploit similarities among parts or products for efficiencies in design and manufacturing. Production Flow Analysis (PFA) seeks to exploit similarities in processes. Applications of GT in Computer Aided Process Planning (CAPP), and GT and PFA in cellular manufacturing (CM) have produced very significant, and well-publicized productivity improvements compared to traditional methods; yet a vast majority of firms are yet to use them.

To explain this paradox one must understand the practical limitations to GT's application in design, process planning, and cellular manufacturing from both a technical and attitudinal standpoint. This chapter traces the technological development and application of GT from 1940 to the present through several illustrations from the authors' experience.

Another factor is the perceived cost-benefit equation in an era of "downsizing." Not surprisingly, the viewpoint of engineers and managers differ, but even the most modern systems (designed with the help of experienced GT and CAPP users) are finding widespread acceptance for CAPP - but not GT. There is hope, however. Successful field trials of a radically new software technology promise to achieve the goals of GT and PFA, but in ways not anticipated by GT pioneers.

2. TECHNOLOGICAL DEVELOPMENT OF GT

GT played an important role in World War II. In what is believed to be the largest industrial migration in history, the Soviets (Manchester 1981) moved 600 strategic industries endangered by the German onslaught of 1941 out of the Ukraine to safety East of the Ural mountains (Turkel 1984). They used GT to organize parts and processes and resumed full production within 15 months. The Soviet T-34 tank, built to a standard design in one of the relocated factories, entered the conflict in 1942 and never lost a major engagement thereafter. German armor was also hampered by the proliferation of designs from Dr. Porsche's studio (Panzer I, II, III, IV, Leopard, Panther, Tiger, Elephant, etc.), some good and others not, which complicated spare parts logistics.

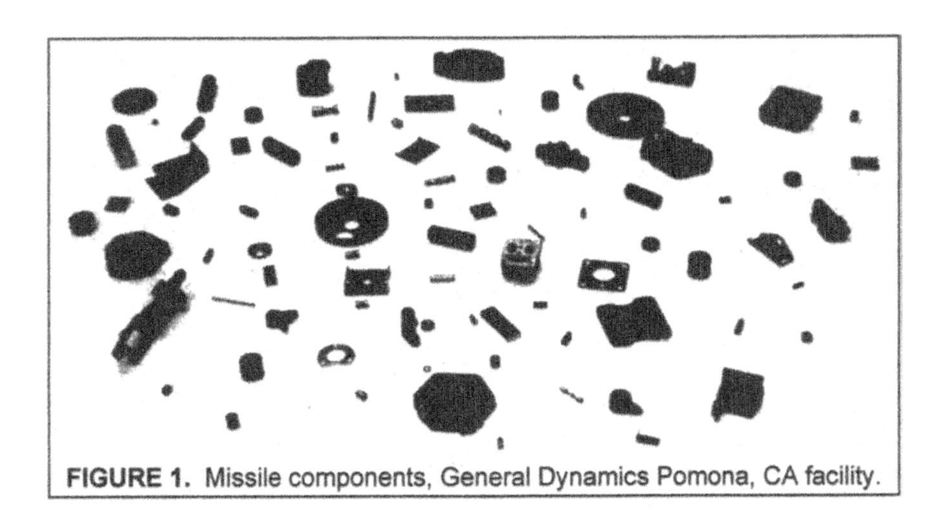

FIGURE 1. Missile components, General Dynamics Pomona, CA facility.

Significant part numbers, or manual GT systems more often achieved somewhat limited success. Such systems were popular in the US and Europe in the 1960s, but their application was generally limited to drawing retrieval. Because humans have trouble with strings longer than eight or nine characters in length, manual classification is limited to a few features; appropriate for simple parts but inadequate for complex ones. Moreover, manual retrieval is cumbersome. Few manual GT systems pass the path-of-least-resistance test. Is it easier to look up an existing part than it is to process a new one?

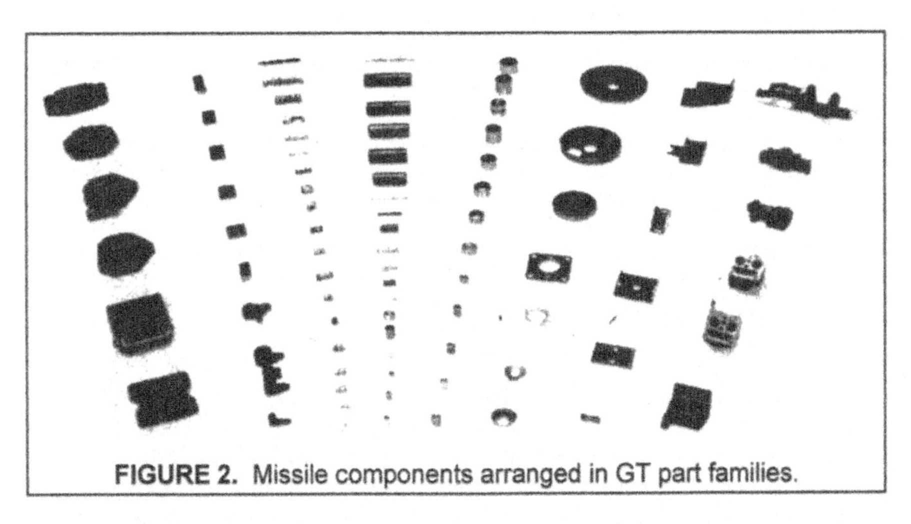

FIGURE 2. Missile components arranged in GT part families.

The first computerized GT system to enjoy commercial success in both Europe and North America was developed in the early 1970s by B. A. Schilperoort of TNO, the Institute for Applied Research, in Apeldoorn, The Netherlands. Known as MICLASS for *Metaalinstituut* Classification, it was used for process planning, design retrieval, and cellular manufacturing applications in 200 installations worldwide.

Asked to determine how to organize production to maximize return on investment for CNC machine tools, Schilperoort recognized the importance of setup reduction, one-piece lot sizes, and design for manufacturability years before most of his colleagues in the West. Standing on the shoulders of Europe's GT pioneers and alert to CM work in Japan (Samuel 1982, Schonberger 1992, and Russell 1990) and at Volvo in Sweden, he developed the tools to do so, creating a menu-driven, interactive GT, CAPP, and CM system with on-line help in 1973. Schilperoort combined ideas from Germany's Professor Dr. Opitz (round parts), Dieter Zimmerman (sheet metal), Koloc, a Czech, who previously classified tens of thousands of parts with a four digit schema of his own design, and Massberg, who developed a comprehensive, 85-digit schema. Schilperoort achieved success by avoiding the pitfalls that claimed previous attempts (Nolen and Schilperoort 1991). Specifically:

- He did not presume to know what features were important. The universe of features was determined by classifying **every** feature on tens of thousands of parts produced at companies where Schilperoort's engineering students were interned. A code 150 digits long was required (Nolen and Keus 1991) to capture everything.
- A valid taxonomy 12 digits in length was distilled from the initial 150 digits by frequency analysis.
- A set of rules insured consistent classification.
- *Definitions* made part feature classification an objective - not subjective - activity.
- *Conventions*, or default conditions, covered instances when the engineering drawing was silent with regard to raw material or feature information, a common occurrence with older drawings.
- Computers automated classification. Through interactive menu-driven classification, the computer - not the user - quickly created long GT codes that were accurate and easily manipulated. Retrieval was no longer cumbersome.
- He recognized the prevalence of design and process proliferation, and devised tools to deal with it. Similar parts had different machined features, tolerances, finishes, raw materials and so forth. Worse yet, they had completely different routings! Process proliferation frustrated PFA. He used GT to identify families, see FIGURE 4, followed by PFA, see FIGURE 5, to find the latent standard routing (LSR) for the family. The LSR became CAPP's parametric master process plan and CM's dominant flow path. Later he added machine tool

capacity analysis to the software's capabilities to calculate machine load and balance CM flow.

While companies such as John Deere in the US made significant, and well publicized returns from their GT classification investment in the form of standard processes, tooling, and cellular manufacturing, most companies balked at the effort required. Appeals to limit classification to a representative, random sample (Barnett 1994) of 10% of **active** parts (usually three months of effort for one person), enough to establish both parametric master process plans and cell layouts, fell on deaf ears.

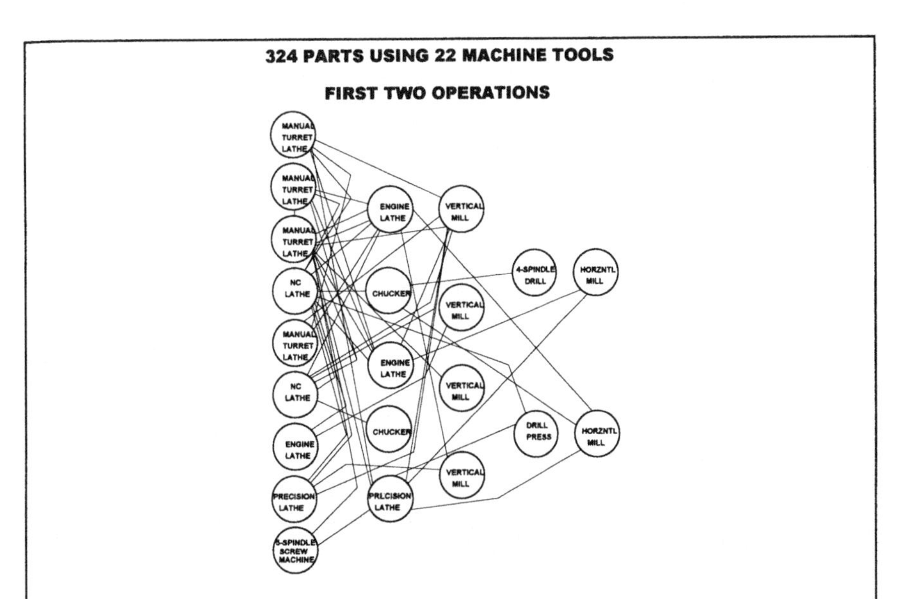

FIGURE 3. Process Proliferation.

MICLASS was intended to address Process Proliferation, illustrated graphically in this diagram from a case study ***("EG&G Sealol (A)," Harvard Business School, 9-686-045 (March, 1987) pp 1-18)*** at a plant where one of the authors (Nolen) installed cellular manufacturing. Proliferation of designs and processes is costly (Dallas 1976) and commonplace in industry. It is often aggravated by uncontrolled copy-and-edit CAD and CAPP systems. Proliferation frustrates Production Flow Analysis because even similar parts depicted here have very different routings. Schilperoort realized that isolating families of parts with GT and then using PFA to identify the latent standard routing for parametric master process plans and cell layouts produced superior results.

Copy-and-edit-CAPP gets work out of engineering quickly but usually creates the burden of proliferation for the shop. Used properly, parametric master process plans built around GT part families make CAPP an engineering productivity tool that also improves shop operations. Standardized process plans become the basis for cells when volume along a flow path justifies dedicated equipment.

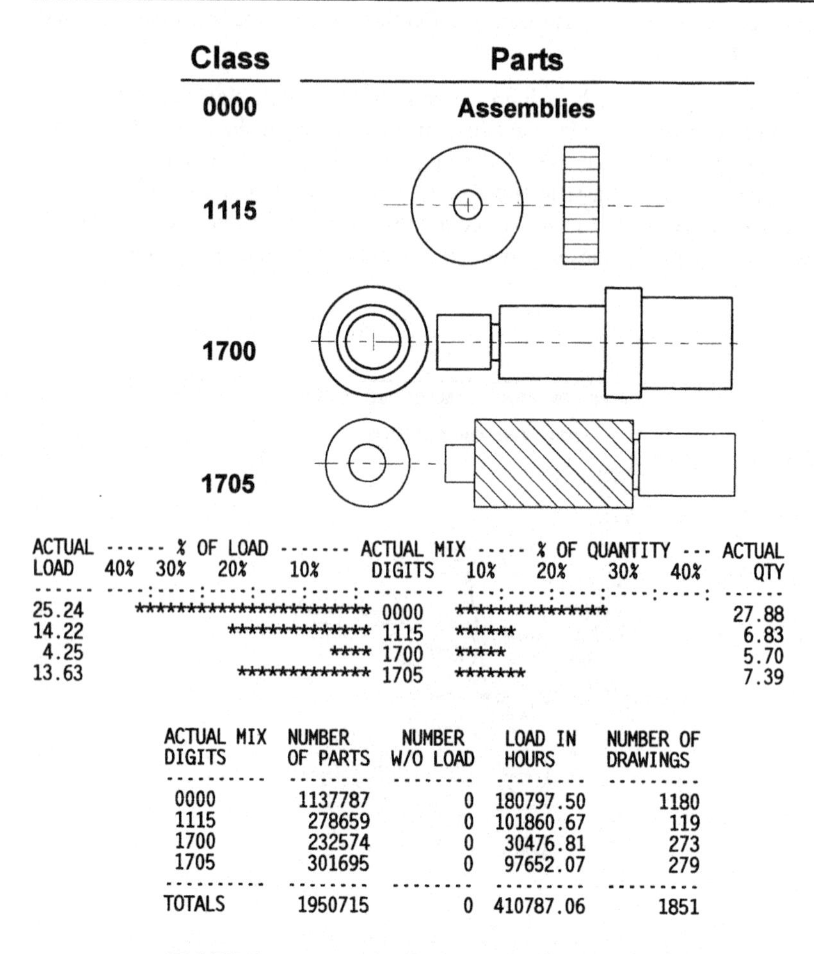


```
ACTUAL ------ % OF LOAD ------- ACTUAL MIX ----- % OF QUANTITY --- ACTUAL
LOAD   40%  30%   20%    10%      DIGITS   10%    20%    30%   40%    QTY
------ ----:----:----:----:----:-------- ----:----:----:----:----: ------
25.24     ************************ 0000  ***************               27.88
14.22              ************** 1115  ******                         6.83
 4.25                        **** 1700  *****                          5.70
13.63               ************* 1705  *******                        7.39
```

ACTUAL MIX DIGITS	NUMBER OF PARTS	NUMBER W/O LOAD	LOAD IN HOURS	NUMBER OF DRAWINGS
0000	1137787	0	180797.50	1180
1115	278659	0	101860.67	119
1700	232574	0	30476.81	273
1705	301695	0	97652.07	279
TOTALS	1950715	0	410787.06	1851

FIGURE 4. GT analysis at Bodine Electric Company.

GT analysis performed by one of the authors (Lyman-Cordes) to identify the machine tool load associated with various families of parts. Bodine makes many models of motors, but gear box components are shared. Dedicating a cell to the best selling model (a common seat-of-the-pants method of cell design) means other models may starve for gear box components' machine capacity. Those who dedicate equipment to products often run out of equipment before they run out of products.

The GT analysis depicted here prevents such capacity crises. Four-digit shape codes appear opposite graphic representations of the parts. Next, a histogram displays each family's percentage of machine tool load and percentage of the part population. Finally, the actual part count and hours are presented in tabular form

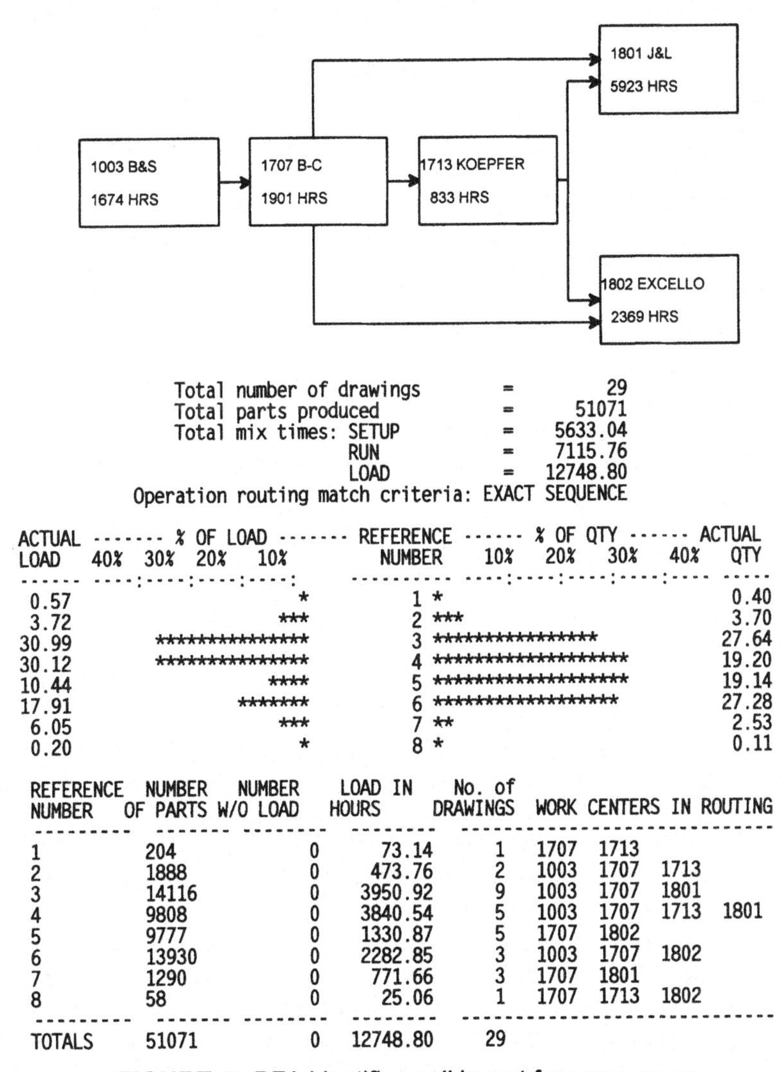


```
Total number of drawings        =        29
Total parts produced            =     51071
Total mix times: SETUP          =   5633.04
                 RUN            =   7115.76
                 LOAD           =  12748.80
Operation routing match criteria: EXACT SEQUENCE

ACTUAL ------- % OF LOAD ------- REFERENCE ------ % OF QTY ------ ACTUAL
LOAD   40%  30%  20%  10%           NUMBER    10%   20%   30%   40%   QTY
------ ----:----:----:----:     ---------- ----:----:----:----:---- -----
 0.57                    *            1 *                            0.40
 3.72                  ***            2 ***                          3.70
30.99      ***************            3 *****************           27.64
30.12      ***************            4 ********************        19.20
10.44                 ****            5 ********************        19.14
17.91              *******            6 *******************         27.28
 6.05                  ***            7 **                           2.53
 0.20                    *            8 *                            0.11
```

REFERENCE NUMBER	NUMBER OF PARTS	NUMBER W/O LOAD	LOAD IN HOURS	No. of DRAWINGS	WORK CENTERS IN ROUTING
1	204	0	73.14	1	1707 1713
2	1888	0	473.76	2	1003 1707 1713
3	14116	0	3950.92	9	1003 1707 1801
4	9808	0	3840.54	5	1003 1707 1713 1801
5	9777	0	1330.87	5	1707 1802
6	13930	0	2282.85	3	1003 1707 1802
7	1290	0	771.66	3	1707 1801
8	58	0	25.06	1	1707 1713 1802
TOTALS	51071	0	12748.80	29	

FIGURE 5. PFA identifies cell layout for worm gears.

PFA analysis by Bodine Electric Company engineers under the supervision of Chris Hargan and one of the authors (Lyman-Cordes) produced the layout for the primary machine tools in the worm gear cell shown above. Flows through the cell are captured as parametric master process plans in CAPP. Process proliferation within worm gears meant eight different methods of manufacture shown in the histogram and table above were in use. Each method must be evaluated to select the most capable one.

MICLASS applications remained limited almost exclusively to variant process planning in the metalworking arena. To the consternation of those following in Schilperoort's footsteps, copy-and-edit CAPP almost always **increased** process proliferation. It did, however, get work out of the Manufacturing Engineering department faster, and that was the basis for its economic justification.

The reluctance to invest in feature identification is well known to Mark Dunn of United Technologies Research Center, Hartford, CT. Dunn is the former Chairman of the PDES Form Features Committee, and in 1982 he and Charles Sack developed Computer Managed Process Planning (CMPP), a generative process planning system, for the United States Army. For cylindrical parts, CMPP totally automated:

- manufacturing instructions,
- tolerance stack-up,
- inspection instructions,
- machine tool instructions and,
- created two dimensional graphics of part geometry, stock, and cutter location paths.

Although CMPP automated 70% of the process planning effort at United Technologies divisions such as Sykorski, its use was limited by the "input problem", a term coined by Dunn. CMPP simply required more information than users were willing to provide. Users abandoned it to return to (inconsistent and much less specific) manual methods, especially for simple parts. CMPP was a technical triumph that did not pass the path-of-least-resistance test.

With the advent of Relational Data Base Management Systems (RDBMS) and the acceptance of Structured Query Language (SQL) by the International Standards Organization (ISO), CAPP and GT's horizons expanded. SQL made communication between different hardware platforms transparent for the user. GT code strings could now be replaced with graphics, words and dimensions. Electronic CAPP documents were easily tailored to user needs.

Attitudinal barriers sometimes pay no attention to technological advances. When Alexander Houtzeel and Rob McNiff set out to reinvent CAPP (Houtzeel 1990) in 1990, they enlisted the help of 40 companies that arguably comprised the world's most experienced CAPP user group. In two three-day working meetings, one in Texas, USA for North American companies and the other in Coventry, UK for European companies, they forged the specification for a new CAPP system to take advantage of advances in computing and network technology. One of the authors (Nolen) led the GT break-out session at both meetings. The GT consensus:

- engineers understand GT benefits, but they do not have the time or staff to mount a classification effort, and
- they do not believe management is convinced of its merits or financial justification.

- Individually, however, they will use a CAPP system having GT capabilities to maximum advantage.

Where CAPP was concerned, the attendees on both sides of the Atlantic agreed:

- Commercial CAPP systems helped the Manufacturing Engineering activity but do not communicate with other applications i.e. CAD, MRP etc.
- Home-grown systems communicated with other applications but were simply an "electronic pencil" for Manufacturing Engineers.
- A CAPP system was needed that improves Manufacturing Engineering productivity, communicates with CAD and MRP, and transfers information (text, graphics and machine instructions) electronically to and **from** the shop floor.

Houtzeel Manufacturing Systems' HMS CAPP was written around these specifications. Today HMS CAPP claims to be the most popular commercial system in the world. Aerospace companies including Boeing, McDonnell Douglas and British Aerospace are its largest installations.

In 1991, an object-oriented PC-based GT system called Group Technology Assistant (GTA) was developed with the help of one of the authors (Nolen) under a project sponsored by the United States Air Force. Using accepted terminology from Industry Handbooks to describe features eliminated the need for definitions and conventions. Precise retrieval with little or no training became possible for all users familiar with metalworking terminology and engineering drawing conventions. GTA is available from Southwest Research Institute, San Antonio, TX through a Cooperative Research and Development Agreement under the provision of public law 96-48 Stevenson - Wydler Technology Innovation Act.

GTA features include:

- A point-and-click graphical user interface (GUI)
- Classification using graphics, industry terminology, numerical data and a streamlined set of on-line definitions and conventions
- On-line commercial standards from industry handbooks for screw threads, gears, raw material gages, alloy, etc.
- An Object Oriented AI shell
- A relational database supporting the ISO SQL Instruction Set

In 1994, engineers under the supervision of Jack Petry at the Giddings & Lewis Assembly Automation Division in Janesville, WI used HMS CAPP's GT capabilities (designed from the GTA specification and identical to it all important respects) to successfully automate 70% of the process planning effort. What makes this accomplishment remarkable is the nature of the business environment. G&L

manufactures custom engineered assembly systems. Examples include the Saturn assembly line in Spring Hill, TN and the assembly line for the Ford Expedition chassis manufactured by A O Smith. These complex systems are one-of-a-kind designs that must be maintained for 30 years.

Extensive part run-offs at G&L and again at the customer's plant are required prior to acceptance. Final assembly and run-off generate many engineering changes initiated by both G&L and customer personnel which must be documented to capture the system's as-built configuration. The CAPP system's relational database, client-server network, user-defined electronic documents, and support for text, graphics and CNC machine instructions provided the capabilities G&L needed for two-way communication between Manufacturing Engineering and the shop floor. Process plans are dispatched electronically to the shop and feedback in the form of red-lined engineering drawings and process instructions returns, documenting product configuration.

Market demands are compressing the delivery time for assembly systems from 18 months to 6 months. Five manufacturing engineers process the output of 125 design engineers. According to Petry, Manufacturing Engineers no longer had the time to reinvent process instructions for similar parts.

The solution was straightforward. A random sample of 1,200 active parts (10% of the population) was classified with a GT taxonomy developed by the authors that draws heavily form Schilperoort but takes advantage of modern computer technology.

The taxonomy divides the universe of parts into mutually exclusive groups that correspond to:

- cylindrical parts
- block parts from either solid raw material or near-net-shape raw material such as castings or forgings
- sculptured parts from either solid raw material or near-net-shape raw material such as castings or forgings
- parts from sheet or coil

These groups are subdivided by the size and proportions of the part envelope dimensions. The cylindrical part envelope is defined by diameter and length. Diameter includes the largest swing about the centerline for asymmetrical parts. Lengths of cylindrical parts are divided into ranges that correspond to commercial raw materials. Within each range, parts are subdivided according to their proportions, the L/D ratio. An L/D of 3 is a fireplug, easily machined without support. An L/D of 20 is a noodle requiring a moving steady rest. An L/D of 0.1 is pancake requiring chucking. A gear in a wristwatch and a gear in a drawbridge may both have an L/D of 0.1. But the L value for the wristwatch is so small it can only be from sheet metal while the L value for the drawbridge is so large it can only be cast. A user retrieving a small gear will be shown only process plans for parts of similar size and proportions.

In a similar fashion, block and sculptured parts are subdivided into cubic, columnar, or flat proportions based on the X, Y, and Z dimensions of their part envelope. The part envelope is oriented in the Cartesian coordinate system such that $X \geq Y \geq Z$. Block parts have more material present in the XY plane than do sculptured parts.

Parts from sheet or coil are subdivided in a similar fashion. The thickness of the sheet or coil, usually the Z value of the part envelope, is subdivided into ranges that correspond to commercial light gage sheet, heavy gage sheet, plate and slab stock.

Classification or retrieval takes 3 to 6 minutes. Category and size envelope are entered on the first screen, perimeter and symmetry on the next, then machined features and tolerances, and finally raw material alloy, heat treatment, and plating or coatings. Extensive libraries form industry handbooks are on-line menu selections. All terminology and graphic prompts conform to accepted standards as published in industry handbooks. Precise retrieval can be performed by individuals with no system training, provided they understand metalworking terminology and engineering drawings. Classification requires three days of training to learn a streamlined set of on-line definitions and conventions.

G&L engineers and the authors analyzed the 1,200-part GT and CAPP database to identify part families. Next each family was evaluated to determine "best practice" manufacturing technology. This resulted in the creation of 32 parametric master process plans that eliminated 70% of the process planning effort.

Even though successful, three years later the G&L GT system remains the only one installed. Meanwhile, CAPP sales have not only increased, they continue to accelerate. As technology continues to shorten product life cycles, economic pressure to reduce both batch sizes and time-to-market will increase. Yet GT is variously perceived as a failed 20-year-old technology, or an activity that cannot be economically justified in today's environment. The authors' experience indicates this view is almost universally held.

How does manufacturing cope with the strain proliferation puts on its ability to meet market demands? Most use sensible strategies such as increased use of flexible automation, augmented by tooling and component standardization generally implemented through brute force and limited to important products. The emphasis on setup reduction and mistake-proofing also has a positive effect, although improvements are spotty. Many plants have self-directed teams empowered to implement improvements. Collectively, however, these efforts usually lack coherent direction from a plant-wide perspective.

Is there a solution that addresses the problem systematically, on a factory-wide basis, and is economically justifiable in today's climate? It appears there is. The authors have recently completed alpha field trials using the Bodine Electric CAPP database to evaluate a radical software technology that appears capable of achieving generative process planning, and automating the collection of both part feature and process data for concurrent engineering and CM. This software is known as the Autognome. It is a self-contained system that automatically builds the knowledge stores required for generative process planing by reading an existing CAPP database.

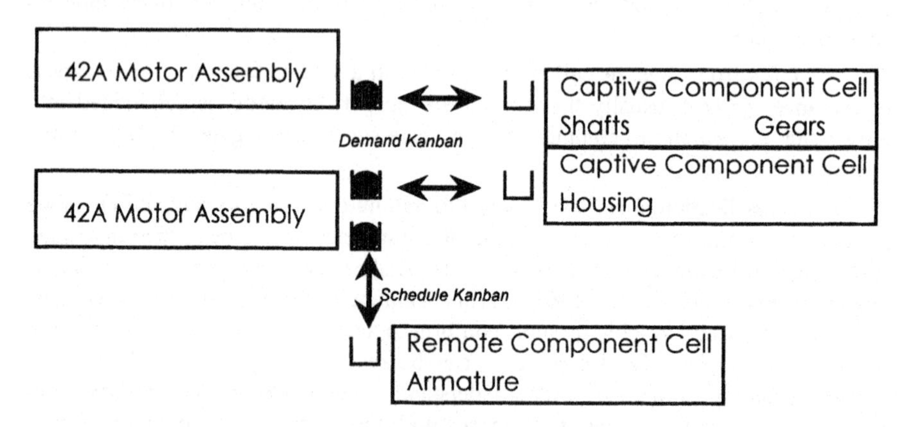

FIGURE 6. Assembly cells linked to component cells by *kanban* at Bodine Electric Company

Cells designed by Bodine Electric Company engineers under the supervision of Chris Hargan and one of the authors (Lyman-Cordes) replaced traditional batch manufacturing methods and layout. The 42 A motor is 35% of Bodine's volume but is now produced in 5% of its floor space on an order-for-order basis. The remote Armature cell feeds many other models in addition to the 42 A. Due to the high cost of winding equipment, dedicating it to one model could not be justified.

Bodine Electric in Chicago, IL, volunteered a copy of its cell routings to test the Autognome. Bodine's *kanban*-linked cells are very successful, making 35% of the companies electric motors in 5% of its floor space. Because cell routings conform to rules known to Bodine and one of the authors (Lyman-Cordes), they were a ready-made test of the Autognome's ability. The Autognome read a database of 375 parts representing 6 types of gears, then it created rules which generate appropriate machine tool sequences.

3. THE AUTOGNOME

The Autognome (Greek: self-knowing) is software originally designed to solve one of the toughest computing problems: translating written language. In route to that goal, development took a detour into manufacturing applications: specifically generative process planning. The Autognome is not an expert system - it learns the way we do. It experiences reality, formulates rules to fit its experience, and keeps the rules that help it perform best, constantly refining them as its experience grows. The Autognome is not a neural network - its knowledge stores are generated and maintained in a self-contained system. It does not have to be trained with correct responses.

The Autognome is elegant in its simplicity, but its genius lies in the way it implements learning theory to automate knowledge acquisition and use. Early research in the 1960s was directed at learning the rules of human languages to support translation from one language to another. The research was successful, but 16 kilobyte computers with tape drives were not capable of practical applications

Language is semiotic; a sign-process. Signs do two things: they stand for something, and they affect the behavior of an observer. In the language of manufacturing routings (much simpler syntax than written English) part features are signs. Hardened gear teeth, for example, are meaningful signs to an engineer; they signal hobbing followed by heat treatment. The Autognome builds knowledge-stores from either written language or a CAPP database by the process described below.

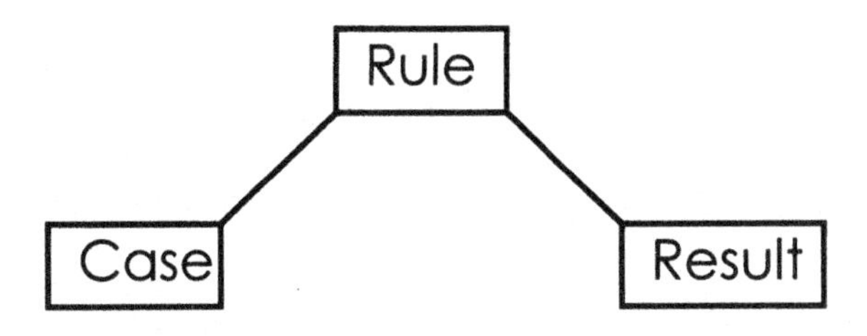

FIGURE 7. An Act.

First, let us establish some terminology. The Autognome seeks to establish and maintain a knowledge store comprised of a system of meaningful acts. An act, FIGURE 7, is a *case*, a *rule*, and a *result*. Acts are built from experience and form a pyramid-like network explained in the paragraphs that follow. An act can be the result of another act that preceded it. Likewise it can be the case for an act that follows it. This pyramid-like network is stored in tabular form illustrated in

FIGURE 8. *Induction* searches the knowledge store evaluating all potential *rules* associated with a given *case* and *result. Deduction* searches the knowledge store evaluating all potential *results* associated with a given *case* and *rule. Abduction* is hypothesizing. When none of the acts in the knowledge store have a satisfactory probability of success, abduction forces the generation of new acts that are potentially correct.

A conceptual understanding of how the knowledge store is built is fundamental. A simple example of this process can be seen in FIGURE 8. Let us imagine an Autognome with no experience - its knowledge stores are empty - that begins to read an ASCII character English language text such as the million-word Brown University Corpus. The Autognome is indifferent to language. The corpus could be French, Russian, manufacturing process instructions or CAD solid models.

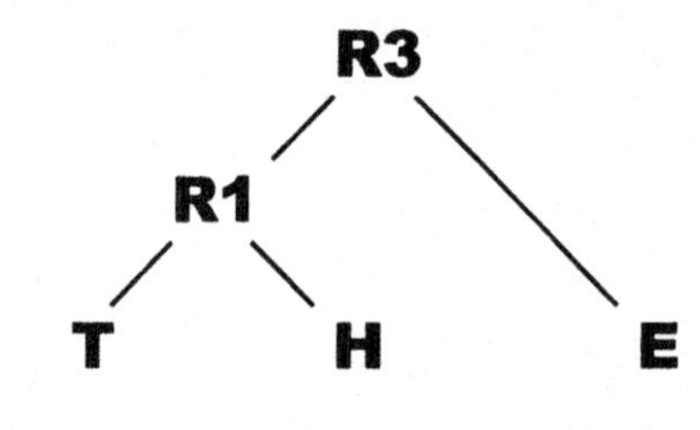

Acts		
Rule	**Case**	**Result**
R1	T	H
~~R2~~	~~H~~	~~E~~
R3	R1	E
~~R4~~	~~T~~	~~R2~~

FIGURE 8. Pyramid of acts builds knowledge store.

The Autognome gains experience as it reads English. If "the" is the first word in a corpus and the Autognome has no experience, it will hypothesize the acts in the table above. However, 1,000,000 words later it will strike acts from the knowledge store, retaining only those that perform successfully. The syntax of process plans is much simpler than written English. The Autognome quickly learned acts generating correct routing sequences after reading 300 routings from the gear cells at Bodine Electric Company.

The Autognome forms rules to explain surprising new experiences. At first, everything is a surprise. Let us assume the first word of the corpus is "the." The Autognome will form rules to fit its experience, which at this point is limited to a one-word vocabulary. It will make all the combinations of rules that potentially explain its experience: if T then H, if H then E, if TH then E, and if T then HE. At this point, all these rules are equally plausible; the Autognome cannot predict which ones will perform best.

One-million words later the pyramid is **very** large, but only two of the first four rules remain. The Autognome uses mathematics to identify the most elemental meaningful objects and the sequence (syntax) in which they are used. Because they are always found together, phonemes (letters representing elemental sounds of a language) are eventually recognized. TH is a phoneme that is followed by one of the vowels, in this case E. In the Bodine trial, work center numbers were defined to be the most elemental meaningful objects.

Similar Acts	
Cases	**Results**
CH,SH,TH	A,E,I,O,U

FIGURE 9. Generalization.

The Autognome will eventually group H-phonemes into a class and relate them to a class of vowels. Generalization lets the Autognome successfully predict appropriate acts in a class even if it has not experienced every possible combination of Case-Result pairs. In trials with Bodine Electric Company data, the authors observed the Autognome correctly generalizing classes of acts for "green" gear cutting (prior to heat treatment), heat treatment, hard gear cutting, and inspection.

The Autognome uses a proprietary technique to *generalize*; it can recognize similarities among classes of acts. This means that, unlike expert systems, the Autognome can generate correct acts even if those acts were not previously encountered. For example, it will eventually combine the H-sounds in FIGURE 9 into a class of acts that have an H-sound as a case and a vowel as a result. The Autognome will not be surprised when it encounters the words "chat", "shot" or "shut" even if it has not encountered them before. In manufacturing it means that every combination of hobbing, heat treating and involute grinding are potential acts, even if some combinations have not been used before.

In the Bodine trial, the Autognome correctly formed classes of gear cutting operations, heat treatment operations, inspection operations and finishing operations such as grinding. These classes of similar acts are indicated by *Predicates,* and may be seen in FIGURE 10 with designations such as P 1040. Even though predicates represent a class of acts, they behave as a single act. In use, all of the acts that make up the predicate class are evaluated against the case at hand, and the one with the best probability of success is chosen.

The process of building knowledge stores is computationally intensive, but within the capability of modern, multi-processor workstations. For example, in the Bodine trial the first family of 150 gears generated 95,000 acts, 38,000 of which were kept. These were organized into 4,300 classes. A Pentium™ PC using a relational database to store acts required almost 24 hours to build a knowledge store. A modern multi-processor workstation running Windows NT and optimized for the Autognome will require about 15 minutes for the same task. This means roughly 16 hours are needed to build the knowledge store for the average plant's 10,000 parts. The success of the Bodine trial has set plans in motion for an extensive feasibility study sponsored by the National Center for Manufacturing Sciences to be conducted at an Eastman Kodak facility.

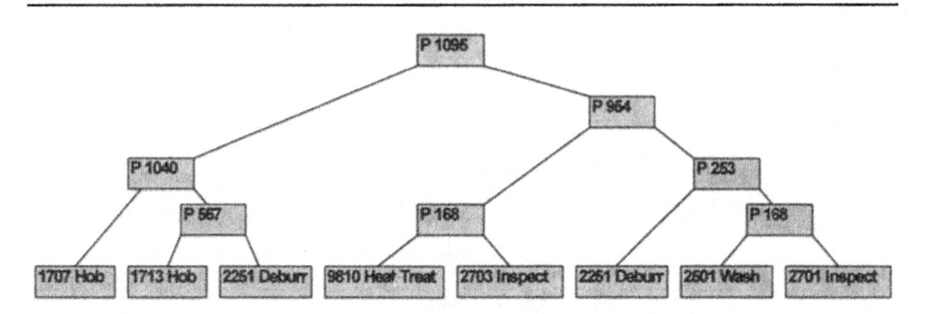

FIGURE 10. Typical Worm gear knowledge store from Bodine Electric Co.

The Autognome runs on a Pentium™ PC and trains itself in a three-step process by reading an existing routing database. First, it **makes the strange familiar** by creating new rules to explain new routing sequences. By the end of this process it has established a syntax of routing sequences i.e. green gear cutting is followed by heat treatment is followed by grinding. Next, to **make the familiar meaningful,** it learns how rules combine to generate routings. In this step it is learning to distinguish meaningful (read actual) routings from the set of all syntactically correct routings; the way we distinguish a meaningful sentence from a syntactically correct nonsense sentence. Finally, it **values the meaningful** by learning what routings do, associating features such as gear teeth with actions such as hobbing.
The knowledge-store grows rapidly at first, then shrinks as some actions are discarded and specific rules are combined into general ones. The Autognome generalizes by grouping similar actions into classes called Predicates. In this example P 168 is used twice to generate inspection operations after both heat treat and washing. For 150 gears the Autognome created 95,000 acts but kept only 38,000 of them which it grouped into 4,300 classes. The knowledge store is said to converge when the Autognome cannot make additional rules without additional experience.

Thus far, the acts described have captured the rules of syntax - rules that constrain what may follow something else. Sentences that are syntactically correct, however, may not make sense i.e. "Green shadows dream awkwardly."

To discover the meaning in sentences, one must understand the rules used to give them meaning. In other words, the underlying meaning reveals the rules (intelligence) that directed the creation of the corpus. This is done in a process almost identical to that used to build the syntactic knowledge store. In this process the syntactic acts are used as input to build a second, higher-level, *representational* knowledge store. When it is complete, Autognome has captured rules that created sentences in the corpus or the routings in the CAPP database. The set of meaningful sentences or routings is minuscule compared to the set of syntactically correct ones.

In performance mode, the Autognome uses induction and deduction to generate the next operation. Each completed operation becomes the case for the next rule. If worm gear then hob. If hob then heat treat. If heat treat then Rockwell hardness test.

Step	Input	Output
Make the strange familiar	Experience	Syntactic acts
Make the familiar meaningful	Syntactic acts	Representational Acts
Value the meaningful	Representational acts and real-world objects	Intelligent Response

FIGURE 11. Knowledge building process.

Autognome's three-step knowledge building process is seen here. First it learns the relationships between the most elemental meaningful objects. In English these objects are letters or phonemes. In process planning they are work centers. The syntactic knowledge store captures relationships among objects. In process planning this means rules about allowable work center sequences, or routing syntax. The representational knowledge store is more abstract, capturing relationships among syntactic acts that produced the routings actually observed. When this knowledge store is complete, the Autognome can distinguish meaningful or actual routings from the much larger set of all syntactically correct routings. Finally, associating real-world objects i.e. worm gear teeth, with representational acts, i.e. hobbing, heat treatment and grinding, generates process plans.

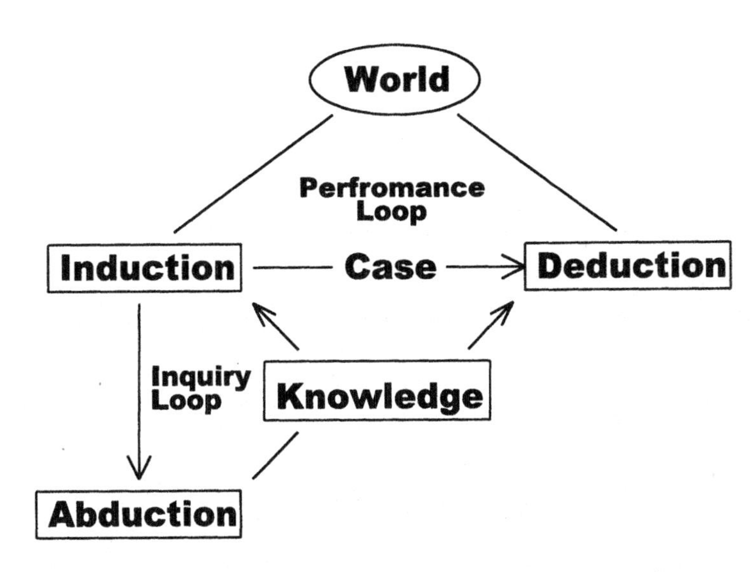

FIGURE 12. The Autognome in performance mode.

The performance loop uses induction (what rule governs the current situation—the case) and deduction (given a case, what result can be expected) to generate meaningful actions: what should be done next? When the case and/or result are doubtful, i.e. low probability of success, the Autognome performs Abduction. It creates new acts that enrich the knowledge store, providing alternative ways to deal with the case at hand.

When it is unsuccessful, the Autognome *inquires* by statistically examining acts in its knowledge store are deficient or inadequate. It then generates new acts to address these shortcomings. Recall the hypothetical acts created to explain the word "the" at the start of this discussion. Over time only the successful acts created by abduction are retained.

One step remains. Part features must be associated with representational rules that govern routing creation. The part feature, i.e. worm gear, is the real-world sign that triggers representational acts generating worm gear routings.

Dunn would be quick to point out that most companies do not have computer sensible part feature databases. Indeed, Kodak was selected by NCMS as the site for the Autognome's feasibility study because they **have** an electronic part feature database. Almost everywhere else, engineering drawings contain part feature data that only humans can interpret. CAD models are simply set theory applied to lines and circles in 3-D space; most do not know if a cylinder is solid (an OD) or void (an ID). Dunn would probably add that people are reluctant to build part feature databases.

What are the alternatives? Long term, the Autognome may become robust enough to read CAD solid model data and learn features. How computer intensive that activity is can only be guessed. Short term, we can seed the Autognome with part feature data from a random sample of 10% of the routings a la G&L. Using its generalizing abductive capability, the Autognome will generate correct process plans for new parts provided they do not contain features different form those in the sample.

REFERENCES

Barnett, A., 1994, How numbers can trick you, *Technology Review* , October, 38-40.

Dallas, D. B., (Ed.)., 1976, *Tool and Manufacturing Engineers Handbook*, 3rd ed., New York, NY, 35-13.

Houtzeel, A., 1990, *Advanced CAPP Project Specification* (Proprietary).

Manchester, W., 1981, *Arms of Krupp*, Bantam Books, Inc., 484-495.

Nolen, J. and Schilperoort, B. A., 1991, *Beekbaghen*, Netherlands, 16 March.

Nolen, J. and Keus, J., 1991, *Instituut voor Toegepast Natuurwetenschappelijk Onderzoek* (TNO), Apeldoorn , The Netherlands, March 17.

Russell, D., 1990, The truth about big business in Japan, *Business Tokyo*, April, 22-28.

Samuel, R. A., 1982, Productivity: The Japanese Experience, *MIT REPORT*, X, 5, May, 1-2.

Schonberger, 1982, *Japanese Manufacturing Techniques*, The Free Press, New York, NY, 221-233.

Turkel, S., 1984, *The Good War: An Oral History of World War II*, Pantheon, New York, NY, 680.

AUTHORS' BIOGRAPHY

James Nolen is President of James Nolen & Company, a manufacturing management consulting firm conducting engagements world-wide. Previously, Mr. Nolen managed a US/Japanese joint venture company and held positions in manufacturing engineering with EG&G and in materials management with Andersen Consulting. He contributes to the *Tool and Manufacturing Engineers Handbook* and the trade press in North America, Europe and Japan. He is the author of *Computer-Automated Process Planning for World-Class Manufacturing* (Marcel Dekker, New York, 1989), selected for translation by *Japan Management Association* in 1991. The Society of Manufacturing Engineers video series *Cellular Manufacturing in a Global Marketplace* draws heavily on Mr. Nolen's experience. He is a senior member of the Society of Manufacturing Engineers and holds three United States Patents. Mr. Nolen received a B.S. degree (1970) in Industrial Management from the University of Maryland, completed a metal working apprenticeship at Warner and Swasey Co. (1971), and earned a M.S. degree (1974) in Information Systems from Case Western Reserve University, Cleveland, Ohio.

Carey Lyman-Cordes is Director of James Nolen & Co.'s Chicago office. She is experienced in managing, designing, and implementing cellular manufacturing facilities. At Deere and Co., she designed and implemented a successful hydraulic cylinder cell using group technology and cellular manufacturing methods. She has been a consultant for a variety of manufacturing and distribution environments as a manager for Arthur Young's Consulting Division. She was Manager of Manufacturing Engineering for Lovejoy, Inc., where she was responsible for redesigning the entire facility into manufacturing cells. Most recently, She was Manager of Manpower Planning for United Airlines' Mileage Plus, Inc. Ms. Lyman-Cordes holds a B.S. degree in Industrial Engineering from University of Illinois (1981). She is a Past President of the O'Hare Chapter of the Institute of Industrial Engineers and is active in Society of Manufacturing Engineers.

D1

Part-Machine-Labor Grouping: The Problem and Solution Methods

S. M. Shafer

1. INTRODUCTION

It is safe to say that no topic in the cellular manufacturing (CM) field has been as widely researched as has the topic of cell formation. Generally speaking, cell formation is concerned with grouping machines and labor into cells. Once formed, the cells are dedicated to the production of a set of parts that have similar processing requirements called part families.

Over the years, a great deal of research has been devoted to the cell formation problem. Unfortunately, as Vakharia (1986) pointed out, the focus of this research has been on the development of new techniques to solve the problem, with little research being directed at determining the appropriateness of these techniques to particular situations.

As the first of five chapters in topic area D, this chapter serves as an introduction to the design of cellular layouts. We begin by formally defining the cell formation problem in the next section. Then, a taxonomy for cell formation procedures is presented. This is followed by a review of the cell formation literature organized on the basis of the taxonomy previously presented. Following this, cell formation performance measures and studies that have compared a variety of cell formation techniques are reviewed, respectively. Finally, the chapter is concluded with some guidelines for future research in this area.

2. THE CELL FORMATION PROBLEM

CM is based on the group technology (GT) philosophy of achieving efficiencies by exploiting part similarities. More specifically, the objective of CM is to realize many of the benefits associated with mass production in less-repetitive job shop environments. To achieve this objective, sets of parts that have similar processing requirements are grouped together into part families. The equipment and labor requirements for each part family are determined simultaneously with, or subsequent to the identification of the part families. The equipment and labor are then located together to form a manufacturing (or machine) cell.

The procedures used to identify and define the part families and/or machine cells are referred to as cell formation procedures. In general, cell formation procedures seek to partition or cluster a given shop's population of parts (machines and labor) into part families (cells) in order to improve the shop's operating efficiency and effectiveness.

More formally, define the set $M = \{m_1, m_2, .., m_m\}$ to represent the m machines available in a particular shop, set $P = \{p_1, p_2, ... , p_n\}$ to represent the n parts produced in the shop, and set $W = \{w_1, w_2, ..., w_r\}$ to be the r shop floor workers employed in the shop. Further, define the set $C = \{c_1, c_2, ... c_k\}$ to be a partition of the set M, $F = \{f_1, f_2, ..., f_k\}$ to be a partition of set P, and $T = \{t_1, t_2, ..., t_k\}$ to be a partition of set W. In effect, set C corresponds to k subsets (cells) of set M, F to k subsets (part families) of set P, and T to k subsets (worker teams) of set W. It is worth noting that many cell formation procedures require the *a priori* specification of k. Finally, define the set $U = \{u_1, u_2, ...u_k\}$ where element $u_i = \{c_i, f_i, t_i\}$.

The objective of cell formation procedures is to find the set U^* and the parameter k^* that optimizes the criteria specified by the decision maker subject to various constraints. To illustrate, Shafer and Rogers (1991) developed a goal programming model primarily to define the cell formation problem. In this model four objectives are explicitly considered: 1) minimizing equipment setup times, 2) minimizing intercellular movements, 3) minimizing investment in new equipment, and 4) maintaining acceptable machine utilization levels. In addition, the model includes a number of constraints such as upper and lower limits on part family and machine cell sizes. Other objectives that have been considered in the literature include: maximizing part similarity; minimizing production costs; balancing workload; minimize tooling cost; improving on-time delivery performance; reducing work-in-process levels; and, improving quality. Of course, as the Shafer and Rogers formulation illustrates, the cell formation problem typically requires the consideration of multiple objectives.

3. A TAXONOMY OF CELL FORMATION TECHNIQUES

A number of researchers have offered taxonomies for cell formation techniques. King and Nakornchai (1982) classified cell formation techniques as similarity coefficient methods, set-theoretic methods, evaluative methods, and other analytic techniques. Set-theoretic methods use a union operation "to build up supersets of machines and components which can be represented as a path along the edges of a lattice diagram." Evaluative methods are procedures that require the analyst to make a series of judgments as systematic listings of the machines and components are generated. Production flow analysis (PFA) is an example of a evaluative method. Other analytic approaches include methods that can be implemented on a computer and therefore not requiring careful inspection by the analyst. The rank order clustering algorithm (ROC) and the direct clustering algorithm (DCA) are examples of these analytic approaches.

Wemmerlöv and Hyer (1986) classified cell formation procedures into four categories: 1) Approaches that identify part families without the help of machine routings; 2) Approaches that identify machine groups; 3) Approaches that identify part families using routings; and, 4) Approaches that identify part families and machine groups simultaneously. Vakharia (1986) classified cell formation techniques as descriptive methods (e.g., PFA), block diagonal matrix methods (e.g., ROC and DCA), similarity coefficient methods, and other analytical methods (such as mathematical programming formulations).

Ballakur and Steudel (1987) offered a particularly comprehensive taxonomy for cell formation techniques. According to this taxonomy, cell formation techniques are initially classified as those based on part family grouping, machine grouping, or machine-part grouping. Subsequently, the part family grouping category is further divided into classification and coding methods, and statistical cluster analysis methods. In a similar fashion, the machine grouping category is further divided into algorithmic procedures and non-algorithmic procedures. Finally, machine-part procedures are divided into combinatorial analysis methods, algorithmic techniques, and manual techniques.

Chu (1989) classified cell formation techniques into six categories: array-based approaches, agglomerative approaches, non-agglomerative approaches, mathematical programming methods, graph theoretic methods, and other heuristic procedures.

Singh (1993) classified cell formation procedures into the following eight categories: coding and classifications, machine-component group analysis, similarity coefficients, knowledge-based, mathematical programming, fuzzy clustering, neural networks, and heuristics. Kaparthi and Suresh (1994) extended Chu's taxonomy with three additional categories: operations sequence-based methods, construction algorithms (i.e., methods based on seed selection procedures), and biologically-inspired methods (e.g., genetic algorithms and neural network algorithms).

In an effort to integrate the contributions made by these and other researchers, the taxonomy shown in Figure 1 was developed. The basis of this taxonomy is the methodology the cell formation procedures are based on. According to the taxonomy, the six basic methodologies used in cell design including manual methods, classification and coding approaches, statistical clustering techniques, algorithms for sorting the machine-component matrix, mathematical techniques, and approaches based on artificial intelligence.

4. A SURVEY OF CELL FORMATION LITERATURE

In this section relevant contributions to the cell formation literature are reviewed based on the taxonomy presented in the previous section. In addition, the cell formation categories are briefly defined and examples are included to illustrate the mechanics of a representative techniques.

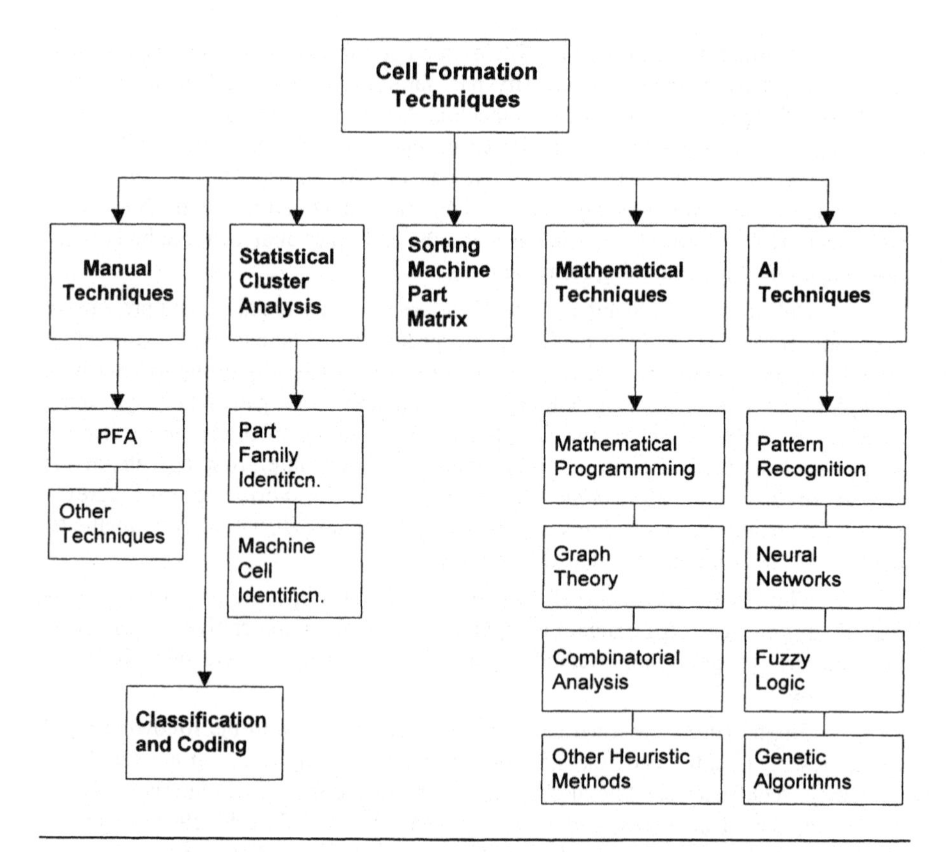

FIGURE 1. Taxonomy of cell formation techniques.

4.1. Manual Techniques

Manual techniques are defined to be those techniques that require the analyst to make a series of judgments. Typically, these techniques identify part families and their corresponding machine cells iteratively. Further, because they depend on human judgment, they do not lend themselves to being implemented on a computer.

4.1.1. Production Flow Analysis.

Production Flow Analysis was developed by Burbidge (1963, 1971, 1975, 1977) as a manual procedure for identifying part families and machine cells based on part routing information. It is based on the premise that the majority of components and machines belong to clearly defined families and cells. It consist of the following four basic steps: 1) Finding the best division of plant and parts between major groups; 2) Finding the best division into families and groups inside major groups; 3) Finding the best arrangement for the layout of machines in the group; and, 4) Dividing the parts into tooling families.

PFA is undoubtedly one of the most comprehensive cell formation procedures developed to date. Indeed, many, if not most, of the more recently developed techniques rely on the work done by Burbidge. For example, a popular research area has been the development of algorithmic procedures for sorting Burbidge's machine-component matrix [e.g., King (1980), Chan and Milner (1982), and Kusiak and Chow (1987)]. Chapters D2 and I2 of this book discuss PFA applications.

4.1.2. Other Manual Techniques.

El-Essawy and Torrance (1972) developed a cell formation procedure similar to PFA called component flow analysis. In the first stage of component flow analysis, parts are sorted into groups based on their manufacturing requirements. In the second stage the groups are manually analyzed to obtain machining groupings taking into account various situational factors. Finally, a detailed analysis of flow patterns of the cells is performed and appropriate adjustments are made to ensure an acceptable design is achieved.

A variety of other manual procedures have been reported in published case studies. For example, the Langston Division of Harris-Intertype Corporation took Polaroid pictures of randomly selected parts and then placed the pictures on grids (Gettelman 1971). Based on a visual inspection of the pictures, it was determined that over 93% of the parts could be placed into one of five part families. As another example, at the Naval Avionics Center in Indianapolis, numerically controlled (NC) machines formed simple "hubs" around which other machines were easily assigned so that parts could be completed within the cell (Allison and Vapor 1979). More recently, a case study investigating the practices of 14 companies that utilized cellular manufacturing found that the great majority of the cases used some type of manual cell formation procedure typically based on part shape or nomenclature (Shafer, Meredith, and Marsh 1997). For example one company based its part families on the drawing name such as "plate" or "shaft" while another firm based their cells on product features and dimensional envelopes.

4.2. Classification And Coding Systems

Classification and coding approaches involve assigning alphanumeric codes to each part based on design characteristics and/or processing requirements. Parts with similar codes are identified and grouped into part families. Classification and coding approaches were considered earlier, in topic area B.

4.3. Statistical Cluster Analysis

Statistical cluster analysis is another approach that can be used to identify part families or machine cells. Typically, a measure of association (or proximity) that quantifies either the similarity or distance (dissimilarity) between two parts or machines is developed. Such a measure of association is referred to as either a similarity or distance coefficient. These similarity or distance coefficients are calculated for each pair of parts or machines and are stored in a proximity matrix. The proximity matrix is used as input by clustering procedure such as single linkage clustering (SLC) or average linkage clustering (ALC) to identify part families or machine cells.

With statistical cluster analysis (also referred to as hierarchical cluster analysis), each part or machine is initially placed in its own separate cluster. These clusters are successively combined together based on the clustering algorithm specified until all parts or machines are grouped into a single cluster. If a similarity coefficient is used, separate clusters are successively combined starting with the clusters that are most similar and ending with the combination of the two clusters that are the least similar. In contrast, if a distance measure is used, the clusters are successively combined starting with the two clusters that are the least dissimilar and ending with the combination of the two clusters that are the most dissimilar.

SLC and ALC are two of the more popular clustering algorithms used. With SLC, combining two clusters is based on the strongest single link between the two clusters. Alternatively, with ALC, combining two clusters is based on the average value of all links between two clusters. To illustrate, let cluster A contain the elements $\{a_1, a_2, a_3\}$ and cluster B the elements $\{b_1, b_2\}$. Then the similarity between A and B with SLC is $\max\{a_1\text{-}b_1, a_1\text{-}b_2, a_2\text{-}b_1, a_2\text{-}b_2, a_3\text{-}b_1, a_3\text{-}b_2\}$ where $a_i\text{-}b_j$ represents the pairwise similarity between a_i and b_j. On the other hand, the similarity between A and B using ALC is $(a_1\text{-}b_1 + a_1\text{-}b_2 + a_2\text{-}b_1 + a_2\text{-}b_2 + a_3\text{-}b_1 + a_3\text{-}b_2)/6$.

4.3.1. Part Family Identification.

Carrie (1973) illustrated how numerical taxonomy can be applied to CM using the Jaccard similarity coefficient. With the Jaccard similarity coefficient, the similarity between two parts is calculated as the ratio of the number of machines the two parts require in common to the number of machines either or both parts together require. More formally, the Jaccard similarity coefficient (JSC) is defined as:

$$JSC_{jk} = \frac{N_{jk}}{N_{jj} + N_{kk} - N_{jk}} \qquad (1)$$

where, JSC_{jk} = the Jaccard similarity coefficient between parts j and k; N_{jk} = the number of machines that components j and k have in common in their production; and, N_{jj} = the number of machines component j requires in its production.

To illustrate the calculation of the JSC, consider the machine-part matrix shown in Figure 2. From Figure 2 it can be seen that part 1 requires machines A, C and E in its processing. Likewise, part 2 requires machines B and D in its processing. The similarity between parts 1 and 2 is calculated to be equal to zero. Similarly, similarities are computed for every part pair, as shown in Figure 2.

	Machines				
Parts	A	B	C	D	E
1	1	0	1	0	1
2	0	1	0	1	0
3	1	0	1	0	1
4	0	1	0	1	0
5	1	0	1	0	0

	Parts				
Parts	1	2	3	4	5
1	--	0	1.0	0	0.67
2		--	0	1.0	0
3			--	0	0.67
4				--	0
5					

FIGURE 2. Machine-part matrix; Part-part similarities.

Statistical cluster analysis begins with each component in its own cluster and successively combines the clusters until all components are combined into a single cluster. Referring to the machine-component matrix shown in Figure 2, the statistical cluster procedure would begin by placing the five components into their own cluster. Next, it can be observed from the proximity matrix shown in Figure 3 that components 1 and 3, and components 2 and 4 have the highest pairwise similarity. We arbitrarily select the cluster with component 1 and the cluster with component 3 to be combined. After doing this there would be four clusters: {1, 3}, {2}, {4}, and {5}.

The next step is to update the proximity matrix based on the new cluster. How the proximity matrix is updated depends on which clustering procedure is being used. If SLC is used then the similarity between the new cluster {1, 3} is based on the maximum similarity between either component 1 or component 3 and the components in the other clusters. Alternatively, if ALC is used then the similarity between the new cluster {1, 3} is based on the average similarity of both component 1 and component 3 to each of the components in the other clusters.

More specifically, the similarity between the clusters {1, 3} and {5} based on the SLC is calculated as the max(JAC_{15}, JAC_{35}). Alternatively, the similarity between these clusters based on ALC would be ($JAC_{15}+JAC_{35}$)/2. For the purpose of this example, we use SLC. The updated proximity matrix after components 1 and 3 are clustered together is shown in Figure 3.

	Component Clusters			
Component Clusters	{1, 3}	{2}	{4}	{5}
{1, 3}	--	0	0	0.67
{2}		--	1	0
{4}			--	0

FIGURE 3. Updated proximity matrix after clusters {1} and {3} are combined.

Values in Figure 3 indicates that clusters {2} and {4} should next be combined. After combining these clusters, the proximity matrix would be updated as shown in Figure 4.

Component Clusters	{1, 3}	{2, 4}	{5}
{1, 3}	--	0	0.67
{2, 4}		--	0

FIGURE 4. Updated proximity matrix after clusters {2} and {4} combined.

From Figure 4, it is clear that clusters {1, 3} and {5} should be next clusters combined. Finally, in the last step, clusters {1, 3, 5} and {2, 4} would be combined into a single cluster.

The results of statistical cluster analysis are typically reported in the form of a dendogram as shown in Figure 5. The dendogram shows graphically the similarity level at which the various clusters were combined. For example, according to Figure 6, clusters {1, 3} and {5} were combined at a similarity level of 0.67. By specifying a threshold value for the similarity level, the analyst can use the dendogram to determine which parts should be clustered together to form part families. In essence, the threshold value is the smallest similarity value acceptable for two clusters to be combined. Referring to Figure 6, if a threshold value of 0.75 is specified then three part families would be created: {1, 3}, {5}, and {2, 4}. Likewise, at a threshold level of 0.50 two part families would be created: {1, 3, 5} and {2, 4}.

Numerous other similarity coefficients have been proposed in the cell formation literature. Choobineh (1988) presented a similarity measure that extends the Jaccard similarity coefficient to consider the sequence of operations. Selvam and Balasubramanian (1985) developed a dissimilarity measure based on operations sequences. Dutta, Lashkari, Nadoli, and Ravi (1986) employed the following dissimilarity coefficient to cluster parts into part families:

$$DIS_{jk} = N_{jj} + N_{kk} - 2N_{jk} \tag{2}$$

This basic measure of dissimilarity was embedded in an algorithm that consisted of a greedy-style heuristic to reallocate parts to part families in order to reduce the value of the overall dissimilarity measure of all j, k pairs in all families. Dutta *et al.* also presented coefficients for: 1) measuring the average dissimilarity of a part with the parts in its part family, 2) the dissimilarity introduced into a family due to the inclusion of each part in the family, and 3) the contribution each part family makes to the overall dissimilarity.

Vakharia and Wemmerlöv (1990) developed a similarity coefficient that considers the within-cell machine sequence and the machine loads. This similarity coefficient measures the proportion of machine types that are used by two parts in the same order. The Jaccard similarity coefficient, in contrast, only considers the set of machine types used by two parts, not the sequence of usage.

Wu, Venugopal, and Barash (1986) and Tam (1990) suggested another similarity coefficient based on operation sequences. Specifically, they noted that by employing combinations of three transformations, namely, substitutions, deletions, and insertions, the operation sequence for one part can be derived from the operation sequence of any other part. Based on this, a similarity measure for a pair of parts is defined as the minimum number of transformations required to derive operation sequence of one part from sequence of the other part.

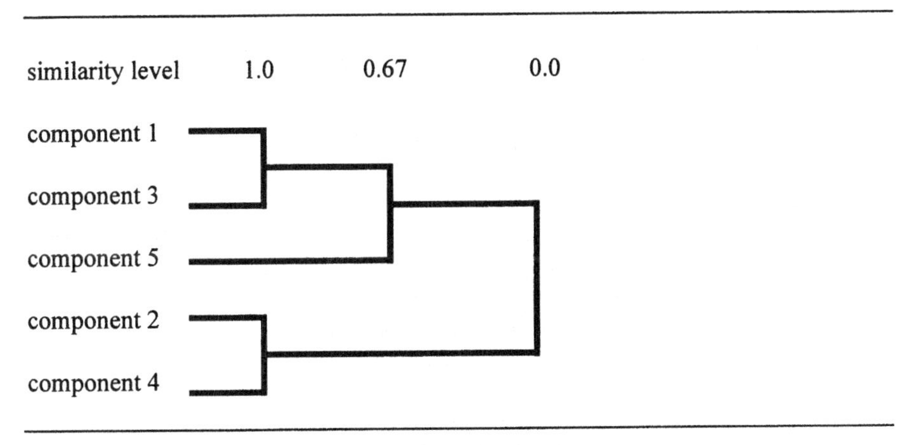

FIGURE 5. Sample dendogram.

4.3.2. Machine Cell Identification.

McAuley (1972) was the first to apply the Jaccard similarity coefficient to the cell formation problem. Specifically, McAuley calculated the Jaccard similarity coefficient for each machine type pair. Thus, this approach is similar to the approach Carrie (1973) employed (discussed earlier) except that the similarity between machine types was being calculated as opposed to the similarity between parts.

Waghodekar and Sahu (1984) proposed a cell formation technique called MACE (machine-component cell formation). MACE generates three outputs based on three different similarity coefficients. The Jaccard similarity coefficient is the first one used. The second similarity coefficient used is the product of two ratios (SCP). The first ratio is the number of components that visit both machine types over the total number of components that require the first machine type. The second ratio is the number of components that visit both machine types over the number of components that require the second machine type.

The third similarity coefficient considered is based on the total flow (SCTF) of common components processed by a pair of machines types. Two ratios are also needed for this measure. The first ratio is the number of components that require both machine types over the number of components that the first machine type has in common with all machine types. The second ratio is the number of components that require both machine types over the number of components that the second machine type has in common with all machine types. These two ratios are multiplied to calculate SCTF as follows:

$$SCTF_{il} = (M_{il} / TFC_i)(M_{il} / TFC_l) = (M_{il}^2) / (TFC_i \times TFC_l) \quad (3)$$

where

$$TFC_i = \sum_{\substack{l=1 \\ i \neq l}}^{N} M_{il} = \text{the total flow of common components processed by} \quad (4)$$

machine type i, with respect to all remaining machine types.

De Witte (1980) proposed three similarity coefficients used to measure the absolute relation between machines types, the relative mutual interdependence between machine types, and the relative single interdependence between machine types. Additional information not contained in the machine-component matrix is needed to calculate these similarity measures. This information includes the number of available machines of each machine type and the part routings and quantity (e.g., annual volume) for each component. This information is used to derive the 'combination' matrix **C**. In this matrix the rows and columns correspond to the machine types. The diagonal elements in the matrix represent the number of times a particular machine type occurs in the part routings and the off-diagonal elements represent the number of times the components travel between the corresponding machine types (ignoring directional considerations).

The divisibility number of the machines is also needed in calculating the similarity measures proposed by De Witte. The divisibility number for a single machine is usually equal to one. However, if a particular machine can be used in more than one location (e.g., a hand drill) then its divisibility number may be set to the number of possible locations. Typically, the divisibility for a machine type is set equal to the number of machines of that type available.

The similarity coefficient showing the absolute relation (SA) is the ratio of the number of times parts move between the two machine types being compared with the average number of parts a machine types processes. Thus, $SA < 1$ indicates that the number of parts that move between the two machine types being compared is less than the number of parts a machine type processes on average. Likewise, $SA > 1$ indicates that the number of parts that move between the two machines being compared is greater than the number of parts a machine type processes on average.

Ballakur and Steudel (1987) developed a similarity coefficient measure call the cell bond strength (CBS). This measure consists of the sum of two terms, one fore each machine type, where each term is the proportion of production of that machine type that is common to the other machine type. Each of the terms of CBS will result in values on the interval [0, 1] and thus the CBS measure will result in values on the interval of [0, 2]. The CBS measure is used in a dynamic programming procedure to generate an optimum chain of machine types, i.e., the chain that maximizes the sum of bonds between adjacent machine types in the chain. A second procedure is used to partition the chain into machine type cells subject to cell size restrictions.

Gunasingh and Lashkari (1989) developed a similarity coefficient that considers the common tooling between the parts and machines (SCCT). As a final example, Gupta and Seifoddini (1989) presented a similarity coefficient that incorporates production volume data, routing sequence information, and unit operation times.

4.3.3. Concluding Remarks.

It is interesting to note how similarity and dissimilarity measures have evolved over time. The first application of similarity measures to the cell formation problem dates back to the early 1970s where the Jaccard similarity coefficient was used to identify machine cells (McAuley 1972) and part families (Carrie 1973). The Jaccard similarity coefficient is a simple measure and only considers the parts (machines) two machines (parts) process (require) in common.

In the 1980s additional information began to be incorporated in the calculation of similarity coefficients. In the early 1980s part volume data was incorporated (De Witte 1980, Mosier and Taube 1985). By mid-1980s part sequencing information was being incorporated into the calculation of similarity coefficients (Selvam and Balasubramanian 1985). In the late 1980s tooling information was used in the calculation of a similarity coefficient (Gunasingh and Lashkari 1989a). Finally in the early 1990s similarity in machine setups was incorporated in the assessment of part similarity (Shafer and Rogers 1991). It is clear from this trend that proximity measures are becoming more sophisticated and becoming more closely linked to the objectives associated with adopting cellular manufacturing.

A final issue relates to determining whether statistical cluster analysis should be used to cluster the machines into cells or the parts into families. In a study conducted by Shafer and Meredith (1990) it was found that using statistical cluster analysis to identify part families outperformed similar procedures to identify machine cells largely due to the greater flexibility the analyst has in assigning duplicate machines to alternative cells with part family identification.

4.4. Algorithms For Sorting Machine-Component Matrix

A number procedures for sorting the rows and columns of the machine component matrix in order to simultaneously identify the part families and machine cells have been developed. Most often, these approaches seek to rearrange the rows and columns of the machine-component matrix to form independent blocks of ones along the diagonal of the matrix. Given this objective, these procedures are often referred to as block diagonal methods (BDM).

The rank order clustering algorithm developed by King (1980) begins by reading each row and column of the machine-component matrix as a binary word.

Next, the decimal equivalent of each row and column is calculated. The rows and columns are then sorted based on their decimal equivalents. King also illustrates the application of the bond energy algorithm developed by McCormick, Schweitzer, and White (1972) to the cell formation problem. King and Nakornchai (1982) present a more efficient version of the ROC algorithm and discuss a new relaxation procedure for bottleneck machines (i.e., machines required in two or more cells).

Chan and Milner (1982) developed a similar approach to the ROC algorithm called the direct clustering algorithm (DCA) for reordering the machine-component matrix. This procedure sequentially moves rows with the left-most positive cells to the top and columns with the top-most positive cells to the left of the matrix. In repeated iterations the positive cells will be "squashed" toward the diagonal of the matrix. This method, like the ROC algorithm, can effectively deal with bottleneck machines.

4.5. Mathematical Techniques

Mathematical techniques include a variety of analytical cell formation techniques including mathematical programming, graph theory, and combinatorial analysis. Because these types of techniques are discussed in more detail in later chapters, they are only briefly reviewed here.

4.5.1. Mathematical Programming.

Mathematical programming techniques use linear programming or goal programming to identify part families and their corresponding manufacturing cells. These approaches can consider a variety of objectives such as minimizing machine setup times, minimizing part transfers between the cells, minimizing investments in new equipment, and maintaining acceptable utilization levels. Likewise, these models can include a number of problem limitations such as limits on part family and cell sizes. Because these procedures require sophisticated algorithms to solve the mathematical models developed, they have tended to not be widely used in practice.

Kusiak (1987) employed the *p*-median 0/1 integer programming formulation to identify part families. The *p*-median approach involved initially selecting p of the parts to serve as medians or seeds for the clusters. Subsequently, the remaining parts were assigned to the seed parts such that the sum of part similarity in each family was maximized. Part similarity between two parts was defined as the number of machines the two parts had in common in their processing. A significant contribution of this approach was that it was one of the first procedures developed to process a proximity matrix using mathematical programming as opposed to statistical cluster analysis. Kusiak also illustrated how the formulation could be extended to consider situations for which alternate part routings existed.

Choobineh (1988) developed a 0/1 integer programming formulation for which the cost of producing a part in each cell and the cost of purchasing new equipment was minimized. A major contribution of this formulation was that the funds available for the purchase of new equipment was considered. Also, the formulation contained constraints to ensure that the capacity of each machine in each cell was not exceeded. A major limitation of the model was is that some of the coefficients used in the model's objective function would not be known until after the model is solved.

Co and Araar (1988) presented a 0/1 integer programming model to minimize the deviation between workload assigned to each machine and the capacity of each machine. This model is formulated and solved individually for each machine type. The most significant limitation of this model is that two parts that have similar processing requirements in terms of tooling and setups could be assigned to different machines in order to achieve a balance of capacity for each machine type.

Gunasingh and Lashkari (1989b) presented a 0/1 integer programming model with two alternative objective functions: 1) to maximize the similarity between machine and parts, and 2) to minimize the cost of machines less savings in intercellular movements. The model proposed by these researchers made two important contributions. First, similarity was based on tooling. Second, the authors distinguished between two situations: 1) reorganizing an existing system, and 2) setting up an entirely new system. Additionally, cell size limits were considered in the model. One noteworthy limitation, however, was than an upper limit on purchases was specified for each machine type instead of specifying a total budget for all new equipment purchases. With this approach, unused funds available for one machine type could not be used to purchase other machine types.

Shtub (1989) illustrated how the cell formation problem can be formulated as a general assignment problem. Specifically, the formulation was for minimizing the cost of assigning parts to cells such that minimum and maximum usage levels fore each cell were achieved.

Wei and Gaither (1990) presented the first 0/1 integer programming model for minimizing the cost associated with intercellular transfers. Intercellular transfers occur for two reasons: 1) the part was an exceptional part, or 2) the cell did not have sufficient capacity to process all parts assigned to it. Cell size limits are also included in this model. The major limitations of the formulation were occasional oversimplifications. For example, parts were randomly assigned to machines when multiple machines were available.

Askin and Chiu (1990) presented a 0/1 integer programming formulation that considered four cost categories: 1) machine costs, 2) group cost, i.e., the overhead associated with establishing a cell, 3) tooling cost, and 4) intercellular transfer cost. In addition, cell size limits and limits on worker hours were also included in the model.

Finally, Shafer and Rogers (1991) developed three goal programming models corresponding to three unique situations: 1) setting up an entirely new system and purchasing all new equipment, 2) reorganizing the system using only existing equipment, and 3) reorganizing the system using existing equipment and some new equipment. The formulations directly address four important design objectives associated with cellular manufacturing, namely: 1) reducing setup times, 2) producing parts cell complete, 3) minimizing investment in new equipment, and 4) maintaining acceptable machine utilization levels. Shafer and Rogers also presented a heuristic solution procedure that involved partitioning the goal programming formulations into two subproblems and solving them in successive stages.

4.5.2. Graph Theory Methods.

Rajagopalan and Batra (1975) developed a cell formation approach based on graph theory. With this approach, the Jaccard similarity coefficient is first calculated between each machine pair of machines. Next, a graph is defined where each vertex represents a machine and the edges represent the relationships between the machines. An edge connects two vertices in the graph only if the similarity coefficient for the machine pair is greater than some threshold value. Finally machine cells are formed by locating maximal "cliques" (a maximal complete subgraph) or near cliques and merging them together. In CM terminology, a clique is a group of machines in which every pair is related with a similarity coefficient greater than the specified threshold value.

As another example, Vannelli and Kumar (1986) investigated the problem of identifying the minimal number of bottleneck machines and/or exceptional parts so that through either duplication of machines or subcontracting of parts perfect part-machine groupings could be developed. The approach taken to solve this problem is to identify minimal cut-nodes in either partition of the bipartite machine-part graph.

4.5.3. Combinatorial Analysis.

Purcheck (1974) presented a combinatorial method for grouping parts into families and grouping machines into cells. The method maximizes the percentage of processing requirements the machine cells can complete in the manufacture of the part families. The interested reader is also referred to Purcheck (1975a, 1975b).

4.5.4. Other Heuristic Methods.

The task of categorizing mathematical cell formation techniques is complicated by the wide variety of techniques that have been developed and by the fact that some procedures employ multiple methodologies. Representative of these other heuristic methods are *capacitated cell formation procedures* that consider factors such as capacity and volume in the formation of cells.

As one example, Kang and Wemmerlöv (1993) presented a new heuristic procedure based on allocating operations to machines as opposed to the traditional approach of allocating parts to machines. The heuristic specifically considers capacity constraints and employs cluster analysis based on a new similarity coefficient to form the cells.

As another example, Suresh, Slomp, and Kaparthi (1995) developed a three -phase hierarchical procedure for capacitated cell formation. In the first phase, neural network methods are employed to identify part families and associated machine cells. Neural networks are utilized given their capability of clustering large industry-size data sets rapidly. In the second phase, a goal program is solved to create independent cells based on multiple objectives. The solution is further refined interactively by reducing intercell transfers on a part-by-part basis in the final phase.

4.6. Artificial Intelligence Techniques

The techniques of syntactic pattern recognition, neural networks, fuzzy logic, and genetic algorithms have been included under artificial intelligence (AI) techniques. This is an emerging and rapidly growing area.

Venugopal (1985) proposed the use of a syntactic pattern recognition approach to the cell formation problem. Syntactic pattern recognition attempts to represent complex patterns in terms of simpler subpatterns and relations among the subpatterns. Presumably, the simplest subpatterns or pattern primitives will be easily recognized. Venugopal's procedure involves the following four steps: 1) primitive selection, 2) cluster analysis, 3) grammar inference, and 4) syntactic recognition.

The use of artificial neural network methods and fuzzy methods are discussed in chapter D3, and the use of genetic algorithms is discussed in chapters D4 and D5.

5. MEASURING PERFORMANCE

The fundamental criterion for evaluating performance measures is how closely related the performance measures are to the basic objectives associated with adopting CM. As was mentioned earlier, Shafer and Rogers (1991) identified the following four key design objectives associated with CM: 1) setup time reduction, 2) producing parts cell complete, 3) minimizing investment in new equipment, and 4) maintaining acceptable machine utilization levels. Other important objectives associated with the adoption of CM include improving product quality, reducing inventory levels, and shortening lead times. In the remainder of this section, a number of performance measures that have been used in previous studies are briefly summarized and compared.

Mosier (1989) used four performance measures including a simple matching measure, a generalized matching measure, a product moment correlation coefficient measure, and an intercellular transfer measure. The first three performance measures simply gauge how well the clustered solution matches the original machine-component matrix and thus do not allow for the possibility that the clustered solution obtained may be in some sense better than the configuration of the original matrix. Indeed, the original configuration may not even be a good choice. Although these three measures are easy to calculate, they are not related to the design goals associated with adopting CM. The last measure used, the intercellular transfer measure, is the volume weighted number of parts that require processing in more than one cell. This measure does indirectly address the CM design objective of producing parts cell complete. However, because operation sequences are not considered, it is only a surrogate measure of the actual number of intercellular transfers.

The seven performance measures used by Shafer and Meredith (1990) required the development of computer simulation models. The measures used included average and maximum work-in-process levels, average and maximum flow times, part travel distances, extra-cellular operations, and the longest average queue. Using these measures is computationally more complex but they are all directly related to CM design objectives.

Four performance measures were used in the cell formation comparison studies conducted by Chu and Tsai (1990) and Shafer and Rogers (1993). These measures included the total bond energy, proportion exceptional elements, machine utilization, and the grouping efficacy. While total bond energy is not related to the design goals associated with CM, the other three measures are surrogate measures for the extent that parts are produced cell complete and that acceptable machine utilization levels are maintained. Kumar and Chandrasekharan (1990) developed a grouping efficacy measure (GE).

Numerous other authors have implemented performance measures implicitly through the objective functions of the mathematical models that they have developed (see Section 4.5.1). In Table 1 a number of performance measures are categorized on the basis of whether or not they require part volume data and/or operation sequence data. In general, the performance measures tend to better reflect the design objectives associated with CM as part volume and operation sequence data is included.

6. STUDIES ON COMPARISON OF CELL FORMATION TECHNIQUES

In this section studies that have focused on comparing alternate cell formation procedures will be reviewed. Many of these studies focused on comparing algorithms for sorting the machine-component matrix (see Section 4.4) and statistical cluster analysis techniques (see Section 4.3).

Mosier (1989) used a mixture model experimental approach and compared seven similarity coefficients and four statistically-based hierarchical clustering procedures. The degree of cluster definition, i.e., the ratio of the nonzero density outside the clusters to the nonzero density inside the clusters of the machine-component matrix were additional factors included in the study. As mentioned in the previous section, the performance measures used were a simple and a generalized matching measure, a product moment correlation coefficient measure and an intercellular transfer measure. Based on the results, the Jaccard similarity coefficient showed promise in spite of its simplicity.

Shafer and Meredith (1990) utilized computer simulation to compare cell formation procedures taken from all three of the major categories of cell formation procedures defined in Ballakur and Steudel's (1987) taxonomy. These cell formation techniques were then applied to data collected from three companies. Computer simulation models were developed based on the solutions generated with the cell formation procedures. The performance measures included average and maximum work-in-process levels, average and maximum flow times, part travel distances, extra-cellular operations, and the longest average queue. It was found that using clustering algorithms to first form part families and then assigning machines to the families provided more flexibility and worked best for the scenarios studied.

Chu and Tsai (1990) compared the rank order clustering algorithm, the direct clustering algorithm, and the bond energy algorithm using the following performance measures: 1) total bond energy, 2) proportion exceptional elements, 3) machine utilization, and 4) grouping efficiency. Eleven data sets from the literature were used for their study and the bond energy algorithm performed best for these problems. However, most of these data sets were extremely small example problems and not enough of the remaining problems were large enough to viably conclude the dominance of any one particular method.

Miltenburg and Zhang (1991) performed one of the more extensive studies comparing cell formation procedures based on the machine-component matrix as input. They tested nine different clustering procedures, a modified rank order clustering algorithm, single and average linkage clustering, the bond energy algorithm, and an ideal seed no-hierarchical clustering algorithm. The objective of the study was to form cells such that there was a high usage of machines by the parts in each cell (i.e., a high machine utilization) and to form cells that do not tend to allow exceptional elements. The primary performance measure used combined these objectives together into a single measure. Two secondary measures used were a measure of closeness of nonzero elements to the diagonal and average bond energy. Results were obtained for 544 solved problems and no significant differences were detected among most of the procedures. The ideal seed non-hierarchical clustering algorithm, however, showed the most promise for these particular randomly generated problems and performance measures. However, these procedures impose a significant computational burden for larger problems as compared to the other cell formation procedures.

Shafer and Rogers (1993) investigated 16 measures of similarity or distance and four hierarchical clustering procedures that are applicable to the cell formation problem. The same performance measures as used by Chu and Tsai (1990) were employed except that grouping efficiency was replaced by the improved grouping efficacy. A total of 704 solutions were derived and it was found that little difference typically existed among the results from different similarity coefficients. Single linkage clustering was, in general found to be statistically inferior to average linkage, complete linkage, and Ward's clustering methods for the bond energy, machine utilization, and grouping efficacy criteria. However, single linkage clustering was statistically superior when considering the proportion of exceptional elements.

TABLE 1. Categorization Of Performance Measures (from Rogers and Shafer 1995).

Part Volumes & Sequencing not Considered	Part Volumes Considered	Part Sequencing Considered	Part Volumes & Sequencing Considered
Simple Matching Generalized Matching Product Moment Correlation Coefficient Total Bond Energy Proportion Exceptional Elements Machine Utilization Grouping Efficiency Grouping Efficacy Grouping Measure Clustering Measure	Weighted Intercellular Transfers	Global Efficiency Group Efficiency Group Technology Efficiency	Average Work-In-Process Maximum Work-In-Process Average Flow Time Maximum Flow Time Part Travel Distances Extra-Cellular Operations Longest Average Queue

Kaparthi and Suresh (1994) evaluated the performance of 10 cell formation procedures for a wide range of data set sizes. Four categories of cell formation procedures were included in the study, including array-based methods, non-hierarchical clustering methods, augmented machine matrix methods, and neural network algorithms. In addition to data set size and cell formation procedure, the degree of data imperfection was included as an experimental factor. Three performance measures were used including a Rand Index, the bond energy recovery ratio, and the computational requirements. One conclusion of the study was that, for realistic, large-sized data sets, the augmented linear clustering algorithm and artificial neural networks had the best overall performance.

7. CONCLUSIONS

This chapter introduced the topic of design of cellular layouts. The major topics included formally defining the cell formation problem, providing a comprehensive taxonomy of cell formation procedures, reviewing the cell formation literature, overviewing cell formation performance measures, and reviewing studies that compared cell formation techniques.

As this chapter illustrates, a large amount of research has been directed toward the development of cell formation procedures. Somewhat less clear, however, is how useful practitioners are finding this stream of research. Anecdotal evidence would seem to indicate that practitioners favor the use of much less sophisticated techniques such as forming part families on the basis of drawing names, part features, or dimensional envelopes.

This potential gap between theory and practice highlights how little is actually known about the practice of cell formation and thus provides several avenues for future research. For example, research is needed to determine how practitioners actually form cells. If it turns out that practitioners are not using the sophisticated procedures being developed by the academic community, the reasons for this can be further investigated. For example, it may be determined that the cell formation techniques presented in the academic literature are perceived as being overly complex, that simple informal procedures provide satisfactory results, and even that the solution quality of the sophisticated techniques is poor when the complexities of an actual plant are considered. To illustrate, Miltenburg and Montazemi (1993) found that many of the cell formation procedures they applied to a manufacturer of farm machinery with 5498 parts and over 600 machines provided poor solutions and required excessive amounts of computational time.

If the existence of a gap between practice and theory is verified, future research should focus on closing this gap. For example, if it is determined that practitioners have a preference for simple informal techniques, studies can compare these techniques to more sophisticated techniques to determine whether there is any performance differential. If a significant performance differential is discovered in favor of the more sophisticated cell formation techniques, additional research can be devoted to developing user friendly decision support systems that would make sophisticated cell formation techniques more assessable to practitioners. Also, research is needed related to the development of guidelines pertaining to the selection of appropriate cell formation procedures based on the goals, environment, and other characteristics of the organization. Ideally, this intelligence would also be embedded in cell formation decision support systems.

REFERENCES

Allison, J.W. and Vapor, R., 1979, GT approach proves out, *American Machinist*, February, 86-89.

Askin, R.G. and Chiu, K.S., 1990, A graph partitioning procedure for machine assignment and cell formation in group technology, *International Journal of Production Research*, 28, 8, 1555-1572.

Ballakur, A. and Steudel, H.J., 1987, A within-cell utilization based heuristic for designing cellular manufacturing systems, *International Journal of Production Research*, 25, 5, 639-665.

Burbidge, J.L., 1963, Production flow analysis, *Production Engineer*, December, 742-752.

Burbidge, J.L., 1971, Production flow analysis, *Production Engineer*, April/May, 139-152.

Burbidge, J.L., 1975, *The Introduction of Group Technology*, John Wiley & Sons.

Burbidge, J.L., 1977, A manual method of production flow analysis, *Production Engineer*, October, 34-38.

Carrie, A.S., 1973, Numerical taxonomy applied to group technology and plant layout, *International Journal of Production Research*, 11, 399.

Chan, H.M. and Milner, D.A., 1982, Direct clustering algorithm for group formation in cellular manufacturing, *Journal of Manufacturing Systems*, 1, 1, 65-75.

Choobineh, F., 1988, A framework for the design of cellular manufacturing systems, *International Journal of Production Research*, 26, 1161-1172.

Chu, C.H., 1989, Clustering analysis in manufacturing cell formation, *OMEGA: International Journal of Management Science*, 17, 289-295.

Chu, C-H. and Tsai, M., 1990, A comparison of three array-based clustering techniques for manufacturing cell formation, *International Journal of Production Research*, 28, 1417-1433.

Co, H.C. and Araar, A., 1988, Configuring cellular manufacturing systems, *International Journal of Production Research*, 26, 9, 1511-1522.

De Witte, J., 1980, The use of similarity coefficients in production flow analysis, *International Journal of Production Research*, 18, 503-514.

Dutta, S.P., Lashkari, R.S., Nadoli, G., and Ravi, T., 1986, A heuristic procedure for determining manufacturing families from design-based grouping for flexible manufacturing systems, *Computers and Industrial Engineering*, 10, 193-201.

El-Essawy, I.F.K. and Torrance, J., 1972, Component flow analysis - an effective approach to product systems' design, *Production Engineer*, May, 278-283.

Garcia, H. and Proth, J.M., 1986, A new cross-decomposition algorithm: the GPM. Comparison with the bond energy method, *Control and Cybernetics*, 15, 2, 155-165.

Gettelman, K.M., 1971, Organize production for parts -- not processes, *Modern Machine Shop*, November, 50-60.

Gunasingh, K.R. and Lashkari, R.S., 1989a, The cell formation problem in cellular manufacturing systems -- a sequential modeling approach, *Computers and Industrial Engineering*, 16, 469-476.

Gunasingh, K.R. and Lashkari, R.S., 1989b, Machine grouping problem in cellular manufacturing systems - an integer programming approach, *International Journal of Production Research*, 27, 9, 1465-1473.

Gupta, T. and Seifoddini, H., 1990, Production data based similarity coefficient for machine-component grouping decisions in the design of cellular manufacturing system, *International Journal of Production Research*, 28, 1247-1269.

Kang, S-L. and Wemmerlöv, U., 1993, A work load-oriented heuristic methodology for manufacturing cell formation allowing reallocation of operations, *European Journal of Operational Research*, 69, 292-311.

Kaparthi, S. and Suresh, N.C., 1994, Performance of selected part-machine grouping techniques for data sets of wide ranging sizes and imperfection, *Decision Sciences*, 25, 515-539.

King, J.R., 1980, Machine-component grouping in production flow analysis: an approach using a rank order clustering algorithm, *International Journal of Production Research* 18, 2, 213-232.

King, J.R. and Nakornchai, V., 1982, Machine-component group formation in group technology: Review and extension, *International Journal of Production Research*, 20, 2, 117-133.

Kumar, C.S. and Chandrasekharan, 1990, Grouping efficacy: A quantitative criterion for goodness of block diagonal forms of binary matrices in group technology, *International Journal of Production Research*, 28, 233-243.

Kusiak, A., 1987, The generalized group technology concept, *International Journal of Production Research*, 25, 4, 561-569.

Kusiak, A. and Chow, W.S., 1987, Efficient solving of the group technology problem, *Journal of Manufacturing Systems*, 6, 2, 117-124.

McAuley, J., 1972, Machine grouping for efficient production, *Production Engineer*, 51, 53-57.

McCormick, W.T. Jr., Schweitzer, P.J., and White, T.W., 1972, Problem decomposition and data reorganization by a clustering technique, *Operations Research*, 20, 993-1009.

Miltenburg, J. and Montazemi, A.R., 1993, Revisiting the cell formation problem: Assigning parts to production systems, *International Journal of Production Research*, 31, 2727-2746.

Miltenburg, J. and Zhang, W., 1991, A comparative evaluation of nine well-known algorithms for solving the cell formation problem in group technology, *Journal of Operations Management*, 10, 1, 44-72.

Mosier, C., An experiment investigating the application of clustering procedures and similarity coefficients to the GT machine cell formation problem, *International Journal of Production Research*, 27, 1811-1835.

Mosier, C. and Taube, L., 1985, Weighted similarity measure heuristics for the group technology machine clustering problem, *Omega*, 13, 577-579.

Purcheck, G.F.K., 1974, Combinatorial grouping - a lattice-theoretic method for the design of manufacturing systems, *Journal of Cybernetics*, 4, 3, 27-60.

Purcheck, G.F.K., 1975a, A mathematical classification as a basis for the design of group technology production cells, *Production Engineer*, January, 35-48.

Purcheck, G.F.K., 1975b, A linear-programming method for the combinatorial grouping of an incomplete power set, *Journal of Cybernetics*, 4, 4, 51-76.

Rajagopalan, R. and Batra, J.L., 1975, Design of cellular production systems: a graph-theoretic approach, *International Journal of Production Research*, 13, 6, 567-579.

Rogers, D.F. and Shafer, S.M., 1995, Measuring cellular manufacturing performance, *Planning, Design, and Analysis of Cellular Manufacturing Systems*, (Eds. A.K. Kamrani, H.R. Parsaei, and D.H. Liles), Elsevier, 147-165.

Selvam, R.P. and Balasubramanian, K.N., 1985, Algorithmic grouping of operation sequences, *Engineering Costs and Production Economics*, 9, 125-134.

Shafer, S. M., 1988, A comparison of manufacturing cell formation techniques, Unpublished Doctoral Dissertation, University of Cincinnati., United States.

Shafer, S. M. and Charnes, J.M., 1993, Cellular versus functional layouts under a variety of shop operating conditions, 24, 3, 665-681.

Shafer, S.M. and Meredith, J.R., 1990, A comparison of selected manufacturing cell formation techniques, *International Journal of Production Research*, 28, 661-673.

Shafer, S.M., Meredith, J.R., and Marsh, R.F., 1997, A comparison of cellular manufacturing research assumptions with practice, Working Paper, Auburn University, United States.

Shafer, S.M. and Rogers, D.F., 1991, A goal programming approach to the cell formation problem, *Journal of Operations Management*, 10, 1, 28-43.

Shafer, S.M. and Rogers, D.F., 1993, Similarity and distance measures for cellular manufacturing. Part I. A survey, *International Journal of Production Research*, 31, 5, 1133-1142.

Shtub, A., 1989, Modeling group technology cell formation as a generalized assignment problem, *International Journal of Production Research*, 27, 5, 775-782.

Singh, N., 1993, Design of cellular manufacturing systems: An invited review, *European Journal of Operational Research*, 69, 284-291.
Suresh, N.C., Slomp, J., and Kaparthi, S., 1995, The capacitated cell formation problem: A new hierarchical methodology, *International Journal of Production Research*, 33, 1761-1784.
Tam, K.Y., 1990, An operation sequence based similarity coefficient for part families formations, *Journal of Manufacturing Systems*, 9, 55-68.
Vakharia, A.J., 1986, Methods of cell formation in group technology: A framework for evaluation, *Journal of Operations Management*, 6, 257-271.
Vakharia, A.J. and Wemmerlöv, U., 1990, Designing a cellular manufacturing system: A materials flow approach based on operations sequences, *IIE Transactions*, 22, 84-97.
Vannelli, A. and Kumar, K.R., 1986, A method for finding minimal bottle-neck cells for grouping part-machine families, *International Journal of Production Research*, 24, 2, 387-400.
Wei, J.C. and Gaither, N., 1990, An optimal model for cell formation decisions, *Decision Sciences*, 21, 2, 416-433.
Wemmerlöv, U. and Hyer N.L., 1986, Procedures for the part family/machine group identification problem in cellular manufacturing, *Journal of Operations Management*, 6, 2, 125-147.
Waghodekar, P.H. and Sahu, S., 1984, Machine-component cell formation in group technology: MACE, *International Journal of Production Research*, 22, 937-948.
Wu, H.L., Venugopal, R., and Barash, A., 1986, Design of a cellular manufacturing system: A syntactic pattern recognition approach, *Journal of Manufacturing Systems*, 5, 81-88.

AUTHOR'S BIOGRAPHY

Scott M. Shafer is an Associate Professor of Operations Management in the Department of Management at Auburn University. He received a B.S. in Industrial Management (1984), a B.B.A. in Marketing (1984), and a Ph.D. in Operations Management (1989) from the University of Cincinnati. His current research interests are in the areas of cellular manufacturing, operations strategy, business process design, production scheduling, and microcomputer applications. A recent study investigating the productivity of 738 researchers in the field of operations management ranked Dr. Shafer in the top 20 in terms of both publication quality and research productivity. Additionally, Dr. Shafer is Certified in Production and Inventory Management (CPIM) by the American Production and Inventory Control Society.

D2

Design of Manufacturing Cells: PFA Applications in Dutch Industry

J. Slomp

1. INTRODUCTION

Cellular manufacturing (CM) can be defined as the grouping of people and processes into specific areas dedicated to the production of a family of parts. There is a growing interest for CM in industry. In the last two decades many firms in the Netherlands as well as in other countries have implemented CM. Though the principle of CM in each firm will be basically the same, there may be considerable differences with respect to the characteristics of the cells. Table 1 presents seven firms in the Netherlands which have completely cellularized their manufacturing departments (see Slomp, Molleman and Gaalman 1993). As can be seen, these departments differ in terms of product types, the number of employees, number of manufacturing cells, number of different part types, and the number of bottleneck parts (i.e., parts that have to be manufactured in more than one cell). It is furthermore informative to know that in cases A, B, D, E, and F the manufacturing activities are based on forecast and planning. In cases C and G the manufacturing activities are customer order driven. This distinction has an impact on the performance objectives in each case. Table 2 lists some main characteristics of the manufacturing cells of the firms. These illustrate the variety of cells with respect to the number of machines, the type of material flow, the degree of automation, and the nature of the bottleneck capacity.

The variety of firms and manufacturing cells indicates the need for a `flexible' cell formation method which can be used in various situations. Basically, cell formation can be seen as a clustering problem which requires a `matrix handling technique'. In Figure 1a an illustration is given of a small part-machine matrix (A=(IxM), I refers to the number of parts, M is the number of different machines). A cell (a_{im}) in this matrix indicates whether part i requires machine m (a_{im}=1) or not (a_{im}=0). Figure 1b shows what the matrix will look like after basic clustering. In effect, clustering comes down to a shifting of rows and columns. Cell formation in reality, however, has certain complexities which are not dealt with in the example of Figure 1. First, the size of the problem will always be larger (see e.g., the number of part types in Table 1). This reduces the possibilities of mathematical programming methods. Second, there may be more than one machine of each type. This relaxes the need to create a block-diagonal structure in Figure 1, but also poses the question of how to divide the machines among the cells. Third, the demand for each part type may vary significantly as well as the utilization of each machine type. This creates differences in the importance of each of the intergroup relations (see Figure 1b) and therefore complicates the design of a clustering algorithm ('how to minimize the cumulative effect of the intergroup relations?'). Finally, certain objectives may have an important impact on the design of a clustering method, for instance, the objective to design manufacturing cells of a certain size.

TABLE 1. Firms And Their Manufacturing Cells.

Firm	final products of the firms	# direct labour (# shifts)	# cells	# different part types	bottleneck parts
A	Industrial Instrumentation	75 (2)	5	5000	<10%
B	Centrifugal Pumps (small)	54 (2)	6	450	0%
C	Centrifugal Pumps (large)	30 (1)	5	8000	20%
D	Parts for Electrical Supply Instal.	150 (2)	14	40000	5%
E	Machines for the Textile Industry	100 (2)	8	6000	5%
F	Large Pumps, Compressors	100 (2)	8	1200	>20%
G	Injection Moulding Mach.	70 (2)	5	130	<20%

part-> mach.	A	B	C	D	E
1	1			1	
2			1		1
3	1		1		
4		1			1

(a) part-machine matrix, not clustered

part-> mach.	A	D	C	B	E
1	1	1			
3	1		1		
2			1		1
4				1	1

(b) part-machine matrix, clustered (1 intergroup relation)

FIGURE 1. Part-Machine matrices.

TABLE 2. Characteristics Of The Manufacturing Cells.

Firm	# machines per cell (average, min., max.)	material flow within cells	FMC in a cell	bottleneck capacity
A	(15, 9, 20)	flow line & job shop	yes	machine and labour capacity
B	(3, 2, 6)	flow line	no	machine and labour capacity
C	(5, 4, 6)	flow line & job shop	no	labour capacity
D	(10, 6, 16)	job shop	yes	machine and labour capacity
E	(5, 1, 10)	flow line & job shop	yes	machine and labour capacity
F	(6, 3, 12)	flow line & job shop	yes	machine capacity
G	(5, 1, 10)	flow line & job shop	yes	machine and labour capacity

This chapter shows the applicability and limitations of group analysis which is a major element in Burbidge's production flow analysis (Burbidge 1991). Burbidge (1991) argues that group analysis is a general technique for cell formation which can be applied in all types of manufacturing departments. §2 briefly summarizes all the steps of production flow analysis. Next, in §3, the major steps of group analysis are presented on the basis of the data of a real firm. The steps differ slightly from the ones presented by Burbidge (1991). New elements are the integration of a cluster algorithm (Bond Energy Analysis, Mc.Cormick, Schweitzer and White 1975) and the addition of a method for an initial assignment of people to the manufacturing cells. §4 briefly presents three other cases in which Production Flow Analysis was used to reorganize the production floor. This is followed by conclusions. The reader is also referred to chapter I2, by Karvonen, Holmström and Eloranta, which refers to PFA application in a broader context in Finland

2. PRODUCTION FLOW ANALYSIS

Production Flow Analysis is an approach to cell formation which relies on information about the routes of products, modules and parts through the company. According to Burbidge (1991), production flow analysis consists of five subtechniques used sequentially to simplify the material flow in an enterprise. *Company Flow Analysis* analyses the flow of material between the different factories of a large company so as to develop a system in which the factories are as independent as possible. *Factory Flow Analysis* studies the flow of material in each factory. It attempts to define departments in such a way that the various departments are connected through a unidirectional flow system only. *Group Analysis* divides each department into manufacturing cells. Each cell is responsible for the entire making of a family of part types. *Line Analysis* studies the flow of material between the machines of a manufacturing cell. This information is needed for the design of the layout of a cell. *Tooling Analysis*, finally, focusses on each of the major machines in a manufacturing cell and attempts to find `families' of parts which require the same machine setup (or tools). This information can be useful for the planning and control of the manufacturing cells.

Company flow analysis, factory flow analysis and line analysis can usually be performed by drawing a network of elements (factories, departments or machines) with material flows in between. The advantages and disadvantages of a seperation of activities in different factories, departments or machines need to be studied carefully. As a result, it may be desirable to simplify the network. This can be done by combining elements and/or by re-designing the tasks of some elements. For instance, it may be possible to reduce the material flow between two departments by moving a machine from the one department to the other. Group analysis and tooling analysis are usually more complex and require, as mentioned before, matrix resolution. The next section illustrates the steps in group analysis.

3. GROUP ANALYSIS

This section presents the various steps in group analysis by means of a case study. The group analysis method presented here can be seen as an adapted version of group analysis procedure of Burbidge (1991) and consists of the following steps: (i) formulation of objectives, (ii) collecting information, (iii) formation of articles/product modules, (iv) integration of modules, and (v) allocation of workers to cells. The section ends with a case-specific evaluation of the method and its solution.

3.1. The Case

The firm DSW-Metaal employs about 125 workers and is specialized in the manufacturing of pipes and pipe constructions. In the course of time, much know-how, expertise and practical experience has been built up in this area. Both copper and steel pipes are processed according to the specifications of the customer. This can involve welding, bending, drilling, and more specific operations such as flairing and trumping. The customers come from the heating, building, chemical and cooling industries.

In early 1991 the management identified several external and internal problems. Externally, problems were felt regarding the long delivery times/throughput times and the poor controllability of the product quality. Internally, the high degree of coordination effort that was necessary and the poor orientation towards labour quality were identified as problems. According to the management, this was all mainly due to the production structure of the company. Up to 1991 there had been a functional product situation consisting of six functional groups: (i) a copper smith's workshop, (ii) pipe works, (iii) drilling, (iv) welding, (v) punching/sheet metal processing, and (vi) tool manufacture/finishing. Between 1991 and 1993 the firm adopted a CM layout. As a result, company performance improved significantly. For example, the throughput times of products were reduced from 2-12 weeks to 2-5 weeks. There was also a strong reduction in the number of procedures within the company, an indication of the required control efforts. Shorter control loops were realized which simplified quality control.

For the grouping of products, machines, and workers DSW-Metaal at the time made use of a spreadsheet program (Quatro-Pro) containing the following product information: articles (i.e. products consisting of one or more parts), machines, routings, operating times and the annual demand of each article. In all, 297 articles and 40 machine types were stored into the program. A technical specialist applied the information to group the articles and machines into autonomous groups. He then `played with the speadsheet program' aiming at a balanced grouping in which several constraints and objectives were taken into account. In doing so, no explicit grouping procedure or method was used.

The spreadsheet data of DSW-Metaal (concerning 1991) were used in a research project to test and improve a more systematic grouping method (van Burken et al., 1995). Afterwards, the method and the solution were evaluated by the technical specialist.

3.2. Step (i): Formulation Of Objectives

When forming groups, the company objectives should be reckoned with. A critical attitude towards these objectives is important as well as a verification of whether they are really necessary. The principle of `minimal critical specification' (see Cherns, 1987), which (merely) specifies essential objectives, plays a key role in this. In the study presented here, the objectives of DSW-Metaal, as formulated in 1991, are adopted without further discussion. In doing so, the results can more easily be compared with the actual solutions. The objectives of DSW-Metaal include:

a) Machines, workers and articles should be clustered into six cells. In the initial situation there were six heads of departments (of the various functional departments). In the new situation they will become foremen of the cells. The choice of six cells furthermore is based on group size considerations;

b) There should be a balanced loading of cells. It is important to maintain a certain stability of the order mix to avoid a high degree of work in a cell during one particular period only. Furthermore, the management of DSW-Metaal stressed the importance of loading cells in a balanced way so as to create equal cells;

c) The number of intergroup relations should be kept to a minimum. This implies that each article should be manufactured as much as possible by one autonomous group only. In doing so, the coordination efforts between the groups will be minimal;

d) The extra machine investments necessary should be minimal. To avoid intercell movements it may be necessary to purchase additional machines;

e) The required additional schooling and training of workers should be minimal. Cell manufacturing is more vulnerable to absenteeism of workers than a functional structure. Within a functional structure, the workers of one department can easily replace one another. This is harder in a CM layout in which each cell has various machines. A certain multifunctionality of workers is needed and this, not surprisingly, may involve further or new training of workers;

f) Workers should be assigned as much as possible to their present functional foreman. This objective is based on the idea that this ensures a minimum of commotion on the shopfloor. Further, it will remain possible to benefit from the current experience of working together.

The above objectives are to a certain extent contradictory. For example, nothing much will change at DSW-Metaal when priority is given to objectives d, e and f. The various objectives will have to be incorporated into the following steps of group analysis.

3.3. Step (ii) Collecting Information

The availability of sufficient information is crucial for the effectiveness of a grouping method. Most firms make use of MRP-software to control production. The data generated from such production control systems usually suffice to apply a grouping method. Vital data include article codes, machine types, routings, processing times and the annual demand of each article. Information concerning the capabilities of the workers are usually available at the production manager's office. As mentioned, DSW-Metaal had already stored vital data into a spreadsheet program (Quatro Pro).

The available data for group analysis merely reflect the situation at a given moment. In actual practice the available machinery, the product mix, and product volume are liable to changes. This can be anticipated by formulating and using the data of various future scenarios. A scenario can be represented by the expected annual demand for each product type. A grouping solution should be acceptable in all possible scenarios. In case of seasonal demand fluctuations, it is useful to check the acceptability of a grouping solution through the year. Frenzel (1993) applied group analysis on the manufacturing data of one year. After the cell formation, he analysed the seasonal fluctuations in the loading of the cells by using the data over 3-month periods.

3.4. Step (iii): Formation Of Articles/Product Modules

As mentioned earlier, the article-machine matrix is the starting point for cell formation. The size of an article-machine matrix can be fairly large. Wissink (1990) describes a situation with about 2500 part types and 36 machine types, Frenzel (1993) a situation with about 1000 part types and 35 machine types, and Rolefes (1995) a situation with some 4000 part types and 36 machine types. Table 1 also gives an indication of the sizes of the matrices which can be found in industry. The size of the article-machine (or: part type - machine type) matrix complicates the clustering of articles and machine types into autonomous groups. It becomes difficult to cluster manually; a spreadsheet does not give the required overview anymore. Burbidge (1991) describes a procedure for grouping articles/part types into product families or modules, as a result of which the article-machine matrix can be reduced. EXHIBIT 1 gives a formal description of the procedure.

First, Burbidge classifies machines into five categories. These categories (S, I, C, G, and E) are based upon the number and the exchangeability of machine types. The Special (S) machines concern special, mostly expensive machines of which there is only one of each type. It is difficult to transfer work to other machines. Intermediate (I) machines are similar to those of category S, but there is more than one machine of each type. Of the common (C) machines there are several of each type. It is furthermore easy to transfer work to related machine types. G-type machines are more or less unique. They are used for many of the part types (e.g., a saw for cutting pipes). It is not logical to locate these machines in manufacturing cells. Finally, Burbidge defines the so-called Equipment (E) category which consists of items to assist manual operations.

Second, after categorization, the machine types are re-ordered into the SICGE-sequence, and within each category in the sequence of decreasing number of part types which makes use of the machine type. The modules are then formed around the so-called key machines which are sequentially picked from the re-ordered SICGE-list. Each module consists of a number of part types which all require the key machine. After the forming of a module, the connected parts as well as the key machine type are removed from the database. The next module is created by using the smaller database. This process is continued until all part types are grouped into modules.

In the DSW-Metaal case, the modules are formed after removing the machine types with little annual load (a few hours per year) and the articles with little annual demand. These machine types and articles are excluded from playing a role in the forming of modules. Figure 2 gives an example of a module created in the DSW-Metaal case. Figure 3 gives a summary of all modules, the module-machine type matrix.

EXHIBIT 1. Formal Description Of Group Analysis.

m	=	machine type	m=1...M (sequence after step 0)
I	=	part type	i=1...I
k	=	module	k=1...M
I_k	=	part types that fit in module k	
$\{I\}_k$	=	set of part types in module k	
A_{im}	=	1 , if part type i requires machine type m 0 , otherwise	
x_{mk}	=	value of cell (m,k) in module-machine type matrix	

Step 0. Initiation:
- Apply SICGE to categorize the machine types. After categorization, the machine types are positioned in the SICGE-sequence, and within each category in the sequence of decreasing number of part types which make use of the machine type;
- Start with k:=1, m:=1 en $I_k := I$

Step 1. If $a_{im}=1$ then $i \in \{i\}_k$ $(i \in I_k$, m=m...M)

Step 2. $I_{k+1} := I_k - \{i\}_k$

Step 3. If $I_{k+1} = \varnothing$ then stop, else k:=k+1, m:=m+1, and go to step 1

The cells in the module-machine type matrix X (MxM) can be calculated as follows:

$$x_{mk} = \Sigma_{i \in Ik} \Sigma_{k=m...M} a_{im}$$

3.5. Step (iv): Integration Of Modules

The format of a module-machine type matrix (maximally MxM) is such that it is sometimes possible to create independent groups of modules and machines manually by joining modules and machines (e.g., from Figure 1a to 1b). Combining modules is not an easy task in all situations. The interrelations between the various modules often pose a problem and there is a tendency to continuously extend the combined modules. This wil eventually lead to one large module, which is not the desired result. In brief, combining modules could benefit from a more formal procedure. - Burbidge (1991) has not formulated such a procedure, but only gives some guiding rules. The study presented here has made use of the clustering algorithm referred to as Bond Energy Algorithm (BEA) of McCormick et al. (1972). In this method, modules and machine types are rearranged so as to create clusters (`clouds') of related modules and machine types. The algorithm is only applied for part of the matrix, that is for the S-machines and the modules that make use of these machines. These special machines were chosen in view of their uniqueness so basically they can be assigned to one group only. Appendix 2 gives a formal description of the algorithm. The result of the algorithm is presented in Figure 4.

machine type (# machines) Article number	351 (1)	341 (1)	241 (1)	212 (9)	821 (5)	831 (3)
508	*		*	*		
509	*		*	*		
500013	*	*		*		
1000001	*	*				
1300001	*	*		*	*	*
# operations	5	3	2	4	1	1
Total workload (hours)	174	42	159	221	45	194

FIGURE 2. Article module (Number 6).

Figure 4 is a useful starting point for combining modules and machine types into cells. First, the remaining product modules (of non-S-machines) and machine types are added to the matrix. Next, modules and machine types are combined. This requires an evaluation of the possibilities to combine modules and adjoining machine types. The company objectives are a guideline for the type of clustering. For example, DSW-Metaal intended to form six equal cells. To realize this, a `target' cumulative process time for each cell was first calculated (=total cumulative process time divided by six). Based on this, modules were linked to the cells with the aim of creating a minimal number of relations between the cells. However, it is not possible to reduce the number of intergroup relations to nil.

Eventually the following six cells were formed: cell 1 (modules 3, 7 and 14, total process time:5671 hours), cell 2 (modules 9 and 10, total process time:9738 hours), cell 3 (module 11, total process time:11867 hours), cell 4 (module 12, total process time:10631 hours), cell 5 (module 13, total process time:11086 hours) and cell 6 (modules 1, 2, 4, 5, 6, 8, 15 and 16, total process time:5891 hours). Next, machines and articles that had initially been excluded from consideration (because of little annual demand and use) will have to be assigned to the various cells. To minimize the intergroup relations it is recommended to additionally purchase - machines 525, 231 and 251, which are a refurbishing machine, an automatic cutting machine and a sawing machine, respectively. Total investment costs are estimated at NLG 130,000.

EXHIBIT 2. Bond Energy Analysis For Forming Groups.

I	=	counter;
α_{mk}	=	1 , if module k requires machine type m 0 , otherwise;
A_{mk}	=	Number of part types in module k that need machine type m (i.e., the value of cell (k,m) in the module - machine type matrix).

Step 0 Create a matrix with only S-(special)-machines and the modules (K^*) which require these machines;

Step 1 Place one of the columns (i.e., a module) arbitrarily. Set i:=1;

Step 2 Try placing individually each of the remaining N-i columns in each of the i+1 possible positions (to the left and right of the i columns already placed), and compute each column's contribution to the Measure of Effectiveness (ME). Place the column that gives the largest incremental contribution to the ME in its best position. Increment i by 1 and repeat until i:=N. The ME is calculated as follows:

$$ME = \tfrac{1}{2} \Sigma_x \Sigma_y A_{xy} (\alpha_{x,y-1} + \alpha_{x,y+1} + \alpha_{x-1,y} + \alpha_{x+1,y})$$

$(x=1..K^*,\ y=1..i+1,\ \alpha_{x0}=\alpha_{0y}=0)$

Step 3 When all the columns have been placed, repeat the procedure on the rows. (The row placement is not really needed).

Instead of applying the value A_{mk} in the expression of ME, it would probably be better to use the number of visits of (part types in) module k to machine m. This requires information about the number size of the transport batches of each part type. In the case of DSW-Metaal described in this chapter the assumption is made that it is sufficient to do the calculations with A_{mk} in order to find related modules.

On the basis of the grouping result, a first, global layout can be designed in which each group gets a certain area of the workfloor. Intergroup relations, those which cannot be removed by rerouting or redesign of the articles, may have a decisive impact on the mutual positions of the groups.

3.6. Step (v): Allocation of Workers To Cells

In the previous step cells were designed consisting of articles/products and machines. To these operators will have to be assigned. One of DSW-Metaal's objectives was to assign workers as far as possible to their original foreman. It was also desirable to reduce the extra training necessary. To this end a three-step assignment procedure (Sjoberg and Slomp, 1995) was designed. In the first step the functional cells are provisionally assigned to the product cells that were created. In the second step the qualified workers within the product cells are assigned to machines/tasks. The third step finally smoothes out any surplus or shortage of labour between the various cells. The assignments in steps 2 and 3 are such that the actual capabilities of workers are optimally used.

ma./mo->	1	2	3	4	5	6	7	8	9	10	11	12	13	14	15	16	b	u
931	1																1	180
901		1															1	36
571			2														2	743
951				4													4	395
202					5												5	358
351						5											5	174
171			1				5										6	228
341					5	3		11									19	708
161									11								11	325
261									8	22							30	458
525							1		2	15	24						42	1145
231							5		3		24	36					68	2330
241				3		2		2	1			1	55				64	1840
312							4				3		5	1			13	7208
251				1			3	3			15	18	30		1		71	5951
212			2		5	4	4	11	1	2	11	14	14	1	1	3	73	7478
135		1					2	1	1			11	32			1	49	2568
961				1					5	4			5				15	1257
821						1			6	16	1						24	1411
831						1	2		5	14	4						26	3403
221							3	1	2		9	19	25		1	1	61	10685
841								1	1			4					6	1141
811									8	21	11	2					42	1703
591									8	1	1	1					11	3159
TB	1	2	5	9	15	16	29	30	62	95	103	106	166	2	3	5		
TWL	180	43	1582	603	914	835	2628	2404	4470	5268	11867	10631	11086	1461	377	535		

mo=module, ma=m/c type, TB= total # operations, TWL=tot. annual workload (hrs); b=# opns. on m/c type, u=annual workload on m/c type.

FIGURE 3. Module-machine type.

mod. Mach.	3	7	10	9	11	12	13	8	5	6	4	2	1
571	2												
171	1	5											
161				11									
261			22	8									
525		1	15	2	24								
231		5		3	24	36							
241				1		1	55	2		2	3		
341								11	5	3			
202									5				
351										5			
951											4		
901												1	
931													1

FIGURE 4. Module-machine matrix (after BEA-application).

3.7. Evaluation

Evaluation of the stepwise approach in the DSW-Metaal case is done by comparing it with the `methods' used in practice and by asking comments from the technical specialist who was responsible for the cell formation. It appeared that the outcome of the group analysis is almost similar to the one earlier created by the technical specialist (note that there was no communication beforehand about the details of the latter outcome!). The stepwise approach, however, does offer some major advantages, according to the technical specialist. First and foremost the overview that is generated and next, the possibilities to think up alternative solutions. Furthermore, it reduces the time needed for `puzzling'.

The assignment of operators to the cells at the time was done in a meeting of the foremen of the functional ánd product cells. Afterwards, the technical specialist was not too satisfied with the results. It appeared that each of the foremen kept his best workers. As a consequence, the remaining cells got the least skilled workers for the not-dominant, but essential functions in the cells. This caused quality problems. A formal assignment procedure (step v) would have been helpful in reducing the impact of subjective judgement of the foremen.

In 1993, the firm adopted a CM layout consisting of six manufacturing cells, each with a size of 20 operators. In spite of the positive results, the firm recently decided to reduce the number of cells to three. This will increase the functionality in each cell which gives opportunities to further specialization. Workers who are skilled in the same function can more easily learn from each other. Another argument of adopting fewer cells is a financial one: less foremen are needed to organize the groups. Basically, the move of going from six to three manufacturing cells can be seen as an adaption of the firm's objectives (step 1). It emphasizes the importance of thinking through the objectives carefully before grouping starts.

4. THREE APPLICATIONS OF PFA

This section briefly describes three applications of production flow analysis. The first case illustrates the importance of factory flow analysis as a required step before group analysis. The second case shows how group analysis can be a tool for evaluation of the material flow through the manufacturing cells of a firm. The third case illuminates the fact that group analysis can also be applied to redesign a cellular manufacturing situation into a more functional layout.

Case I (Schurink 1991) concerns a firm which designs, produces and sells machines for the textile industry. The manufacturing department of the firm is organized in manufacturing cells, each responsible for the complete manufacturing of a family of part types. In 1991, the firm decided to investigate to what extent part manufacturing and subassembly (modules) can be integrated in the manufacturing cells. The investigation was occasioned by problems in the assembly department due to a cumulation of acceptable tolerances of parts. Furthermore, the management of the firm expected several other advantages of the integration, such as a shorter throughput time, a less complex production control, less quality procedures required, and an improvement of customer service (faster response times). The wish to integrate the manufacturing and the assembly departments can be seen as an outcome of a factory flow analysis in which the relations between departments are evaluated carefully. The investigation on the merging of the two departments was performed on one final product of the firm, the so-called `steamer', consisting of 36 modules (single parts excluded), and showed, using clustering techniques (Bond Energy Analysis, McCormick et al., 1971), how all the modules could be produced, almost completely, in two cells and one support cell. Few modules had to be split. On the basis of the intercell relations, a basic layout of the shop floor was designed. Within the analysis, single parts (which are not part of a subassembly) were not taken into account, because it was thought that these parts can be easily assigned to one of the manufacturing cells, without creating many extra intercell relations. Furthermore, after the analysis, it appeared that several single parts can be manufactured in both cells, which will simplify the balancing of the loads of the cells. In spite of the positive results of the study, management of the firm decided to postpone further investigations because of the presence of unpredictable changes on the market, the development of a new textile printing machine, and the organizational complexity of merging two departments.

Case II (Wissink 1990) refers to the part manufacturing department of a firm which produces, among other things, oil tank-gauging equipment. Group analysis and line analysis were performed to design manufacturing cells. In step (iv) six alternatives were developed and evaluated using the following criteria: (a) transparancy, or the unambiguity of assigning parts to cells; (b) independency, or the number of intercell movements; (c) size of the cells; (d) flexibility, or the possibility to manufacture parts in more than one cell; (e) stability of the workload of the cells in time; (f) the need for additional training of employees; and (g) required investments in new equipment. Some of the criteria are more or less contradictory, for instance (a) and (d). This illustrates the fact that a cell layout cannot be optimal with respect to all criteria. In 1991, the firm implemented four manufacturing cells, with 9 to 14 persons each. For a few years, the cellular structure was evaluated using the group analysis software developed by Wissink (1990) and applying the data of the previous year. This gave insight in the required adaptations of the cell structure. In 1993, the firm merged with another company. The manufacturing department of this company was functionally organized. The new, merged manufacturing department adopted cellular manufacturing. In 1996, however, the management of the firm decided to shift to a functional layout arguing that is was difficult to maintain the right level of multifunctionality among workers.

Case III (Rolefes 1995, and Posthumus 1997) concerns a factory that manufactures parts, or sometimes small subassemblies, which can be used in the metal-electrical industry. Most of the parts and subassemblies (almost 95%) are produced for the other business units of the company. About 140 employees are directly involved in the manufacturing process. Until 1993, the manufacturing department was organized in 14 relatively autonomous manufacturing cells. Important manufacturing processes include machining/turning operations, sheet-metal processing, strip and tool bound punching, electrostatic painting, galvanic plating, and construction and assembly. Since 1993 several changes have been implemented in the CM system of the firm. Basically, the firm moved from a cellular manufacturing system to more functionally organized cells. Important drive for this major change was mainly instigated by a combination of market developments, technological developments, and a change in management philosophy. The competition on the market asked for more specialization regarding the manufacturing operations. New manufacturing equipment, such as some advanced CNC-machines, lead to a reduction in the number of manufacturing steps of many part types. This fact reduced the need for cell manufacturing. Furthermore, management philosophy has been altered from a focus on the quality of labour and organization to a strong focus on financial results (a shift from a more internal focus towards an external focus). Within this context two `group analysis' studies have been performed in the firm. Rolefes (1995) developed software for group analysis and redesigned the machining/turning area into two cells. Originally, five cells were responsible for the machining and turning operations. Important element in the group analysis approach of Rolefes was the decision to deal with some important machine types (more or less identical CNC-machines) in the I-category as if they were S-machines. The layout of the cells was designed in cooperation with the workers. The two cells were implemented satisfactorily in 1995. Posthumus (1997) has performed a similar study in the sheet-metal processing area of the firm. By means of group analysis, he generated four alternatives for the six cells responsible for sheet-metal processing. In a meeting of the managers of the cells involved, a choice was made for the most functional alternative. This alternative concerns two autonomous manufacturing cells, each responsible for the complete manufacturing of a family of part types, and two functional departments. The new layout will be implemented next year. A new project is started to adapt the production control system for functional manufacturing.

5. CONCLUSIONS

The variety of firms and manufacturing cells in industry indicates the need for a `flexible' cell formation method which can be used in various situations. This chapter has presented an example and some adaptations of `group analysis', the cell formation method presented by Burbidge (1991). The steps involved are: (i) formulation of objectives; (ii) collecting information; (iii) formation of modules; (iv) integration of modules; and (v) allocation of workers to cells. Strong elements of the approach are its simplicity, its flexibility to deal with various firm specific characteristics, and the ease by which alternative groupings can be generated. Group analysis is not only a useful tool for cell formation. A small case in this chapter has indicated that group analysis can also be used as a tool for a regular evaluation of the material flow through and between the cells of a firm. Another case showed the application of group analysis in the redesign of a cellular manufacturing situation into a more functional layout.

Several elements of group analysis require further research. First, group analysis basically assumes that the information it uses remains unchanged after transition from one to another manufacturing layout. In practice data will change after transition to, for instance, a cell layout. Setup times probably will reduce and processing times may increase (less specialization in a cell layout). This may have an impact on the quality of the final solution. Knowing the changes in data beforehand another production structure would perhaps be selected earlier. Second, group analysis, as described by Burbidge (1991), only clusters parts and machines and does not include the allocation of workers to cells. As indicated in this chapter, the assignment of workers to cells should be seen as a major element of cell formation. This issue requires more empirical and analytical research. It is conceivable to integrate `the allocation of workers to cells' earlier in the stepwise procedure. Third, group analysis attempts to divide a manufacturing department into little `autonomous factories'. It is likely that these autonomous units will be more sensitive to changes in the environment. Therefore, some type of sensitivity analysis is needed in order to study the stability of the cells.

Group analysis is a simple technique helpful in the formation of cells. It is easy to explain the steps to managers and consultants. Suresh et al. (1995) describes a more advanced approach with several algorithms to support the steps of group analysis. It does not offer the advantage of transparency, but may lead to a better solution more easily. It is a challenge to make such an advanced approach applicable in industry.

REFERENCES

Burbidge, J.L., 1991, Production flow analysis for planning group technology, *Journal of Operations Management*, 10, 1, January.

Frensel, M., 1993, *De mogelijkheden voor groepsgewijze produktie in de mechanische afdeling van een bedrijf*, Afstudeerrapport Faculteit Bedrijfskunde, Rijksuniversiteit Groningen, (in Dutch).

McCormick, W.T., Schweitzer, P.J. and T.W.White, 1972, Problem decomposition and data reorganisation by a clustering technique, *Operations Research*, 20, 993-1009.

Posthumus, G., 1997, *Herontwerp productiestructuur*, Afstudeerrapport Faculteit Bedrijfskunde, Rijksuniversiteit Groningen, (in Dutch).

Rolefes, S., 1995, *Functioneel of groepsgewijs produceren - is er een gulden middenweg*, Afstudeerrapport Faculteit Bedrijfskunde, Rijksuniversiteit Groningen, (in Dutch).

Schurink, J., 1992, *Groepsgewijze produktie en bouwgroepen*, Afstudeerrapport Faculteit Werktuigbouwkunde, Universiteit Twente, (in Dutch).

Sjoberg, T. and J. Slomp, 1995, *The assignment of operators to clusters of machines and part types*, working paper, Faculteit Bedrijfskunde, Rijksuniversiteit Groningen.

Slomp, J., Molleman, H.B.M. and Gaalman, G.J.C., 1993, Production and operations management aspects of cellular manufacturing - A survey of users, in: *Advances in Production Management Systems (B-13)*, (Eds. I.A. Pappas and I.P. Tatsiopoulos), Elsevier Science Publishers B.V., 553-560.

Suresh, N.C., Slomp, J. and Kaparthi, S., 1995, The capacitated cell formation problem: A new hierarchical methodology, *International Journal of Production Research*, 33, 6, 1761-1784.

Wissink, R., 1990, *Vorming en besturing van produktie-units in de groepsgewijze onderdelenfabricage van Enraf-Nonius Delft B.V.*, Afstudeerrapport Faculteit Werktuigbouwkunde, Universiteit Twente, (in Dutch).

AUTHOR'S BIOGRAPHY

Dr. Jannes Slomp is Associate Professor at Faculty of Organization and Management at University of Groningen, Netherlands. He received his M.Sc. in Mechanical Engineering and Ph.D. in Industrial Engineering from the University of Twente, The Netherlands. His research interests are in cellular manufacturing, flexible manufacturing systems, and concurrent engineering. He is involved in several industrial projects concerning the implementation of flexible manufacturing systems and the transition of manufacturing departments from functional to cellular layout, and vice versa. He has published papers in many international journals, besides numerous articles in Dutch journals, and conference proceedings.

D3

Artificial Neural Networks and Fuzzy Models: New Tools for Part-Machine Grouping

V. Venugopal

1. INTRODUCTION

This chapter focuses on two, relatively new approaches for the part-machine grouping problem: artificial neural networks and fuzzy models. Neural networks and fuzzy models were introduced in group technology (GT) literature during 1990s to overcome some of the limitations of conventional approaches.

Fuzzy models are good at measuring and expressing the fuzziness in any system. Traditional approaches such as cluster analysis, mathematical programming and heuristics assume that a given part (machine) can be a member of only one part-family (machine-cell). Also, whenever new parts (machines) are introduced in the production system, traditional approaches require redoing the entire problem with the whole set of data. Neural networks can classify any new parts (machines) to existing groups without considering the entire data set again. This presents a major advantage of being able to handle large data sets. Neural networks also have the *ability to learn* complex patterns and to *generalize* the learned information *faster*. Also, they have the *ability to work with incomplete information* due to their parallelism. These characteristics have made both NN and fuzzy models as the most relevant tools for solving part-machine grouping problem.

This chapter provides a state-of-the-art review to synthesize the literature pertaining to neural network and fuzzy models for part-machine grouping to benefit both researchers and practitioners. The rest of this chapter is divided into five major sections. §2 gives an overview of neural network models; §3 examines GT literature and discusses application of neural network models to part-machine grouping problem. §4 and §5 do the same for fuzzy models. Finally, directions for future research and conclusions are provided in §6.

2. OVERVIEW OF ARTIFICIAL NEURAL NETWORKS

Artificial neural networks (ANNs) are distributed information processing systems that simulate the biological learning processes. It basically consists of different components such as processing unit (PU), connections, propagation rule, activation/transfer function and learning rule. The PUs are densely interconnected through directed links (connections). PUs take one or more input values, combine them into a single value using propagation rule, then transform them into an output value through activation/transfer function.

On the basis of types of connectivity, three types of neural networks can be constructed. A first type of neural network architecture is called hierarchical feedforward (Figure 1 a). The input layer and output layer are physically separated by the feed-forward and unidirectional connections in between. In the second type of architecture (Figure 1b), called recurrent neural networks, this physical separation between input and output does not exist, which means that the information of the input signal is reverberated through the network until stability occurs. The third type of architecture (Figure 1c), called hybrid neural networks combine the power of the above two architectures. This results in very powerful neural network paradigms, both in terms of information processing and plausibility.

A neural network learns from a set of training patterns. Through training, they generalize the features within the training patterns and stores these generalized features internally in its architecture. The learning takes place mainly through the readjustment of the weights. ANN can be classified as either supervised or unsupervised based on the learning type. Supervised learning requires the pairing of each input value with a target value representing the desired output and a "teacher" who provides error information. In unsupervised learning, the training set consists of input vectors only. The unsupervised learning procedures construct internal models that capture regularities in the input values without receiving additional information.

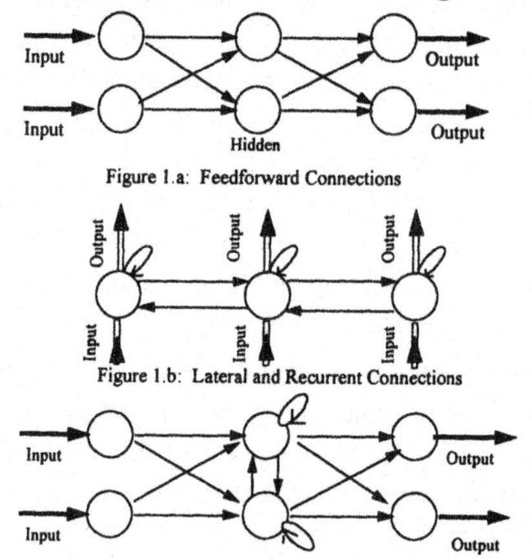

Figure 1.a: Feedforward Connections

Figure 1.b: Lateral and Recurrent Connections

FIGURE 1. a) Feedforward; b) Lateral and recurrent; c) Hybrid systems.

Different types of neural network models have been developed in the neural network literature. A taxonomy of neural network models that can be used as classifiers are given by Lippman (1987). For further details on neural networks, readers are referred to Wasserman (1989) and Kosko (1989). The next section gives an overview on the applications of different neural network models for part-machine grouping problem.

3. NEURAL NETWORK APPROACHES TO PART-MACHINE GROUPING

Use of neural networks for part-machine grouping is relatively new and started in late '80s and early '90s. Research in this area is still in the early stage. However, the few works that have been published so far can be classified based on three dimensions, *viz.*, type of input, the learning mode (supervised / unsupervised) and type of models used (Table 1). The taxonomy shown in Table 1 clearly indicates the focus of the existing literature, current trend and the gaps. Following sections provide a brief overview of the several research works as shown in the taxonomy.

3.1. Processing Requirements Of Parts As Input

This stream of research has used parts' processing requirements as an input to the neural network to find part-machine grouping. The details of some of the methods under this stream are given below.

3.1.1. Unsupervised Learning.

3.1.1.1. Competitive Learning (CL) Model

Competitive Learning model uses a network consisting of two layers - an input layer and an output layer (competitive layer). The competitive learning model minimizes the dissimilarity between the input (a vector from part-machine incidence matrix) and the group center represented by the weights of the network. During the training, the PUs in the bottom layer receive the input patterns and the output node computes the dissimilarity between its weight and the input pattern. The output node with the smallest dissimilarity is designated as the winner and its weight vector is adjusted towards the input pattern. This process is repeated for all input vectors. After training, the network can classify the inputs into groups.

Malave and Ramachandran (1991), Venugopal and Narendran (1992, 1994) and Chu (1993) simulated the *basic* CL model and reported good results on a few relatively smaller problems. Malakooti and Yang (1995) modified the CL algorithm by using the generalized Euclidean distance, and a momentum term in the weight vector updating equation to improve stability of the network.

The advantage of CL model is its simpler structure. The disadvantage is that it is very sensitive to learning rate. Instead of indicating the need for a new group for significantly different parts(machines), the model appears to force an assignment to one of the existing groups.

3.1.1.2. Interactive Activation and competition (IAC) Model

In the interactive activation and competition model, the processing units are organized into several pools (layers). Each pool represents specific characteristics of the problem. In each pool, all the units are mutually inhibitory. Between pools, units may have excitatory connections. The model assumes that these connections are bi-directional, so that whenever there is an excitatory connection from unit i to unit j, there is also an excitatory connection from unit *j* back to unit *i*. The steps involved in this procedure are summarized in McClelland and Rumelhart (1988).

TABLE 1. A Taxonomy of NN Methods For Part-Machine Grouping.

INPUT TYPE	LEARNING MODE: Unsupervised	LEARNING MODE: Supervised
Parts' Processing Requirements	**1. Competitive Learning / modification** • Malave and Ramachandran (1991) • Venugopal and Narendran (1992, 1994) • Chu (1993) • Malakooti and Yang (1995) **2. Interactive Activations and competition (IAC)** • Moon 1990 • Moon and Chi (1992) • Currie (1992) **3. Self-Organising Feature Map / modification** • Venugopal and Narendran (1992, 1994) • Lee et. al. (1992); Rao and Gu (1994) • Kiang et.al., (1995) • Kulkarni and Kiang (1995) **4. ART-1 / modification** • Kusiak and Chung (1991) • Dagli and Huggahalli (1991,1995) • Kaparthi and Suresh (1992,1993) • Venugopal and Narendran (1992,1994) • Dagli and Sen (1992); Liao and Chen (1993) • Rao and Gu (1995); Chen and Cheng (1995) **5. Fuzzy ART** • Burke and Kamal (1992,1995) • Karparthi and Suresh (1994) • Suresh and Kaparthi (1994); Suresh et.al., (1995) • Kamal and Burke (1996)	**1. Back Propagation** • Jamal (1993) **2. Stochastic Learning** • Arizono et.al., (1996)
Parts' Shape, images & features	**1. Self-Organising Feature Map / Modification** • Moon and Roy (1992) **2. ART-1 / modification** • Liao and Lee (1994)	**1. Back-Propagation** • Kamarthi and Kumara (1990) • Kao and Moon (1991) • Kaparthi and Suresh (1991) • Moon and Roy (1992) • Chakraborty and Roy (1993) • Chung and Kusiak (1994) • Kusiak and Lee (1996) **2. Hopfield Network** • Venugopal and Narendran (1992)

Moon (1990a, 1990b) and Moon and Chi (1992) investigated the performance of IAC by constructing a neural network consisting of three pools of neurons: one for parts, one for machines and one for instances (hidden layer). They used the entries in the machine similarity matrix as connection weights between units within the machine pool. Similarly, they use the entries in the part-similarity matrix as connection weights between units within the part pool. Since the connection weights are determined in advance and not changed, the *method does not use the idea of learning in neural network.* However, for taxonomy purpose, this chapter considers this work under this category.

Currie (1992) used the activation levels of IAC model as similarity measure of parts and then used bond-energy algorithm to reorganize the activation level matrix (similarity matrix) to find part families. The use and performance of IAC model for part-machine grouping were not clearly brought out in the literature.

3.1.1.3. Self-organizing Feature Map Model

Self-organizing feature map (SOFM) was proposed by Kohonen (1984). This model implements a clustering algorithm that is very similar to the classical K-means clustering algorithm (Lippman 1987). The SOFM network uses *two dimensional output layer (Kohonen layer).* Each output node has an associated topological neighborhood of other output nodes. During the training process, the weights for the winning output as well as the output nodes in the neighborhood of the winner are updated. The size of the neighborhood is decreased as training progresses until each neighborhood has only one output node.

Venugopal and Narendran (1992) simulated and compared SOFM model with that of competitive learning algorithm and ART-1. Only in some cases, the performance of SOFM model was comparable with those of ART-1 and CL. Kiang et al., (1995) and Kulkarni et al. (1995) also used self-organizing feature map to find part families. Unlike the above SOFM applications, they have used the operations sequenced based similarity coefficient as input to the network. The other related works in this area are due to Lee et al. (1992) and Rao and Gu (1994).

The advantage of SOFM model is that it gives the flexibility of deciding and fixing the number of groups. The disadvantage of this model is the time required to train the network.

3.1.1.4. Adaptive Resonance Theory Model (ART-1)

This model implements a clustering algorithm that is very similar to the simple sequential leader algorithm (Lippman 1987). As the inputs (binary) are presented to the network, the model selects the first input as the specimen for the first group. If the next input is similar to this specimen, within a specified vigilance threshold limit, then it is treated as the member of the first group. The weights connected to this group is also updated in the light of the new input vector. If the new input is not similar to the specimen, it becomes the specimen for a new group, associated with the second PU in the output layer. This process is repeated for all inputs. The vigilance parameter and the weight updating procedures distinguishes ART-1 from competitive learning and SOFM.

Dagli and Huggahalli (1991) analyzed the performance of ART-1 on smaller problems represented by binary part-machine matrix. They observed *category proliferation problem* in the fast learn mode. To overcome this problem, they suggested to present inputs in decreasing order of the number of "1"s and to update the weights in a different way. While, these improvement suggestions solved their problems, it was not clear how it alleviated the problem. In the subsequent work (Dagli and Sen (1992), they investigated the performance of the above mentioned improvement schemes on a relatively large size problem (1200x200 and 2000x200 incidence matrices). In a recent paper, Dagli and Huggahalli (1995) had suggested additional improvements. They suggested a procedure to determine near-optimal value of vigilance parameter. They had also proposed supplementary procedures to improve the solution obtained based on near-optimal vigilance parameter.

Venugopal and Narendran (1992, 1994) used the ART-1 model with the learning factor ρ = 0.95. They found number of groups depended upon threshold value and the number of groups growing with time.

Kaparthi and Suresh (1992) investigated the capabilities of ART-1 in handling large industry-size data sets. Their experiments on large data sets involving 10000 parts and 100 machines revealed that the model can identify the groups relatively fast with a good degree of perfection. In a subsequent work (Kaparthi and Suresh (1993), they have showed that performance of basic ART can be improved significantly by merely reversing the zeros and ones and making a corresponding change in similarity coefficient (step 5 of the basic ART-1 model). They also showed that reverse notation can resolve category proliferation problem.

Rao and Gu (1995) considered part-machine grouping problem with constraints. They embedded the constraints in the threshold functions (values) and connections of the network. Their procedure avoids constraint violation by introducing an inhibitory bottom-up connection which is large enough to more than nullify the other connection weights. Chen and Cheng (1995) indicated that the solution given by ART-1 depends on the initial disposition of the machine-part incidence matrix. They proposed supplementary procedures which can eliminate this weakness by providing smooth weights change in the ART-1 model.

The disadvantages of ART-1 are: a) The quality of solution depends on choice of vigilance parameters; b) The solution is dependent on the order in which input vectors are presented; c) The number of groups grows with time.

3.1.1.5. Fuzzy ART Model

Fuzzy ART model is an improvement over ART-1 as it incorporates fuzzy logic into ART-1. Burke and Kamal (1992 & 1995) demonstrated the use of Fuzzy ART and have shown that Fuzzy ART is superior to ART-1 and other algorithms. In addition, they concluded that the complement coding option (for normalizing the inputs) for fuzzy ART need not result in superior solutions, at least for the binary part-machine incidence matrices. Their study was confined to *small data sets.*

Suresh and Kaparthi (1994) and Kaparthi and Suresh (1994) investigated the performance of Fuzzy ART on large data sets with different degrees of imperfection and showed that performance of Fuzzy ART is better and more consistent. Subsequently, Suresh et al. (1995) incorporated this within a capacitated cell formation procedure to address some other practical objectives and constraints.

In a recent work, Kamal and Burke (1996) introduced a method, FACT, an improved version of Fuzzy ART. Their proposed method can be used to group parts-machines under a multiple objective environment.

The advantages of Fuzzy ART are: a) it is capable of rapid stable clustering of binary and non-binary input patterns; b) it can overcome the problem of category proliferation in fast learn/slow recode option. The disadvantage is that the quality of the solutions depends on the choice of the parameters as in other models.

3.1.2. Supervised Learning

3.1.2.1. Back-Propagation Model

In the back-propagation network structure, the output is obtained by propagating forward the input vector and is compared with the targeted one. The error is measured and transmitted back to the input layer. The weights of the interconnections are iteratively modified in order to reduce the measured error.

Jamal (1993) used back-propagation model for solving part-machine grouping problem. Their results indicated that the efficiency of the network will depend on the number and type of examples with which it was trained.

3.1.2.2. Stochastic Neural Network Model

Deterministic neural network models do not have the capability to escape from local optimal solution. Stochastic neural network models avoid to local optimal solution. The well-known stochastic neural network models are Boltzmann machine model and Gaussian machine model (Arizono et al. 1995). Arizono et al. (1995) formulated the part-tool grouping problem as a 0-1 integer programming problem. They used a stochastic NN, whose system dynamics are similar to Boltzman Machine, to solve the 0-1 linear integer program.

The disadvantage of stochastic neural networks is that specifying the values of various parameters and weights is more complicated than in the deterministic neural network model.

3.2. Part Size, Shape/Image And Design Features As Input

This stream of research has used parts' size, shape/images and design features as inputs to the neural network to find part-machine grouping. Under unsupervised category, Moon and Roy (1992) have used SOFM model and Liao and Lee (1994) have used ART-1 model. Under the category of supervised learning, several authors have applied back-propagation model (Kao and Moon 1991; Kaparthi and Suresh 1991; Moon and Roy 1992; Chakraborty and Roy 1993; Chung and Kusiak 1994; Kusiak and Lee 1996). For details on these, readers are referred to Moon (1998).

3.3. General Observations On Neural Network Models For Grouping

A careful analysis of the literature presented above (§3) reveals the following:

- A bulk of research papers considered unconstrained single objective part-machine grouping problem in the light of binary input and uses the binary input to the neural network
- Many of the applications (using parts' processing requirements as input) have focused on unsupervised learning models
- Many of the applications (using parts' size, shape/image and features as input) have focused on supervised learning models
- There is a growing research interests in applying advanced neural network models for constrained multi-objective part-machine grouping problem.
- Evolution of NN to part-machine grouping problem has seen four stages:

  Stage 1. Basic neural network models are directly applied to part-machine grouping problem. (e.g., Moon 1990; 1992; Kaparthi and Suresh 1991; 1992; Venugopal and Narendran 1992; 1994; Chu 1993).

  Stage 2. Suitable improvements are embedded within the basic neural network model to solve grouping problem. The improvements have taken place in terms of modifying the weight change dynamics and the process of presenting the input to the network (e.g., Dagli and Huggahalli 1992; Kaparthi and Suresh 1993)

  Stage 3. Supplemental procedures are developed to improve the solution obtained from the neural network model. (e.g., Chen and Cheng 1995; Suresh et al. 1995)

  Stage 4. Objectives/constraints of part-machine grouping problem are embedded within the neural network model. (e.g., Rao and Gu 1995).

4. OVERVIEW OF FUZZY MODELS

The term fuzzy model is used here to refer to fuzzy sets/mathematics, fuzzy data analysis such as fuzzy clustering and classifications. The foundation of any fuzzy model is fuzzy set. Fuzzy set as its name suggests, is basically a theory of classes with unsharp boundaries. In a conventional set concept (crisp set), element is either a member of the set or not. Fuzzy sets on the other hand, allow elements to be partially in a set. Each element is given a degree of membership in a set. This membership value can range from 0 (not an element of the set) to 1 (a member of the set). A membership function is the relationship between the values of an element and its degree of membership in a set. As mentioned before, fuzzy models are suitable for dealing with impreciseness that may exist in the parameters of any system.

Impreciseness does exist while grouping parts and machines. For example, if parts are to be described based on the attribute "length" as the set of short and long parts. Let $A = \{$ set of short parts $\}$. Since, length starts at 0 unit (say, centimeter), the lower range of this set is known clearly. The upper range is rather hard to define. Suppose if the upper range is set at 3 *cm*, then $A=[0,3]$. Now the problem is, why is some part whose length is 3 *cm* is considered to be short and the one whose length is 3.1 cm is considered to be long. Obviously, this is a structural problem. Even if the upper bound is moved to 4 cm, the problem still exists. This may lead to inaccurate part-machine grouping. A more natural way to construct the set *A* would be to relax the strict separation between short and long. To be more concrete, the set of small part can be defined by a membership function as defined in Figure 2.

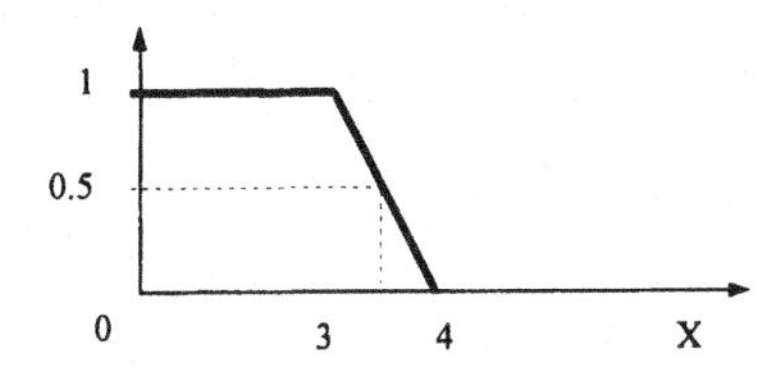

Figure 2

This way, a part of length 3.5 cm would still be short to a degree of 0.5. Mathematically, this membership function can be expressed as:

$$\mu_A(x) = \begin{cases} 0 \text{ if } x > 4 \\ (4.0 - x) \text{ if } 3 < x \leq 4 \\ 1 \text{ if } x \leq 3 \end{cases} \quad (1)$$

Likewise, fuzziness arises with respect to other attributes of the parts such as shape and routing. It is important to note here that fuzzy sets are not probability.

One of the difficulties in fuzzy modeling is choosing an appropriate membership value / function. For more details on possible membership functions forms and the associated theory, readers are referred to Zimmerman (1991).

5. FUZZY MODELS FOR PART-MACHINE GROUPING

The suitability of fuzzy model to GT problem was first demonstrated by Rajagopalan (1975) and Batra and Rajagopalan (1977). Since then, only a few researchers have investigated the applications of fuzzy models to part-machine grouping. The few works that have been published so far can be classified based on the dimensions, *viz.*, type of input used and the approaches. Fuzzy approaches can be distinguished further into two types. The first type is fuzzified versions of classical/conventional methods, such as fuzzy c-means clustering, fuzzy single linkage clustering and fuzzy mathematical programming. The second type is fuzzified versions of modern/intelligent approaches such as fuzzy neural networks and fuzzy expert systems (Table 2). The taxonomy shown in Table 2 clearly indicates the focus of the existing literature, current trend and the gaps.

One of the first issues in applying fuzzy models to part-machine grouping is to quantify the fuzziness through membership functions. Considering this fact, Ben-Arieh and Triantaphyllou (1992), Ben-Arieh et al. (1996) and Narayanaswamy et al. (1996) have proposed methods to quantify the fuzziness in part features and machining processes. Following sections provide a brief overview of the different fuzzy models for part-machine grouping as shown in the taxonomy (Table 2).

5.1. Processing Requirements Of Parts As Input

This stream of research has used fuzziness in parts' processing requirements as an input and developed the fuzzy models to find part-machine grouping. The details of some of the methods under these two groups are given below.

5.1.1. Fuzzified Version of Conventional Methods.

5.1.1.1. Fuzzy C-means Clustering

Fuzzy C-means clustering is the fuzzified version of conventional K-means (C-means) clustering method. The conventional methods implicitly assume that disjoint groups exist in the data set and therefore a part (machine) can belong to only one groups. The grouping results can thus be expressed as a binary matrix: $U = \left[u_{ij}\right]_{cxn}$.

Where c is the number of groups; n is the number of parts (machines);

- $$u_{ij} = \begin{cases} 1 & \text{the } j\text{th part (machine) belong to the } i\text{th group ;} \\ 0 & \text{the } j\text{th part (machine) does not belong to the group} \end{cases} \qquad (2)$$
- $$\sum_{i=1}^{c} u_{ij} = 1 \qquad \text{for } j = 1, 2, ☞, n \qquad (3)$$
- $$\sum_{j=1}^{n} u_{ij} > 0 \qquad \text{for } i = 1, 2, ☞, c \qquad (4)$$

In fuzzy clustering , the grouping results can be expressed as in matrix U, where $0 \le u_{ij} \le 1$ and still satisfying the constraints (3) and (4). The number of groupings satisfying these conditions is infinite. Fuzzy C-means aims to minimize the distance (Euclidean) between the data set and cluster center with specified degree of fuzziness.

Chu and Hayya (1991) generated non-binary incidence matrix (U) from a standard binary part-machine incidence matrix. They used divisive fuzzy C-means clustering algorithm (Bezdek 1981) for part-machine grouping. Their study observed that the solution depends on the parameters such as desired number of cluster, the degree of fuzziness.

Gindy et al. (1995) indicated that during the iterative process of group formation, new parts tended to gravitate towards the groups which already had the largest number of parts. Hence, they proposed an extended version of the fuzzy C-means clustering algorithm by incorporating a "harder" definition of cluster centroids. They also proposed a new validity measure based on cluster compactness and machine repetition factors. Other related works in this area are due to Wen et al.

(1996) and Leem and Chen (1996). Fuzzy C-means clustering shares the same advantages and limitations as that of conventional K-means (C-means) clustering.

TABLE 2. A Taxonomy of Fuzzy Models For Part-Machine Grouping.

INPUT TYPE	APPROACHES: Fuzzified version of conventional methods	APPROACHES: Fuzzified version of Intelligent methods
Parts' Processing Requirements	1. Fuzzy clustering • Fuzzy C-means Clustering • Chu and Hayya (1991) • Gindy et.al, (1995) • Wen and Smith (1996) • Leem and Chen (1996) • Fuzzy Single Linkage Clustering • Zhang and Whang (1992) 2. Fuzzy Heuristics • Fuzzy Rank Order Clustering • Zhang and Whang (1992)) 3. Fuzzy Mathematical Programming • Tsai et.al., (1997)	1. Fuzzy ART neural network • Burke and Kamal (1992,1995) • Karparthi and Suresh (1994) • Suresh and Kaparthi (1994); • Suresh et.al., (1995) • Kamal and Burke (1996)
Parts' Shape & features	1. Fuzzy clustering • Rajagopalan (1975) • Rajagopalan and Batra (1978) • Li et.al., (1988) • Xu and Whang (1989)	

5.1.1.2. Fuzzy Single Linkage Clustering

Fuzzy single linkage clustering is the fuzzified version of conventional single linkage clustering method. The fuzziness factor is to be incorporated in the similarity coefficient. The use of fuzzy single linkage clustering for part-machine grouping problem has been illustrated by Zhang and Wang (1992).

5.1.1.3. Fuzzy Rank Order Clustering

Zhang and Wang (1992) suggested and illustrated fuzzy rank order clustering to find part-machine groups. This method is nothing but a King's (1980) rank order clustering algorithm except that this method uses U (a non-binary matrix) as a primary data rather than the binary part-machine matrix.

5.1.1.4. Fuzzy Mathematical Programming

The fuzzification can be incorporated in conventional mathematical programming at the objective function level and/or constraints level (Kacprzyk and Orlovski 1987). Recently, Tsai (1997) formulate the part-machine grouping problem as fuzzy mixed integer programming (FMIP). They proposed a new fuzzy operator and examined the impact of different membership functions and operators in solving FMIP.

5.1.2. Fuzzified Version Of Modern Methods.

5.1.2.1. Fuzzy-Neural Methods

As discussed in §3, the cross fertilization of fuzzy logic and artificial neural nets has grown stronger and stronger since 1990s. §3.1.1.3 described the research works on the application of such a hybrid model (Fuzzy ART) to part-machine grouping. Fuzzy models' capabilities are not fully exploited in these works to capture the fuzziness in part features and machining processes. These research works have incorporated fuzzy logic only within the network and not at the input level. Hence, this is one of the promising areas for further research.

5.2. Part Size, Shape/Image And Features As Input

The other stream of research has used fuzziness in parts' features as inputs and applied the fuzzy models to find part-machine grouping. Rajagopalan (1975) and Batra and Rajagopalan (1977) introduced for the first time, fuzzy clustering in the GT literature. They used an iterative algorithm to form fuzzy part-families and to decide the membership grade of each component with reference to every family. Xu and Wang (1989) used the method of fuzzy equivalence to find suitable number of part families. They then used fuzzy classification to modify the results obtained from fuzzy equivalence analysis. They developed a system incorporating these two methods which allowed the user to modify the membership function and other parameters dynamically. Another related work in the area of application of fuzzy cluster analysis and pattern recognition is due to Li et al. (1988).

5.3. General Observations On Fuzzy Models For Grouping

The taxonomy shown in Table 2 and a careful analysis of the relevant literature presented above (§5) reveals the following:

- Only few works have been done in this area compared to neural network applications to part-machine grouping.
- There is a growing research interest on the aspect of quantifying the fuzziness in part features and machining processes. Also, there is a need for better understanding in choosing an appropriate membership function to measure the fuzziness of part features and machining processes.
- Many of the applications have focused on using fuzzified version of conventional methods, in particular fuzzy clustering.
- Investigation on the performance of fuzzified version of **other** conventional methods such as multi-objective mathematical programming problem still needs to be done.
- The existing works on the applications of fuzzy ART do not exploit the full capabilities of fuzzy models. Investigation on the performance of fuzzified versions of other modern approaches such as fuzzy knowledge-based approaches and fuzzy genetic approaches still needs to be done.

6. FUTURE DIRECTIONS AND CONCLUSIONS

The application of neural networks and fuzzy models to part-machine grouping problem have produced encouraging results. However, the powers and limitations of neural networks and fuzzy models to GT problem remain to be charted fully. For example, in spite of many papers reporting good performance of ANNs and fuzzy models, they do not explain adequately why or when such approaches are likely to perform better. As indicated separately in §3.3 and §5.3, enough opportunities exist for further research in both the areas. The promising area for future research lies in the form of applying hybrid models to part-machine grouping. Future research should aim to complement each other's weaknesses, thus creating more powerful approaches to solve practical GT/CM problems. The current stage of such an application of hybrid model (fuzzy ART) is at preliminary stages. Fuzzy model with genetic approach is another promising area, as genetic algorithm is particularly well suited for tuning membership functions and handling multi-objective problems.

REFERENCES

Arizono, I., Kato, M., Yamamoto, A. and Ohta, H., 1996, A new stochastic neural network model and its application to grouping parts and tools in flexible manufacturing systems, *International Journal of Production Research,* 33, 6, 1535-1548.

Batra, J.L. and Rajagopalan, R., 1977, A composite components through graphs and fuzzy clusters, *Proceedings of the 18th International MTDR Conference,* London.

Ben-Arieh, D. and Triantaphyllou, E., 1992, Quantifying data for group technology with weighted fuzzy features, *International Journal of Production Research,* 30, 6, 1285-1289.

Ben-Arieh, D., Lee, S.E. and Chang, P.T., 1996, Fuzzy part coding for group technology, *European Journal of Operational Research,* 92, 637-648.

Bezdek, J.C., 1981, *Pattern Recognition with fuzzy objective function and algorithms,* Plenum Press, New York.

Burke, L.I. and Kamal, S., 1992, Fuzzy ART and cellular manufacturing, *ANNIE 92 Conference Proceedings,* 779-784.

Burke, L.I. and Kamal, S., 1995, Neural networks and the part family/machine group formation problem in cellular manufacturing: A framework using fuzzy ART, *Journal of Manufacturing Systems,* 14, 3, 148-159.

Chakraborty, K. and Roy, U., 1993, Connectionist Models for part-family classifications, *Computers and Industrial Engineering,* 24, 2, 189-198.

Chen, S.-J. and Cheng, C.-S., 1995, A neural network-based cell formation algorithm in cellular manufacturing, *International Journal of Production Research,* 33, 2, 293-318.

Chu, C.H., 1993, Manufacturing cell formation by competitive learning, *International Journal of Production Research,* 31, 4, 829-843.

Chu, C.H. and Hayya, J.C., 1991, A fuzzy clustering approach to manufacturing cell formation, *International Journal of Production Research,* 29, 8, 1475 - 1487.

Chung, Y. and Kusiak, A., 1994, Grouping parts with a neural network, *Journal of Manufacturing Systems,* 13, 4, 262-275.

Currie, K.R., 1992, An intelligent grouping algorithm for cellular manufacturing, *Computers and Industrial Engineering,* 23, 1-4, 109-112.

Dagli, C. and Huggahalli, R., 1991, Neural network approach to group technology, *Knowledge-based Systems and Neural Networks,* Elsevier, New York, 213-228.

Dagli, C. and Sen, C.F., 1992, ART1 neural network approach to large scale group technology problems, in *Robotics and Manufacturing: Recent trends in research, education and applications,* 4, ASME Press, New York, 787-792.

Dagli, C. and Huggahalli, R., 1995, Machine-part family formation with the adaptive resonance theory paradigm, *International Journal of Production Research,* 33, 4, 893-913.

Gindy, N.N.Z., Ratchev, T.M. and Case, K., 1995, Component grouping for GT applications - a fuzzy clustering approach with validity measure, *International Journal of Production Research,* 33, 9, 2493-2509.

Jamal, A.M.M., 1993, Neural network and cellular manufacturing, *Industrial Management & Data Systems,* 93, 3, 21-25.

Kamal, S. and Burke, L.I., 1996, FACT: A new neural network based clustering algorithm for group technology, *International Journal of Production Research,* 34, 4, 919-946.

Kacprzyk, J. and Orlovsky, S.A., 1987, *Optimisation Models Using Fuzzy sets and Possibility Theory,* Dodrecht, Boston.

Kao, Y. and Moon, Y.B, 1991, A unified group technology implementation using the backpropagation learning rule of neural networks, *Computers and Industrial Engineering,* 20, 4, 425-437.

Kaparthi, S. and Suresh, N.C., 1991, A neural network system for shape-based classification and coding of rotational parts, *International Journal of Production Research,* 29, 9, 1771-1784.

Kaparthi, S. and Suresh, N.C., 1992, Machine-component cell formation in group technology: neural network approach, *International Journal of Production Research,* 30, 6, 1353-1368.

Kaparthi, S. and Suresh, N.C., 1993, An improved neural network leader algorithm for part-machine grouping in group technology, *European Journal of Operational Research,* 69, 3, 342-356.

Kaparthi, S. and Suresh, N.C., 1994, Performance of selected part-machine grouping techniques for data sets of wide ranging sizes and Imperfection, *Decision Sciences,* 25, 4, 515-539.

Kiang, M.Y., Kulkarni, U.R. and Tam, K.Y., 1995, Self-organizing map network as an interactive clustering tool - An application to group technology, *Decision Support Systems,* 15, 4, 351-374.

King, J.R., 1980, Machine-component grouping in production flow analysis: an approach using rank order clustering algorithm, *International Journal of Production Research,* 18, 2, 213-232.

Kohonen, T., 1984, *Self-organisation and associative memory,* Springer-Verlag, Berlin.

Kosko, B., 1992, *Neural Networks and Fuzzy Systems,* Prentice-Hall International Inc.

Kulkarni, U.R. and Kiang, M.Y., 1995, Dynamic grouping of parts in flexible manufacturing systems - A Self-organizing neural networks approach, *European Journal of Operational Research,* 84, 1, 192-212.

Kusiak, A. And Chung, Y., 1991, GT/ART: Using neural networks to form machine cells, *Manufacturing Review,* 4, 4, 293-301.

Kusiak, A. and Lee, H., 1996, Neural computing based design of components for cellular manufacturing, *International Journal of Production Research,* 34, 7, 1777-1790.

Lee, H., Malave, C.O. and Ramachandran, S., 1992, A self-organizing neural network approach for the design of cellular manufacturing systems, *Journal of Intelligent Manufacturing,* 3, 325-332.

Leem, C-W. and Chen, J.J., 1996, Fuzzy-set based machine-cell formation in cellular manufacturing, *Journal of Intelligent Manufacturing,* 7, 5, 355-364.

Li, J., Ding, Z., and Wei, W., 1988, Fuzzy cluster analysis and fuzzy pattern recognition methods for formation of part families (NAMRC), Society of Manufacturing Engineers, 558-563.

Liao, T.W. and Chen, J.L., 1993, An evaluation of ART1 neural models for GT part family and machine cell forming, *Journal of Manufacturing Systems,* 12, 4, 282-289.

Liao, T.W. and Lee, K.S., 1994, Integration of feature-based CAD system and an ART1 neural model for GT coding and part family forming, *Computers and Industrial Engineering,* 26, 1, 93-104.

Lippman, R.P., 1987, An introduction to computing with neural nets, *IEEE ASSP Magazine,* 4-22.

Malakooti, B. and Yang, Z., 1995, A variable-parameter unsupervised learning clustering neural network approach with application to machine-part group formation, *International Journal of Production Research,* 33, 9, 2395-2413.

Malave, C.O. and Ramachandran, S., 1991, A neural network-based design of cellular manufacturing system, *Journal of Intelligent Manufacturing,* 2, 305-314.

McClelland, J.L. and Rumelhart, D.E., 1988, *Explorations in parallel distributed processing: A Handbook of models, programs and exercises,* MIT Press, MA.

Moon, Y.B., 1990 a, An interactive activation and competition model for machine-part family formation in group technology, *Proceedings of International Joint Conference on Neural Networks,* Washington D.C., 2, 667-670.

Moon, Y.B., 1990 b, Forming part families for cellular manufacturing: A neural network approach, *International Journal of Advanced Manufacturing Technology,* 5, 278-291.

Moon, Y.B. and Roy, U., 1992, Learning group technology part families from solid models by parallel distributed processing, *International Journal of Advanced Manufacturing Technology,* 7, 109-118.

Moon, Y.B. and Chi, S.C., 1992, Generalised part family formation using neural network techniques, *Journal of Manufacturing Systems,* 11, 3, 149-159.

Moon, Y.B., 1998, Part family identification: New pattern recognition technologies, in *Group Technology and Cellular Manufacturing - A State-of- the-art synthesis of research & practice,* Kluwer Academic Publishers.

Narayanaswamy, P., Bector, C.R. and Rajamani, D., 1996, Fuzzy logic concepts applied to machine-component matrix formation in cellular manufacturing, *European Journal of Operational Research,* 93, 88-97.

Rajagopalan, R., 1975, *Design Retrieval and Group Layout Through Fuzzy sets and Graphs,* Unpublished Ph.D Thesis, Indian Institute of Technology, Kanpur, India.

Rao, H.A. and Gu, P., 1994, Expert self-organizing neural network for the design of cellular manufacturing systems, *Journal of Manufacturing Systems,* 13, 5, 346-358.

Rao, H.A. and Gu, P., 1995, A multi-constraint neural network for the pragmatic design of cellular manufacturing systems, *International Journal of Production Research,* 33, 4, 1049-1070.

Suresh, N.C. and Kaparthi, S., 1994, Performance of Fuzzy ART neural network for group technology cell formation, *International Journal of Production Research,* 32, 7, 1693-1713.

Suresh, N.C., Slomp, J. and Kaparthi, S., 1995, The capacitated cell formation problem: a new hierarchical methodology, *International Journal of Production Research,* 33, 6, 1761-1784

Tsai, C-C., Chu, C.H. and Barta, A.T., 1997, Modelling and analysis of manufacturing cell formation problem with fuzzy mixed-integer programming, *IIE Transactions,* 29, 7, 533-547.

Venugopal, V. and Narendran, T.T., 1992, A neural network approach for designing cellular manufacturing systems, *Advances in Modelling and Analysis,* 32, 2, 13-26.

Venugopal, V. and Narendran, T.T., 1992, Neural network model for design retrieval in manufacturing systems, *Computers in Industry,* 20, 11-23.

Venugopal, V. and Narendran, T.T., 1994, Machine-cell formation through neural network models, *International Journal of Production Research,* 32, 9, 2105-2116.

Wasserman, D., 1989, *Neural Computing - Theory and Practice,* Van Nostrand Reinhold.

Wen, H.J., Smith, C.-H. and Minor, E.D., 1996, Formation and dynamic routeing of part families among flexible manufacturing cells, *International Journal of Production Research,* 34, 8, 2229-2245.

Xu, H. and Wang, H., 1989, Part family formation for GT applications based on Fuzzy mathematics, *International Journal of Production Research,* 27, 9, 1637-1651.

Zhang, C. and Wang, H., 1992, Concurrent formation of part families and machine cells based on the fuzzy set theory, *Journal of Manufacturing Systems,* 11, 1, 61-67.

Zimmerman, H.J., 1991, *Fuzzy Set Theory,* Kluwer, Boston.

AUTHOR'S BIOGRAPHY

Dr. V.Venugopal is Associate Professor of Operations and Logistics Management at Center for Supply Chain Management, The Netherlands Business School, Nijenrode University, The Netherlands. Venugopal's work is primarily in the field of Operations Management. He has been teaching courses such as Operations Management, Logistics and Distribution Management and Total Quality Management in both graduate and executive MBA programs. His current research interests are in the area of Modeling Manufacturing and Supply Chain Management issues. He has published widely in international Journals such as *International Journal of Production Research, European Journal of Operational Research, Decision Support Systems, Computers and Industrial Engineering, Journal of Systems Management* and *Computers in Industry.*

D4

Cell Formation Using Genetic Algorithms

J.A. Joines, R.E. King and C.T. Culbreth

1. INTRODUCTION

With increasing computer speeds, researchers are increasingly applying artificial and computational intelligence techniques to the cellular manufacturing problem (i.e., neural networks, fuzzy reasoning, evolutionary techniques, etc.). Many of these methods are patterned after non-hierarchical clustering methods, mathematical programming techniques, array-based clustering methods, etc. (Joines, Culbreth and King 1996b). However, they offer several advantages over traditional cell formation methods. Fuzzy and neural systems are convenient and efficient means of modeling complex systems by simulating the approximate reasoning of humans and the learning capability of the brain (Joines et al. 1996b).

Genetic algorithms (GAs) and simulated annealing are efficient stochastic search algorithms that also emulate natural phenomena. They have been used successfully to solve a wide range of optimization problems, especially combinatorial problems. Because of the NP-completeness of the grouping problem and the existence of local minima, these algorithms (Lenstra 1974; Kumar, Kusiak and Vannelli 1986; Venugopal and Narendran 1992b) offer promise in solving large scale problems. Simulated annealing mimics the process of cooling a physical system slowly in order to reach a state of globally minimum potential energy. (Harhalakis, Proth and Xie 1990; Boctor 1991; Alfa, Chen and Heragu 1992; Venugopal and Narendran 1992b; Vakharia, Chang and Selim 1994; Shargal, Shekhar and Irani 1995). The stochastic nature of the algorithm allows it to escape local minimum, explore the state space, and find optimal or near-optimal solutions. Boctor (1991) and Venugopal and Narendran (1992b) used simulated annealing to solve integer programming formulations of the cell formation problem.

GAs, on the other hand, mimic Darwin's evolutionary process by implementing a "survival of the fittest" strategy. GAs solve linear and nonlinear problems by exploring all regions of the state space and exponentially exploiting promising areas through mutation, crossover, and selection operations. They have proven to be an effective and flexible optimization tool that can find optimal or near-optimal solutions. Unlike many other techniques, GAs do not make strong assumptions about the form of the objective function (Michalewicz 1996).

This chapter demonstrates the potential of using GAs for the manufacturing cell design problem owing to the algorithms efficiency and flexibility. The next section briefly reviews various GA approaches to the cell formation problem. §3 summarizes our GA cell design methodology and illustrates the efficiency and flexibility of the approach in solving different cell formation models. The next section (§4) develops a promising new area of research that utilizes local improvement procedures in combination with GAs to form an effective global optimization routine. Finally, we conclude with a summary of the results to date and plans for future work in this area.

2. LITERATURE REVIEW

Several researchers have applied GAs to the manufacturing cell design problem. These studies can be broken into two distinct categories: GAs that solve integer programming formulations directly and those that determine the best permutation of the machine or part indices. The methods that solve IP formulations utilize a set notation representation described in §3.2. Joines et al. (1996c, 1996b) developed a genetic algorithm approach to solve various integer programming formulations of the cell design problem, with the ability to include/exclude various objective functions and constraints at design time. This methodology can determine cell and family composition simultaneously. For a given set of literature data sets, the algorithm was able to find as good solutions as, if not better than, those in the literature. This methodology is described in detail in the next section.

Venugopal and Narendran (1992a) used a GA to solve a multi-objective integer programming formulation of the cell formation problem which minimizes the volume of intercell moves and the total within-cell load variation. Gupta et al. (1995) used a similar GA to minimize a weighted total number intercell and intracell movements.

GAs utilizing an order-based representation are essentially non-classic, array-based clustering techniques (i.e., they determine the best ordering of the machines(parts)). Joines (1993) used two order-based GAs to determine the best permutation of the parts and machines that would maximize the total bond energy. Daskin (1991) used the GA to determine a good starting permutation. Machines where assigned to groups based on the order of the machines in the permutation which minimized the combined capital cost of the machines in all groups and the material handling cost of moving parts between cells. Kazerooni et al. (1996) developed two new similarity coefficients for machines and parts that included production volume and process sequence. To maximize each of the similarities, two independent order-based GAs were used to determine the best sequence of machines and parts separately. Utilizing order-based GAs to solve the problem usually requires visual inspection of the part/machine incidence matrix to determine the cell/family composition which maybe very difficult in large, complicated data sets (Joines et al. 1996b). Billo et al. (1994; 1995) used an order-based GA to minimize the number of cells and maximize the similarity of parts. However, their algorithm modified the representation to include cell dividers (i.e., the sequence was broken up to determine cell compositions automatically). It was shown to outperform a hierarchical clustering algorithm. Most these methods either do not consider developing part families or determine the part families independently.

3. INTRODUCTION TO GENETIC ALGORITHMS

Commonly, the objective of cell design is to form a set of autonomous manufacturing units that eliminates any inter-cell movement of parts. To meet the need for a flexible, efficient, and effective clustering technique, a cell formation design methodology using genetic algorithms has been developed (Joines et al. 1996c). The formulation uses a unique representation scheme (set notation) for individuals (part/machine partitions) that reduces the size of the cell formation problem and increases the scale of problems that can be solved. This approach offers improved design flexibility by allowing a variety of evaluation functions to be employed and by incorporating design constraints during cell formation. Unlike most clustering algorithms, this methodology can identify all naturally occurring clusters as well as find solutions with a constrained number of clusters owing to management constraints. The effectiveness of the GA approach has been demonstrated on several problems from the literature and industry. This section explores the flexibility and effectiveness of this methodology.

3.1. Genetic Algorithm Fundamentals

GAs maintain and manipulate a family, or population of solutions, in the search for an optimal solution by simulating the evolutionary process. This is different from other optimization methods which maintain a single solution and improve it until an optimal solution is found. In general, the fittest individuals of any population tend to reproduce and pass their genes to the next generation, thus improving successive generations. However, some of the worst individuals do, by chance, survive and reproduce. A more complete discussion of GAs, including extensions to the general algorithm and related topics, can be found in books by Davis (1991), Goldberg (1989), Holland (1992), and Michalewicz (1996). The algorithm used in this research is shown in FIGURE 1.

Each solution or individual in the population is described by a vector of variables (chromosome representation). The first step is to initialize the population either randomly or by seeding. Once the initial population is generated, each individual, i, is evaluated using the objective function to determine its fitness or value, F_i. A subset of the population is selected to parent the next generation. An individual in the population can be selected to be a parent more than once. A probabilistic selection is performed such that the fittest individuals have an increased chance of being selected. These parents then undergo reproduction using genetic operators to produce a new population. To complete the new population, a subset (arbitrary or otherwise) of the old population is added to the new population. The GA moves from generation to generation until some specified stopping condition is met.

1. Set generation counter $i = 0$.
2. Create the initial population, Pop(i), by randomly generating N individuals.
3. Determine the fitness of each individual in the population by applying the objective function to the individual and recording the solution found.
4. Increment to the next generation, $i = i + 1$.
5. Create the new population, Pop(i), by selecting N individuals stochastically based on the fitness from the previous population, Pop(i-1).
   a) Randomly select R parents from the new population to form the new children by application of the genetic operators.
   b) Evaluate the fitness of the newly formed children by applying the objective function.
6. If $i <$ the maximum number of generations to be considered, go to Step 4.
7. Print out the best solution found.

FIGURE 1. A simple genetic algorithm.

3.2. Integer Program Formulation And GA Solution Technique

Mathematical programming approaches for the clustering problem are nonlinear or linear integer programming problems (Kusiak 1987; Boctor 1991). These formulations suffer from three critical limitations. First, because of the resulting nonlinear form of the objective function, most approaches do not concurrently group machines into cells and parts into families (Boctor 1991). Second, the number of machine cells must be specified *a priori*, which affects the grouping process and potentially obscures natural cell formations in the data. Third, since the variables are constrained to integer values, most of these models are computationally intractable for realistically sized problems (Joines 1993; Lee and Garcia-Diaz 1993). A common integer programming which ensures that each machine and part is assigned to only one cell and family, respectively:

$$x_{uk} = \begin{cases} 1, & \text{if machine } i \text{ is assigned to cell } l \\ 0, & \text{otherwise} \end{cases}$$

$$y_{jk} = \begin{cases} 1, & \text{if part } j \text{ is assigned to family } l \\ 0, & \text{otherwise} \end{cases}$$

$$\sum_{i=1}^{k} x_{il} = 1, \quad i = 1, \ldots, m$$

$$\sum_{j=1}^{k} x_{jl} = 1, \quad j = 1, \ldots, n$$

$k =$ number of clls(families) specified

$m =$ number of machines

$n =$ number of parts

As the number of parts and machines increases, the models become too large to be stored in memory or become computationally intractable (Lee and Garcia-Diaz 1993). To overcome these limitations, an integer programming model with the following set variable declarations has been proposed independently by Joines et al. (1993) and Venugopal and Narendran (1992a).

$$x_i = l \quad \text{if machine } i \text{ is assigned to cell } l$$

$$y_j = l \quad \text{if part } j \text{ is assigned to family } l$$

This model reduces the number of variables by a factor of k and eliminates the constraints by incorporating them into the set representation. Each part and machine variable is equal to the number of its assigned family or cell. For example, y_2=1 indicates that part 2 is assigned to part family 1 while x_3 = 2 indicates that machine 3 is assigned to machine cell 2. A standard integer programming solution technique cannot be employed without significant modification because of the objective function's inability to decode this variable representation. However, for the genetic algorithm, the objective function is a computer procedure that can easily decode the set representation and evaluate a solution. In building a genetic algorithm, six fundamental issues that affect the performance of the GA must be addressed: chromosome representation, initialization of the population, selection strategy, genetic operators, termination criteria, and evaluation measures. In the following subsections, these issues are introduced and described briefly for the cell formation genetic algorithm.

3.2.1. Chromosome Representation.

For any GA, a chromosome representation is needed to describe each individual in the population of interest. The representation scheme determines how the problem is structured in the GA and also determines the genetic operators that are used. Each individual or chromosome is made up of a sequence of genes from a certain alphabet. An alphabet could consist of binary digits (0 and 1), floating point numbers, integers, symbols (i.e., A, B, C, D), matrices, etc. In Holland's original design, the alphabet was limited to binary digits. Since then, problem representation has been the subject of much investigation. It has been shown that more natural representations are more efficient and produce better solutions(Michalewicz 1996). An integer representation is used where the first m variables represent the machines while the last n variables are associated with the parts (Joines et al. 1996c). Therefore, each individual is a vector of $m+n$ integer variables on the range of one to the maximum number of cells or families (k_{max}).

3.2.2. Initialization Of The Population.

For any GA, it is necessary to initialize the population. The most common method is to randomly generate solutions for the entire population. Since GAs iteratively improve existing solutions, the beginning population can be seeded with individuals from other cell formation algorithms or from an existing solution to the problem. The remainder of the population is then seeded with randomly generated solutions.

3.2.3. Selection Strategy.

The selection of parents to produce successive generations plays an extremely important role in the GA. The goal is to allow the "fittest" individuals to be selected more often. However, all individuals in the population have a chance of being selected to reproduce the next generation. There are several selection schemes, e.g., roulette wheel selection and its extensions, scaling techniques, ranking methods, elitist models, and tournament selection (Goldberg 1989;Michalewicz 1996).

Ranking methods only require the evaluation function to map the solutions to a totally ordered set, thus allowing for minimization and negativity. Ranking methods assign the probability of selecting individual i, P_i, based on the rank of solution i after all solutions are sorted. A normalized geometric ranking selection scheme employing an elitist model is used for the cell design problem described in this chapter. The elitist model injects the best individual from the previous population into the new population while normalized geometric ranking defines P_i to be $q'(1-q)^{r-1}$; where q is the probability of selecting the best individual, r is the rank of the individual with one being the best, N is the population size, and q' equals $q/(1-(1-q)^N)$.

3.2.4. Genetic Operators.

Genetic operators provide the basic search mechanism of the GA. Two types of operators (mutation and crossover) are used to create new solutions based on existing solutions in the population. Mutation operators tend to make small random changes in one parent to form one child in an attempt to explore all regions of the state space. Crossover operators combine information from two parents to form two offspring such that the children contain a "likeness" (a set of building blocks) from each parent. The application of these two basic types of operators and their derivatives depends on the chromosome representation used.

Seven float operators described by Michalewicz (1996) were modified to work with the integer representation: uniform mutation, multi-uniform-mutation, non-uniform mutation, multi-non-uniform mutation, boundary mutation, simple crossover, arithmetic crossover. Two problem-specific genetic operators (cell-swap crossover and cell-two-point crossover) based on the proposed cell formation representation were also developed and shown to enhance the GA's performance (Joines et al. 1996c). Let $X=(x_1, x_2,..., x_{m+n})$ and $Y=(y_1, y_2,..., y_{m+n})$ be two n-dimensional integer row vectors denoting individuals (parents) from the population. Each of these operators were used in the experiments of this chapter and are described briefly below. Let $a_i = 0$ and $b_i = k_{max}$ be the lower and upper bound, respectively, for each variable i in the cell formation problem.

Uniform Mutation: Randomly select one variable, j, and set it equal to a truncated uniform random number, $\lfloor U(a_i, b_i) \rfloor$ where $\lfloor x \rfloor$ is the largest integer less than or equal to x.

$$x_i' = \begin{cases} \lfloor U(a_i, b_i) \rfloor, & \text{if } i = j \\ x_i, & \text{otherwise} \end{cases}$$

Multi-Uniform Mutation: Apply equation () to all of the variables in the parent.

Non-Uniform Mutation: Randomly select one variable, j, and set it equal to an non-uniform random number based on equation (). The new variable is equal to the old variable plus or minus a random displacement.

$$x_i' = \begin{cases} \lceil x_{i,} + (b_{i,} - x_{i,})f(G) \rceil, & \text{if } r_1 < 0.5,\ i = j, \\ \lfloor x_{i,} - (x_{i,} - a_{i,})f(G) \rfloor, & \text{if } r_1 \geq 0.5,\ i = j, \\ x_{i,} & \text{otherwise,} \end{cases}$$

where $f(G)$ equals $(r_2(1 - \frac{G}{G_{max}}))^b$, r_1 and r_2 are uniform random numbers between (0,1), G is the current generation, G_{max} is the maximum number of generations, b is a shape parameter, and $\lceil x \rceil$ represents the smallest integer than or equal to x.

Multi-Non-Uniform Mutation: Apply equation () to all of the variables in the parent.

Boundary Mutation: Randomly select one variable, j, and set it equal to either its lower or upper bound, where $r = U(0,1)$.

$$x_i' = \begin{cases} a_i, & \text{if } r < 0.5,\ i = j, \\ b_i, & \text{if } r \geq 0.5,\ i = j, \\ x_{i,} & \text{otherwise,} \end{cases}$$

Simple Crossover: Generate a random number r from a discrete uniform distribution from 2 to $(m+n-1)$ and create two new individuals (X' and Y') according to equations () and ().

$$x_i' = \begin{cases} x_i, & \text{if } i < r \\ y_i, & \text{otherwise} \end{cases} \qquad y_i' = \begin{cases} y_i, & \text{if } i < r \\ x_i, & \text{otherwise} \end{cases}$$

Arithmetic Crossover: Produce two complimentary linear combinations of the parents, where $r = U(0,1)$.

$$X' = rX + (1-r)Y \qquad Y' = (1-r)X + rY$$

To achieve the necessary integer representation, the following is performed:

$$X' = (\langle ax_1 + by_1 \rangle, \langle ax_2 + by_2 \rangle \langle ax_3 + by_3 \rangle \langle ax_4 + by_4 \rangle \langle ax_5 + by_5 \rangle),$$

where

$$\langle ax_i + by_i \rangle = \begin{cases} \lceil ax_i + by_i \rceil, & \text{if } x_i > y_i, \\ \lfloor ax_i + by_i \rfloor, & \text{otherwise.} \end{cases}$$

Cell-Swap Crossover: The previous genetic operators work for any integer programming formulation. The next two crossover operators only work with the cell formation representation. Let X=($x_1, x_2, \ldots, x_m, y_1, y_2, \ldots, y_n$) and W=($w_1, w_2, \ldots, w_m, z_1, z_2, \ldots, z_n$) be two $m+n$-dimensional cell formation individuals (parents). It differs from the simple float crossover in that, instead of randomly selecting a cut point, the cut point is generated between machines and parts, effectively exchanging their part variables.

$$x_i' = \begin{cases} x_i, & \text{if } i < m \\ z_i, & \text{otherwise} \end{cases} \qquad w_i' = \begin{cases} y_i, & \text{if } i < m \\ w_i, & \text{otherwise} \end{cases}$$

Cell-Two-Point Crossover: The cell-two point crossover operator also works in a manner similar to the simple float crossover but rather than randomly selecting a single cut point, two cut points are generated. Generate two random number r_1 and r_2 from a discrete uniform distribution from 2 to (m-1) and (m+2) to (m+n-1), respectively and create two new individuals (X′ and Y′) according to equations () and ().

$$x_i' = \begin{cases} x_i, & \text{if } i < r_1 \\ w_i, & \text{if } r_1 \le i \le m \\ y_i, & \text{if } m < i < r_2 \\ z_i, & \text{otherwise} \end{cases} \qquad w_i' = \begin{cases} w_i, & \text{if } i < r_1 \\ x_i, & \text{if } r_1 \le i \le m \\ z_i, & \text{if } m < i < r_2 \\ y_i, & \text{otherwise} \end{cases}$$

3.2.5. Termination Criterion.

The GA moves from generation to generation selecting and reproducing parents until a termination criterion is met. Several stopping criteria can be used. For example, a specified maximum number of generations, convergence of the population to "virtually" a single solution, lack of improvement in the best solution over a specified number of generations, etc. can be used.

3.2.6. Evaluation Functions.

Evaluation functions of many forms can be used in a GA, subject to the minimal requirement that the function can map the population into a partially ordered set. As stated, the evaluation function is independent of the GA (i.e., stochastic decision rules). The "grouping efficacy" measure was used in the original model as the evaluation criterion to test the integer-based genetic algorithm on several data sets from the literature. Grouping efficacy was chosen as the initial evaluation measure because it has been used frequently in the literature and results were available for comparison (Joines et al. 1996c). It generates block diagonal cell formation and practically viable solutions (Kumar and Chandrasekahran 1990; Ng 1993). Grouping efficacy seeks to minimize the number of exceptional elements and the process variation (i.e., the number of voids (zeros) in the diagonal blocks). More specifically, grouping efficacy attempts to minimize the number of voids plus the number of exceptional elements divided by the total operational zone. The operational zone is defined by the number of operations (exceptional elements plus operations along the diagonal) plus the number of voids. Grouping efficacy has a value of one when there are no exceptional elements and no voids and a value of zero if the number of exceptional elements equals the total number of operations. Joines et al. (1996c) and Ng (1993) have shown that grouping efficacy is not necessarily a good measure from an algorithm point. Therefore, Ng developed a weighted grouping efficacy that was also used. Several other grouping measures have also been developed.

3.3. Extending The Original Model

In this chapter, the flexibility/extensibility of the GA approach with respect to its ability to incorporate new representations and other manufacturing information is explored.

3.3.1. Alternative Operations.

The ability to analyze the ordering of operations within routing sequences is important not just for material flow considerations, but also because cell throughput depends on setup times, which are, in turn, usually sequence dependent. Most cell formation algorithms use the binary part/machine matrix, A, consisting of elements a_{ij}=1, if part j requires processing on machine i, otherwise a_{ij}=0. This format leads to fixed routing sequences with no machine substitutions. Several researchers have proposed the following representation, where the operational sequence is embedded in the matrix (Nagi, Harhalakis and Proth 1990):

$$a_{ij} = \begin{cases} k, & \text{the } k\text{th operation of part } j \text{ is required on machine } i \\ 0, & \text{no processing on machine } i \end{cases}$$

Using this modified part/machine matrix, alternative operations and machine redundancy can be incorporated into the part/machine incidence matrix by specifying that several machines can perform the kth operation of a particular part. Because the evaluation function is independent of the decision rules, the GA provides the flexibility to interchange various objective functions without changing the algorithm. The evaluation function needs to take into account only the new part/machine incidence matrix representation. Specifying alternative operations has the advantage of allowing the algorithm to determine the most appropriate routing sequences in the context of minimizing intercell flows. Joines et al. (1997) demonstrated the usefulness of the GA utilizing this representation on several data sets including a real case study which is discussed in §5. This extended model has been used to develop cells for real, industry problems (Joines et al. 1997).

3.3.2. Alternative Fixed Routings.

A generalized cell formation problem that can consider complete fixed alternative routings can be formulated from the classic IP formulation (Joines et al. 1996c) by adding set of assignment variables, z_{jh}, and constraints to ensure that each part is assigned one and only one route where z_{jh} =1, if part j uses route h, otherwise z_{jh}=0.

The GA approach can be extended by including a new set of variable declarations that allow evaluation of fixed alternative routings. Specifically, let z_j = h, if part j uses route h. If there is only a single routing for part j, this variable is unnecessary and is not be included in the programming model. With this new variable definition, a new chromosome representation can be developed easily by adding the z variable component to individuals in the population.

Again, because of the flexibility of the GA, the algorithm remains unchanged and only the evaluation function and representation is modified to accommodate this new representation. In comparison to the alternative operations model in the previous section, this formulation offers two distinct advantages. First, it forms cells considering complete alternative routings in which preferred sets of machines can be specified and preserved through the clustering process. A second advantage of this model is the ability to assign priorities to certain process plans. This would allow the cell designer to generate cells that balance work loads, achieve minimum investment in tooling costs, take full advantage of manpower skills, etc.

3.3.3. Complete Alternative Routings With Machine Redundancy.

Even though the complete alternative routings offers the ability to specify a preferred set of machines or to assign priorities to certain routes, it can also be limiting. The operational sequences model offers advantages by allowing machine redundancy as well as determination of the true effect of intercell movement. These two models can be combined into one representation by allowing the complete alternative routings to contain their operation sequences as well.

3.3.4. Incorporating Demand And Flow.

The approach can be extended to include other manufacturing information like alternative operations, complete alternative routings, and functionally identical machinery. Most of the techniques in the literature assume that part demands are equal as does the genetic search methods described in the previous chapters. However, when the demands among parts are unequal, the machine cells and part families formed using an equal demand assumption may be very inefficient. Therefore, cells should be formed based on the part demands (i.e., machines in the cells should satisfy more demand of parts within the cell than outside). Gupta and Seifoddini (Gupta and Seifoddini 1990) developed a similarity coefficient that incorporates production requirements, the actual sequence of operations, the average production volume for each part, and the unit processing time for each of the part's operations. Recently, several researchers have included part demand in their formulations when calculating the cost associated with intercell movements (i.e., an intercell movement is weighted based on the parts-volume demand) (Heragu and Kakuturi 1994; Vakharia et al. 1994; Suresh, Slomp and Kaparthi 1995).

Therefore, the grouping efficacy measure (Γ) used in the previous GA models has been modified to include volume demand by weighting the exceptional elements by the part demand divided by the total demand. The new measure tries to minimize the demand satisfied outside the cells and the voids along the diagonal blocks. This measure is still unitless and reduces to the original Γ if the demands are all equal. Cells can now be formed based on the demands of all the parts. The measure has been tested on several real data sets and shown to be an effective cell design tool (Joines 1996a).

All the previous models utilized a grouping measure, we have also included cost issues into the model (Joines 1996a). The original model that employed only the part/machine incidence matrix assumed there was no precedence among operations while the extended models assumed complete precedence. However, in many instances, parts may require precedence in only the first and/or the last few operations while he operations in the middle nay be performed in any order (i.e., no precedence). The algorithm has easily been modified to include precedence graphs to handle these situations.

4. HYBRIDIZING GAs WITH LOCAL IMPROVEMENT PROCEDURES

The previous section demonstrated the flexibility of GA to incorporate additional information while this section will explore the promising new area of hybrid-GAs (i.e., the combination GAs and local improvement procedures). GAs are efficient at exploring the entire search space; however, they are relatively poor at finding the precise local optimal solution in the region at which the algorithm converges. Local improvement procedures, e.g., two-opt switching for combinatorial problems and gradient descent for unconstrained nonlinear problems, quickly find the local optimum of a small region of the search space, but are typically poor global searchers. Because these procedures do not guarantee optimality, in practice, several random starting points are generated and used as input into the local search technique and the best solution is recorded. This global optimization technique (multi-start) has been used extensively but is a blind search technique since it does not take account past information (Houck, Joines and Kay 1996). Genetic algorithms, unlike multi-start, utilize past information in the search process. Therefore, local improvement procedures (LIPs) have been incorporated into GAs in order to improve their performance through what could be termed "learning." Such hybrid-GAs have been used successfully to solve a wide variety of problems (Davis 1991; Chu and Beasley 1995; Renders and Flasse 1996; Houck, Joines and Kay 1997). Houck, Joines, and Kay (1996) showed that for the continuous location-allocation problem, a GA that incorporated a LIP outperformed multi-start and a two-way switching procedure, where both methods utilized the same local improvement procedure as the hybrid-GA.

4.1. Utilizing Baldwin Effect And Lamarckian Evolution

Local improvement procedures have been incorporated into GAs as the evaluation function in order to improve the algorithm's performance through learning. (Recall, the GA makes no assumptions on the form of the objective; only that it map the individuals in the population into a totally ordered set.) Incorporating a LIP as an evaluation function gives rise to the concepts of the Baldwin Effect and Lamarckian evolution.

4.1.1. Baldwin Effect.

The Baldwin Effect allows an individual's fitness (phenotype) to be determined based on learning, i.e., the application of local improvement. Like natural evolution, the result of the improvement does not change the genetic structure (genotype) of the individual, it just increases the individual's chances of survival. Whitley et al. (1994) demonstrated that using a LIP can, in effect, change the landscape of the fitness function into flat landscapes around the local basins showed that utilizing either form of learning is more effective than the standard GA approach without the LIP. Whitley et al. (1994) argued that, while Lamarckian learning is faster, it may be susceptible to premature convergence to a local optimum as compared to Baldwinian learning.

4.1.2. Lamarckian Evolution.

Lamarckian evolution, in addition to using learning to determine an individual's fitness, changes the genetic structure of an individual to reflect the result of the learning. This results in the inheritance of acquired or learned characteristics that are well-adapted to the environment. The improved individual is placed back into the population and allowed to compete for reproductive opportunities. Lamarckian evolution inhibits the schema processing capabilities of genetic algorithms and may led to premature convergence (Gruau and Whitley 1993; Whitley et al. 1994).

However, Baldwinian learning certainly aggravates the problem of multiple genotype to phenotype mappings. A genetic algorithm works on both genotypes and phenotypes. A genotype refers to the composition of the values in the chromosome or individual in the population, whereas a phenotype refers to the solution that is constructed from a chromosome. In a direct mapping, there is no distinction between genotypes and phenotypes, but, for some problems, a direct mapping is not possible or desired (Nakano 1991). However, it has been noted that having multiple genotypes map to the same phenotype may confound the GA (Hinton and Nolan 1987;Nakano 1991). This problem occurs when an LIP is used in conjunction with a GA. Consider the example of maximizing $\sin(x)$. Suppose a simple gradient-based LIP is used to determine the fitness of a chromosome. Then any genotype between $[-\pi/2, 3\pi/2]$ will have the same phenotype value of 1 at $\pi/2$.

4.1.3. Partial Lamarckian.

Hybrid genetic algorithms need not be restricted to operating in either a pure Baldwinian or pure Lamarckian manner. Instead, a mix of both strategies, or what is termed "partial Lamarckianism"(Houck et al. 1996) could be employed. For example, a possible strategy is to update the genotype to reflect the resulting phenotype in 50% of the individuals. While this 50% partial Lamarckian strategy has no justification in natural evolution but, for simulated evolution, it provides for an additional search strategy. This section will examine the effectiveness of partial Lamarckianism with regard to the cell formation problem.

4.2. Grouping Efficacy Of Local Improvement Procedure

To investigate the capabilities of the hybrid-GA approach for the cell formation problem, several local improvement procedures have been developed for the grouping efficacy measure (Γ) (Joines 1996a). The LIP in Figure 2**Error! Reference source not found.** has been developed for the alternative operations or machine redundancy model of §3.3.1. and uses the following notation.

N_l, M_l = set of parts (machines) assigned to part family (machine cell) l

p_{il} = number of parts needing processing by machine i in cell l

q_{jil} = number of machines that operate on part j in family l

c, c^* = index of the current (best) cell / family assignment for a machine / part

α_{il}, α_{jl} = decrease in the number of exceptional elements if a switch from c to l occurs

β_{il}, β_{jl} = decrease in the number of voids if a switch from c to l occurs

The improvement procedure is a single switching algorithm which utilizes switching rules defined by Ng (1993). Algorithm in Figure 2 is a steepest descent algorithm since, in Steps 2(b) and 3(b), the switch which yields the largest increase is used. Given an initial machine cell and part family assignment, the first step of the algorithm is to determine the grouping efficacy measure. From this calculation, the size and composition of each machine cell M_l and part family P_l is determined. Steps 2 and 3 perform the actual machine and part switching for all machines and parts, respectively. For the alternative operations or machine redundancy model, no new variables were introduced. The part/machine incidence matrix now contains the sequence of operations rather than just indicating that a part needs processing by a machine. Therefore, the switching rules developed by Ng remain the same. The only exception is the way p_{il} and q_{jl} are calculated. When determining the number of parts p_{il} that need processing by a particular machine i in cell l in Step 2(a), part operations that can be performed by another machine in the same cell l are not included in the count. When determining the number of machines, q_{jl}, that process part j in cell l in Step 3(a), only the unique number of operations that can be performed in the cell l are used (i.e., two machines performing the same operation in the cell are only counted once).

5. ALTERNATIVE OPERATION: LOCAL IMPROVEMENT EXPERIMENTS

To investigate the trade-off of disrupted schema processing or premature convergence in Lamarckian learning and of multiple genotype mapping to the same phenotype in Baldwinian learning, a series of experiments using LIPs as evaluation functions were performed on several different cell formation test problem instances. For each test problem instance, the GA was run with varying levels of Lamarckianism from 0% (pure Baldwinian) to 100% (pure Lamarckian) to determine if there was a trend between the two extremes, or if combinations of Baldwinian learning and Lamarckian learning were beneficial. In these experiments, individuals were updated to match the resulting phenotype with a probability of 0, 5, 10, 20, 30, 40, 50, 60, 80, 90, 95, 100%. Also, the GA was run using no local improvement to determine if a pure genetic approach was better in terms of computational efficiency and/or quality of solution.

Each run of the GA was replicated 20 times, with common random seeds. The GA was terminated when either the optimal or best known solution was found or after one million function evaluations were performed. Using function evaluations allows a direct comparison between the hybrid GA methods and the pure GA. Six different problem instances were used during the experiments. For each of these instances, three different values of k_{max} were used to determine if varying values influenced any of the methods in terms of convergence to the optimal. The first three data sets are taken from Heragu and Kakuturi (1994), a 12 machine by 19 part problem (Heragu2), a 20 machine by 20 part problem (Heragu3), and a 27 machine by 40 part problem (Heragu4). The fourth data set is the 20 machine by 20 part Nagi problem (Nagi et al. 1990). The last two instances are real industry data sets used to test the ability of these methods to solve realistically sized problems.

1. For a given machine cell and part family assignment, compute $\Gamma=(e-e_O)/e+e_V)$ which determines P_l and M_l for each cell l.
2. For each machine i from 1 to m, do,
   3. Determine the number parts needing process in by machine l for each family pil, excluding those parts whose operation can be performed by other machines in the same cell.
   4. Determine the best cell switch for machine i:

   $$c^* \leftarrow \arg\max_l \{\beta_{il}\Gamma - \alpha_{il} | c \neq l\},$$

   where $c \leftarrow x_i$, $\alpha_{il} \leftarrow p_{ic} - p_{il}$, and $\beta_{il} \cdot |P_c| - |P_l|$.
   (c) If $\beta_{ic^*}\Gamma - \alpha_{ic^*} > 0$ (i.e., switching machine i to cell c^* increases Γ) then $M_{c^*} \leftarrow M_{c^*} + \{i\}$,. $M_c \leftarrow M_c - \{i\}$, $x_i \leftarrow c^*$ and $\Gamma=(e-e_O-\alpha_{ic^*})/e+e_V-\beta_{ic^*})$
3. For each part j from 1 to n, do,
   4. Determine the number unique operations that can be performed in each cell l for part j, qjl.
   5. Determine the best cell switch for part j:

   $$c^* \leftarrow \arg\max_l \{\beta_{il}\Gamma - \alpha_{il} | c \neq l\},$$

   where $c \leftarrow y_j$, $\alpha_{jl} \leftarrow q_{jc} - q_{jl}$, and $\beta_{il} \cdot \leftarrow |M_c| - |M_l| - \alpha_{jl}$.

FIGURE 2. Grouping efficacy based LIP.

The first industry data set is a 22 machine by 148 part problem (Real1) from a division of a large international company. This particular division manufactures and machines rotary unions. Due to product changes over the past few years, the current layout had become very inefficient in terms of material flow (i.e., distance traveled by parts) and space utilization. In addition, any new machines added over the past few years were placed in empty space with out regard to material flow. Management wanted an improved layout to be designed. In particular, management was interested in determining if changing the machinery area to a cellular layout would be beneficial in reducing material flow and work in progress. Since attempts to cluster the data manually were ineffective, the GA approach was used to cluster the data. In addition, there were several functional identical machines which could be exploited using the alternative operations model. The objective was to identify the machine cells and the various parts assigned to those cells. Based upon the clusters found, a layout was to be generated and evaluated.

The last data set, a 115 machine by 2557 part problem (Real2), is taken from a high-end furniture case-goods plant in Hickory, North Carolina. Because of the results achieved from one small trial cell, management is enthusiastic in developing more cells. Typically, furniture companies build to inventories (i.e., they have very large inventories of finished goods with little turnover). Also, it is not uncommon to find lead times of eight weeks or greater where they assume two days or more per operation of a lot in the furniture industry. Immediately, it is quite easy to determine why their shop floors are full of work in progress (WIP). Usually, it is often difficult see across the shop floor and some times even to the next machine because of the amount of WIP in most furniture plants. Therefore, in an attempt to reduce their lot sizes, lead times and WIP, many furniture companies are looking for new and innovative approaches like cellular manufacturing. To confound the issue, these data sets tend to be rather large (e.g., greater than 2000 parts and over 100 machines) with very complex routings (i.e., 15 to 25 operations per part). Also, most furniture companies have a range of machine technology from manual machines to the latest CNC technology. Owing to capacity issues, there are often multiple functional identical machines for most of the machine types which needs to be included in the various models. The size and complexity of these data sets would exclude many cell formation techniques. Therefore, the evolutionary approach employing alternative operations offers several distinct advantages.

6.1. Alternative Operation Results

For each of the instances and k_{max} values, both the mean number of function evaluations and the final function value were examined using analysis of variance (ANOVA). Since the ANOVA shows a significant effect at an $\alpha > 0.0001$ for all six problem sets in terms of function evaluations and final function value except where indicated on the summary Tables 1 and 2, the means of each of the 12 different search strategies considered (no local improvement, referred to as N; pure Baldwinian, referred to as 0, pure Lamarckian, 100; and nine levels of partial Lamarckian, referred to as 5, 10, 20, 40, 50, 60, 80, 90, and 95, respectively) were compared using the Student-Newman-Keuls (SNK) multiple means comparison tests.

The main results of the multiple means comparison are concisely presented in Tables 1 and 2 which show a ranking of each search strategy. This ranking was constructed with respect to the fitness of the final result and number of function evaluations required. The rankings were determined optimistically as follows: for each strategy, the group was found which yielded the best results for each test problem instance, as determined by the SNK means analysis, i.e., if the strategy was placed into a single group, that group's rank was used; however, if the method was placed into several overlapping groups, the method was placed into the group with the best rank. Therefore, these rankings represent how many groups of means are significantly better for each test problem.

Table 1 shows the results of these rankings for the final fitness of the solutions returned by each of the search strategies, where 1 represents the best rank and is shaded. All of the strategies employing at least 40% Lamarckian learning were better in terms of final fitness value. For the real data sets, pure Baldwinian (0) produced statistically the worst solutions. Since the real data sets tend to be big, a large number of the function evaluations are used to initialize the problem. Therefore, Baldwinian learning does not have a lot time to overcome the problem of a one-to-one genotype to phenotype mapping.

Table 2 shows the results of the rankings with respect to number of function evaluations required to locate the final functional value returned by the search strategy. This table demonstrates that for the literature data sets any level of Lamarckian learning consistently yielded the quickest convergence, with a couple of exceptions as noted. This is due in part to the fact that the literature data sets were extremely easy when employing the LIP. However, for the real data sets, a level of updating of greater than 50% is required.

7. CONCLUSIONS

The models developed in this chapter illustrate the flexibility of the genetic algorithm to incorporate various evaluation functions, representations, etc. with out effecting the underlying stochastic decision rules. This methodology has been used to from cells for real data sets. For example, the four cell design generated by the GA for the first real data set will be implemented in 1997 and has a projected 38.5% reduction in material flow per year.

The techniques utilizing genetic algorithms (GAs) and LIPs offer a very effective stochastic global optimization method for solving variants of the cell formation problem, as well as, other types of problems. In the empirical investigation conducted a general trend was observed: increasing use of Lamarckian learning led to the quicker convergence of the genetic algorithm to the best known solution than both Baldwinian learning and no LIP. By forcing the genotype to reflect the phenotype, the GA converges more quickly and to better solutions than by leaving the chromosome unchanged after evaluating it. This may seem counterintuitive since forcing the genotype to be equal to the phenotype might have forced the GA to converge prematurely to a local optimal. For the cell formation problems, a higher percentage of Lamarckian learning gives the best mix of computational efficiency and solution quality. When employing the LIP, the problems from the literature turned out to be very easy and a multi-start procedure utilizing the same LIP would be as effective. However, for the real data sets, this would not be case since GA takes a number of generations to reach the best known solution. With the improved efficiency, the GA approach is now capable of solving more complicated versions of the cell formation problem without a significant increase in computational cost.

TABLE 1. Summary of Fitness Value (SNK).

Problem Instance	*N*	*0*	*5*	*10*	*20*	*40*	*50*	*60*	*80*	*90*	*95*	*100*
Herag2-2	1	1	1	1	1	1	1	1	1	1	1	1
Herag2-3	2	1	1	1	1	1	1	1	1	1	1	1
Herag2-4	4	1	1	1	1	1	1	1	1	1	1	1
Herag3-4	2	1	1	1	1	1	1	1	1	1	1	1
Herag3-5	1	1	1	1	1	1	1	1	1	1	1	1
Herag3-6	3	1	1	1	1	1	1	1	1	1	1	1
Herag4-4	3	1	1	1	1	1	1	1	1	1	1	1
Herag4-5	2	1	1	1	1	1	1	1	1	1	1	1
Herag4-6	2	1	1	1	1	1	1	1	1	1	1	1
Nagi-4	2	1	1	1	1	1	1	1	1	1	1	1
Nagi-5	1	1	1	1	1	1	1	1	1	1	1	1
Nagi-6	2	1	1	1	1	1	1	1	1	1	1	1
Real1-4	2	3	1	1	1	1	1	1	1	1	1	1
Real1-5	2	3	1	1	1	1	1	1	1	1	1	1
Real1-6	2	3	1	1	1	1	1	1	1	1	1	1
Real2-6	4	5	3	2	1	1	1	1	1	1	1	1
Real2-7	2	3	3	1	4	1	1	1	1	1	1	1
Real2-8	3	4	2	1	1	1	1	1	1	1	1	1

†-Anova did not show a significant effect at $\alpha = 0.01$

TABLE 2. Summary of Convergence (SNK).

Problem Instance	*N*	*0*	*5*	*10*	*20*	*40*	*50*	*60*	*80*	*90*	*95*	*100*
Herag2-2	1	1	1	1	1	1	1	1	1	1	1	1
Herag2-3	2	1	1	1	1	1	1	1	1	1	1	1
Herag2-4	3	1	2	1	1	1	1	1	1	1	1	1
Herag3-4	2	1	1	1	1	1	1	1	1	1	1	1
Herag3-5	1	1	1	1	1	1	1	1	1	1	1	1
Herag3-6	2	1	1	1	1	1	1	1	1	1	1	1
Herag4-4	2	1	1	1	1	1	1	1	1	1	1	1
Herag4-5	2	1	1	1	1	1	1	1	1	1	1	1
Herag4-6	3	2	1	1	1	1	1	1	1	1	1	1
Nagi-4	2	1	1	1	1	1	1	1	1	1	1	1
Nagi-5	2	1	1	1	1	1	1	1	1	1	1	1
Nagi-6	2	1	1	1	1	1	1	1	1	1	1	1
Real1-4	4	4	3	2	1	1	1	1	1	1	1	1
Real1-5	2	2	2	1	1	1	1	1	1	1	1	1
Real1-6	2	2	2	2	2	1	1	1	1	1	1	1
Real2-6	3	3	3	2	2	2	1	1	1	1	1	1
Real2-7	3	3	3	2	3	2	2	1	1	1	1	1
Real2-8	4	4	3	3	3	2	1	1	1	1	1	1

†-Anova did not show a significant effect at $\alpha = 0.01$

The cells for the furniture data set are currently being evaluated by management. Owing to the complexity of the number of operations (15-25 operations), a hybrid cell approach was found to be more effective where the first few operations are done in one cell, the next few operations are done in another, etc. Future work will include looking for a better hybrid cell approach to this problem. In the furniture industry, there is generally a mixture of both precedent and non-precedent operations. Generally, only the first and last few operations require a strict precedent of operations. To solve this problem, precedent graphs need to be included into the models which can be utilized by the GA cell design methodology developed in this chapter (only the evaluation function needs to change).

The first phase of the research has been to develop a very flexible and effective tool that encompasses readily accessible data. For example, the part routing, demand information, and types and quantities of machines are generally available from an MRP system. However, to include more factors like machine duplication, operator control, labor rates, etc., a true cost model needs to be developed including the cost of obtaining this additional information. For example, part routings in the furniture industry have been developed to optimize the current functional layout. Complete alternative routes or even alternative operations are not readily accessible. Only in the event of a problem with the primary route (i.e., machine capacity or breakdown, etc.) is a secondary route even developed. Researchers should be concerned with not only developing effective cell formation techniques given perfect information but also concerned about the information that is necessary to develop good cells.

REFERENCES

Alfa, A., Chen, M., and Heragu, S., 1992, Integrating the grouping and layout problems in cellular manufacturing systems, *Computers & Industrial Engineering* 23, 1-4, 55-58.

Billo, R., Tate, D., and Bidanda, B., 1994, Comparison of a genetic algorithm and cluster analysis for the cell formation problem: A case study. *3rd Industrial Engineering Research Conference*, Atlanta, GA, 538-548.

Billo, R., Tate, D., and Bidanda, B., 1995, A genetic cluster algorithm for the machine component grouping problem, working paper, University of Pittsburgh.

Boctor, F., 1991, A linear formulation of the machine-part cell formation problem, *International Journal of Production Research,* 29, 2, 343-356.

Chu, P. C. and Beasley, J. E., 1995, A genetic algorithm for the generalised assignment problem, The Management School, Imperial College, London.

Daskin, M. S., 1991, An overview of recent research on assigning products to groups for group technology production problems., Northwestern University, Chicago.

Davis, L., 1991, *The Handbook of Genetic Algorithms,* Van Nostrand Reingold, New York.

Goldberg, D. E., 1989, *Genetic Algorithms in Search, Optimization, and Machine Learning,,* Addison-Wesley, NJ.

Gruau, F. and Whitley, D., 1993, Adding learning to the cellular development of neural networks, *Evolutionary Computation* 1, 4, 213-233.

Gupta, T. and Seifoddini, H., 1990, Production data based similarity coefficient for machine-component decisions in the design of a cellular manufacturing system, *International Journal of Production Research,* 28, 7, 1247-1269.

Gupta, Y., Gupta, M., Kumar, A., and Sundram, C., 1995, Minimizing total intercell and intracell moves in cellular manufacturing: A genetic algorithm approach, *International Journal of Computer Integrated Manufacturing*, 8, 2, 92-101.

Harhalakis, G., Proth, J. M. and Xie, X. L., 1990, Manufacturing cell design using simulated annealing: an industrial application, *Journal of Intelligent Manufacturing* 1, 1, 18.

Heragu, S. S. and Kakuturi, S., 1994, Grouping and placement of machine cells, working paper, Rensselaer Polytechnic Institute, New York.

Hinton, G. E. and Nolan, S. J., 1987, How learning can guide evolution, *Complex Systems* 1, 495-502.

Houck, C., Joines, J., and Kay, M., 1996, Comparison of genetic algorithms, random restart, and two-opt switching for solving large location-allocation problems, *Computers and Operations Research,* 23, 6, 587-596.

Houck, C., Joines, J., and Kay, M., 1997, Empirical Investigation of Partial Lamarckianism, *Evolutionary Computation* 5, 1.

Joines, J. A., 1993, Manufacturing cell design using genetic algorithms, MS thesis, North Carolina State University.

Joines, J. A., 1996a, Hybrid Genetic Search for Manufacturing Cell Design, Ph.D. thesis, North Carolina State University.

Joines, J. A., Culbreth, C. T., and King, R. E., 1996b, A comprehensive review of production oriented cell formation techniques, *Internation Journal of Factory Automation and Information Management*, 3, 3&4, 225-265.

Joines, J. A., Culbreth, C. T., and King, R. E., 1996c, Manufacturing cell design: An integer programming model employing genetic algorithms, *IIE Transactions* 28, 1, 69-85.

Joines, J. A., King, R., and Culbreth, C., 1997, Moving beyond the parts incidence matrix:, *Computers & Industrial Engineering,* in 2nd review, .

Kazerooni, M., Luong, L., and Abhary, K., 1996, Cell formation using genetic algorithms, *International Journal of Factory Automation and Information Management*, 3, 3&4, 283-301.

Kumar, K. R. and Chandrasekahran, M. P., 1990, Grouping efficacy: a quantitative criterion for goodness of block diagonal forms of binary matrices in group technology, *International Journal of Production Research,* 28, 2, 233-243.

Kumar, K. R., Kusiak, A., and Vannelli, A., 1986, Grouping of parts and components in flexible manufacturing systems, *European Journal of Operational Research,* 24, 387-397.

Kusiak, A., 1987, The generalized group technology concept, *International Journal of Production Research,* 25, 4, 561-560.

Lee, H. and Garcia-Diaz, A., 1993, A network flow approach to solve clustering problems in group technology, *International Journal of Production Research,* 31, 3, 603-612.

Lenstra, J. K., 1974, Clustering a data array and the traveling-sales problem, *Operations Research,* 22, 413-414.

Michalewicz, Z., 1996, Genetic Algorithms + Data Structures = Evolution Programs, Springer-Verlag, New York.

Nagi, R., Harhalakis, G., and Proth, J. M., 1990, Multiple routings and capacity considerations in group technology applications, *International Journal of Production Research,* 28, 12, 2243-2257.

Nakano, R., 1991, Conventional genetic algorithm for job shop problems, 4th ICGA: 474-479.

Ng, S., 1993, Worst-case analysis of an algorithm for cellular manufacturing systems, *European Journal of Operational Research,* 69, 3, 384-398.

Renders, J.-M. and Flasse, S., 1996, Hybrid methods using genetic algorithms for global optimization, *IEEE Transactions on Systems, Man, and Cybernetics,* B, 26, 2, 243-258.

Shargal, M., Shekhar, S., and Irani, S. A., 1995, Evaluation of search algorithms and clustering efficiency measures for machine-part matrix clustering, *IIE Transactions* 27, 1, 43-59.

Suresh, N. C., Slomp, J., and Kaparthi, S., 1995, The capacitated cell formation problem: A new hierarchical methodology, *International Journal of Production Research,* 33, 6, 1761-1784.

Vakharia, A. J., Chang, Y., and Selim, H. M., 1994, Cell formation in group technology: A combinatorial search approach, working paper, University of Arizona, Tucson.

Venugopal, V. and Narendran, T. T., 1992b, Cell formation in manufacturing systems through simulated annealing: An experimental evaluation, *European Journal of Operational Research*, 63, 3, 409-422.

Venugopal, V. and Narendran, T. T., 1992a, A genetic algorithm approach to the machine-component grouping problem with multiple objectives, *Computers & Industrial Engineering*, 22, 4, 469-480.

Whitley, D., Gordon, S., and Mathias, K., 1994, *Lamarckian Evolution, the Baldwin effect and the function optimization. Parallel Problem Solving from Nature* PPSN III, Springer-Verlag, 6-15.

AUTHORS' BIOGRAPHY

Jeffrey A. Joines is a Research Associate for the Furniture Manufacturing and Management Center (FMMC) in the Dept. of Industrial Engineering at North Carolina State University, Raleigh, NC. He received his B.S. in Industrial Engineering, B.S. in Electrical Engineering, M.S. in Industrial Engineering, and Ph.D. in Industrial Engineering from NCSU. His interests include object-oriented simulation, manufacturing cell design, and genetic algorithms. He is a member of INFORMS, IEEE, IIE, Phi Kappa Phi and Alpha Pi Mu.

Russell E. King is a Professor of Industrial Engineering at NCSU. He received a B.S in Systems Engineering, Master of Science in Industrial Engineering and Ph.D. in IE from the University of Florida. His research and teaching interests are in the production scheduling and control, real-time control of automate systems, and a stochastic processes. He is a member of ASEE, IIE, APICS, SME, Alpha Pi Mu and INFORMS and is an associate faculty member of the Integrated Manufacturing Systems Engineering Institute (IMSE) and the OR Program at NCSU. He is an associate editor IIE Transactions.

C. Thomas Culbreth is a Professor of Industrial Engineering and Director of the Furniture Manufacturing and Management Center at NCSU. He received B.S. degree in Furniture Manufacturing and Management, a master of Economics degree, and Ph.D. in Industrial Engineering from NCSU. His current research interests focus on the application of cellular manufacturing and flexible automation in furniture manufacturing. His a member of ASEE, IIE, SME, and Alpha Pi Mu.

D5

A Bi-Chromosome Genetic Algorithm For Minimizing Intercell and Intracell Moves

C.H. Cheng, W. H. Lee and J. Miltenburg

1. INTRODUCTION

In the ideal case, cell design produces perfectly independent machine cells. That is, all operations of parts in a part family are completed within a single machine cell. However, the ideal case is rarely realized in practice. Very often, some of the parts in a part family have to move between machine cells to use machines in different cells. Consequently, the degree of machine cell independence is reduced by intercell moves.

Many cell design algorithms have been proposed, for example, McAuley (1972), King (1980), Seifoddini and Wolfe (1986), Askin and Subramanian (1987), Chandrasekharan and Rajagopalan (1987), Kusiak and Chow (1987), Gunasingh and Lashkari (1989), Askin and Chiu (1990), Harhalakis, Nagi and Proth (1990), Boctor (1991), Boe and Cheng (1991), Srinivasan and Narendran (1991), Venugopal and Narendran (1992), Srinivasan (1994), Chen, Cotruvo and Baek (1995), and Chen and Cheng (1995). As discussed by other researchers [Ballakur and Steudel (1987), Logendran (1990, 1991), Stanfel (1985), and Gupta, Gupta, Kumar and Sundaram (1995, 1996)], these algorithms suffer four limitations.

First, many cell design approaches do not consider the sequence of operations. Calculations of intercell moves, which consider the sequence of operations, will better reflect the actual intercell moves happening in the shop floor. Second, a majority of design algorithms ignore intracell moves. Logendran (1990) argues that intracell moves are as important and unproductive as intercell moves. Third, some design approaches disregard machine workload. Machine workload imbalance is one of the major criticisms of cellular manufacturing. A good design should be one having a reasonable workload for each machine in the system. Finally, most algorithms do not take layout into consideration when grouping machines and parts into machine cells and part families. Layout often determines the distance that a part has to travel between machine cells.

This chapter develops a solution methodology to address the four limitations discussed above. This research follows, extends and improves previous work done by Stanfel (1985), Ballakur and Steudel (1987), Logendran (1990, 1991), and Gupta et al. (1995, 1996). Stanfel (1985) proposes a mathematical model that includes the concept of extraneous machine transitions to reflect intracell processing. However, he does not consider machine workloads. Ballakur and Steudel (1987) develop a workload-based model for cell design. Logendran (1990) considers the work of Stanfel (1985) and Ballakur and Steudel (1987), and proposes a workload-based model focusing on minimizing total moves, which includes both intercell and intracell moves. In his follow-up work, Logendran (1991) includes the consideration of layout in the workload-based model. In an attempt to improve the computational efficiency of Logendran's workload-based models, Gupta et al. (1995) propose the use of genetic algorithms (GA). Gupta et al. (1996) use a multi-objective model so that in addition to minimizing the movement of parts, other objectives can be included such as machine workload and utilization, cell load variation, and so on.

In this chapter, we propose a GA with a bi-chromosome representation for the cell design problem. The bi-chromosome GA corrects a problem with the single (standard) chromosome GA representation of Gupta et al. (1995 & 1996). The problem is the generation of illegal offsprings, which requires additional computational effort to correct.

The remainder of the chapter is organized as follows. A mathematical model of the problem is presented in § 2. A bi-chromosome GA for solving the problem is developed in § 3. A computational study of the effectiveness of the algorithm follows in § 4. A summary is given in § 5.

2. THE MODEL

The model proposed by Logendran (1991) serves as our basic model. "If a part is required to visit n cells ($n \geq 1$) to either partially or completely process its requirements, then it contributes to ($n - 1$) intercell moves in the total intercell moves equation. ... If a part is required to visit m machines ($m \geq 1$) dedicated to a cell as a portion of its processing requirements or in its entirety, then it contributes to ($m - 1$) intracell moves in the total intracell moves equation. Logendran (1990, pp. 914)"

The objective function seeks to minimize the total intracell and intercell moves. The sequence of operations is an important factor in evaluating the movement of parts. For example, consider the three cells in Figure 1. Machines 1 and 3 are assigned to cell 1, machine 2 is assigned to cell 2, and machine 4 is assigned to cell 3. Suppose that a part must visit the four machines in the sequence M1→M2→M3→M4 to complete its processing requirements. If account of the sequence is not taken, then the cost is one intracell move and two intercell moves. However, when account of the sequence is taken, the actual cost is found to be three intercell moves and no intracell move. This example clearly demonstrates the importance of operation sequences.

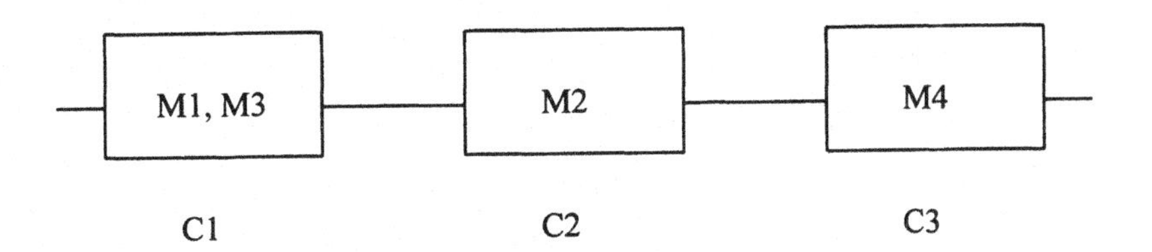

FIGURE 1. Linear single-row cellular layout.

The layout of cells is also an important factor in the design of the system. Consider again the linear single-row layout in Figure 1. If the distances between cells are equal, then the distance travelled from cell 1 to cell 3 is twice the distance from cell 1 to cell 2. Now consider the linear double-row layout in Figure 2. If the distance between cells 1 and 2 is equal to the distance between cells 2 and 3, then distance between cells 1 and 3 will be $\sqrt{2}$ times the distance between cells 1 and 2.

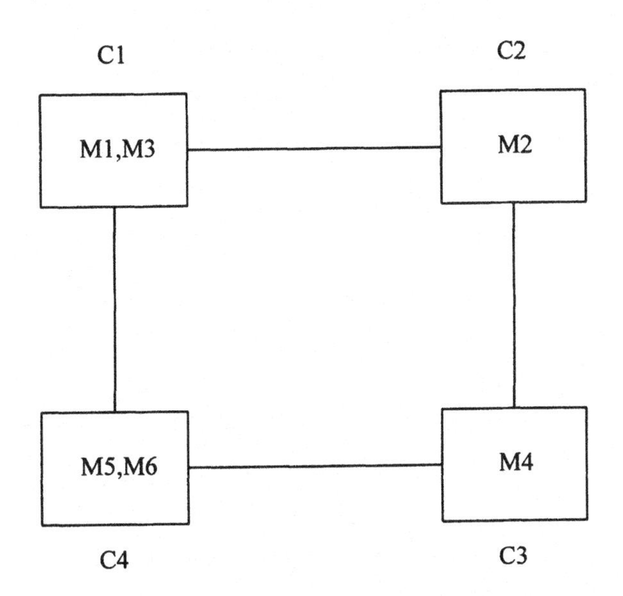

FIGURE 2. Linear double-row cellular layout

Define now the following notation for cell design problems where cells are organized in single-row or double-row layouts:

Indices: i : part ; j : machine ; k : operation

Parameters

P	total number of parts
M	total number of machines
$N \geq 2$	total number of cells
A_j	available capacity of machine j
K_i	total number of operations to be performed on part i
$0 \leq \theta_1 \leq 1$	weighting for intercell moves
$\theta_2 = 1 - \theta_1$	weighting for intracell moves
$t_{i,j,k}$	processing time of part i at machine j for operation k

Decision variables

$\alpha_i^{(part)}$	cell to which part i is assigned
$\alpha_j^{(machine)}$	cell to which machine j is assigned
$c_{i,k}$	cell in which operation k on part i is performed (where k = 1, 2, ... is in the proper sequence of operations)

$$L = \begin{cases} 1 & \text{if layout is single-row} \\ 0 & \text{if layout is double-row} \end{cases}$$

Derived decision variables

$$d_{i,k,k+1}^{(L)} = \begin{cases} 0 & \text{if } |c_{i,k} - c_{i,k+1}| = 0,\ L = 1 \text{ or } L = 2 \\ 1 & \text{if } |c_{i,k} - c_{i,k+1}| = 1,\ L = 1 \text{ or } L = 2 \\ \sqrt{2} & \text{if } |c_{i,k} - c_{i,k+1}| = 2,\ L = 2 \\ 2 & \text{if } |c_{i,k} - c_{i,k+1}| = 2,\ L = 1 \end{cases}$$

distance travelled by part i between operations k and $k+1$

m_i total number of intracell moves for part i

Define a cost T to be the weighted sum of the intercell moves and the intracell moves. Then

$$T = \theta_1 \sum_{i=1}^{P} \sum_{k=1}^{K_i - 1} d_{i,k,k+1}^{(L)} + \theta_2 \sum_{i=1}^{P} m_i \tag{1}$$

and we select values of the decision variables, $\alpha_i^{(part)}$, $\alpha_j^{(machine)}$, $c_{i,k}$, L and to minimize T.

In practice, the weight assigned to intracell moves should not be as high as that assigned to intercell moves. So $\theta_1 > 0.5$. In this research, we use $\theta_1 = 0.7$ and $\theta_2 = 0.3$. Another model parameter is the number of cells N. In general, as the number of cells increases, intercell moves increase and intracell moves decrease. We vary the number of cells N = 2, 3, 4, 5, 6, ... to provide different scenarios for the designer so that the final choice among them can be made on the basis of intangible factors not incorporated into the model. As pointed out by Gupta et al. (1995), most manufacturing firms in the U.S. use six or less cells, and hence, it is possible to evaluate all the alternatives.

As discussed by Logendran (1991), workstation utilization is another important factor in determining cell configuration. Once the machine cells are formed, each part is assigned to the cell where the largest portion of its processing time is spent, with ties broken arbitrarily, and then the machine utilizations are calculated. Define total machine utilization to be the ratio of the workload to the available machine capacity:

$$U_j^{(total)} = \frac{\left(\sum_{i=1}^{P} \sum_{\substack{k=1: \\ c_{i,k}=\alpha_j^{(machine)}}}^{K_i} t_{i,j,k} \right)}{A_j} \tag{2}$$

Define also intracell machine utilization to be the ratio of the workload for parts assigned to the cell in which the machine is located to the available machine capacity:

$$U_j^{(intracell)} = \frac{\left(\sum_{\substack{i=1: \\ \alpha_i^{(part)}=\alpha_j^{(machine)}}}^{P} \sum_{\substack{k=1: \\ c_{i,k}=\alpha_j^{(machine)}}}^{K_i} t_{i,j,k} \right)}{A_j} \tag{3}$$

If the workload assigned to a machine exceeds its capacity, i.e., $U_j^{(total)} > 1.0$, then we allow a duplicate machine to be allocated to the cell. To avoid confusion between duplicate machines, we will call each a workstation in the remainder of the chapter.

3. THE SOLUTION METHOD

The cell design problem is a combinatorial problem [Levine (1993)]. Heuristic approaches are required to solve it efficiently. GAs seem to be a good candidate to solve combinatorial problems due to their ability to perform a parallel search and their robustness [Levine (1993)]. Applying GAs to solve the workload model was first proposed by Gupta et al. (1995 & 1996).

In Gupta et al.'s implementation, standard (or group number) representation was employed. The value of a gene represents the cell number with the position of the gene corresponding to the workstation number. For example, chromosome (2,3,1,1,2) signifies that workstation 1 is assigned to cell 2, workstation 2 is assigned to cell 3, workstation 3 is assigned to cell 1, and so on. The length of the chromosome is the total number of workstations. The crossover operator used is the single point crossover. The crossover point in a chromosome is chosen randomly. The portions of the chromosomes after the crossing point are exchanged to produce the offsprings.

One weakness of this representation is the occurrence of empty cells and therefore illegal offsprings. Consider for example the chromosome shown in Figure 3. The number of cells is three. However, after the crossover, cell 3 is empty in offspring 1 and cell 2 is empty in offspring 2. This violates the requirement that each cell must contain at least one workstation.

To overcome this problem, Gupta et al. (1995) used a mutation operator. Two random integers r_1 and r_2 are selected such that $1 \le r_1 \le M$ and $1 \le r_2 \le N$ where M and N are the total number of machines and cells, respectively. The algorithm then assigns workstation number r_1 in the illegal offspring to cell r_2. This process is repeated until no empty cell exists. This mutation operator increases the computational requirements and hence affects the performance of the GA.

Using more than one chromosome to represent a solution overcomes this difficulty. Let a solution to the cell design problem be represented by two chromosomes:

1. machine sequence chromosome, length = M,
2. cell boundary chromosome, length = M - 1.

Each gene in the machine sequence chromosome represents a machine number and therefore is integer-valued. The length of a chromosome is the number of machines. The second chromosome is the cell boundary chromosome for a given machine sequence chromosome. Its task is to partition the machines in the machine sequence chromosome into cells. Gene 1 in the second chromosome identifies the edge between cells 1 and 2 in the first chromosome. Gene 2 identifies the edge between cells 2 and 3, and so on. Consequently each gene in the cell boundary chromosome is also integer-valued. To illustrate these concepts, consider the example in Figure 4 where $M = 9$ machines are to be assigned to $N = 3$ cells.

The third and sixth gene in the cell boundary chromosome contain the values 1 and 2, respectively and so identify the boundaries between cells 1 and 2, and 2 and 3, respectively (as shown in the figure). As a result, machine 2, 3, and 5 are assigned to cell 1, machines 1, 6, and 8 are assigned to cell 2, and cell 3 contains machines 4, 7, and 9. Notice this approach will never generate illegal offsprings. The computational requirements for the second chromosome are modest and, in our experience, much less than the computational requirements associated with correcting illegal offsprings.

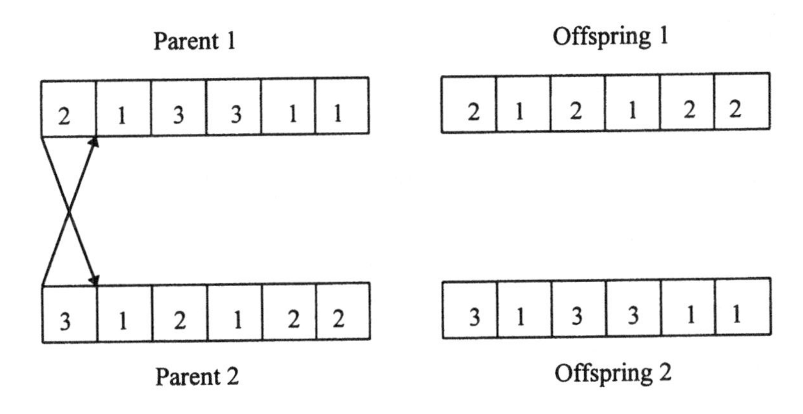

FIGURE 3. Occurrence of empty cell after crossover.

The crossover operator used is the edge-recombination operator. Edge-recombination operator transfers 95% of edge (information) from parent to offspring and is considered to be the most efficient operator for general sequencing problems [Michalewicz (1992)].

Another advantage of this approach is the ease with which different numbers of cells, N, can be considered. That is we may compute the value of T^* from $T^* = \underset{N=2,3,\ldots,M}{\text{MINIMUM}}\{T(N)\}$ where $T(N)$ is the value of T for a specified value of N. This is different from the approach of Gupta et al. where the number of cells, N, is fixed.

Our implementation is based on a GA package called GENITOR [Whitley (1989)]. The fitness function is equation (1) from § 2. Both chromosomes are randomly generated in the initial population. The parents are selected using fitness ranking. The processes of selection, crossover, and replacement are continued until a maximum number of generations are completed. Finally, we note that other researchers have used more than one chromosome to represent a solution. For example, Juliff (1993) uses a multi-chromosome GA to solve the pallet loading problem.

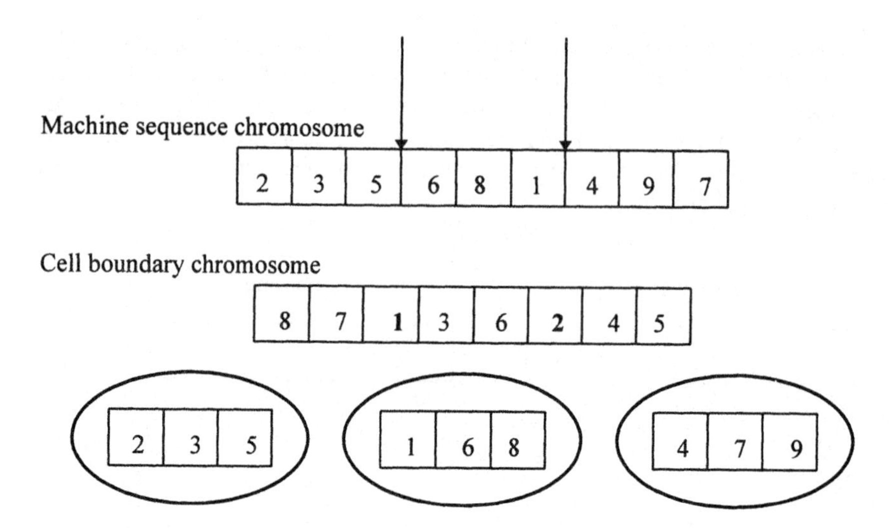

FIGURE 4. Interpretation of information in chromosomes.

4. COMPARATIVE STUDY

We solve and analyze four well-known problems from the literature to gauge the effectiveness of our algorithm. The first problem is from Ballakur and Steudal (1987). The second is from Tabucannon and Ojha (1987), the third is from King and Nakornchai (1982) and the fourth is from Gupta et al. (1996). Logendran (1991) modified the second and third problems by adding part processing times, making them useful for the problem considered in this chapter.

Table 1 shows the workstation-part load matrices for the four problems. Each element in the matrices is the processing time in hours required by a part at a workstation. The third section of columns in the table shows the total processing time required at each workstation. In this study we assume that the available capacity at a workstation is eight hours and whenever the required capacity at a workstation exceeds eight hours an extra workstation is added. For example, workstation 2 in problem 1 processes parts 1, 2, 4 and 6, and requires 2.5+2+4.5+1.5 = 10.5>8 hours to do this. So a second workstation is added. The next column shows the total utilization of each workstation calculated from equation (2). For example, the total utilization of workstation 2 is 10.5/8 = 1.31. Note that account is not taken of the number of workstations in this example. This is done so that we can more easily compare our results with those of other researchers. In the last column of the table we calculate the average total utilization for the problem. For problem 1 this was (0.88+1.31+0.88+0.5+0.75)/5 = 0.863.

Table 2 shows the results from applying the algorithm of § 3 to these problems. The best solution to problem 1 is shown in the first row of the table. N=2 cells, the layout may be single-row or double-row (from L=1 or 2), and the value of the cost function is T=3.8, which is calculated from equation (1) with θ_1=0.7 and $\theta_2 = 1-\theta_1$=0.3. The first cell consists of workstations 1 to 4 with parts 1, 2 and 4 to 7. The second cell is much smaller with only one workstation and one part. Also shown in the table are the solutions for N=3, 4 cells for each of the two layouts, so that we may compare our results with those of other researchers where the number of cells may be fixed before the problem is solved.

The next section of the table shows the workstation intracell utilizations calculated from equation (3). This utilization is a measure of the available workstation capacity that is used by parts assigned to the cell in which the workstation is located. In this research we set 50 percent of the total utilization as the minimum acceptable intracell utilization. When a workstation spends less than this amount of its capacity processing parts assigned to the cell in which it is located, then the assignment of machines and parts to cells should be re-examined to see whether some improvements may be made. To illustrate, consider again the first row in Table 2. The intracell utilizations of the five workstations are 0.875, 1.31, 0.875, 0.5 and 0.375, while the total utilization (from Table 1) are 0.875, 1.31, 0.875, 0.5 and 0.75. So the first four workstations spend 100 percent of their total utilization processing parts assigned to the cells in which they are located. That is, they do not process parts assigned to other cells. However workstation 5, which is assigned to cell 2, only spends 0.375/0.75 = 50 percent of its total utilization processing cell 2 parts. The other 50 percent is spent processing parts assigned to cell 1. This suggests that other assignments of workstations and parts should be examined to see whether this low intracell utilization can be improved. The next four rows of the table examine three- and four-cell assignments. However, there is no improvement in the intracell utilization for this workstation. The last column in Table 2 shows the average intracell utilizations for the solutions to each problem. For example, the average intracell utilization for the solutions to Problem 1 range from 0.79 for the two-cell solution in the first row to 0.626 for the four-cell solution. These values may be compared to the best possible value, 0.863, which was given in the last column of Table 1.

Table 3 shows the comparison of the solutions from Table 2 with those produced by two other algorithms, one from Logendran (1991) and the other from Gupta et al. (1996). Logendran's algorithm is a heuristic consisting of four steps; 1. cell representation where key workstations are selected to be seeds around which cells are formed, 2. clustering where other workstations are assigned to the cells identified in the first step, 3. improvement where exchanges of workstations between cells are considered, and 4. assignment where parts are assigned to cells. Two variations the first step are considered in our study. The first variation selects as seeds those workstations having the highest total workload, while the second variation selects those workstations that process the largest number of parts. The algorithm of Gupta et al. (1996) is GA-based which was reviewed briefly in §3.

TABLE 1. Workstation - Part Load Matrix

Problem(1)	Workstation, j	Part, i 1	2	3	4	5	6	7	8	9	10	11	12	13	14	15	16	17	18	19	20	21	22	23	24	25
1	1		0.5			5		1.5																		
	2	2.5	2		4.5		1.5																			
	3		2.5			3.5	0.5	0.5																		
	4	2.5			1		0.5																			
	5			3		1		2																		
2	1		0.69	?		2.42			2.44	2.48					2.72											
	2	0.5		?	0.61	0.9	2.09	1.35																		
	3			?							2.5	3.03	0.71	1.61												
	4		3.1	?				1.35		1.03		0.58	0.99													
	5		1.22	?					4.45						3.84											
	6	0.5		?			4.55		2.26																	
	7	0.55		?	4.74	3.61	1.47			3.87	4.68															
3	1		0.55		4.74			1.35																		
	2			1.22		3.61																				
	3	0.5	1.69		2.42			1.35																		
	4	0.51		3.1			4.55																			
	5			0.61	0.9	2.09	1.47																			
4	1		0.8			1.2		0.5		1.7	1.8		2										0.5			
	2	0.1							1.7		1.3				1.7					0.8						1.1
	3	0.9	0.8			1.8	0.7					0.5					0.2			1.4	0.3	0.3				
	4			0.4					0.6			0.8									1.1	1.7	1.4	1.7		
	5				0.9			1.5	0.9	0.1					0.6			0.5			0.6			1.2		
	6	1.2	1.1					1						0.6		1.9	0.4				2					
	7	0.1			1.3	0.2											1.9									
	8	0.6	1.2	1.7				1.4				2								1.8		1				
	9						0.9	0.4		1.3			0.8	1.2											1.3	
	10							1.5		1.4				1.8					0.7		0.2		1.7			1.8
	11	0.3	0.4																					1.7		
	12								0.1	1.5			1.5	0.2								1.1			1.4	
	13				1.4			0.3				1.7	1.6	1.5			1.3	1.2	0.4	0.4		0.7	0.1			
	14		1.6							1.3						1										
	15					0.3				0.4		1.3		0.3	0.5								1.9		1.5	0.2

Notes:
1. Problem 1 is from Balakur and Steudel (1987).
   Problem 2 is from Tabucannon and Ojha (1987) and was modified by Logendran (1991).
   Problem 3 is from King and Nakornchai (1982) and Waghodekar and Sahu (1984).
   Problem 4 is from Gupta et al. (1996).
2. Total utilization is calculated from equation (2). For example, the total utilization for workstation 1 in Problem 1 is (0.5+5+1.5)/8 = 0.875.

TABLE 2. GA Results for Problems from the Literature

Problem	No. of cells, N	Layout, L	Cost, T		Workstations and parts in cell c									
					$c=1$	2	3	4	5	6	$j=1$	2	3	4
1	2	1or2	3.8	workstation, j	1to4	5					0.88	1.31	0.88	0.5
				part, i	1,2,4to7	3								
	3	1	5	workstation, j	5	1,2,3	4				0.88	1	0.88	0.31
				part, i	3	2,4to7	1							
	3	2	5	workstation, j	4	1,2,3	5				0.88	1	0.88	0.31
				part, i	1	2,4to7	3							
	4	1	6.9	workstation, j	4	2	1,3	5			0.88	0.75	0.81	0.31
				part, i	1	4,6	2,5,7	3						
	4	2	6.49	workstation, j	1,3	2	4	5			0.88	0.75	0.81	0.31
				part, i	2,5,7	4,6	1	3						
2	2	1or2	6.3	workstation, j	1,2,4to7	3					1.34	0.68	0.98	0.88
				part, i	1to10,12,14,15	11,13								
	3	1or2	7.5	workstation, j	1,5	2,4,6,7	3				0.73	0.68	0.58	0.81
				part, i	2,8,14	1to7,9,10,12	11,13							
	4	1	9.4	workstation, j	3	2,4,6,7	1	5			0	0.68	0.58	0.81
				part, i	11,13	1,3to7,9,10,12	?	2,8,14						
	4	2	8.88	workstation, j	2,6,7	1,5	3	4			0	0.68	0.58	0.81
				part, i	1,4to7,9,10	2,8,14	11,13	3,12						
3	2	1or2	3.5	workstation, j	1,2,3,4	5					0.83	0.6	0.56	1.02
				part, i	1,2,4to7	3								
	3	1	4.3	workstation, j	1,3	4,5	2				0.83	0.45	0.56	1.02
				part, i	2,4,7	1,3,6	5							
	3	2	4.3	workstation, j	2	4,5	1,3				0.83	0.45	0.56	1.02
				part, i	5	1,3,6	2,4,7							
	4	1	5.5	workstation, j	2	4,5	3	1			0.76	0.45	0.09	1.02
				part, i	5	1,3,6	2	4,7						
	4	2	5.5	workstation, j	2	1	3	4,5			0.76	0.45	0.09	1.02
				part, i	5	4,7	2	1,3,6						
4-Jan	4	1	37.3	workstation, j	14	11	2to10,12,13	1			0.23	0.78	1.08	1
				part, i	15	25	1to9,11to14, 16to24,26to30	10						
	4	2	34.7	workstation, j	11	14	15	1to10,12,13			1.06	0.94	1.08	1
				part, i	25	15	24	1to14,16to23,26to30						
	5	1	44.2	workstation, j	14	11	3to10,12,13,15	1	2		0.23	0.66	1.08	0.89
				part, i	15	25	1to7,9,11to13, 16to24,26,28to30	10	8to14,27					
	5	2	37.1	workstation, j	1,3to10,12,13	11	14	15	2		1.06	0.66	1.08	0.89
				part, i	1to7,9to13, 16to23,28to30	25	15	24	8,14,27					
	6	1	54.9	workstation, j	1	2	3to8,11	9,10,12,13	15	14	0.23	0.66	0.99	0.71
				part, i	10	8,14,27	1to5,7,11,16,19to21, 23,25,26,28,29	6,9,12,13,17,22,30	24	15				
	6	2	39.5	workstation, j	2	1	15	14	11	3to10,12,13	0.23	0.66	1.08	0.89
				part, i	8,14,27	10	24	15	25	1to7,9,11to13, 16to23,26,28to30				

Notes: 1. Intracell utilization is calculated from equation (3).

TABLE 3. Comparison of Algorithm

Problem	No. of cells. N	Layout. L	Cost. T Logendran	Cost. T Gupta et al.	Cost. T GA	Upper bound	Average Intracell Utilization Logendran	Average Intracell Utilization Gupta et al.	Average Intracell Utilization GA
1	2	1or2	4.6. 3.8	3.8	3.8	0.863	0.694. 0.659	0.657	0.659
	3	1	6.8. 7.1	5	5	0.863	0.62. 0.559	0.559	0.588
	3	2	5.98. 5.87	5	5	0.863	0.62. 0.559	0.585	0.588
	4	1	10.4. 13.2	7	6.9	0.863	0.52. 0.52	0.52	0.55
	4	2	7.36. 7.36	6.78	6.49	0.863	0.52. 0.52	0.434	0.55
2	2	1or2	6.3. 6.7	6.3	6.3	1.194	0.72. 0.699	0.702	0.72
	3	1	8.9. 8.9	7.5	7.5	1.194	0.676. 0.609	0.635	0.635
	3	2	8.08. 8.08	7.5	7.5	1.194	0.676. 0.609	0.635	0.635
	4	1	12.3. 13.7	9.4	9.4	1.194	0.527. 0.550	0.583	0.583
	4	2	10.66. 9.39	8.88	8.88	1.194	0.527. 0.550	0.592	0.592
3	2	1or2	3.5. 3.5	3.5	3.5	0.742	0.604. 0.604	0.604	0.604
	3	1	6.8. 6.8	4.3	4.3	0.742	0.496. 0.494	0.522	0.522
	3	2	5.57. 5.57	4.3	4.3	0.742	0.496. 0.494	0.522	0.522
	4	1	9.8. 11.2	5.5	5.5	0.742	0.362. 0.36	0.414	0.414
	4	2	7.64. 7.64	5.5	5.5	0.742	0.414. 0.412	0.414	0.414
4	4	1	na	38.8	37.3	1.021	na	0.523	0.581
	4	2	na	36.34	36.47	1.021	na	0.523	0.553
	5	1	na	48.7	44.2	1.021	na	0.501	0.549
	5	2	na	43.77	37.07	1.021	na	0.418	0.547
	6	1	na	59.7	54.9	1.021	na	0.431	0.44
	6	2	na	42.16	39.47	1.021	na	0.437	0.495

Notes: ___ denotes those instances when an algorithm found the best known solution.

Two performance measures are reported in Table 3, the cost *T* calculated from equation (1) and the average intracell utilization calculated from equation (3). Values of *T* are reported in the middle columns of the table. The lower the value of *T* the better is the solution. Values of the average intracell utilization are reported in the last columns of the table. Higher average utilizations represent better solutions. The upper bound on the average intracell utilization is also shown. Recall that it is the average total utilization calculated in Table 1. The best values of the performance measures for each problem instance are underlined to show at a glance the effectiveness of each algorithm. In this study the two GAs clearly outperformed the algorithm of Logendran, which is not GA-based. With respect to the two GAs, the one presented in this chapter was more effective than the algorithm of Gupta et al. Of the 24 instances considered in the study, the algorithm of Gupta et al. found the best solution 17 and 8 times for the cost and utilization performance measures respectively, while the algorithm presented in this chapter found the best solution 23 and 18 times for two measures respectively.

5. SUMMARY

In this chapter, we consider a cell design problem where cell layout, intercell and intracell part movement, sequence of operations, and machine workload are all taken into account. A GA employing two chromosomes to represent a solution to the problem is outlined. The GA is compared with two algorithms from the literature, one of which is also a GA. That GA uses a single chromosome to represent a solution. The other algorithm is a four-step heuristic. The two GAs outperform the heuristic, and the GA presented in this chapter outperforms the GA from the literature. We conclude that a bi-chromosome representation is superior to a single chromosome representation for this cell design problem.

REFERENCES

Askin, R.G. and Subramanian, S.P., 1987, A cost-based heuristic for group technology configuration, *International Journal of Production Research*, 25, 101-113.

Askin, R.G. and Chiu, K., 1990, A graph partitioning procedure for machine assignment and cell formation, *International Journal of Production Research*, 28, 1555-1572.

Ballakur, A. and Steudel, H.J., 1987, A within-cell utilization based heuristic for designing cellular manufacturing systems, *International Journal of Production Research*, 25, 639-665.

Boctor, F.F., 1991, A linear formulation of the machine-part cell formation problem, *International Journal of Production Research*, 29, 343-356.

Boe, W.J. and Cheng, C.H., 1991, A close neighbor algorithm for a designing cellular manufacturing system, *International Journal of Production Research*, 29, 2097-2116.

Bruns, R., 1993, Direct chromosome representation and advanced genetic operators for production scheduling. *Proceedings of the Fifth International Conference on Genetic Algorithms*, 352-359.

Burbidge, J.L., 1989, *Production Flow Analysis*, Clarendon Press, Oxford.

Chandrasekharan, M.P. and Rajagopalan, R., 1987, ZODIAC – an algorithm for concurrent formation of part-families and machine-cells, *International Journal of Production Research*, 25, 835-850.

Chen, C.L, Cotruvo, N.A., and Baek, W., A simulated annealing solution to cell formation problem, *International Journal of Production Research*, 33, 2601-2614.

Chen, S.J. and Cheng, C.S., 1995, A neural network-based cell formation algorithm in cellular manufacturnig, *International Journal of Production Research*, 33, 293-318.

Del Valle, A.G., Balarezo, S., and Tejero, J., 1994, A heuristic workload-based model to form cells by minimizing intercellular movements, *International Journal of Production Research*, 32, 2275-2285.

Fonseca, C.M. and Fleming, P.J., 1993, Genetic algorithms for multiobjective optimization: formulation, discussion and generalization, *Proceedings of the Fifth International Conference on Genetic Algorithms*, 416-423.

Gunasingh, R.K. and Lashkari, R.S., 1989, The cell formation problem in cellular manufacturing systems – a sequential modelling approach, *Computers and Industrial Engineering*, 16, 469-476.

Gupta, Y.P., Gupta, M.C., Kumar, A., and Sundaram, C., 1995, Minimizing total intercell and intracell moves in cellular manufacturing: a genetic algorithm approach, *International Journal of Computer Integrated Manufacturing*, 8, 92-101.

Gupta, Y.P., Gupta, M.C., Kumar, A., and Sundaram, C., 1996, A genetic algorithm-based approach to cell composition and layout design problems. *International Journal of Production Research*, 34, 447-482.

Harhalakis, G., Nagi, R., and Proth, J.M., 1990, An efficient heuristic in manufacturing cell formation for group technology applications. *International Journal of Production Research*, 28, 185-198.

Juliff, K., 1993, A multi-chromosome genetic algorithm for pallet loading, *Proceedings of the Fifth International Conference on Genetic Algorithms*, 467-473.

King, J.R., 1980, Machine-component group formation in production flow analysis: an approach using a rank order clustering algorithm, *International Journal of Production Research*, 18, 213-232.

King, J.R. and Nakornchai V., 1982, Machine-component group formation in group technology: review and extension, *International Journal of Production Research*, 20, 117-133.

Kusiak, A. and Chow, W.S., 1987, Efficient solving of the group technology problem, *Journal of Manufacturing Systems*, 6, 117-124.

Levine, D.M., 1993, A genetic algorithm for the set partitioning problem, *Proceedings of the Fifth International Conference on Genetic Algorithms*, 481-487.

Logendran, R., 1990, A workload based model for minimizing total intercell and intracell moves in cellular manufacturing, *International Journal of Production Research*, 28, 913-925.

Logendran, R., 1991, Impact of sequence of operations and layout of cells in cellular manufacturing, *International Journal of Production Research*, 29, 375-390.

McAuley, A., 1972, Machine grouping for efficient production, *Production Engineer*, 51, 53-57.

Michalewicz, Z., 1992, *Genetic Algorithm + Data Structure = Evolution Programs*, Springer-Verlag, Hong Kong.

Seifoddini, H. and Wolfe, P.M., 1986, Application of the similarity coefficient method in group technology, *IIE Transactions*, 18, 271-277.

Srinivasan, G., 1994, A clustering algorithm for machine cell formation in group technology using minimum spanning trees, *International Journal of Production Research*, 32, 2149-2158.

Srinivasan, G. and Narendran T.T., 1991, GRAFICS – a nonhierarchical clustering algorithm for group technology, *International Journal of Production Research*, 29, 463-478.

Stanfel, L.F., 1985, Machine clustering for economic production, *Engineering Costs and Production Economics*, 9, 73-81.

Tabucanon, M.T. and Ojha, R., 1987, ICRMA – a heuristic approach for intercell flow reduction in cellular manufacturing systems, *Material Flow*, 4, 189-190.

Venugopal, V. and Narendran, T.T., 1992, A genetic algorithm approach to the machine-component grouping problem with multiple objectives, *Computers and Industrial Engineering*, 22, 469-480.

Waghodekar P.H. and Sahu S., 1984, Machine-component cell formation in group technology: MACE, *International Journal of Production Research*, 22, 937-948.

Wei, J.C. and Gaither, N., 1990, A capacity constrained multiobjective cell formation method, *Journal of Manufacturing Systems*, 9, 222-232.

Whitley D., 1989, The GENITOR algorithm and selection pressure: why rank-based allocation of reproductive trails is best, *Proceedings of the Third International Conference on Genetic Algorithms*, 116-121.

AUTHORS' BIOGRAPHY

Dr. Chun Hung Cheng is an Associate Professor of Engineering Management at Department of Systems Engineering and Engineering Management, the Chinese University of Hong Kong, Hong Kong. Professor Cheng obtained his Ph.D. in Management Information Systems from the University of Iowa in 1990, and started his teaching career at Kentucky State University, U.S.A. He returned home in 1995 and joined the Chinese University of Hong Kong. He has conducted research in Information Systems and Operations Management, and published 22 articles in refereed journals.

Mr. Wai Hung Lee was a M.Phil. student at Department of Systems Engineering and Engineering Management, the Chinese University of Hong Kong, Shatin, N.T., Hong Kong. He is now with Mass Transit Railway Corporation, Hong Kong.

Dr. John Miltenburg is a Professor of Operations Management in the School of Business at McMaster University, Canada. This chapter was written while he was on sabbatical leave at the Chinese University of Hong Kong. Professor Miltenburg obtained his Ph.D. from the University of Waterloo. He also worked for three years at General Motors in a Manufacturing Engineering Department. He has published 40 articles in refereed journals and written a book, *Manufacturing Strategy*, (Productivity Press, 1995).

E1

Design / Analysis of Manufacturing Systems: A Business Process Approach

N. Viswanadham, Y. Narahari and N. R. S. Raghavan

1. INTRODUCTION

Manufacturing has gone through periods of great changes time and again. New materials, new information and production technologies, new planning and control techniques, and new bases for competition have all contributed to these changes. In recent years, global competition, more demanding customers, economic liberalization, more stringent regulations on environment, emergence of common markets, disintegration of large States, etc. have added to the complexity of managing manufacturing firms. To address these new complexities, computer aided automation, flexibility management, strategic alliances, management of end-to-end business processes, especially supply chain and new product development processes, are particularly important.

Traditionally, the manufacturing system has been viewed as a sequential arrangement of functions such as design, manufacture, R&D, marketing, finance, etc. However, the recent trend has been to view manufacturing as *a collection of value-delivering processes*. The *traditional, functional or hierarchical structure* typically presents responsibilities and reporting relationships, whereas the *new, process structure* is a dynamic view of how the organization delivers value to the customer. In this chapter, the design and analysis of manufacturing organizations are approached as business processes and their interfaces. Several performance measures are defined for end-to-end business processes such as order-to-delivery, new product development, etc. These metrics attempt to summarize directly, product and system performances as well as customer satisfaction levels.

§2 provides a total systems view of manufacturing in accordance with recent perspectives. The notion of the business process is introduced, and two important business processes, new product development and supply chains are discussed. §3 introduces appropriate performance measures, while §4 conceptualizes the business process approach to performance analysis and brings out appropriate methodologies. §5 and §6 focus on lead time performance measure and describe queuing network models for new product development process (§5) and supply chain process (§6). These models serve to capture several critical process parameters and also enable rapid experimentation to compress lead times.

2. THE MANUFACTURING ENTERPRISE

Traditionally, manufacturing meant primarily the factory floor, and during 1970's and 80's much effort was directed at implementation of new techniques and technologies such as flexible manufacturing systems (FMS), computer integrated manufacturing (CIM), just-in-time (JIT) systems, total quality management (TQM), etc. It is generally recognized, however, that these efforts have not led to expected gains, and they have prompted new ways of approaching the manufacturing enterprise as a total system. Accordingly, a manufacturing system here refers to the total enterprise and consists of several interconnected subsystems, all of which need to act in cohesion so that customer desired products are delivered, on time.

2.1. Functional vs. Process Organizations

Contemporary trend is to view a manufacturing system in terms of *value delivering, business processes*. The traditional, functional or hierarchical structures present responsibilities and reporting relationships, whereas the process structure is a dynamic view of how the organization delivers value to the customer.

Hierarchical organization structures based on functional divisions have several problems. Each function tends to act as a "silo" and hands over its output over the wall to the next function. Turf-wars and dominance by functions result in slow progress of work through the system. Lack of proper communication between functions results in work going back and forth with long iteration periods. The hierarchical arrangements in the functions require that decisions are to be sought from the top and work processes move up and down the ladder. Thus one finds that the ratio of cycle-time to processing-time is large, most of the cycle time comprising non-value adding times, devoted to move, wait, and information collection.

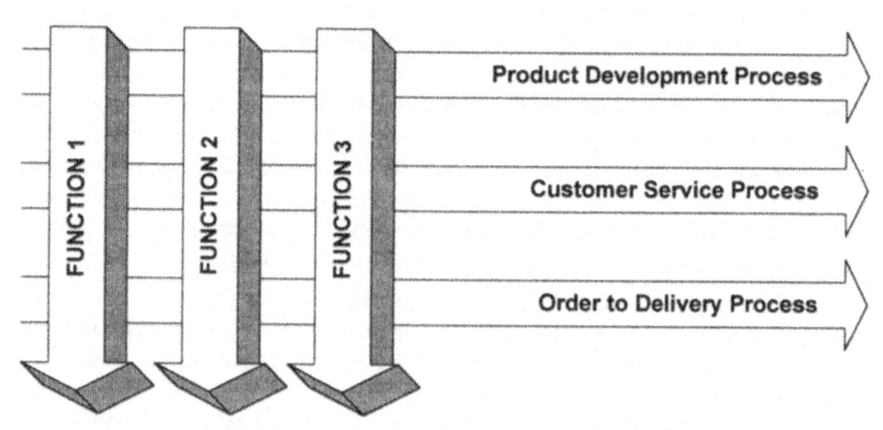

FIGURE 1. Functions and business processes.

Despite the above criticisms, the advantages of grouping work on a functional basis, by skill or specialty of work have been recognized for some time. Grouping by function enables resources to be pooled among different work activities. It promotes certain types of specialization, more easy transfer of skills, and efficient management of similarly-skilled personnel. Thus, even within focused factories and cellular manufacturing, organizations are at times functionally structured. In contrast, the business process perspective, believed to have originated with total quality movement, views a manufacturing system as a collection of business processes which deliver value to customers (Figure 1), and manages them using cross-functional teams in a coordinated manner (Davenport 1993).

2.2. Business Processes

A business process can be defined as *a structured, measured set of activities ordered in time and space, designed to produce a customer-desired output*. Examples of typical processes include: order-to-delivery process, new product development process, factory floor process, and supply chain process. A process perspective is a horizontal view, with inputs at the beginning and customers at the end. Subscription to this view means de-emphasizing functional views of the business. A primary issue in a vertical organization is the ill-management of hand-off between functions. Process orientation either eliminates hand-offs or coordinates them effectively. Processes are typically cross-functional and might be cross-organizational. A process is a mechanism for creating outputs in a systematic way, which have greater economic value than inputs. As the workpiece or job moves through various work processes, a transformation occurs, adding value to it. The transformation could be physical, locational, transactional or informational.

Thus, products are produced, and services performed, by means of *business processes*. Businesses are composed of several interrelated processes. They should be efficient and effective for gaining competitive advantage. Competition will then be in terms of processes against processes, rather than functions versus functions. Thus, it is not enough to be a world-class manufacturer of a certain product; the product should also reach the customers in time, and fulfill customer needs with respect to looks, features, performance, durability, usability, etc. We next consider two important business processes in manufacturing enterprises.

2.3. New Product Development Process

A product development process (PDP) is the sequence of steps or activities that an enterprise employs to conceive, design, and commercialize a product (Ulrich and Eppinger 1995). Some organizations define and follow a precisely defined PDP while others may not even be able to describe their processes. Also, processes of different organizations may have differences and unique characteristics. According to Cooper (1983), who constructs a process model for a typical PDP, a well-defined process enables better control and tracking of new product development projects. A generic PDP consists of five phases (Ulrich and Eppinger 1995): *concept development, system-level design, detail design, testing and refinement, and production ramp-up.*

The PDP is critical to a company's success. A cross-functional, cross-organizational team with a process owner is ideal for managing this process. Each work process can be analyzed for inputs, outputs, transformation mechanism, variabilities involved, and their reduction, etc. Time to market, costs, flexibility to generate a variety of products, time to next product, defect rates (or reliability), etc. become important performance measures. Benchmarking PDP with competitors and other industry leaders will help identify best practices and the enablers for speeding up the process. Some of the enablers in a PDP include: quality function deployment (QFD), rapid prototyping and computer aided design (CAD) tools, groupware, project management, concurrent engineering, Internet, etc.

2.4. Supply Chain Process

The supply chain process encompasses the full range of intra-company and inter-company activities from material procurement, through manufacturing and distribution, to proper delivery of products to customers. The speed of the supply chain process (SCP) determines the delivery time in make-to-order environments. Supply chain cost, performance, lead time variability, etc. depend on all constituents of the supply chain. One can define supply chain management as the coordination or integration of all activities involved in procuring, producing, delivering and maintaining products/services to customers in various geographical locations.

In the short term, SCP involves proper production and distribution of multiple products in the order-to-delivery process. Long term issues in SCP involve location of production and inventory facilities, choice of alliance partners such as suppliers and distributors in the logistics chain, make-to- order or make-to-stock policies, degree of vertical integration, capacity decisions for various plants, etc.

Traditionally, marketing, distribution, planning, manufacturing and purchasing activities are performed independently, each with its own functional objectives. But SCP is a process-oriented approach to coordinating all functional units involved in the order-to-delivery process.

3. PERFORMANCE MEASURES

Performance measures are needed to evaluate, control, and improve business processes. They are also useful for comparison with processes utilized in other companies for benchmarking purposes.

Performance measures defined in the past for organizations have tended to have a financial orientation, e.g., return on sales or investment. The use of these measures has been criticized on many counts, and in recent years, there have been efforts to augment these with new, non-financial performance measures that may lead to more effective business strategy and performance. These include lead time, quality, reliable delivery, customer service, rapid product introduction, and flexible capacity. It is our contention however, that these new, non-financial measures should also be defined to be amenable to business process approaches, as indicated below:

- Lead time: the interval between start and end of a business process; it is the concept-to-market in the case of PDP, for instance.
- Customer service: Delivery of customer-desired products, at the right time, in the right place and in right quantities, every time; Customers could be external customers or internal customers.

Similarly, the important performance measures of quality, cost, dependability, flexibility, capacity, asset utilization, etc. should all be redefined in a process-oriented approach.

4. PERFORMANCE ANALYSIS BASED ON A PROCESS APPROACH

Traditional analysis has assumed that the enterprise is weakly coupled, i.e., various subsystems work almost independently. Paper-based communication, high inventories and arms-length relationships with suppliers, distributors, etc. justified this assumption. Analysis and optimization were confined to individual subsystems, and each function focused on optimizing its own measures. This has generally resulted in sub-optimization at the total system level.

The business process-oriented approach squarely addresses this issue of sub-optimization. It is also becoming more appropriate, given that present-day manufacturing systems are becoming more strongly coupled in terms of material and information flows. Stake holders in manufacturing enterprises are increasingly connected via Internet and Intranets, they share common databases, inventories are made low, with customers, both internal and external, insisting on more synchronized just-in-time modes of operation, etc.

For process-oriented performance analysis, three types of models may be considered: static models, dynamic models and software models.

4.1. Static Models

A business process is cross-functional and involves complex interactions among various entities. Constructing high-fidelity analytical models is quite a challenge. However, to understand the effects of critical and dominant constituents of a business process, it may often be enough to consider simple, static models. Influence diagrams, process flow graphs, and data flow diagrams are some static models that have been used in business process modeling.

4.2. Dynamic Models

To describe a business process in a dynamic and stochastic environment, one can view a business process as a discrete event dynamic system (Viswanadham and

Narahari 1992). Such a system has a state space wherein each state is a vector of components, each component describing an individual entity of the process. The state of the system changes due to the occurrence of an event such as the entry of a new customer order or the delivery of a product to a customer or the completion of design of a new product, etc. Under the usual assumption of Markovian dynamics, one can visualize a business process as a Markov chain (Viswanadham and Narahari 1992). A higher-level formalism such as a stochastic Petri net (SPN) (Viswanadham and Narahari 1992) can be used to automatically generate this Markov chain. An SPN model has the added advantage of directly representing features of a business process such as concurrency and asynchronism of activities, and conflicting and random nature of interactions.

Queuing models are also quite relevant in this context, and the next two sections are devoted to performance analysis of business processes using queuing network (QN) models. Several researchers have used queuing models in related problems. Adler, Mandelbaum, Nguyen, and Schwerer (1995) have developed a QN model of a product development process and used the model to analyze and predict development lead time. Buzacott (1996) has used certain well-known queuing models to assess the impact of the nine principles of reengineering enunciated by Hammer and Champy (1993). The máin conclusion of his study is that the reengineering principles do improve the performance in a marked way, especially under significant variability of activity times. Malone and Smith (1988), in their study, have looked at organizational and coordination structures, which constitute a key element of any business process. Again using simple queuing models, they have compared the efficacy of various organizational structures under a variety of conditions.

Dynamic models can also be used to capture certain key features of business processes at a higher level of abstraction, enabling rapid performance analysis. We believe there is a high potential for SPN and QN in modeling of business processes. Design issues can also be addressed by combining these models with mathematical programming or heuristic optimization techniques.

4.3. Software Models

Static and dynamic models discussed above are generally high abstraction models, based on simplifying assumptions such as Markovian dynamics. To obtain high-fidelity models, one has to represent many other details, through detailed specification and description in a programming language.

Object-oriented design and analysis, which has emerged in recent times as an important development in software engineering, seems to provide a natural paradigm for design and analysis of competitive business processes. There have been some efforts in this direction, e.g., the case studies in Booch (1994) and Jacobson (1995). Mujtaba (1994) describes enterprise modeling and simulation in an object-oriented environment; Bailetti, Callahan, and DiPetro (1994) present an object-oriented approach to represent coordination structures in the management of projects.

In an object-oriented model of a business process, the various constituents or entities of the process are natural candidates for being the objects. For example, the coordination structure, the information processing network, the physical layout, the resource allocation function, the technology adopted, etc., can all be considered as objects derived from various basic objects through inheritance. Using a generic library of such objects, one can quickly and modularly construct a model of an enterprise or an entire business process, and use the model in various ways to, for instance, derive detailed software specification and implementation to realize the business process; and, to simulate and assess the performance of the process.

5. LEAD TIME MODELS FOR NEW PRODUCT DEVELOPMENT

We now describe briefly a product development organization (PDO) studied by Adler et al. (1995) which serves as a typical example of a multi-project PDO, involved in development of plastics products which are either new products or reformulations. Much of the effort is spent on new products, which is the main area of concern here for modeling purposes. The main resources in the PDO are the product and process engineers and technicians, in addition to product management personnel, manufacturing engineers, marketing and sales personnel, etc.

In this PDO, at any given time, many new product development projects are in progress in different phases. This causes contention for engineering/human resources and results in delays at various points. Often times, different phases of the same product development project could be contending for a given resource.

An important aspect of a PDO is the need for feedback and rework at most stages of the process. This is necessitated because manufacturability and other problems can get unearthed at various stages, which calls for repeating a subset of PDP activities all over again. One can characterize this iteration structure using feedback probabilities.

The activities in this PDO can be categorized into five phases: Phase 1: Concept and Feasibility (CF); Phase 2: Project Plan and Team Formation (PP); Phase 3: Product Development; Phase 4: Manufacturing Standardization/Product Launch, and Phase 5: Continuous Improvement. Phase 5 is not considered here as in [5] since very little data can be collected on this issue. Phase 3 of the process contains the bulk of the work and again, as in [5], is chosen for a detailed study here. Phase 3 involves 8 activities: Manufacturing Process Development-1 (MPD1); Material Prototype and Testing (MPT); Product Prototype and Testing in Laboratory (PPTL); Product Prototype Testing and Manufacturing (PPTM); Manufacturing Process Development-2 (MPD2); Sales; Specifications; and Field Trials (FT). Figure 2, derived from the data available in Adler et al. (1995), shows the iteration structure typical in new product development.

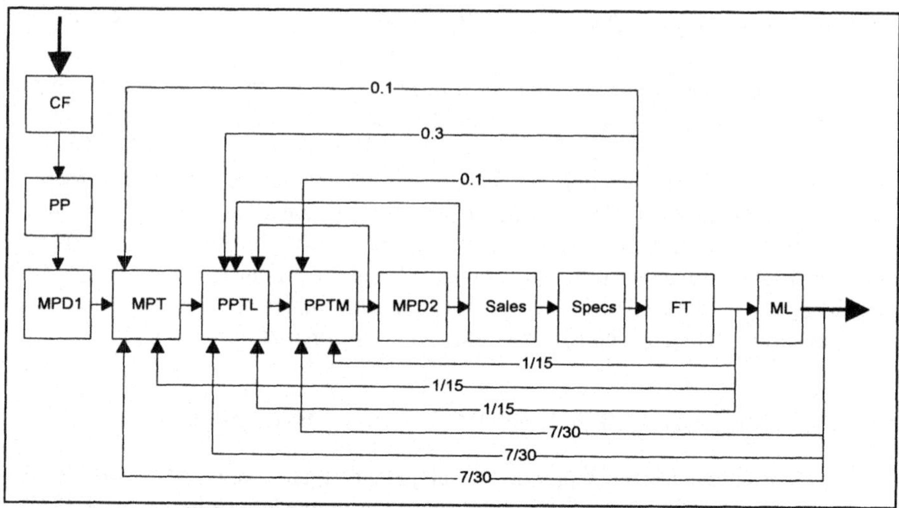

FIGURE 2. New product development process.

5.1. A Single Class Queuing Network (QN) Model

Figure 3 shows a coarse, single class QN model of the PDO in consideration. It is an 8-node open Jackson network with each node containing one server. It is an aggregated model, with MPT, PPTL, and PPTM aggregated into a single stage called PT, and Sales and Specifications into a single stage called SST. The three stages MPT, PPTL, and PPTM of Figure 2 involve essentially prototyping and testing, and involve roughly the same set of resources. Similarly, Sales and Specifications can also be combined into a single stage. Each node thus represents an aggregated, parallel set of activities and the single server in each node is a functional or cross-functional team carrying out this set of activities.

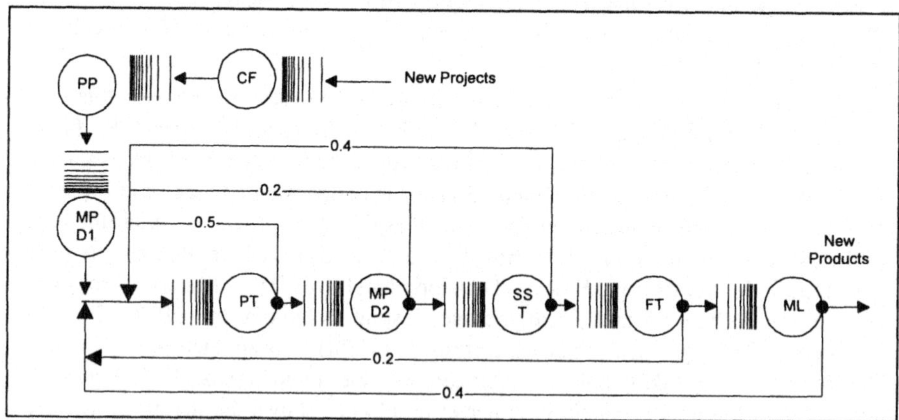

FIGURE 3. A single-class QN for a multi-project PDO.

New product development projects that are in progress in different stages represent the customers or jobs in the network. New projects enter the network into the CF node and successfully completed projects leave the network from node ML. Each project undergoes a sequence of activities in the manner shown in the network. A project can visit a node several times due to rework. The probability of returning to a previous stage after completing service at a particular stage is also shown for all feedback possibilities. It is assumed that the processing time distribution at a node for a project visiting the node for the first time is identical to that corresponding to all subsequent stages. This assumption can be relaxed in the case of the re-entrant line model discussed in the next subsection. The routing is as usual Markovian, i.e., independent of the previous history of the jobs.

A server corresponds to a functional or cross-functional team of human resources who undertake the set of activities corresponding to that node. It is possible that a given engineer/technician is involved in two or more parallel activities corresponding to the given node. The server is thus a conglomerate of all these human resources and the service time corresponds to the most time-consuming activity among the parallel activities.

We assume that inter-arrival times between new product initiations, and service times at various nodes are distributed according to a probability distribution whose mean and variance are known. These values can be obtained using measured data from a PDO. For example, Adler et al. (1995) obtained such data by interviewing the company personnel. The QN model is thus described by the following parameters: 1) number of nodes, 2) mean and variance of inter-arrival time of new projects, 3) mean and variance of service time distribution at each node, 4) routing matrix, and, 5) scheduling policy to be followed at each node. We assume, at each node, a non-preemptive FCFS policy and a buffer with infinite capacity.

5.2. Rapid Performance Analysis For Lead Time Reduction

Our aim here is to analyze the performance using the QN model of Figure 3 and explore opportunities for lead time reduction. The performance measure of primary interest is mean product development lead time (PDLT). Other performance measures such as variance of PDLT, mean number of projects in progress, utilization of resources, can also be computed. Since the model is a single-class open Jackson network with single-server stations, Markovian routing, general inter-arrival times, and general service times, we can use a package such as QNA (Whitt 1983), for which a software tool was developed at the Indian Institute of Science (IISc). The performance measures are computed using analytical formulae and for certain models, it takes less than one tenth of a second even on a primitive PC. Thus, one can use the model to do rapid performance analysis, and to evaluate the performance over a wide range of parameter values and get a comprehensive idea of system performance.

Table 1 shows the mean service times for the eight servers in our model, based on the data provided in Adler et al. (1995). The routing probabilities assumed are shown in Table 2. Assuming a Poisson arrival process with an inter-arrival time of 16 weeks (that is, one new project initiated every four months on an average), the above base model gives a mean development lead time of 82.04 weeks if the processing times are assumed to be exponentially distributed, with the means shown in Table 1. On the other hand, if the processing times are assumed to be deterministic with the above values, the mean lead time drops to 57.68 weeks.

TABLE 1. Mean Service Times for the QN Model.

Station	CF	PP	MPD1	PT	MPD2	SST	FT	ML
Service Time (Weeks)	2.0	1.0	2.5	1.0	2.5	1.5	1.5	4.0

TABLE 2. Routing Probabilities Assumed.

Station	CF	PP	MPD1	PT	MPD2	SST	FT	ML
CF	0	1	0	0	0	0	0	0
PP	0	0	1	0	0	0	0	0
MPD1	0	0	0	1	0	0	0	0
PT	0	0	0	0.5	0.5	0	0	0
MPD2	0	0	0	0.2	0	0.8	0	0
SST	0	0	0	0.4	0	0	0.6	0
FT	0	0	0	0.2	0	0	0	0.8
ML	0	0	0	0.4	0	0	0	0

5.2.1. Lead Time Reduction through Process Control.

Table 3 shows mean development lead times and mean number of ongoing projects in the organization assuming Poisson arrivals over a range of arrival rates. Three different service time scenarios are considered: 1) service times assumed to be exponentially distributed; 2) they are uniformly distributed with coefficient of variation, CV, (= ratio of standard deviation to the mean) of 0.144; and, 3) they are assumed to be constant, all based on the service times provided in Table 1.

The results indicate the increase in mean PDLT as a function of arrival rates of new projects. An item of major interest is the reduction in development lead time when CV of service times reduces from 1.0 (exponential distribution) to 0.144 (uniform distribution), and to 0.0 (deterministic service times). This is a direct consequence of the so-called variability law of queuing theory (Lu, Ramaswamy and Kumar 1994), according to which mean waiting times in a queuing system are positively correlated to variabilities of inter-arrival and service times. The reduction in variability in service times is a consequence of tighter control; in the product development setting, this translates to systematic planning, efficient project management, and a well-defined and well-understood product development process.

TABLE 3. Lead Times With Different Service Time Variabilities.

	Exponential		Uniform		Constant	
Inter-arrival Time (weeks)	Mean PDLT (weeks)	Mean No. of projects	Mean PDLT (weeks)	Mean No. of projects	Mean PDLT (weeks)	Mean No. of projects
24	58.89	2.48	47.65	1.98	46.46	1.93
22	61.84	2.81	50.63	2.31	49.50	2.25
20	65.93	3.26	53.21	2.66	49.59	2.48
18	72.00	4.08	57.95	3.21	55.91	2.92
16	82.04	5.12	64.72	4.04	57.68	3.61
15	90.14	5.84	68.07	4.53	64.69	4.312
14	102.58	7.18	72.18	5.15	66.26	4.732
13	124.76	8.31	84.26	6.48	79.43	6.92
12	179.27	15.72	157.54	13.13	139.70	11.64

5.2.2. Lead Time Reduction Through Input Control.

Waiting times reduce with reduced variability of arrivals, which provides another opportunity for lead time reduction. However, since new product project initiations are often motivated by market opportunities, the arrival process here is subject to the market fluctuations. One way of reducing the arrival fluctuations is to operate the organization in a closed network mode, i.e. initiate a new project only when an existing one finishes. This ensures a constant population of projects inside the PDO and reduces the variance of arrivals. Another way is to admit a new project only when the number of projects is below a threshold or the total workload in the system is below a particular threshold.

Table 4 shows the effect of operating the PDO in a fixed population mode, based on the values shown in Tables 1 and 2. Service times are assumed to be exponentially distributed. The first column gives the current population of the network in the closed network mode; the second gives the mean PDLT for the corresponding population; the third gives the mean inter-arrival times that are consequent on having the corresponding population in the closed network; the fourth column provides the corresponding throughput rate of successful projects completed per year; the final column gives the mean PDLT if the PDO is operated as an open network with these throughput rates.

It must be noted that in a stable open network, the arrival rate is the same as the throughput rate of the network. Upon close observation, the merits of operating in a fixed-population mode become clear. For example, with a population of 5, the mean PDLT is 70.897 weeks; the throughput rate is 3.667 completed projects per year. To obtain this throughput rate using an open mode of operation will entail a mean PDLT of 99.85 which is more than 30% higher. Thus, lower lead times are achieved for a specified throughput rate and, conversely, higher throughput rates can be obtained for specified cycle time. However, a closed mode will entail rejection of some projects and also continuous availability of fresh projects for initiation.

TABLE 4. Lead Time Reduction Through Input Control.

Population For Closed Model	Mean PDLT (weeks)	Associated Mean IAT (weeks)	Throughput Rate (per year)	Mean PDLT for Open Model
1	40.031	40.031	1.299	49.150
2	47.168	23.584	2.201	59.140
3	54.712	18.237	2.851	71.324
4	62.633	15.658	3.320	84.540
5	70.897	14.179	3.667	99.850
6	79.472	13.245	3.926	117.920
7	88.326	12.618	4.121	139.010
8	97.431	12.179	4.271	164.270
9	106.760	11.862	4.383	195.010
10	116.290	11.629	4.472	232.610
11	125.995	11.454	4.540	278.120
12	135.860	11.321	4.593	333.110
13	145.860	11.219	4.634	405.620
14	155.970	11.141	4.667	493.460
15	166.190	11.080	4.693	601.320

6. CONGESTION MODELS FOR SUPPLY CHAIN NETWORKS

Simple QN models can effectively characterize supply chain processes. To illustrate the modeling process, consider a simple supply chain network consisting of two product lines, A and B, assembled in two separate plants and distributed by a centralized distribution network (Figure 4). Product line requires A requires sub-assemblies A1 and A2, while B requires sub-assemblies B1 and B2. Sub-assembly A1 is procured from Supplier S1; sub-assemblies A2 and B1 from Supplier S2; and subassembly B2 from Supplier S3. The logistics and distribution are handled by a single fleet of transporters. This fleet is responsible for transporting sub-assemblies to the plants P1 and P2 and for reaching the finished products to the distributors. At the supply side, a transport vehicle will wait for a batch of A1 from supplier S1 and a batch of A2 from supplier S2 and then transport them to plant P1. Similarly, a vehicle will wait for a batch of B1 from Supplier S2 and a batch of B2 from supplier S3 and transport them to Plant P2. Also the same vehicles may be requisitioned for moving a batch of finished products A and B to the distributors. At a given point, a vehicle can only carry sub-assemblies A1 and A2; or sub-assemblies B1 and B2; finished product A, or finished products B. Batch processing is assumed, with a job representing a batch of sub-assemblies or finished products.

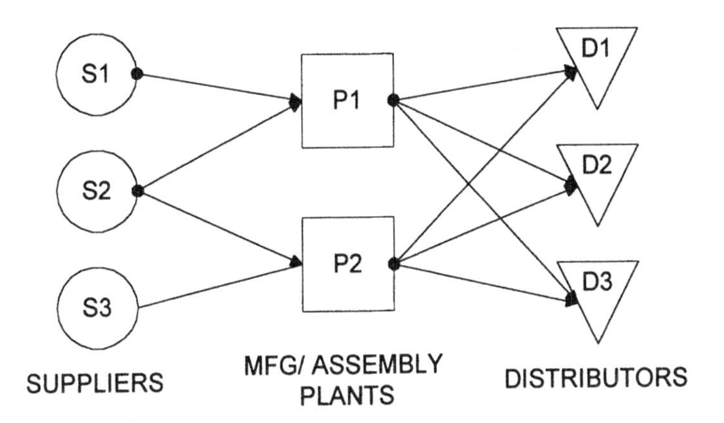

FIGURE 4. Supply chain network considered.

6.1. A Fork-Join QN Model

Figure 5 represents a queuing network model of this system. The model has two classes of jobs, class A and class B, with the routes for these two classes shown in the figure. This model is a fork-join queuing network (Massey, Baccelli and Towsley 1989). Forking denotes simultaneous creation of two or more customers from a single request. For example, a demand for a batch of class A products will simultaneously generate two orders, one for a batch of sub-assemblies A1 and another for sub-assemblies A2. Joining represents synchronization. There are two join elements in this model. The first join occurs before transporting sub-assemblies A1 and A2 to Plant P1; the second join denotes a similar synchronization between sub-assemblies B1 and B2 before being transported to Plant P2.

The model has six nodes, corresponding to the three suppliers, two plants, and the logistics and distribution system. There is queuing space with infinite capacity in front of each of these six nodes. Orders or sub-assemblies or finished products wait in these queues. There are two queues in front of Supplier 2, one containing backorders for sub-assemblies A2 and the other containing backorders for sub-assemblies B1. A sequencing policy has to be specified to decide which of the orders will be taken up next for processing by Supplier 2, and we assume an FCFS policy. There are three queues in front of the logistics and distribution system. This is modeled as a multi-server system comprising identical servers (vehicles). The first queue contains subassembly batches waiting to be transported to Plant P1. The second queue here has subassembly batches waiting to be transported to Plant P2. The third queue has finished products waiting to be moved to the distributors. Interesting scheduling policies can be used here and FCFS will be the usual straw policy employed. The so called buffer priority policies will be of interest here

(Kumar 1993). At the other four nodes (Supplier S1, Supplier S3, Plant P1, and Plant P2), there is a single queue and we assume an FCFS policy at these queues.

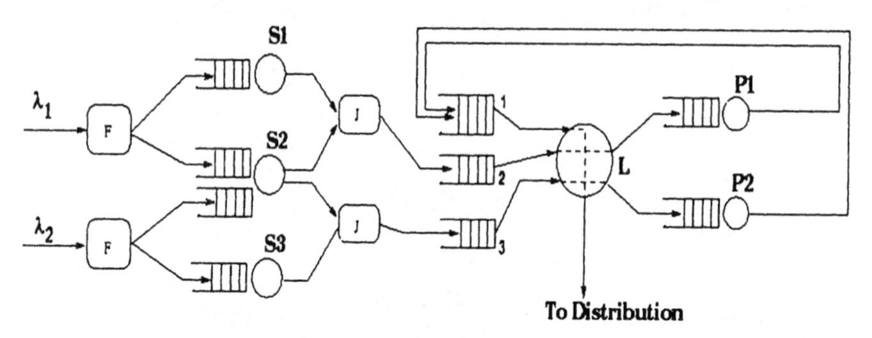

FIGURE 5. An open two-class, fork-join QN model for the supply chain.

The model is an open QN model, with two external arrival processes corresponding to the orders for product lines A and B. These orders are generated by a complex process of demand forecasts, and safety stock calculations in the case of a make-to-stock system; in the case of a make-to-order system, these orders result from a series of feedback steps starting from actual customer orders. Because of the inherent uncertainty of customer demands, we need to model these arrival processes as random processes. In this case, we choose the simplistic, analytically tractable, and real world-oriented Poisson process. Let $\lambda 1$ and $\lambda 2$ be the arrival rates of these two processes. We shall assume that other service times (durations required to prepare the sub-assemblies, transportation times, assembly lead times in the two plants) to be uniformly distributed random variables with appropriate means. We do not assume any setup times nor do we assume any batching of jobs (for example for transportation). If simulation is being used for performance analysis, such features can be included without much difficulty. In this case, we shall keep the model simple and uncomplicated.

The above fork-join QN model is analytically intractable for many reasons, such as non-product-form structure, non-exponential processing times, forking, joining, etc. However, it is amenable for rapid experimentation using simulation.

6.2. Simulation Results

The performance measures of interest that can be computed using simulation are: mean and variance of lead time (from suppliers to distribution), mean waiting time at various stations, mean synchronization time at each join point, and utilization of the various resources. We can study the performance of this supply chain network under various input conditions and scheduling policies. Here, we shall focus on the mean cycle times for the two classes of products and the utilization of the various servers in the network.

Table 5 shows input parameter values chosen for the base model. Tables 6, 7, and 8 show server utilization (steady-state) and the mean cycle times (steady-state) for class 1 and class 2 products, under various scheduling policies at Node L (the logistics and distribution node). Note that there are three buffers at this node: Buffer 1 contains subassembly batches waiting to be transported to Plant P1. Buffer 2 has subassembly batches waiting to be transported to Plant P2. Buffer 3 contains finished products of both classes waiting to be moved to the distributors. Six buffer priority policies are possible here: (1,2,3); (1,3,2); (2,1,3); (2,3,1); (3,1,2); and (3,2,1). The policy (1,2,3) for example, gives highest (non-preemptive exhaustive) priority for jobs in Buffer 1, followed by jobs in Buffer 2, and finally jobs in Buffer 3 (same as the first buffer first serve policy - FBFS). The policy (3,2,1) is the last buffer last serve policy.

Table 6 corresponds to the case when there is only one server (one vehicle) in node L. The table shows utilization for various server nodes and mean cycle times (MCT1 and MCT2) for class 1 and class 2 products. It is clear that the server at L is heavily utilized. Also, there is appreciable difference in the mean cycle times of class 1 and class 2 for some of the buffer priority policies. This can be explained by the exhaustive nature of these policies.

Table 7 shows these performance measures when we have two identical vehicles instead of just a single vehicle. We now see a dramatic improvement in the performance which is due to the reduced congestion at node L. Table 8 presents the results when we have 3 identical vehicles. As can be expected, the performance improves further, but only marginally since the congestion was already just moderate in the case of two vehicles. Thus, considering investments, having two vehicles is much more judicious than three vehicles. In the case of three vehicles, we can also investigate the effect of dedicating a vehicle to each buffer. As to be expected, the performance is better with pooled vehicles than with dedicated vehicles (see last column of Table 8).

TABLE 5. Input Parameters.

Arrival Rate	$\lambda_1 = \lambda_2 =$	~Poisson (0.2) orders/week
Service Rates	server S1	~uniform (1, 2) weeks
	server S2	~uniform (0.5, 0.6) weeks
	server S3	~uniform (1, 1.5) weeks
	L: Buffer 1	~uniform (1, 1.2) weeks
	L: Buffer 2	~uniform (0.8, 1.2) weeks
	L: Buffer 3	~uniform (1, 1.5) weeks
	server P1	~uniform (0.6, 0.8) weeks
	server P2	~uniform (0.6, 0.8) weeks

TABLE 6. Performance With One Server in Node L.

	(1, 2, 3)	(1, 3, 2)	(2, 1, 3)	(2, 3, 1)	(3, 1, 2)	(3, 2, 1)	FCFS
ρ_{S1}	0.2998	0.2998	0.2998	0.2998	0.2998	0.2998	0.2998
ρ_{S2}	0.2201	0.2201	0.2201	0.2201	0.2201	0.2201	0.2201
ρ_{S3}	0.2506	0.2506	0.2506	0.2506	0.2506	0.2506	0.2506
ρ_{L}	0.9205	0.9205	0.9205	0.9206	0.9205	0.9207	0.9206
ρ_{P1}	0.1399	0.1399	0.1399	0.1399	0.1399	0.1399	0.1399
ρ_{P2}	0.1403	0.1403	0.1403	0.1403	0.1403	0.1403	0.1403
MCT1(weeks)	27.0308	7.72051	27.9409	28.4992	7.2490	27.9358	21.3629
MCT2 (weeks)	27.4521	28.8191	26.5452	7.0949	28.2635	6.7599	20.6569

TABLE 7. Performance With Two Servers in Node L.

	(1, 2, 3)	(1, 3, 2)	(2, 1, 3)	(2, 3, 1)	(3, 1, 2)	(3, 2, 1)	FCFS
ρ_{S1}	0.2998	0.2998	0.2998	0.2998	0.2998	0.2998	0.2998
ρ_{S2}	0.2201	0.2201	0.2201	0.2201	0.2201	0.2201	0.2201
ρ_{S3}	0.2506	0.2506	0.2506	0.2506	0.2506	0.2506	0.2506
ρ_{L}	0.4608	0.4609	0.4608	0.4610	0.4609	0.4610	0.4609
ρ_{P1}	0.1399	0.1400	0.1399	0.1399	0.1400	0.1400	0.1399
ρ_{P2}	0.1403	0.1403	0.1403	0.1403	0.1403	0.1403	0.1403
MCT1(weeks)	5.2075	5.1271	5.2236	5.2383	5.1408	5.2142	5.1802
MCT2 (weeks)	4.7428	4.7548	4.7247	4.6399	4.7297	4.6521	4.7043

TABLE 8. Performance With Three Servers in Node L.

	(1, 2, 3)	(1, 3, 2)	(2, 1, 3)	(2, 3, 1)	(3, 1, 2)	(3, 2, 1)	FCFS	Dedicated
ρ_{S1}	0.2999	0.2999	0.2999	0.2999	0.2999	0.2999	0.2999	0.2999
ρ_{S2}	0.2201	0.2201	0.2201	0.2201	0.2201	0.2201	0.2201	0.2201
ρ_{S3}	0.2505	0.2505	0.2505	0.2505	0.2505	0.2505	0.2505	0.2505
ρ_{L}	0.3068	0.3068	0.3068	0.3068	0.3068	0.3068	0.3068	0.2198
ρ_{P1}	0.1399	0.1399	0.1399	0.1399	0.1399	0.1399	0.1399	0.1399
ρ_{P2}	0.1402	0.1402	0.1402	0.1402	0.1402	0.1402	0.1402	0.1402
MCT1 (weeks)	4.9389	4.9381	4.9393	4.9401	4.9388	4.9396	4.9389	5.3395
MCT2 (weeks)	4.4341	4.4347	4.4337	4.4327	4.4341	4.4332	4.4339	4.8488

7. CONCLUSIONS

In this chapter, we have considered an integrated manufacturing enterprise, as a collection of value-delivering, business processes. The business process view contrasts with the traditional functional view, which we submit, is largely inappropriate for contemporary manufacturing systems. We have stressed that a business process approach to modeling manufacturing systems effectively addresses various aspects of system performance. Performance measures need to be defined to suit business processes, and these include lead time, quality, dependability, cost, flexibility, asset-utilization, and capacity. Focusing on lead time, we have presented conceptual lead time models, based on single class and multi-class queuing networks, for two important business processes: the new product development process and the supply chain process. This general approach to modeling may provide valuable insights and direction in current efforts toward reorganization of manufacturing systems into value delivering business processes.

REFERENCES

Adler, P.S., Mandelbaum, A., Nguyen, V. and Schwerer, E., 1995, From project to process management: An empirically-based framework for analyzing product development time, *Management Science*, 41, 3, 458-484.

Bailetti, A.J., Callahan, J.R. and DiPetro, P., 1994, A coordination structure approach to the management of projects, *IEEE Transactions on Engineering Management*, 41, 4, 396-403.

Booch, G., 1994, *Object-Oriented Analysis and Design with Applications*, Benjamin-Cummings, Redwood, CA.

Buzacott, J. A., 1996, Commonalties in reengineered business processes, *Management Science*, 42, 5, 768-782.

Cooper, R.G., 1983, A process model for industrial new product development, *IEEE Transactions on Engineering Management*, 30, 1, 2-11.

Davenport, T. H., 1993, *Process Innovation*, Harvard Business School Press, Cambridge, MA.

Hammer, M. and Champy, J, 1993, *Reengineering the Corporation: A Manifesto for Business Process Revolution*, Harper Business.

Jacobson, I., 1995, The Object Advantage: Business Process Reengineering with Object Technology, Addison-Wesley Publishers.

Kumar, P.R., 1993, Re-Entrant lines, *Queuing Systems: Theory and Applications*, 13, 87-110.

Lu, S.H., Ramaswamy, D. and Kumar, P.R., 1984, Efficient scheduling policies to reduce mean and variance of cycle-time in semiconductor manufacturing plants, *IEEE Transactions on Semiconductor Manufacturing*, 7, 3, 374-388.

Malone, T.W. and Smith, S.A., 1988, Modeling the performance of organizational structures, *Operations Research*, 36, 3, 421-436.

Massey, W.A., Baccelli, F. and Towsley, D., 1989, Acyclic fork-join queuing systems. *Journal of the ACM*, 36, 615-642.

Mujtaba, M.S., 1994, Enterprise modeling and simulation: Complex dynamic behavior of a simple model of manufacturing, *Hewlett-Packard Journal*, December, 80-114.

Ulrich, K.T. and Eppinger, S.D., 1995, *Product Design and Development*, Mc-Graw Hill, NY.

Whitt, W., 1983, The queuing network analyzer, *Bell Systems Technical Journal*, 62(9), 2779-2815.

Viswanadham, N. and Narahari, Y., 1992, *Performance Modeling of Automated Manufacturing Systems*, Prentice Hall, Englewood Cliffs, NJ.

AUTHORS' BIOGRAPHY

Dr. N. Viswanadham is TataChem Professor of Computer Science & Automation, and Chairman, Division of Electrical Sciences, Indian Institute of Science (IISc), Bangalore, India. He has been with IISc since 1967, and obtained his Ph.D. in 1970. He has held visiting appointments at Universities of New Brunswick, Waterloo, Connecticut and Stanford, and General Electric Corporate Research and Development Center. He was a GE Research Fellow in 1989. He is a Fellow of *IEEE, Indian National Science Academy, Indian Academy of Sciences, Indian National Academy of Engineering and Third World Academy of Sciences*. He has authored numerous journal and conference papers. He has co-authored two text books: *Reliability in Computer and Control Systems* (1987), and *Performance Modeling of Automated Manufacturing Systems* (Prentice Hall, NJ, USA, 1992). He has edited or co-edited five books: *Reliability and Fault-Tolerance Issues in Real-Time Systems* (1987), *Systems and Signal Processing* (1991), *Recent Advances in Modeling and Control of Stochastic Systems* (1991), *Artificial Intelligence and Expert Systems Technologies in the Indian Context* (1991), and *Computing and Intelligent Systems* (1993). He is Editor of *Sadhana: Academy Proceedings in Engineering Sciences*, and Associate Editor of *IEEE Trans. Robotics & Automation, Journal of Manufacturing Systems, Intelligent and Robotic Systems, The International Journal of Information Technology* and *Journal of Franklin Institute*.

Dr. Y. Narahari is Associate Professor at Department of Computer Science and Automation, Indian Institute of Science, Bangalore, India. He is currently a visiting research scientist at National Institute of Standards and Technology (NIST), Gaithersburg, MD, USA. He completed his ME in Computer Science in 1984 and his Ph.D. in 1988, both at IISc. He has co-authored, with Dr. Viswanadham, th book *Performance Modeling of Manufacturing Systems*, (Prentice-Hall, NJ, USA, 1992). In 1992, he was awarded Indo-US Science and Technology Fellowship, and he visited the Laboratory for Information & Decision Systems, Massachusetts Institute of Technology to work on dynamic and stochastic scheduling of manufacturing systems. Dr. Narahari's research interests are in performance modeling and scheduling of manufacturing systems and he has been a consultant for several industrial organizations and investigator of many R & D projects.

Mr. N. R. Srinivasa Raghavan is working towards his Ph.D. in Computer Science and Automation, at Indian Institute of Science with Professor Viswanadham. He holds a B.Tech. degree, with top honors, in mechanical engineering in 1993 from S.V.U. College of Engineering, Tirupati, India. He completed M. Tech. in Management Science from IISc in 1995, securing first rank and Institute gold medal. His research interests include modeling and analysis of business processes, scheduling in supply chains, business process reengineering, intelligent agent-based modeling and object-oriented simulation of manufacturing enterprises.

E2

Cellular Manufacturing Feasibility at Ingersoll Cutting Tool Co.

D. J. Johnson and U. Wemmerlöv

1. BACKGROUND

Ingersoll Cutting Tool Company (ICTC) of Rockford, Illinois, USA is a manufacturer and supplier of metal cutting tools and tool inserts to industrial customers. The company had been experiencing growth rates over the last few years substantially higher than the average rate for the industry, as reported by the United States Cutting Tool Institute. ICTC's order receipt to product shipment lead times were shorter than most other firms in the industry and estimated to range from 5 to 20 weeks depending on cutter type. However, demands from key customers, as well as competitive pressures, required that lead times be reduced if ICTC was to maintain or increase the strength of its market position. Although some cost pressures had been felt as a result of new firms entering the industry, response time was still the main competitive priority (especially for made-to-order items). The purpose of this study was to analyze the entire cutter production system to determine changes needed to achieve a 40-50% reduction in delivery lead times.

Before this study began, ICTC had been working on lead time reduction in the office with another University of Wisconsin team (Wayman 1995). The result had been a 75% reduction in customer order processing lead time for the "Modified Standard" cutter product line (cutters similar to make-to-stock cutters with respect to length, width, etc., but redesigned to customer specifications). Much of this success was attributed to a cross-functional team known as Fast Track which handles all design, process planning, NC programming, and raw material ordering for each customer order received. The implementation of this team changed the work organization from one focused on individual processes related to a customer order to one responsible for completing an entire customer order. ICTC was now interested in knowing if the cellular concept could also be successfully applied to cutter manufacturing in order to achieve the desired lead time reduction.

2. PRODUCTS AND PROCESSES

Cutting tools produced by ICTC vary greatly in size and shape, but can be classified as either end mills, face mills, drills, slotting cutters, gear gashers, shell mills, half and full side mills, slab mills, boring cutters, broaches, arbors, adapters, or centering shanks. Approximately 25% of the cutting tools produced (measured by direct labor hours) are made to stock while the remaining 75% are made to order.

The order management/manufacturing/delivery process at Ingersoll can be divided into four parts: order processing, cutter manufacturing, cutter finishing, and delivery (see Figure 1).

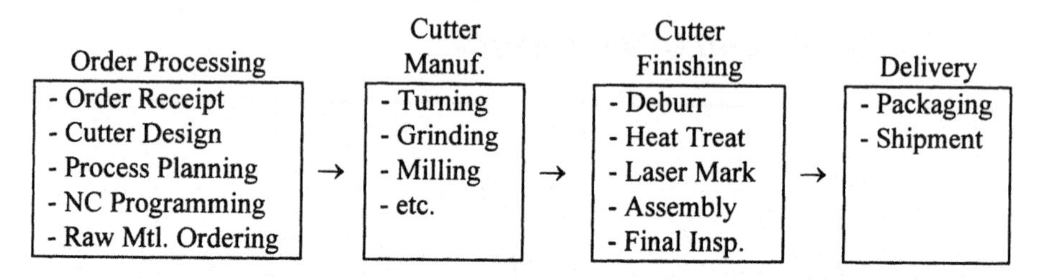

FIGURE 1. Steps in cutter order/manufacturing/delivery process.

Approximately 60% of all customer orders for cutters are handled by the Fast Track team referred to above. This includes orders for Modified Standards, as well as for other, fairly standardized cutters. The average lead time for these orders is approximately two days. The remaining 40% of orders are processed in a functional arrangement where different departments perform the design, process planning, NC programming, and materials ordering. The average lead time for these orders is about 12 days.

At the time this study began, cutter manufacturing was performed in a functional layout with machines performing similar processes grouped together. The number and types of machines in the shop are displayed in Table 1 and a block layout of the manufacturing facility is shown in Figure 2. Most of the machines in the lathe, grinding, and milling machine areas were run by one operator per shift.

TABLE 1. Cutter Manufacturing Equipment.

Machine ID and Type	Quantity	Machine ID and Type	Quantity
4111 Lathe	2	4270 Surface Grinder	1
4112 Lathe	1	4320 5-Axis Mill. Mach.	3
4120 Lathe	1	4330 5-Axis Mill. Mach.	4
4187 Manual Mill. Mach.	2	4340 5-Axis Mill. Mach.	2
4190 Broach	1	4350 6-Axis Mill. Mach.	1
4220 Rotary Grinder	1	4360 5-Axis Mill. Mach.	2
4230 I.D. Grinder	1	4371 Excello Flat Mach.	1
4240 Small O.D. Grinder	1	4373 Excello Gun Drill	1
4250 Large O.D. Grinder	1		

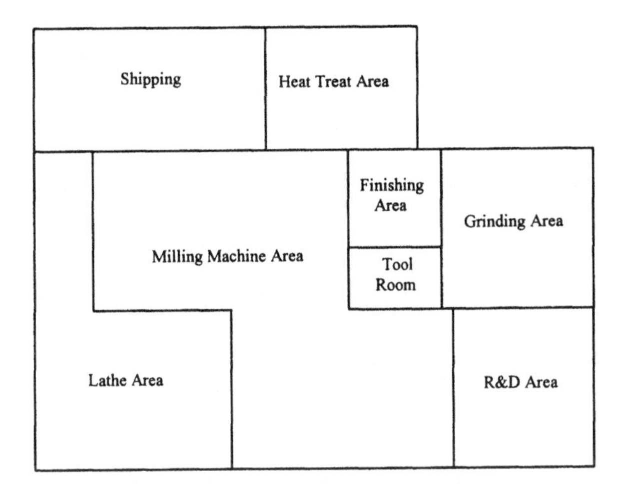

FIGURE 2. Block layout of functional system.

Average made-to-stock and made-to-order batch sizes were 8 and 3 units, respectively. Despite the small lot sizes, lead times were quite long in this part of the manufacturing process, believed by ICTC personnel to range from 20-30 work days. Finishing of all cutters was done in an area referred to as the "finishing cell," with lead times averaging 1½ - 2 days. Lead times in the delivery stage of the process were also fairly short, averaging 1-2 days. Thus, cutter manufacturing contributed the most to the overall lead time and the main analysis was confined to this portion of the process.

3. METHODOLOGY

Discussions were first held with personnel at ICTC to gain an understanding of the existing cutter manufacturing system. These discussions revealed three important facts: (1) a significant amount of lathe work was being outsourced due to insufficient in-house capacity, (2) a large WIP inventory was queued up in front of the 4320 and 4330 milling machines, indicating a potential bottleneck operation, and (3) the reliability of the 4111 lathes and the 4320, 4330, and 4360 milling machines was poor, resulting in sizable amounts of machine downtime.

In order to analyze system behavior, a mathematical model of the existing manufacturing system was developed using the MPX software package (MPX is a registered trademark of Network Dynamics, Inc., 128 Wheeler Road, Burlington, MA 01803; also see Suri 1996). MPX is a queuing theory-based program (as opposed to stochastic/dynamic simulation) that allows the rapid estimation of utilization levels at each machine, WIP inventory levels at each machine, and lead times for each batch. To use MPX, historical data were collected on cutter routings, average production batch sizes for each cutter type, the percentage of released routings that required subcontracted operations, the types of outsourced operations per routing, setup times per batch and run times per piece at each machine group for each cutter type, machine downtime and repair data for each machine group, and demand volumes for each cutter type over a 50 week period. Based on these initial data, we investigated several "what-if" scenarios to determine the impact of changes to the production system.

4. ANALYSIS OF THE EXISTING SYSTEM

The theoretical capacity in ICTC's machine shop was 132 hours/week/machine based on a 22 hours/day, 6 days/week availability. However, this level of machine availability is unrealistic since it does not account for employee training time, meetings, absenteeism, preventative maintenance, plant shutdowns, etc. Consequently, effective capacity was assumed to be 90% of this nominal time, i.e., 118.8 hours/week/machine. Using the MPX software, we then calculated the average utilization for each machine group. The utilization levels for the highest loaded equipment, considering setup time, run time, and machine downtime, are shown in Figure 3.

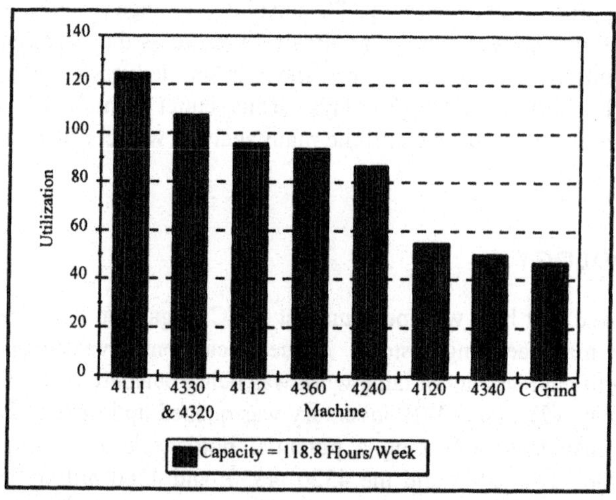

FIGURE 3. Machine utilization based on one year of demand data.

The calculation of utilizations assume that the grinding operations on several product types have been eliminated as a result of a new processing procedure that was being developed. Further, the 4320 and 4330 milling machines were considered completely interchangeable and treated as one machine group for the purpose of this analysis. Finally, Figure 3 shows an estimated utilization for a centerless grinding machine. This operation was outsourced at the time but our analysis led to the recommendation that this subcontracted work should be brought in-house by the purchase of a new machine (see Section 5).

Figure 3 indicates that the two 4111 lathes and the seven 4320 & 4330 milling machines were loaded beyond 100%. However, the calculations were based on the assumption that all load emanating from the historical demand volumes was routed to the machines in the plant. As it turned out, subcontracting of lathe work had been used much of the previous year in order to reduce the load on these machines. Further, work normally intended for the 4320 & 4330 milling machines had also been subcontracted over the last two months of the time period examined in an effort to lower the load. While the actual amount of outsourced work was unknown, an examination of company files showed large WIP levels consistently positioned in front of the 4111/4320/4330 lathes and mills. This indicated that the real utilizations of these machine groups were close to 100%.

Even if capacity was sufficient to process the load on these machine groups throughout the year, the high utilization levels would not provide enough of a capacity cushion to handle spikes in demand without adversely affecting lead time. To illustrate this phenomenon, the net cumulative weekly load (i.e., any unprocessed load from previous periods plus any new load for the current period) on the 4320 & 4330 mills was determined for a 50 week period. Figure 4 demonstrates that demand spikes in excess of capacity that occurred in weeks 29, 30, 32, 36, and 40, resulted in an overload on the system for a total of 16 weeks (weeks 29-44), causing a corresponding increase in the lead times for all cutters produced during this period. Similar results were found for the 4111, 4112, and 4360 machine groups.

This load analysis indirectly points to a fundamental tradeoff that exists in any system, namely that high machine utilizations can only be maintained at the expense of long manufacturing lead times. The specific relationship between work load, WIP inventory , and lead time at the combined 4320 & 4330 machine group is shown in Figure 5. This figure was constructed from queuing theory formulas (see Gross and Harris 1985) using setup and run time data from ICTC and its upper region validated using actual WIP inventory levels. (The reason queuing formulas rather than the MPX software was used to derive Figure 5 is that MPX can not calculate lead times or WIP once utilization levels exceed 95%.)

Two things should be noted. First, the higher the average machine utilization, the longer the average manufacturing lead time. Second, the higher the machine utilization, the more sensitive the system becomes to variations in load. Thus, even though a machine utilization of 96% will result in a lead time of 2 days to get a job through this machine group, this relationship is only true "on the average." Any temporary increase in machine load at this level of utilization quickly causes lead times to increase drastically.

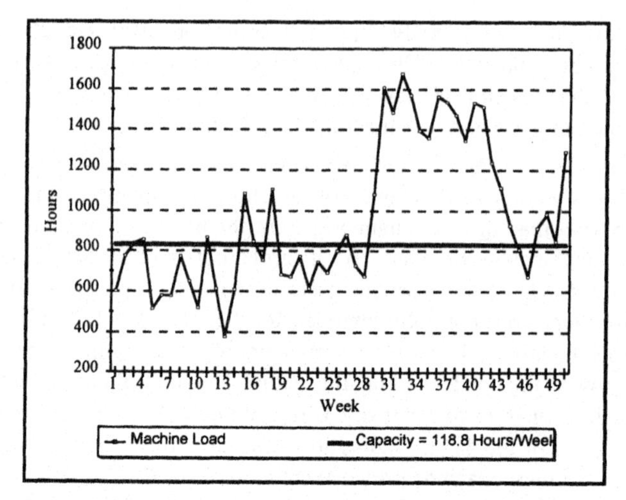

FIGURE 4. Net cumulative weekly load in hours at the 4320 and 4330 milling machine group.

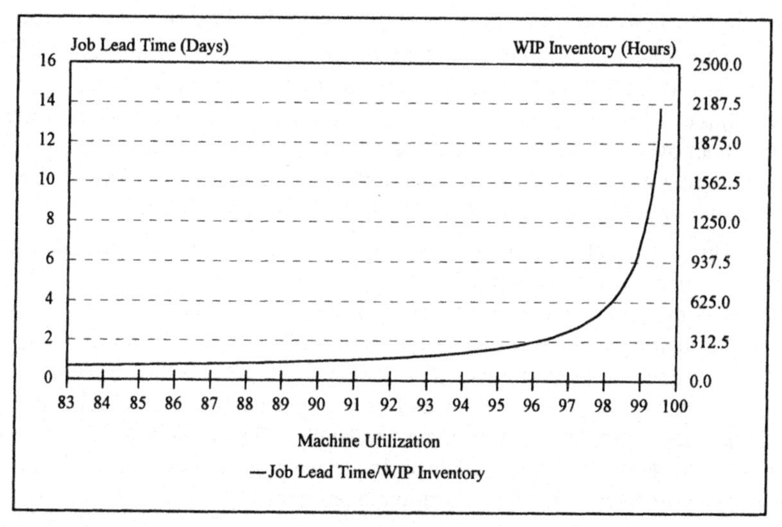

FIGURE 5. Job lead time/WIP inventory versus utilization at the 4320 and 4330 milling machine group.

In addition, high machine utilization levels leave little time for preventative maintenance of equipment. As preventative maintenance is reduced, machine failure incidences and repair times increase, further reducing available capacity and increasing the amount of variability in the system. The result is escalated lead times. Thus, Figure 5 suggests that Ingersoll's highest loaded milling machine group - in order to reduce and to stabilize lead time - should be operated at an average utilization level probably not exceeding 90%.

5. OPTION 1 - RETAINING THE FUNCTIONAL LAYOUT

Given the results of the above analysis, it was obvious that ICTC was constrained by capacity, especially in the lathe and milling machine areas. While subcontracting can occasionally be used to provide a temporary increase in capacity, it also added 3-6 days to cutter lead time and reduced the amount of control which ICTC had over the manufacturing process. Operations that were outsourced due to lack of in-house capabilities, such as centerless grinding, suffered from the same problems. Accordingly, the following steps were recommended to management in order to shorten lead times, and also bring all outsourced milling, lathe, and centerless grinding work back in-house, if the existing functional work organization was to be maintained:

1. Purchase one lathe with capabilities similar to the 4111 lathes (creating a group of three lathes)
2. Purchase at least one mill with capabilities similar to the 4320/4330 machines (creating a group of 8 or more)
3. Purchase a centerless grinding machine in order to perform this operation in-house (first centerless grinder)

Also:

4. Improve the preventative maintenance program in order to reduce machine downtime and further increase capacity
5. Begin monitor utilizations and/or lead times and add capacity as needed. For the shop in general, average machine utilizations should not exceed 75-85% if lead times are to be kept short. Keeping utilization levels within this range usually allows for a large enough capacity cushion to handle demand spikes or other variations without adversely affecting lead time performance.

According to the MPX analysis, adding the proposed capacity would result in an 83% average utilization level at the 4111 lathes and a 94% average utilization level at the 4320 & 4330 milling machines. Figure 6 shows the impact of this capacity infusion on the average job lead time for cutter manufacturing. The left bar shows the average lead time recorded in the shop as part of this study and includes the time required by outsourced lathe, milling, and centerless grinding work. The right bar shows the lead time estimated by MPX and assumes that all lathe, milling, and centerless grinding work is performed in-house. As seen, the additional capacity would reduce average job lead time from 24.5 to approximately 2.5 days if the shop operated under the same 50-week historical load pattern as used in our analysis.

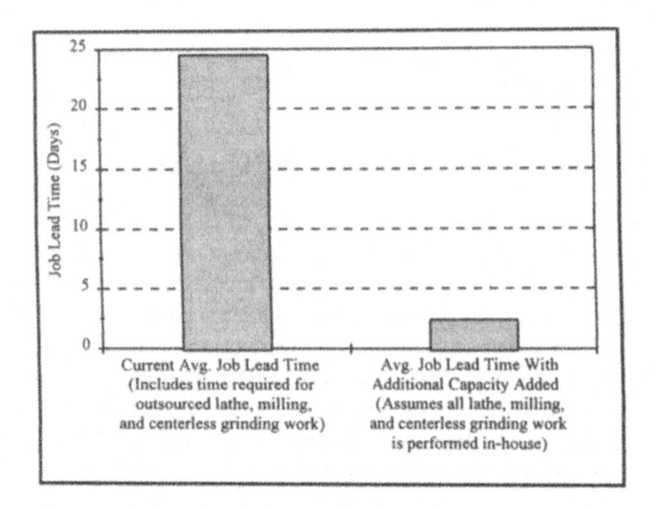

FIGURE 6. Impact of capacity infusion on average job lead time for cutter manufacturing.

6. THE POTENTIAL FOR CELLULAR MANUFACTURING

The next step in our analysis was to investigate the feasibility of implementing cellular manufacturing at ICTC. From a technical or part-machine perspective, a manufacturing cell is a group of functionally dissimilar machines or work stations, located closely together and dedicated to the (complete) manufacture, assembly, and/or packaging of a set of similar parts or products (called a part or product family). From a work organization perspective, a manufacturing cell is an organizational entity, consisting of one or more operators under the guidance of a supervisor or team leader and assigned responsibility for a total manufacturing process (unbroken sequences of operations). This includes responsibility for such tasks as scheduling of work, quality control and improvement, simple preventative maintenance, and other "indirect" tasks. Thus, a manufacturing cell adopts a product-oriented philosophy where work is organized to complete similar parts or products and the focus is on materials flow and system efficiency (speed, quality, and total manufacturing cost).

6.1. Data Analysis

We initially explored the possibility of forming cells dedicated to the complete manufacture of single cutter types (i.e., end mills, face mills, drills, etc.). For this part of the study, released part routings for a two-month period were first sorted to determine the kind of equipment required for each type of cutter. The examination of the routings revealed, first of all, that the majority of the cutters manufactured at ICTC required six or fewer in-house manufacturing operations. Second, it was unlikely that cells could be dedicated to cutter types without duplicating a significant amount of equipment, since several cutter types required the same pieces of unique equipment (i.e., the 4112 and 4120 lathes, and the 4220, 4230, and 4240 grinding machines).

We next sorted, using a spreadsheet, the routings within each cutter type according to operation sequences. This revealed that end mills could be divided into two independent groups. The same situation was found for the drills. The end mill group with the largest volume all required operations on a 4112 lathe, a 4240 grinding machine or centerless grinder (the 4240 grinding operations could be performed on a new centerless grinding machine, see recommendations for Option 1, leaving the 4240 grinding machine in the functional part of the shop), on the 4330 and 4360 milling machines, and on a 4371 Excello. The drill group with the largest volume required operations on a 4112 lathe, a 4373 Excello, and a 4330 milling machine. At this stage ICTC personnel pointed out that end mills in the large volume group were characterized by their straight shanks and the drills in the large volume group were characterized by both the size and shape of the shank.

Since a 4330 mill and a 4112 lathe were required by both cutter groups, and since the latter is a unique machine (single copy), it was decided to combine the large volume end mills and drills into one part family and process them through a single cell (see Burbidge 1989). This cell, consisting of one 4112 lathe, one centerless grinder, one 4330 and two 4360 milling machines, would process approximately 70% of all drills and 85% of all end mills as measured by product volume. Load calculations indicated a close balance in utilization between the lathe and the milling machines. The groups of low volume end mills and drills would continue to be processed in the functional area of the shop.

As mentioned earlier, discussions with ICTC engineers indicated that a new manufacturing process was being developed for a variety of small cutters less than 9" in diameter. If implemented, this process would eliminate most of the 4220, 4230, and 4240 grinding operations while simultaneously increasing the processing time on the 4111 lathe. When these grinding operations were removed from the routings for these cutters, a second cell became apparent. This cell would require operations on a 4111 lathe, a 4330 milling machine, a 4190 broach or a new EDM machine, and a 4220 grinding machine (the grinding operations could not be completely eliminated). However, load calculations indicated that only the cutter type with the highest volume (face mills) could be processed in this cell without requiring machine duplication. If other cutter types were added to the cell, the purchase of an additional 4111 lathe and an additional 4330 milling machine would be needed (i.e., beyond the recommendations for Option 1). This was not viable due to the constraints on capital at the time. On the other hand, the broach or the EDM machine would not be required if only face mills were assigned to the cell. It was, therefore, suggested that the second cell would process small face mills only. This cell, designed to contain one 4111 lathe, two 4330 mills, and one 4220 grinder, would be able to manufacture 85% of all small face mills.

The two cells just outlined would process 71% of the product volume of all cutters produced by ICTC and would handle about 47% of the total manufacturing load.

6.2. The Impact of Creating Cells

Any time manufacturing cells are implemented in an otherwise functional organization, labor and machines are extracted from larger pools to form the cells. It is important to understand that if the implementation of the proposed cells would not result in any benefits in the form of reduced setup, run, or move times, reduced variability in processing or job arrival times, and/or reduced batch sizes, the lead times for the cell parts would actually be higher in the manufacturing cells than they would be in the functional layout with the additional capacity added. This is a direct result of partitioning a pool of resources (see Suresh and Meredith 1994). An intuitive explanation for this phenomenon is as follows. Suppose a pool of four identical machines is split into two equal sized groups to form cells A and B. If jobs are dedicated to cells and not allowed to transfer from one cell to another, there will be times when a machine in cell A is idle due to lack of work while cell B has a queue of work remaining to be processed. Thus, the net impact of partitioning is increased lead times.

This partitioning effect was shown for ICTC using MPX analysis. As demonstrated in Figure 7, the lead times for the end mills, face mills, and drills through the functional layout would be ¼ - 1 day less than the corresponding lead times in the cells (see the two left-most bars for each cutter type; to clarify: the left bar in Figure 6 is based on the existing functional shop without any capacity added while Figure 7 compares performance when new capacity has been added to both the functional layout and to the cells).

However, improvements usually do occur in the factors mentioned above when converting to cells. Due to the fact that batch sizes were already quite small, move times were fairly short, and the potential for variance reduction was difficult to ascertain, our analysis focused on reductions in setup and run times only. Such improvements may occur as a result of part similarity, common fixturing, teamwork within the cell, quality improvements, workers handling the parts, etc. Figure 7 shows two such cases for each cutter type (see the two right-most bars for each cutter). The first case assumes that setup times per batch and run times per cutter are 50% and 90%, respectively, of those in the functional layout while the second case assumes setup and run times that are 50% and 80%, respectively, of the original times. As the figure shows, either case would result in lead times in the cell that were lower than those in the functional layout, with reductions ranging from 5 to 44% for the case with the least improvement. Based on our observations on the shop floor and discussions with ICTC engineers, we believed that setup and run time reductions assumed by the second case were possible and those assumed by the first case were highly probable.

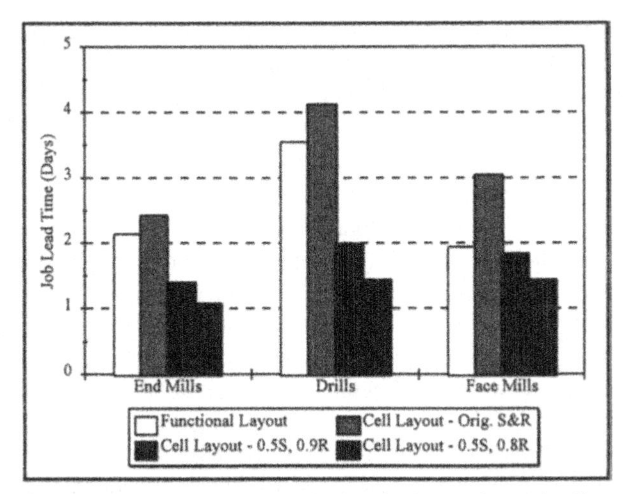

FIGURE 7. Average job lead time through the manufacturing area with all lathe, centerless grinding, and milling work done in-house.

Figure 7 is useful for resolving the choice between Option 1 (adding capacity to the existing layout) and Option 2 (adding the same capacity to cells). However, it should also be clear that if the cell option was to be chosen, the remainder functional layout would be left with less capacity than under Option 1 (and, therefore, suffer from the same type of partitioning effect that the cells do). As a result, the lead times for the non-cell parts would become approximately ¼ - 1 day longer than if the same parts would be processed in the functional layout. However, there are several factors which make this increase in lead time inconsequential. First, many of the non-cell cutters required outside operations which could not be performed in-house due to lack of capabilities. These operations accounted for 3-6 days of the 24.5 day average lead time of the cutters (Figure 6) and an additional ¼ - 1 day would have low impact. Second, the cutters left in the functional shop tended to be more complex to manufacture than those in the cells, were more likely to require customer approval of concept or design drawings, and generally required more design time. Again, an additional ¼ - 1 day would not be of importance. Counteracting these increases is the fact that pulling the cutters to be produced in cells out of the functional layout may lead to reduced variability and processing times, and thereby lower lead times, for the remaining cutters in the functional layout. However, this impact was unknown and, consequently, no estimates were made of this benefit.

6.3. Factors Not Considered in the Cellular Manufacturing Analysis

There are several aspects of adopting cellular manufacturing that were not specifically considered in our analysis. Beginning with the non-financial benefits, these include:

1. Reductions in material handling time in the cells, due to closely located equipment, and their impact on total lead time
2. Possible reduction in variability in the cells and the remaining functional layout, due to increased product similarities, shorter setup times, and controlled order releases, and its impact on lead time
3. Potential improvements in part quality in the cells, due to increased ownership of the product, teamwork within the cell, etc., and the impact this would have on lead time
4. Possible improvements in worker job satisfaction, due to teamwork, expanded scope of job responsibilities, etc., and its impact on absenteeism, employee turnover, and higher productivity in the cells
5. Simplified production planning and control in the cell, due to fewer scheduling and release points and an increased responsibility of order completion put on cell teams

Creating cells, however, is also associated with expenses. The investment in new equipment was not considered since our analysis showed that a capacity expansion would be necessary regardless of the type of work organization adopted. However, the cost of training operators in the cell concept, team work, and additional processes, the cost of moving machinery and preparing the floor foundations, ventilation, etc., are expenses that must be incurred for the cells to be established. However, management did not ask us to assess these costs, and, thus, no estimates were made.

7. OPTION 2 - CELLULAR MANUFACTURING

To summarize, our recommendations to management for the cellular manu-facturing option were as follow:

1. Purchase one lathe with capabilities similar to the 4111 lathes
2. Purchase at least one mill machine with capabilities similar to the 4320/4330 machines
3. Purchase a centerless grinder to bring this operation in-house
4. Implement one manufacturing cell to produce selected straight shank end mills and small straight shank drills
5. Implement a second manufacturing cell to produce small face mills

In addition:

6. Standardize the process routings for parts processed in the cells. In addition, tooling requirements and fixturing should be standardized and taken into account at the cutter design stage.
7. Improve the preventative maintenance program to reduce machine downtime and further increase effective capacity
8. Begin monitoring machine utilizations and/or lead times and add capacity as needed

We also reminded management that implementation of cellular manufacturing at ICTC would represent a fundamental change to the basic principles by which work has been organized for decades and would require effort on behalf of both management and labor. The proposed change would not only involve modifications to the shop floor, but significant cultural and organizational changes as well. Changes of this magnitude require a strong commitment by both management and the workforce to succeed. Management must be willing to invest in training, provide capacity to the cell to prevent it from becoming overloaded as demand increases, give cell workers the authority and responsibility to produce the cutters as required, and to maintain the integrity of the cell by processing only those parts assigned to it. In addition, modifications to the measurement, reward, and planning/control systems must also take place to fit the new work organization. Operators in the cell must be willing to work as a team with the primary goal being the efficient production of cutters. This requires a willingness to provide assistance to other team members as needed, and to take responsibility for the work performed by every member of the cell. If these basic requirements are not met, the cells will not perform as desired.

8. POSTSCRIPT

Management at Ingersoll Cutting Tools readily decided to follow the recommendation to implement cells in the cutter manufacturing area. This may be somewhat puzzling given that the results from the analysis did not show cellular manufacturing providing the spectacular improvements often reported in the academic literature and industrial press (e.g., Wemmerlöv and Hyer 1989; Wemmerlöv and Johnson 1997). Of course, reported improvements due to cells often involve a comparison between an existing shop and one or more cells with capacity added. Given the importance of capacity enhancements in lead time reductions (compare Figures 6 and 7), such results are not unexpected. The comparison we did here, i.e., one where new equipment was added to both the functional and the cellular shop, can be seen as more equitable and relevant than one where capacity is only added to the cells.

It can also be noted that the expected level of performance improvement for ICTC's cells were not high since the factors that often can be improved and contribute to increased material velocity were already at low levels. Specifically, lot size reduction, perhaps the generally most powerful source of lead time reduction, was not available since the lots were already small (in the 3 to 8 unit range). As indicated, move times were not part of our MPX analysis since all equipment in the fairly small plant were already closely located and materials handling was rather efficient. Essentially, and this could be projected without any sophisticated analysis, the advantage of cells over the existing functional layout had to come from variance reduction, lowered setups, and more efficient (and dedicated) processing of jobs. Unfortunately, it was not possible (and is rarely possible) to get any reliable assessments regarding the improvements in these variables. The analysis conducted here could only point to the degree of improvement necessary in order for the cells to outperform the functional layout.

Despite these circumstances, several factors led to management's embracing of this new work organization. First, the goal to achieve a 50% lead time reduction had been formulated at the highest level in the firm. Thus, "no change" was not an option. Second, experimentation with the order processing cell had made management and employees comfortable with organizational change and the idea of team work. A desire to explore the use of teams in the manufacturing area was therefore logical. Third, management was convinced that future improvements in response time could only come from a more integrated organization and a work force with increased responsibilities and involvement with customer orders. The adoption of cells was also seen as an opportunity to increase the skill level and flexibility of the employees through training (most workers were not trained to operate machines in different functional groups). Finally, the fact that, in the face of impending growth, the complexity and costs of the materials handling system and the planning and control efforts would increase was well understood and also contributed to the favorable view of the proposed conversion to cells.

Implementation of the cells was delayed due to a temporary lack of funds for new equipment. Facing a capacity shortage while demand kept increasing, the decision was taken to remove the drills from the end mill/drill cell since drills could more easily be subcontracted. The first cell, producing end mills only, has by now been in operation for a little more than a year. The lead time through the cell has steadily declined, starting at 31 days and now hovering around 21 days (a 32% reduction). Although encouraging, this is far above the predicted average lead time of 2 days (see Figure 7). This outcome can be explained by a continued growth and capacity shortage, and concomitant difficulties of being able to outsource milling work to a sufficient degree to reduce the internal work load. On the positive side, management has seen a great interest among the employees in cell work, and much reduced planning and control efforts. Additional machine tools are now on order, and two more cells, beyond those originally planned, are on the drawing board.

A few final remarks. First, part family/cell formation is not always a simple, mechanical process. It requires an in-depth knowledge of the parts being produced, the potential for using alternative process routings, and the possibility of changing the method of manufacturing in order to eliminate operations. This type of knowledge increases the possibility that part families can be created with volumes large enough to justify the dedication of equipment and people to cells. Second, this study has highlighted the difficulty of knowing with certainty the degree of performance improvement that has been achieved and, therefore, whether the decision to alter the manufacturing system was correct. The conditions at ICTC changed, and changes, continuously. By varying the level of work being outsourced vs. produced in-house, by phasing in new equipment over time, and by reconfiguring part families and cells, it is almost impossible to ascertain lead time improvements from one period to the next. In such situations management has to rely on a variety of strategic and operational considerations - and tangible as well as intangible data - to justify decisions to reorganize.

Acknowledgments. This study was conducted as a part of a larger study on Quick Response Manufacturing (see Suri, Wemmerlöv, Rath, Gadh, and Veeramani 1996). It was sponsored, in part, by a grant from the Industrial and Economic Development Research Program of the State of Wisconsin, by a grant from the College of Engineering at the University of Wisconsin-Madison, and by the Center for Quick Response Manufacturing. We would like to thank Mike Wayman, Dave Magner, Pete Cober, Steve Burggraf, Scott Wenstrom, Jeff Blascoe, Sandy Blankenship, and Mike Dieken for their time and support during this project, as well as all other Ingersoll employees who shared their insights and observations with us.

REFERENCES

Burbidge, J.L., 1989, *Production Flow Analysis*, Clarendon Press, Oxford.

Gross, D. and Harris, C.M., 1985, *Fundamentals of Queuing Theory,* John Wiley, New York.

Suresh, N.C. and Meredith, J.R., 1994, Coping with the loss of pooling synergy in cellular manufacturing systems, *Management Science*, 40, 4, 466-483.

Suri, R., 1996, Using queuing models to support quick response manufacturing, *Proceedings of the International IIE Conference*, Minneapolis, MN, May 1996.

Suri, R., Wemmerlöv, U., Rath, F., Gadh, R., and Veeramani, R., 1996, Practical issues in implementing quick response manufacturing: Insights from fourteen projects with industry, *Proceedings of the Manufacturing and Service Operations Management Conference*, Hanover, NH, June 24-25, 1996.

Wayman, M. J., 1995, Order processing lead time reduction: a case study, *Proceedings of the International IIE Conference*, Nashville, TN, May 1995, 400-409.

Wemmerlöv, U., 1995, Implementation of cellular manufacturing: Change management fundamentals, Presentation to QRM center members, May 17, 1995.

Wemmerlöv, U. and Hyer, N. L., 1989, Cellular manufacturing in the U.S. industry: A survey of users, *International Journal of Production Research*, 27, 9, 1511-1530.

Wemmerlöv, U. and Johnson, D.J., 1997, Cellular manufacturing at 46 user plants: Implementation experiences and performance improvements, *International Journal of Production Research*, 35, 1, 29-49.

AUTHORS' BIOGRAPHY

Danny J. Johnson is a doctoral student in Operations Management at the University of Wisconsin-Madison. His research interests are in the design, implementation, operation, and management of quick response manufacturing systems and the problems faced by firms as they develop and use these systems to improve key performance measures. He has assisted with studies on the implementation of cellular manufacturing systems in industry and is currently conducting an empirical investigation of factors influencing reorganizations to cells. He received his B.S. in Business Administration from Moorhead State University and his MBA in Operations Management from the University of Wisconsin-Madison. Prior to obtaining his B.S., he worked for eight years in the service sector. He is certified in production and inventory management by the American Production and Inventory Control Society (APICS).

Urban Wemmerlöv is the Kress Family Wisconsin Distinguished Professor and Director of the Erdman Center for Manufacturing and Technology Management (*http://www.wisc.edu/bschool/erdman*) at the University of Wisconsin-Madison (School of Business), Associate Director of the Manufacturing Systems Engineering Program, and Executive Committee Member of the affiliated Center for Quick Response Manufacturing (College of Engineering). His teaching and research interests are in the areas of adoption, design, evaluation, and operation of manufacturing systems, and the justification and implementation of new manufacturing technologies and philosophies. Particular areas of interest include Group Technology, cellular work organizations in offices and manufacturing areas, focused factories, and Lean Manufacturing. Professor Wemmerlöv is the associate/area editor of three academic journals and has published over 50 articles and book chapters, and numerous conference papers, in the field of production management.

E3

A Decision Support System for Designing Assembly Cells in Apparel Industry

M. Kalta, T. Lowe and D. Tyler

1. INTRODUCTION

Apparel industries in the Western World have suffered decline over the past few decades; a trend which has been partially attributed to cheap imports and an unwillingness to change in a market that has become increasingly customer-oriented. Fashion-led markets necessitate good design, short lead times, flexibility, and consistent high quality. The major strategic response to this challenge has been to participate in quick response (QR) programs that are reactive to consumer demand (Hunter 1990). The problem for implementing QR is that quick response is a convenient term to communicate direction without explicitly specifying the necessary actions (Tyler 1989). This has led to a variety of approaches to implementation.

Other sectors of industry have faced similar market pressures and, in engineering, the computerised approaches of advanced manufacturing technology (AMT) have played a major role in improving the efficiency of existing batch manufacturing systems. These benefits have been particularly marked when linked to Burbidge's approach (1961) towards reduced quantities, a focus on product families, and the development of flexible low-inventory production cells. The AMT option has not proved to be viable either commercially or practically in the apparel sector (Tyler 1989) and flexibility has had to be sought primarily in the operators and in appropriate choices of system design. An alternative to reprogrammable automation has been to restructure manufacturing through the implementation of just-in-time (JIT), total quality management (TQM) and group technology (GT) concepts. QR, JIT, TQM and GT are all ingredients of teamworking, and all can be seen as building on Burbidge's pioneering work on manufacturing systems.

In most apparel industry implementations, teams comprise a group of employees who do all the operations necessary to complete a garment or a "family" of garments. The apparel industry has adopted two main `reference standards' for implementing cells: the Toyota Sewing System (TSS) and the *kanban* system. Our industrial collaborator uses a modified form of the TSS which is outlined in §2.

An integrated scheduling / decision support system with a common database has been developed by the authors to support apparel industry teamworking systems. The scheduling system enables management to explore a number of possible futures based on the control of essential factory parameters. The system is designed to plan (schedule) team-to-product assignments for teamworking production systems (Lowe et al. 1995). The validated simulation model of the DSS, which is called COATS (Clothing Oriented Animated Teamworking Simulation), has been developed to investigate the dynamic behaviour of assembly teams (Kalta et al. 1997). In addition to being part of the DSS, COATS may be used as a stand-alone simulation system and also by the scheduling system to validate schedules. The purpose of this paper is to describe the Decision Support System (DSS) that provides a user with a set of tools to improve the productive performance of a team on a given style. This is achieved through the use of the Central Database (CDB), the Decision Support Main Module (DSMM) and COATS (Figure 1).

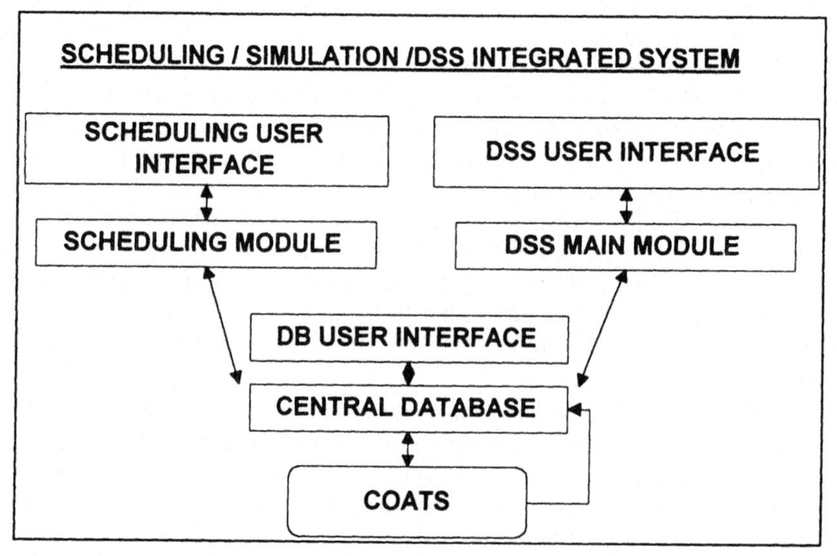

FIGURE 1. Decision support system overview.

2. TEAMWORKING AND THE TOYOTA SEWING SYSTEM

Sips (1993) has relevant background for European perspectives on innovations relating to work groups. Tyler (1989) and Peeters and Pot (1993) relate these trends in production systems specifically to the apparel industry. The drivers for change in the apparel industry were QR-related: responsiveness, quality, customer service and cost. There are relatively few formalised `standards' of teamworking in the apparel context. The Toyota Sewing System (TSS) is one of them: it originated in the Toyota Group in Japan, the pioneer for the implementation of JIT concepts. The characteristics of a TSS production line are as follows (see also Figure 2):

- The line layout is U-shaped with garments progressing around the line.

- Each operator is cross-trained on a different portion of the line (i.e. contiguous operations) depending on skill and operation complexity. Ideally all the operators are cross-trained on all the operations.
- Each operator is assigned at least one operation (normally more than one operation), and the contiguous operations are referred to as the pitch.
- Operators have pitches which overlap with their neighbours by one or more workstations.
- Work is carried out in batch sizes (lots) of one.
- Operators will generally stand at all workstations.
- The first and the last operations are uniquely assigned to the first and the last operator respectively.
- Buffering between workstations is normally not permitted.

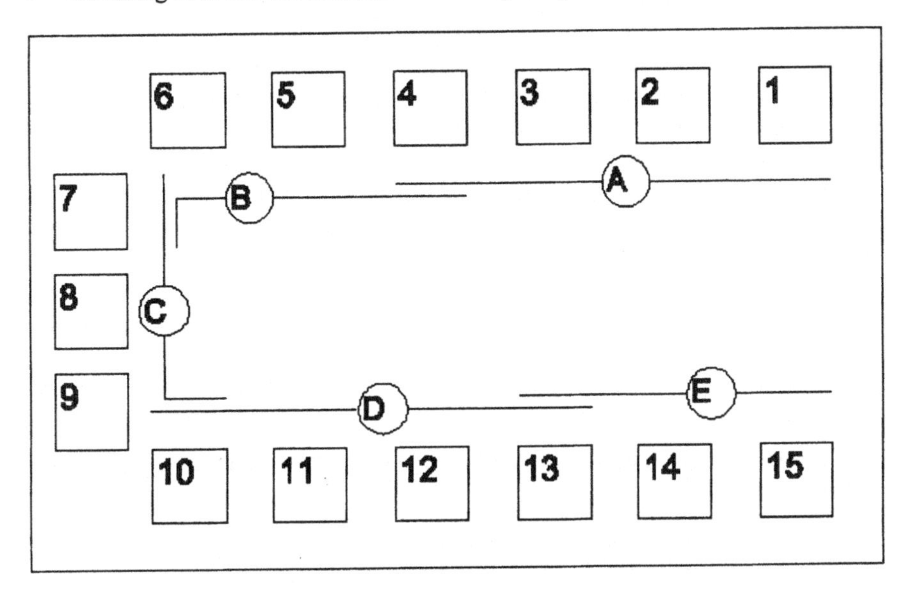

FIGURE 2. Layout for TSS with an overlap of one workstation.

The industrial collaborator in this research, Meritina Ladieswear Ltd (now CV Clothing Ltd), operates a modified version of TSS in their production units. This is referred to as the Meritina Team Approach (MTA, now Manufacturing Team *Approach*, Price, et al. 1990). Although MTA is essentially the same as TSS, there are a number of modifications which have been developed by the Company. These include the option of having small buffers between neighbouring stations and allowing the number of overlap stations between neighbouring operators to be dependent on the skills within the team. Operator work sequences (including the overlap stations and operator assignments) are decided at the outset of production with inputs from both management and team members. A detailed description of both TSS and MTA is given by Kalta et al. (1994). Operators are paid a standard weekly wage with a team bonus for exceeding production targets.

3. DECISION SUPPORT SYSTEM DEFINITION

A Decision Support System (DSS) is a system which provides a user with help in reaching a decision in a domain that has little or no precedent. This type of problem is often defined as ill-structured (Finlay 1989).

A DSS is useful when a fixed goal exists but there is no algorithmic solution. The solutions paths can be numerous and user-dependent (Klein and Tixier 1971). In other words, the goal of a DSS is to improve a decision by better understanding and preparation of the tasks leading up to evaluation and choice. These tasks are collectively called the decision making process.

Structured problems are routine and repetitive, because they are unambiguous (since each such problem has a single solution method). A less structured problem has more alternative solution methods, and the solutions may not be equivalent. A completely "unstructured" problem has unknown solution methods or too many solution methods to evaluate effectively (Klein et al. 1990). From our experience of pre-production planning in the apparel industry, team design as normally carried out is a "less structured" problem, and optimisation is "unstructured" with existing solution methods.

A DSS is defined by Finlay (1989) as an interactive data processing and display system which is used to assist a concurrent decision making process, and which also conforms to the following characteristics:

i) it is sufficiently user friendly to be the preferred choice of support tool by the decision maker;

ii) it displays its information in a format and terminology which is familiar to the user;

iii) it is selective in its provision of information and avoids the decision maker in information overload.

4. THE AIM OF DEVELOPING THE DSS

Having decided on the team to work on a style, the next task is to address issues of cell design. This includes specification of the line layout and the allocation of the team members. This task is generally recognised to be a key factor in a full utilization of team capabilities. To design the best cell, the decision maker should have, in principle, a detailed knowledge of the team which includes:

- the skills matrix of each member (sometimes referred to as a skills inventory);
- the operation learning behaviour of each member (particularly relevant to skills which need to be acquired or developed);

- the productivity history for the team (*i.e.* if there are fluctuations in productivity which are not due to absenteeism or machine breakdown, but are a measure of team dynamics and the aptitude of members to working in a team).

This data is not accessible to decision makers in industry and, in practice, decisions are made on a much more limited data set. Consequently, the task of designing team layouts is considered an ill-structured task. There are many influencing factors, many alternative solutions and these solutions are not equivalent. Therefore, it is concluded, the decision support system approach can be used to assist in optimising the design of cells (Klein et al. 1990).

After discussion with the collaborating company, the specification for the decision support system included the following:

- Support for cell design decisions. This must take place prior to the new style being introduced to the team. Experimentation should be possible to explore different options for work allocation: to identify more productive allocations ("what if?" questions) and to highlight training needs.
- Support for providing realistic targets. Targets are recognised to change during the startup phase, as teams change from one category to another (*e.g.* from semi-skilled to skilled), as individuals acquire skills, as rework reduces, and as the dynamic behaviour of the team changes. Realistic targets are desirable: during the startup phase and afterwards. At present, management has an expectation of performance, but substantial improvements are sought.

To achieve these goals, the following research objectives were identified:

1. The rules needed for the assignment of operator work sequences must be elicited from the personnel involved in the production unit. These rules will provide the knowledge base for the DSS.
2. A methodology for handling the production rules must be developed in such a way that it is possible to incorporate them into appropriate software tools.
3. A methodology for optimisation must be identified, enabling decision makers to explore a variety of cell designs quickly, and leading to predictions of system outputs from the different cell configurations.

5. KNOWLEDGE ELICITATION

A formal procedure for determining operator work sequence allocation was in operation and was the responsibility of the Work Study practitioner based in the factory. This procedure is summarised in §5.1.

The operator sequence allocation was the result of a very ill-structured process, involving management, supervisors and operators. The methodology for identifying the rationale for decisions made was to analyse existing allocations, and to interview those responsible for making the allocations. Practices were observed to vary within our collaborator's factories, notably to the extent to which operators were involved in the allocation process. In the particular factory where this research was based, the operator sequence allocation was decided by the team members at pre-production meetings, with the supervisors only intervening when conflict occurred.

5.1. Operation Sequence Rules

The operation sequences for a cell are decided by management after calculating the pitch time. The pitch time specifies the number of Standard Minutes to be performed by each operator if the work is to be evenly distributed within the team. It is calculated from the following equation:

Pitch Time = (Total Standard Minute Value)/(Number of Team Operators)

The number of operators chosen to work on a style is based on the Standard Hours per dozen. e.g. each operator should have a sequence that is no more that 2 - 2.5 minutes (i.e. pitch time = 2 - 2.5 minutes), otherwise a new operator may be added. If the pitch time is less than two minutes, teams may be split with the introduction of new styles. The size of the split is dependent on the factory styles and the team skills.

After the pitch time is calculated, contiguous operations are allocated on a nearest fit basis to specify the work sequences. Operators are then allocated one of these sequences. Various rules are used in assigning work sequences to operators. These rules have been classified as pre- and in-production rules.

5.2. Operator Work Sequence Allocation Rules

Seven pre-production rules have been identified:

- ***The minimum disruption rule.*** Allocate a style that is similar to the previous style if possible so that the previous operator work sequence order can be adhered to. This rule is used by the DSS.
- ***The compatible skills rule.*** Skills of operators must match allocated sequences. Selection of equally compatible operators (taking into account all the other rules) is random. Selection when a match is not exact is also arbitrary, but training needs are identified. This rule is used by the DSS by using the skill matrices of the operators which are stored in the central data base (described later).
- ***The sequence re-ordering and training rule.*** Mismatches in operation sequences and operator skills may be tackled by re-ordering operation sequences so that a better match may be found. At the moment, this rule is not used by the DSS. This is because the re-ordering of the operations has to done by a work study engineer prior to the introduction of the style. To incorporate this rule in the DSS, more heuristics about re-ordering the sequence of operations have to be elicited.

- ***The performance variation rule.*** All operators are seen as homogeneous in performance terms. In principle, the allocated sequences all take the same time. On this basis, people refer to a balanced line or cell. Although this is not logical, but this how it is done in the factory the work study engineer. The Decision Support Main Module (described later) does not take into consideration the performance variation between operators. But in order to get a realistic production target, the simulation model (COATS) of the DSS considers this performance variation. The performance matrices of the operators are stored in the central data base (described later).
- ***The operator inconsistency rule.*** If any team member is inconsistent (in performance or attendance) then they are not given the lead job (*i.e.* the first sequence of operations). The work variation is considered to cause the operators further down the line to be starved of work. This rule is considered to produce greater line efficiencies. This rule is used by the DSS by using the absenteeism and learning matrices of the operators which are stored in the central data base (described later).
- ***Examination rule.*** To guarantee that faulty garments are passed back to the appropriate operators for repair, the supervisors try to make sure that an operator with a strong personality is assigned to the examining operation. Such operators are simply entered in the database as they do not acquire the examination skill. This rule is used by the DSS.
- ***Critical operation rule.*** A critical operation is often identified and the operators are allocated with operations from this starting point. For example, this operation can be the most difficult operation or the operation which can be performed by only one team member. This rule is used by the DSS.

Three in-production rules have been identified:

- ***The distraction rule.*** If the performance of the team is suffering, this may be due to non-productive talking between some of its members. These operators are separated.
- ***Dynamic reallocation rule.*** If a line imbalance (i.e. big buffer size and/or variation operators utilization) develops, it may lead to a temporary reallocation of new work sequences (*e.g.* to remove a bottleneck).
- ***Resolve locally rule.*** If a conflict occurs within a team, the problem should be addressed internally by the team members. It has been found that transferring operators often breeds resentment within the host team.

None of the in-production rules are used in the DSS, since the DSS is a pre-production tool.

5.3. Skills Classification

There has been significant debate as to whether Japanese manufacturing practices result in a work force with enhanced skills (the upskilling model) or whether there is a degradation of work (the deskilling model). Bratton's (1993) study concluded that a middle position is more defensible: the significance of cellular work structures on manual [engineering] skills is indeterminate. During the research reported here, it was the general perception of management and operators alike that the manual skills within the factory were enhanced as a result of moving from traditional line production to teamworking. This brings an increased need to manage these skills effectively. The practical problem experienced in factory environments is maintaining a realistic assessment of available skills. The Central Database in this research provides local management with the means to assess, record and update a skills inventory.

In our collaborator's factory, 34 operation classes (skills) have been identified by the research team in conjunction with local management. These operation classes have be subdivided into 7 general skills and 27 specific skills. These skills are classified as Complex (A), Average (B) and Simple (C) (Tyler and Fozzard 1987, Price and Tyler 1991) as in Table 1. The view of the operation complexity varies between the line management and the senior management. Of the 29 cases (62%) where comparisons between judgments were possible, 18 cases showed an identical assessment. In the other 11 cases (38%), senior management considered tasks less complex than production line management, as shown in Table 1. (* denotes that a classification was not obtained for the particular operation skill).

The differences in skills classifications are not serious, but they are significant. It is a indication that there is no fully “objective” knowledge base. This is because there is not a satisfactory method of measurement of task complexity. In this case, individual judgments varied depending on the closeness of that individual to the assembly cells: the closer the individual was to operational practice, the more difficult the manual tasks were seen to be. Although either of the classifications could be stored in the Central Database, the authors adopted the line management classification. This was because the classification of an expert independent third party was close to the line management's classification.

5.4. Skills, Absenteeism And Learning Matrices

Based on the skills classification described in 5.3 and considering only the operations which can be performed with performance equal to or higher than 70%, a performance matrix and a learning matrix have been built for each operator. The learning matrix is based on the three performance curves obtained by examining Meritina's production data: advanced (1), average (2) and inexperienced (3).

An absenteeism matrix has been developed for each team utilizing existing factory data. The absenteeism history of each member of a team was determined, and the absenteeism behaviour was classified as follows:

- Class (a): if absence periods are less than or equal 2% of the working periods.
- Class (b): if absence periods are greater than 2% and less than or equal 6% of the working periods.
- Class (c): if absence periods are above 6% of the working periods.

The classifications are pragmatic, based on local management heuristics. This was a necessary consequence of applying the "operator inconsistency rule" in §5.2.

TABLE 1. Skills Classification By Local Management.

OPERATIONS	CLASSIFICATION (SENIOR MANAGEMENT)	CLASSIFICATION (LINE MANAGEMENT)
GENERAL OPERATION CLASSES		
5 THREAD SEAMS	B	B
3 THREAD PANEL	B	B
LOCKSTITCH	*	C
TOPSTITCH	B	B
TWIN-NEEDLE	B	B
COVERSEAM	B	A
CHAINSTITCH	B	B
SPECIFIC OPERATION CLASSES		
EAGLE	C	A
PLEATS & DARTS	C	B
LOCKSTITCH POCKETS	A/B	B
SCROLL POCKETS	B	B
WELT POCKETS	A	B
ZIP POCKETS	A	A
POCKET BAG	C	B
etc.		

6. DESIGN OF THE DECISION SUPPORT SYSTEM

As shown in Figure 1, the DSS consists of three components: Central Database (CDB), Decision Support Main Module (DSMM) and the simulation tool (COATS).

6.1. The Central Database

The central database (CDB) has been developed using Borland's Object Windows Library in C++. It holds the following information:

- *Skill.* The skill records the various skills needed within the factory and are associated with a complexity.
- *Operators.* For each operator, the operator classification, absenteeism classification and the acquired skills with the performance attained.

- *Team*. For each team, a record of the operators that make up the team, the performance classification and the product family it can be assigned to.
- *Style*. Number of operators required, complexity and the operations that make up the style.
- *Operations*. Operation type (skill required), machine type, work content, complexity and whether the operation is a designated change-over station.
- *Orders*. Date assigned, date finished, garments to produce, flow rate week, flow of raw materials.
- *Assignment*. Date assigned, date finished, garment input rate, expected performance, expected production and operator allocations.
- *Operator allocations*. Walk back performance for each operator, start and end stations and the performance at which they work on the given stations.
- *Absenteeism*. The mean time between absences.
- *Machine breakdown*. Inter-arrival mean time for failure; repair time mean and standard deviation for each machine type. (Further details of the machine breakdown analysis is given by Lowe et al. (1994)).
- *Production hours*. Hours worked on each working day.
- *Holidays*. Start and end days for holidays.

The CDB is the central unit of the DSS where the following activities are carried out:

- The user enters and/or accesses the above information.
- The user activates and experiments with the DSMM.
- The user experiments with the DSMM and COATS (*e.g.* "what if?" questions) to identify more productive allocations and to determine realistic targets.

6.2. The Decision Support Main Module (DSMM)

The Decision Support Main Module (DSMM) has been developed using Kappa-PC. Kappa-PC is a windows-based expert system shell with object representation capability. It also has a high-level application development language called KAL. The DSMM uses the rules explained earlier to present all possible options to the user rather than an optimal (*e.g.* most productive) solution. This is possible since the search space is relatively small and the package is designed to allow decision makers to decide which solution is the most productive using COATS (described in 6.3). The complexities of this decision making process have been outlined by Lowder (1991). He describes three initial module (cell) balancing techniques, as follows.

- Touch Module. The touch module compares the skills of the individual team members with the operation sequence for the style. Operators are then assigned a sequence of operations.
- Combined Module. The combined module is able to modify the style process sequence in order to match the team skills (*i.e.* try to find best fit).

- Bid Module. The bid module transfers the decision-making process to the team itself. An analysis of the style for skill content is made and a production goal is agreed on. The operators make bids for work on operations they can perform.

The DSMM produces all possible options for both Touch and Bid Module decisions and, by redefining the operations sequence, decisions relating to the Combined Module can be explored. All initial balancing options have the potential of being improved using COATS.

The DSMM consists of seven Routines: User-Interface, Read, Pre-Pitches-Calculation, Operations-Pitches, Pre-Assignment-Consistency-Checking, Pitches-Allocation and Training-Requirement (Figure 3).

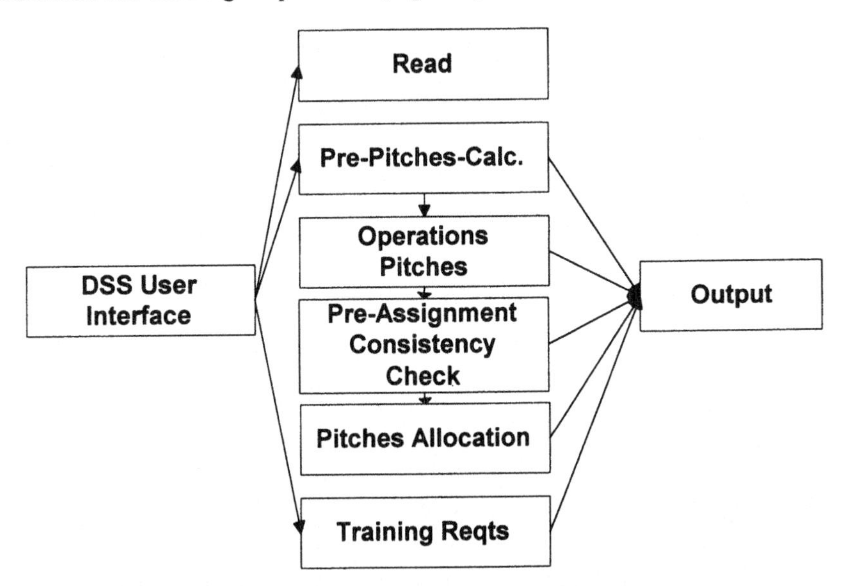

FIGURE 3. Components of decision support main module.

- In the User-Interface Routine, the user can perform the following activities:
  - selecting the specific style/team combination to be investigated.
  - calling the other routines of DSMM.
  - storing the results in a file.
- The Read Routine converts data stored in the central database into Kappa-PC objects.
- The Pre-Pitches-Calculation Routine highlights the missing skills in the team.
- The Operations-Pitches Routine determines the operations of each pitch of the style.
- The Pre-Assignment-Consistency-Checking Routine determines the operators who can work on each pitch. This routine, in addition to the skills acquired by

each team members, it takes into consideration other heuristics (e.g. absenteeism history of the team members).

- Pitches-Allocation Routines determines all the valid pitches/operators combination (possible solutions). It is a backtracking routine with consistency checking to make sure that each combination conforms to the allocation rules described earlier. This backtracking routine has been developed to determine all the possible solutions instead of using the rule-based paradigm in Kappa-PC. This is because:
  - The rule execution in Kappa-PC is relatively slow.
  - As mentioned earlier, the search space for the domain is relatively small (the worst case is a six member team; thus the maximum number of possible solutions is factorial six which is equivalent to 720 combinations). Hence, the search space completeness can be assured using the backtracking routine.
  - The Training-Requirement Routine can be used to guide training during the startup period to make the teams more flexible. This facility must be used where there are no options possible with existing skills and, again, additional management support for the team is needed.

6.3. The Simulation Tool (COATS)

COATS (Clothing Oriented Animated Teamworking Simulation) is a validated simulation model for team working in the apparel industry which has been developed using the simulation software ARENA (Kalta et al. 1997). COATS is configured and accessed by the Central Database. This facilitates experimentation by the rapid recall of relevant simulation parameters (*e.g.* operator performance, change over operations, number of operators).

In COATS, the user is able to select any of five scenarios which capture aspects of dynamic team behaviour. These scenarios are briefly described as follows:

1. Fixed change over operations
2. Partially fixed change over operations (semi-flexible)
3. Partial change over operations (operations can be performed on part of the operation) in addition to the partially fixed mode
4. Fully flexible change over operations without buffers between stations (the Toyota Sewing System)
5. Fully flexible change over operations with buffers between stations

In COATS the processing time at each workstation is modelled by a normal distribution with a standard deviation of 13% of the operator's mean performance (Fozzard 1989). Although Fozzard's research was based on a different system of apparel production (the Progressive Bundle System), this standard deviation figure was confirmed during the validation of COATS to be an acceptable approximation for teamworking (Kalta et al. 1997). However, in a subsequent analysis, using a more rigorous data collection tool, a rather higher level of operator variability in teamworking has been documented (Carter and Tyler 1996). COATS output includes: machine utilization, worker utilization, work in progress in each workstation, number of garments produced, team performance, processing time spent by each operator, the processing time spent by each operator on each operation and the walking back time spent by each operator. Simulated production figures utilizing COATS utilize the performance variation rule documented in §5.2.

			Team 3			
			Absenteeism Category of operator			
			301	**302**	**303**	**304**
			infreq	infreq	infreq	infreq
Style 7905			**Performance of operator (%)**			
Op No	**Skill**	**Skill ID**	**301**	**302**	**303**	**304**
1	Pleats & Darts	9	100	100	100	100
2	5 Thread Seams	2	90	100	110	100
3	5 Thread Seams	2	90	100	110	100
4	Concealed Zip	16	90	110	110	
...	....					
13	Button Hole	30	80	90	110	110
14	Vents	28		80	100	110
15	Blind Hems	25		80	110	110
16	Examination	29				110

FIGURE 4. Team/Style information.

7. DSS OPERATION

The operation of the DSS is explained by reference to a case study. The Scheduling Module allocates Team 3 to style 7905. Figure 4 shows the skill matrices and the absenteeism history of the team members. It also shows the operation sequence and the Standard Times (in seconds) for the style. This information is stored with other factory data in the Central Database. By using the User Interface Routine of the Decision Support Main Routine, this information is read by the Read Routine. Then by calling the other Routines, the Decision Support Main Routine produces the results shown in Figure 5. The system output provides details of the operations for each pitch, the operators who can perform on each pitch, possible allocations and possible training needs.

These possible allocations are fed back to the central database. By using COATS each of these allocations are then simulated. Analysis of the simulation results is carried out by the decision maker to suggest some modification to improve productivity. A summary of the results given by the Decision Support Main Routine, the experimentation with COATS and the resultant suggestions to improve productivity are shown in Figure 6.

The most productive allocation by the Decision Support Main Routine is the first solution with simulated production equal to 158 garments. According to this solution, operator 301 is given the lead job in spite of a relatively low performance. This is because the lead job is the only job the operator can perform on. However, the low performance leads to a reduced flow of work to the other operators (their utilizations are lower). To improve the situation, a suggestion is made to shift the change-over operation to operation 3 instead of 4. By simulating this suggestion, the simulated production increases to 165 garments and the utilizations of the other operators improved. By looking at the simulated result, it can be noticed that the utilization of operator (303) is still low compared with the others. Another suggestion is made to improve productivity. In this suggestion, in addition to the first modification; operation 7 is made a change-over operation as well as operation 8 (*i.e.* there are two change-over operations between operator 302 and operator 303. By simulating this suggestion, the simulated production increases to 170 garments and the utilization of operator 303 increases from 92% to 98%.

8. SUMMARY AND CONCLUSIONS

The decision support system (DSS) has been developed as a post-scheduling pre-production activity. In its most basic form, it carries out an existing function within the Company using rules elicited from the management, but in very much shorter time interval (the processing time is a few seconds).

Having made the team/style assignment, the best cell has to be designed. The DSS provides support for cell design decisions by providing users with a comprehensive statement of possibilities. In some cases, where there are mature teams with many skills within the team, there will be many options. In such cases, management need not spend much time on the problem, as these teams are perfectly capable of taking a prototype layout and adapting it to maximise their own output. However, there will be other cases where there are few options, perhaps only one, and sometimes none. Such cases will occur because of the limited skill base within the teams concerned. The DSS provides management with viable options and guidance when these teams need extra support to ensure that the startup is carried out in a controlled way. The DSS has a "Training Requirements" option, which can be used to focus the training effort undertaken during the startup period to make the teams more flexible. This facility must be used where there are no options possible with existing skills, and again, additional management support for the team is needed.

Style: 7905 / Team: 3

STYLE OPERATIONS PITCHES:

OPERATIONS OF PITCH 1:
Operation 1: PleatsAndDarts
Operation 2: FifthSeam
Operation 3: FifthSeam
Operation 4: ConcealedZip

OPERATORS WHO CAN WORK ON PITCH 1:
Operator: Op301
Operator: Op302
Operator: Op303
(similarly, for operations of every other pitch)

POSSIBLE SOLUTIONS:

SOLUTION NUMBER 1:	SOLUTION NUMBER 2:
Pitch 1 Can Be Allocated to Op301:	Pitch 1 Can Be Allocated to Op301:
Pitch 2 Can Be Allocated to Op302:	Pitch 2 Can Be Allocated to Op303:
Pitch 3 Can Be Allocated to Op303:	Pitch 3 Can Be Allocated to Op302:
Pitch 4 Can Be Allocated to Op304:	Pitch 4 Can Be Allocated to Op304:
(etc.)	

TRAINING HIGHLIGHT

- To Work on Pitch 1, Operator Op304 Should be Trained on: ConcealedZip
- To Work on Pitch 2, Operator Op301 Should be Trained on: AttachFacing
- To Work on Pitch 2, Operator Op304 Should be Trained on: ConcealedZip
- To Work on Pitch 3, Operator Op301 Should be Trained on: UnderPress..

(etc.)

FIGURE 5. Decision support main module output.

Solutions	Operations Allocated to Operator (utilization %)				Simulated Production (Garments)
	301	302	303	304	
1st Solution by DSS	1 to 4 (100)	4 to 8 (86)	8 to 12 (82)	12 to 16 (98)	158
2nd Solution by DSS	1 to 4 (100)	8 to 12 (91)	4 to 8 (88)	12 to 16 (98)	156
1st Suggestion	1 to 3 (100)	3 to 8 (99)	8 to 12 (92)	12 to 16 (98)	168
2nd Suggestion	1 to 3 (100)	3 to 8 (99)	7 to 12 (98)	12 to 16 (98)	170

FIGURE 6. Simulation Results - Team/Style: 3/7905

The research objectives identified in §4 have been achieved:

1. The rules needed for the assignment of operator work sequences were elicited from the personnel involved in the production unit. These rules were found to be relatively unsophisticated conceptually, although considerable complexity was apparent because of the numerous human factors involved in decision making. These rules have provided the knowledge base for the DSS.
2. A methodology for handling the production rules has been developed using an expert system shell. Since the search space was relatively small, it was practical to offer the user all possible options rather than to restrict the range of options by the use of further selection procedures.
3. A methodology for optimisation has been identified, enabling decision makers to explore a variety of cell designs quickly, and leading to predictions of system outputs from the different cell configurations. The output of the Decision Support Main Module (§6) is fed semi-automatically to the simulation system (COATS) to model possible solutions. The simulation results provide a basis for making informed decisions aimed at improving productivity, as well as indicating the production targets that cells ought to achieve if they are to maintain consistency with data in the central database.

Acknowledgments. Meritina Ladieswear Ltd (now CV Clothing Ltd) are thanked for permission to publish factory data. The staff in Meritina's Retford factory are thanked for making available all the resources we required. The Design and Integrated Production

Group (DIPG) of the Engineering and Physical Sciences Research Council (EPSRC) funded the project.

REFERENCES

Bratton, J., 1993, Cellular manufacturing: some human resource implications, *International Journal of Human Factors in Manufacturing*, 3, 4, 381-399.

Burbidge, J.L., 1961, The "new approach" to production, *Production Engineer*, 40, 12, 769-793.

Carter, I. and Tyler, D.J., 1996, A comparison of operator performance behaviours in different clothing production systems, *Journal of Clothing Technology & Management*, 13, 1, 110-129.

Finlay, P., 1989, *Introducing Decision Support Systems*, NCC Blackwell, Manchester, UK.

Fozzard, G., 1989, *Simulation of Clothing Manufacture*, PhD thesis, Manchester Metropolitan University, December.

Hunter, N.A., 1990, *Quick response in Apparel Manufacturing*, The Textile Institute, Manchester.

Kalta, M., Tyler, D., Lowe, T. And Wilson, G., 1994, Prototype simulation models for team working in the clothing industry, *Journal of Clothing Technology and Management*, 11, 2, 16-30.

Kalta, M., Tyler, D., Lowe, T. And Wilson, G. 1997, COATS: Simulation of Dynamic Team Behaviour in the Clothing Industry, in *Restructuring Manufacturing*, (Eds. D. J. Tyler and P. Totterdill), The Textile Institute, Manchester, 75-91.

Klein, M. and Tixier, V., 1971, SCARBEE: A data and model bank for financial engineering and research, *Proceedings IFIP Congress*, North-Holland.

Klein, M. and Methlie, L.B., 1990, *Expert Systems. A Decision Support Approach with Applications in Management and Finance*, Addison-Wesley Publishing Company, UK.

Lowder, R., 1991, Balance: a delicate word in modular manufacturing, *Bobbin*, 33, 3, 132-138.

Lowe, T., Kalta, M., Wilson, G. and Tyler, D., 1994, Analysis of machine breakdown and repair for team working in the clothing industry, *Journal of Clothing Technology and Management*, 11, 2, 16-30.

Lowe, T., Tyler, D., Kalta, M. and Wilson, G., 1995, Artificial intelligence and simulation: Simulation validation for agenda based scheduling, *AISB Quarterly*, (Autumn), 93, 50-53.

Peeters, M.H.H. and Pott, F.D., 1993, Integral organisation innovation in the Dutch clothing industry: The myth of new production systems, *International Journal of Human Factors in Manufacturing*, 3, 3, 275-292.

Price, G., Aspinall, J. and Tyler, D., 1990, Team spirit, *Manufacturing Clothier*, 71, 7, 13-17.

Sips, K., 1993, Autonomous work groups in Flanders and the Netherlands: A critical review of the revival of socio-technical systems in a European context, *International Conference on Self-Managing Work Teams*, (Eds. M. Beyerlein, L. Teal, G. Rust and M. Bullock), September 22-24, Dallas, Texas, 165-174.

Tyler, D.J. and Fozzard, G.J.W., 1987, Simulation incorporating characteristics of manual skill, *Proceedings of the 3rd International Conference on Simulation in Manufacturing, Turin*, November 1987, 95-106.

Tyler, D., 1989, Managing for production flexibility in the clothing industry, *Textile Outlook International*, (September), 63-84.

Price, G. and Tyler, D., 1991, Incentives for abolishing piecework in quick response clothing manufacture, *Textile Outlook International*, (September), 69-81.

AUTHORS' BIOGRAPHY

Dr. Mohamad Kalta has a degree in mechanical engineering and a Ph.D. from UMIST, Manchester, UK. His research interests are in CAD systems and process planning. He then joined Manchester Metropolitan University (MMU) as a Research Fellow where he was involved with simulation of apparel industry assembly teams and with the development of a DSS for pre-production planning of assembly cells. He is currently working in industry as a professional engineer.

Mr. Tim Lowe has a degree in computer science and joined MMU initially to work on simulation software development. He was subsequently a Research Fellow, where his software-writing expertise was used to develop an integrated environment for an industrial database linked to intelligent tools. He was responsible for the scheduling system, which provides management with team-to-style assignment options. This research will provide the basis for his Ph.D. degree thesis. He is currently employed in developing commercial software.

Dr. David Tyler is Senior Lecturer in Department of Clothing Design and Technology, Manchester Metropolitan University, United Kingdom. Starting with a background in the physical sciences, David Tyler's involvement with apparel industry began when he sought to analyze industrial problems from a physicist's perspective. Subsequently, he worked within the industry, initially as a technologist and then as a manager. After joining MMU, he has pursued a number of research interests related to responsive manufacturing and systems modeling, flow line systems and management strategies to optimize performance. The DSS work was part of larger project on teamwork in manufacturing, with Kalta and Lowe as research staff. Current research interests relate to teamworking in product development and its relation to operational practices, and the optimization of performance of textile/apparel supply chains.

E4

Evaluation of Functional and Cellular Layouts Through Simulation and Analytical Models

N. C. Suresh

1. INTRODUCTION

The relative performance of functional layout (FL) and cellular layout (CL) has been analyzed using computer simulation models since early 1970s, and through analytical models mostly since late 1980s. Simulation studies constitute the major portion of the work in this area, due to the complexity and analytical intractability of both functional and cellular manufacturing systems. Analytical modeling has barely begun, and the few models developed so far have been mainly aimed at resolving some of the paradoxical findings from simulation-based research.

Many simulation-based comparisons of FL and CL have, counter-intuitively, indicated that the *performance of FL may be superior to CL under many parameter ranges*. Specifically, it has been shown in many of these studies that the queuing times, flow times and work-in-process (WIP) inventory in FL may be *lower than in CL*. This is, of course, at odds with the message of much of the prescriptive literature on group technology (GT) and cellular manufacturing (CM) over the last three or four decades.

This chapter attempts to provide a background and summary of the body of literature devoted to the relative performance of FL and CL, and the implications of a recent stream of analytical-simulation models. The chapter is organized as follows. Simulation studies devoted to FL-CL comparisons are first summarized in §2. Results from the recent analytical-simulation modeling stream are summarized in §3 and §4, followed by conclusions and future research directions in §5.

2. SIMULATION COMPARISONS OF FL AND CL

Simulation studies of CM were first conducted in the 1970s. Early studies (e.g., Baldwin and Crookall 1972; Athersmith and Crookall 1974; Shunk 1976; Crookall and Lee 1977) investigated the extent of benefits such as reduction in lead time and setup times with the introduction of cellular layouts.

These studies were followed by the works of Leonard and Rathmill (1977a; 1977b) and Rathmill and Leonard (1977), which served to raise some fundamental questions regarding the real extent of reduction in flow time and WIP possible with CM.

Research on CM continued to rely on computer simulation as the primary research tool during the 1980s, with Gupta and Tompkins (1982) investigating the effects of load imbalances, and Ang and Willey (1984) examining the effects of inter-cell movements. There was increasing methodological rigor, beginning with, in particular, the studies of Flynn and Jacobs (1986; 1987).

These later studies may be broadly divided into two sub-categories. One stream of literature, beginning with Mosier, Elvers and Kelly (1984) has addressed the impact of part family-oriented scheduling, while the other, beginning with Flynn and Jacobs (1986), has tended to focus on comparisons of FL and CL systems under various conditions. Studies devoted to the first stream are discussed in chapter F3 by Mahmoodi and Mosier. A taxonomy of this literature is also found in Shambu, Suresh and Pegels (1996). The focus of attention in this chapter is on the second stream that deals with FL-CL comparisons.

Studies On FL-CL Comparisons.

Craven (1973) first sought to highlight the restrictive applicability of cellular manufacturing. Leonard and Rathmill (1977a; 1977b) and Rathmill and Leonard (1977) asserted, rather vehemently, that efficient functional layouts are superior to GT under most parameter ranges. They pointed out that GT has been favorably compared with the poor functional-layout systems that many firms started with; and, efficient functional layouts are capable of providing better flow times and WIP than CL, apart from offering several flexibility-related advantages. Rathmill and Leonard (1977) also offered a queuing theory-based explanation that sought to demonstrate performance *deterioration* with CM.

Ang and Willey (1984) demonstrated that allowing a small amount of inter-cell movements can significantly improve the performance of cells. But, they did not compare CL having inter-cell movements, with the erstwhile FL system, and it is hard to assess the relative improvement in performance. Lee (1985) showed that larger cells, utilizing multi-server facilities, perform better than smaller, excessively partitioned cells.

Flynn (1984) pointed out the need for fair comparisons of GT with job shops on the basis of controlled experimentation. The results from Flynn (1984), and Flynn and Jacobs (1986; 1987), all simulation-based studies using specific shop data, showed that flow time and WIP performance can be inferior in systems with dedicated machines. In Flynn and Jacobs (1986; 1987), a first-come-first-served (FCFS) rule was used. A cellular manufacturing system with a low degree of cell independence was assumed i.e., every part was routed through many cells for processing, in contrast to normal cellular operations. This has turned out be a somewhat distracting feature. They compared four different kinds of layouts: PROCSS, a process layout developed by using CRAFT; PRODED, physically identical to PROCSS but with machines dedicated to the processing of certain part families; and two layouts that relied both on a cellular layout and machine dedication.

In Flynn (1987a), the service discipline was changed to the repetitive-lots sequencing rule proposed by Jacobs and Bragg (1988). As per this rule, priority is given to waiting jobs of the same part type, to capitalize on sequence-dependent setup reduction. It was shown that the performance is improved by using this rule, but, the superiority of FL is seen in this study also. Three layouts were studied: a process, a cellular and a combined layout.

Flynn (1987b) investigated the effect of average setup time on the output capacity of a shop. Two layouts were used in the study: a process layout that used machine dedication, and a cellular layout.

Morris and Tersine (1989; 1990) presented very similar results. These were again simulation-based studies, in which the experimental factors included the ratio of setup to process time, move times between work centers, demand stability and the type of flow within the cells. The simulated shop consisted of 30 machines of 8 different types, and the process layout with 8 departments was compared with a cellular layout with 5 cells. The superiority of the functional layout was once again demonstrated in the parameter ranges simulated, but it was also suggested that CM may compare favorably in a minority of situations characterized by high setup-to-run-time ratio, stable demand, unidirectional flow of work in the cell and high material movement times between process layout departments.

Christy and Nandkeolyar (1986) compared the performance of a job shop with a hybrid cellular shop employing factors such as setup and material handling time reductions, and the proportions of machines and jobs in cells. Garza and Smunt (1991) illustrated the negative impact of inter-cell flow and indicated how changes in other operating factors caused by conversion from a process to cellular layout may counter the negative impact of this flow. The process layout contained eight departments and the cellular layout, six cells.

Burgess (1988) and Burgess, Morgan and Vollmann (1993), using simulation, compared a job shop with a hybrid shop containing a job shop and one cell. The study found that the hybrid system performs better when the cell is operated at higher levels of capacity than the job shop.

Shafer and Charnes (1993, 1995) compared the performance of cellular and functional layouts under a variety of operating conditions, including operation overlapping within cells, and showed superiority of cellular layout under several parameter ranges.

The reader is also referred to the recent works of Morris and Tersine (1994), who extend these studies to machine-and-labor-limited systems. The study of Jensen, Malhotra and Philipoom (1996) provides further insights into the parameter ranges where CL can be inferior or superior to FL.

3. ANALYTICAL MODELS

The conditions appropriate for cellular manufacturing are yet to be comprehensively identified. In addition to empirical research, there is also, as Gupta and Tompkins (1982) pointed out, a clear need for developing analytical tools to evaluate the performance of GT systems. But the complexity of this environment has been a deterrent in this regard.

Analytical models developed so far, in addition to the models presented by Rathmill and Leonard (1977) mentioned earlier, include the works of Karmarkar (1987), Karmarkar, Kekre, Kekre and Freeman (1985), Suresh (1991; 1992; 1993), Agarwal, Huq and Sarkis (1993), Wainwright, Harrison and Leonard (1993), Suresh and Meredith (1994), and Wainwright (1996).

The rest of this chapter focuses on the models that address FL-CL comparisons and attempt to provide an explanation for the performance deterioration shown by simulation-based studies (Suresh 1991; 1992; 1993; Suresh and Meredith 1994). Agarwal et al. (1993) provide further insights based on these models, and the recent work of Johnson and Wemmerlöv (1996) attempts to reconcile the results of past simulation studies in light of these models.

The objective in Suresh (1991) was to analyze the effects of partitioning functional-layout work centers into hybrid cellular manufacturing system components. These models demonstrate the effects of partitioning multi-server work centers. It is shown that partitioning a job-shop work center leads to adverse effects on flow characteristics. This can be countered by setup reduction in the cell components, but the adverse effects in the remainder cell tend to offset much of the benefits obtained, and the overall results for hybrid contexts tend to be along the lines of simulation studies such as those of Flynn and Jacobs (1986) and Burgess et al. (1993). These models were extended further to: 1) develop measures for fair comparison of partitioned and unpartitioned work centers, in Suresh (1992); 2) to explore the impact of various measures to overcome the main source of performance deterioration ("pooling loss") when converting to CM, in Suresh and Meredith (1994); and, 3) to explore these effects in labor-and-machine-constrained situations, i.e., dual-resource-constrained systems, in Suresh (1993). The implications of these models are summarized below.

4. IMPLICATIONS OF ANALYTICAL MODELS

Simulation studies have clearly pointed out one important source of performance deterioration while converting to CM. This factor has come to be addressed squarely only in recent years.

This effect may come as a surprise to many, and can be best understood, perhaps, only by conducting a simulation experiment of a functional and cellular layout. Using the same part or job mix and order streams on both systems, and assuming reduced setup times and move times in the cells, and monitoring queue times, overall lead times, WIP, etc., the performance deterioration in CL will still be evident in several parameter ranges.

This performance deterioration arises due to the fact that when converting to CL, multi-server work-centers get partitioned, which results in an adverse effect that is well known in queuing theory. For instance, in Figure 1, there are ten multi-machine work centers in FL, numbered 1 to 10. These machine pools get partitioned and the machines assigned to five different cells in CL. Likewise, there are workers belonging to ten functionally-specialized labor pools in FL (denoted as types *a* through *j*). These labor pools are also liable to be partitioned and assigned to different cells (and typically cross-trained to perform other functions in CL). This can lead to a considerable *loss of pooling synergy*.

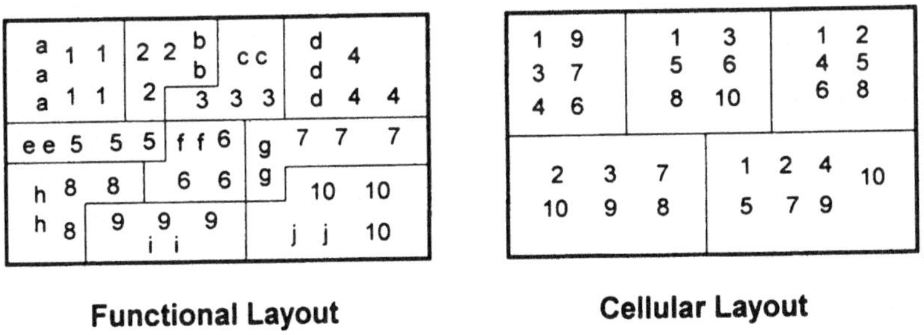

FIGURE 1. Functional and cellular layouts.

Investigating these effects through analytical modeling poses major problems due to the complexity of several interacting multi-server queues. Given this difficulty, the general approach adopted in (Suresh 1991; 1992; 1993; Suresh and Meredith 1994) has been to analyze these effects using a single work center that gets partitioned into dedicated queues. The insights developed from these elementary models are then used to construct several shop-level simulation experiments and test several propositions. This research stream is based on, in effect, a *combined analytical-simulation modeling approach*.

The results of this modeling stream can be best described in a step-by-step approach, considering several effects that take place when converting to CM. We consider these effects in the following order:

1. Effects of lot sizes
2. Effects of partitioning job shop work centers
3. Effects of setup reduction
4. Effects of reduction in variability in job arrivals
5. Effects of reduction in variability of processing times
6. Effects of productivity improvements due to part-family similarities

7. Effects of cross-training workers
8. Effects of labor assignment and dispatching rules.

In practice, all these effects take place simultaneously and the net effects of conversion to CM, even through simulation models, have been somewhat unclear.

These effects are described below in a non-mathematical way. Effects (1) through (6), which deal with loss of pooling synergy due to partitioning machine pools, are discussed in §4.1 (§4.1.1. - §4.1.6.). The loss of pooling on the labor dimension, effects 7 and 8, are addressed in §4.2 (in §4.2.1.- §4.2.2.).

4.1. Loss Of Pooling Synergy Due To Partitioning Machine Pools

In this section, we consider effects 1 through 6, as stated above. Consider an unpartitioned job shop shop work center, and a set of partitioned queues as shown in Figure 2. The unpartitioned system (UPS) is the job shop work center, which is divided into the partitioned system (PS) with dedicated queues. Each of these queues may be in a different cell, and dedicated to a subset of parts (part families).

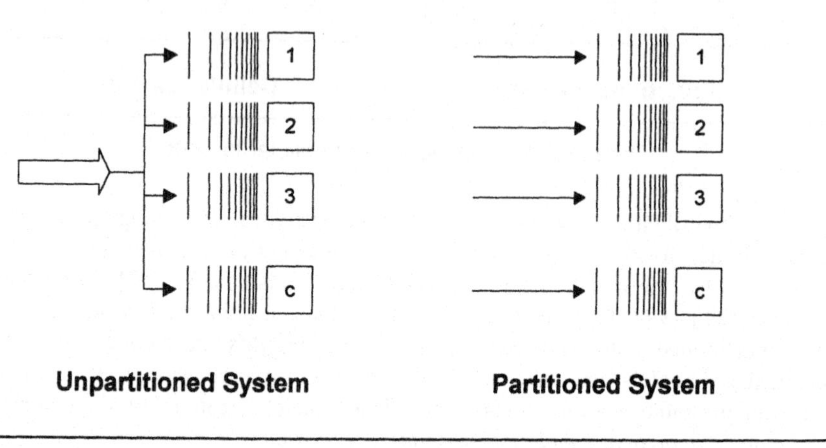

FIGURE 2. An unpartitioned job shop work center and partitioned queues.

4.1.1. Effects Of Lot Sizes.

We first consider the unpartitioned system (UPS), which represents the multi-server work center in FL.

Given stochasticity of arrivals and processing, lot sizes are known to have a *nonlinear* effect on flow time and WIP, as described by Karmarkar (1987). This can be seen in the flow time/WIP curve for UPS in Figure 3a. At low lot sizes, the flow time and WIP (shown in common as a *flow ratio*) are high. This is due to the fact that, at low lot sizes, the number of jobs (setups) is high, which consumes capacity excessively and causes congestion. As lot sizes are increased, the congestion eases swiftly and reduces flow time and WIP. There is an optimum range of lot sizes at which flow time and WIP are minimized; but they are somewhat insensitive to lot size around the optimum point, providing certain amount of flexibility. Beyond the optimum level of lot sizes, they tend to increase somewhat linearly. The net result is a U-shaped function seen in Figure 3a for the UPS system.

4.1.2. Effects Of Partitioning Job Shop Work Centers.

The first physical effect of partitioning this UPS system into a partitioned system, PS1, is to increase flow time and WIP. This is due to the fact that, in the partitioned system, there are random instances when a job has to wait in a queue even though a machine of the required type (or labor with the necessary skill) may be available at other cells. This does not occur in the unpartitioned system which has all similar resources located together. It must be stressed that this pooling loss occurs as long as there is stochasticity, and regardless of the customary assumptions in queuing theory relating to distributions of job inter-arrival times and processing times.

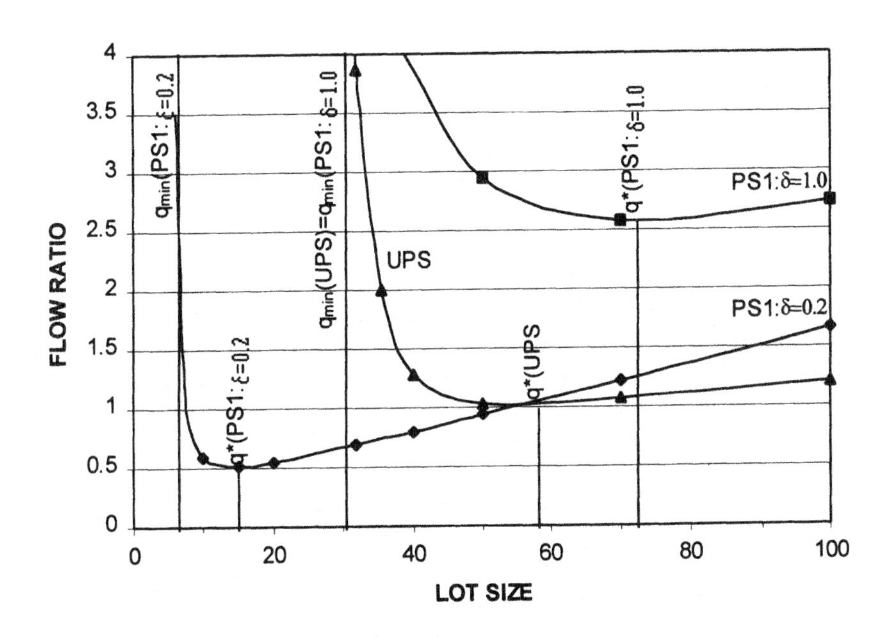

FIGURE 3a. Effects of partitioning and setup reduction.

Considering UPS as a multi-server queuing system (e.g., M/M/c, or GI/G/c), and the partitioned system as a c*single-server system (e.g., c*(M/M/1, M/G/1, or GI/G/1)), the effect of partitioning can be analyzed. This effect is along the lines shown in Figure 3a: the U-shaped curve for PS1 ($\delta = 1$) is seen to be higher than the curve for UPS. ($\delta = 1$ indicates no setup reduction, as yet, which is considered below).

This increase in flow time and WIP can be substantial at times, and it depends on variables such as the lot sizes, arrival and processing times and their distribution. Given the increased flow time and WIP, the general tendency in a partitioned system is to reduce the loading level, which then reduces the *de facto* resource utilization. Thus, partitioning a multi-server resource results in increased lead time and WIP, reduced *de facto* utilization, and also significant, unabsorbed load imbalances.

Therefore, the various measures advocated with CM, such as reduction in setup times and lot sizes, must adequately counter this performance deterioration before leading to the advantages normally associated with CM.

To provide a non-dimensional measure of this adverse effect, the flow time (or WIP) in partitioned systems may be divided by the minimum flow time (or WIP) in the unpartitioned system. This results in the *flow ratio (FR)* measure (Suresh 1992). Ensuring the condition { FR $\leq$ 1 } becomes a major design objective for partitioned systems. In addition to being non-dimensional, two other advantages of using the flow ratio measure include: 1) it is a measure common to flow time and WIP, in accordance with Little's Law; and, 2) it enables a *fair comparison* of partitioned systems with the *best unpartitioned sytem*.

4.1.3. Effects Of Setup Reduction.

We next consider the effects of setup reduction. In the partitioned system, the queues are separate, and each queue, being a part family, consists of jobs that are similar. These similarities are exploited in GT and setup times reduced. The effect of such setup reduction is to cause a leftward and downward shift in the flow time/WIP function. This effect is also seen in Figure 3a for PS1 ($\delta = 0.2$, denoting an average of 80% reduction in setup times).

The higher the degree of setup reduction (i.e., lower the value of δ), the greater the leftward and downward shifts, overcoming the adverse effects of partitioning. For a given lot size, at a certain level of setup reduction, the flow time (or WIP) in the partitioned system equals the lowest level of flow time (or WIP) in the unpartitioned system. This refers to a *threshold level of setup reduction, required to ensure { FR ≤ 1 }*. This is referred to as *breakeven δ.*

Overall, the effect of setup reduction is to improve the performance of the partitioned systems and overcome the negative effects of partitioning.

4.1.4. Effects Of Reducing Variability of Job Arrivals.

A well known result in queuing theory is that reducing the stochasticity in arrivals and processing serves to reduce queue lengths and WIP. It can also be shown that reducing the variability of job arrivals also serves to reduce the adverse effects of partitioning multi-server resources (Suresh and Meredith 1994). Like setup reduction, reducing the variability in job arrivals has beneficial effects, and tends to bring down flow ratios. This is seen in Figure 3b for system PS2 ($\delta = 0.2$).

In cellular manufacturing, several factors may enable reduction in variability in job arrivals: formation of part families and exploiting their similarities, stable arrangements with vendors, better scheduling of incoming materials, proper production planning and control systems, etc. Given the proximity of upstream and downstream work centers, it might be expected that much of the uncertainty and randomness of job arrivals present in a job shop may be reduced. With the benefit of hindsight, it may be observed that such factors have not been considered adequately in earlier simulation studies.

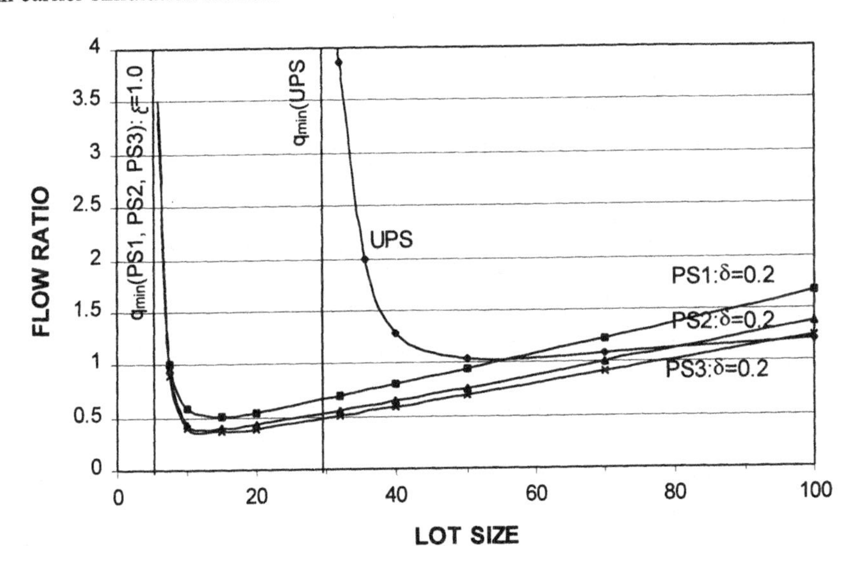

FIGURE 3b. Effects of reducing variability of arrivals and processing.

4.1.5. Effects Of Reducing Variability of Processing Times.

Part-family similarities can again be exploited to reduce variabilities in processing times, through, for example, specialized tooling that are dedicated to each family. This also has a favorable effect on flow time and WIP of the partitioned system, as seen in Figure 3b, for PS3 ($\delta = 0.2$).

4.1.6. Effects Of Productivity Improvements.

GT has traditionally emphasized setup-time reductions, but part-family similarities may also be exploited to effect productivity improvements to reduce operation times, through design of specialized tooling, for instance. There is limited empirical evidence to date for the actual materialization of such benefits. But, the models have shown that, if utilized, they may also have a significant effect in countering the adverse effects of partitioning.

4.2. Loss Of Pooling Synergy Due To Partitioning Labor Pools

We next consider the adverse effects of partitioning labor pools. When converting to CL, like functionally-specialized machine pools, functionally-specialized labor pools also get partitioned. This again leads to adverse effects on flow time and WIP, and other physical measures.

A concerted strategy is called for on the labor dimension to cope with this loss of synergy. This strategy may be based essentially on exploiting several types of flexibility associated with labor. These include, in particular: 1) the ability to be cross-trained, and, 2) the ability to be moved around, inside and outside the cells, wherever there is congestion or bottlenecks.

4.2.1. Effects Of Cross-Training Workers.

When switching to CM, cross-training the work force often becomes a *necessity*. For instance, in Figure 1, it seen that, to ensure a feasible partition of labor pools, some amount of cross-training is necessary after the workers are assigned to cells.

Machine-and-labor-constrained systems are referred to as *dual-resource-constrained (DRC)* systems. A review of this literature for DRC job shops is presented in Treleven (1989). FL-CL comparisons on the basis of DRC systems (Suresh 1993; Morris and Tersine 1994) indicate some useful results for effective operation of CM.

In Suresh (1993), to quantify the level of cross-training, a simple measure, η, has been used, which denotes the average number of functions that each worker is trained in. When there is no cross-training, η equals 1. The results from extending the analysis and experimentation to DRC situations may be summarized as follows:

1. First, a threshold level of cross-training, short of hiring additional work force, may often be *necessary* to ensure a viable partition of existing labor pools and ressaignment into cells.

2. Cross-training has a significant impact on flow time and WIP performance, both in FL and CL (see simulation results for [FL $\eta = 1$ δ=1] and [FL $\eta = 2$, δ=1] in Figure 3c).

3. Cross-training is a very useful measure to counter the adverse of effects of partitioining labor pools, as seen in Figure 3c for [CM $\eta = 2$ δ=1], [CM $\eta = 3$, δ=1] and [CM $\eta = 4$, δ=1]. It is also seen that in the parameter range considered, cross-training by itself does not fully overcome the pooling loss.

4. Even minimal amounts of cross-training (e.g., from $\eta = 1$ to $\eta = 1.1$ or 1.2) lead to significant benefits.

5. The benefits of cross-training are subject to diminishing marginal returns; hence, a large amount of cross-training may not be necessary.

6. With both cross-training and setup reduction, CM seems to more easily outperform FL, as shown in Figure 3d. This relative advantage is also seen to be greater at lower lot sizes, which GT/CM has always emphasized.

7. Cross-training serves to raise the threshold level of setup reduction (breakeven δ). That is, with cross-training, less setup reduction is required to ensure $\{FR \leq 1\}$ i.e., to outperform the best FL.

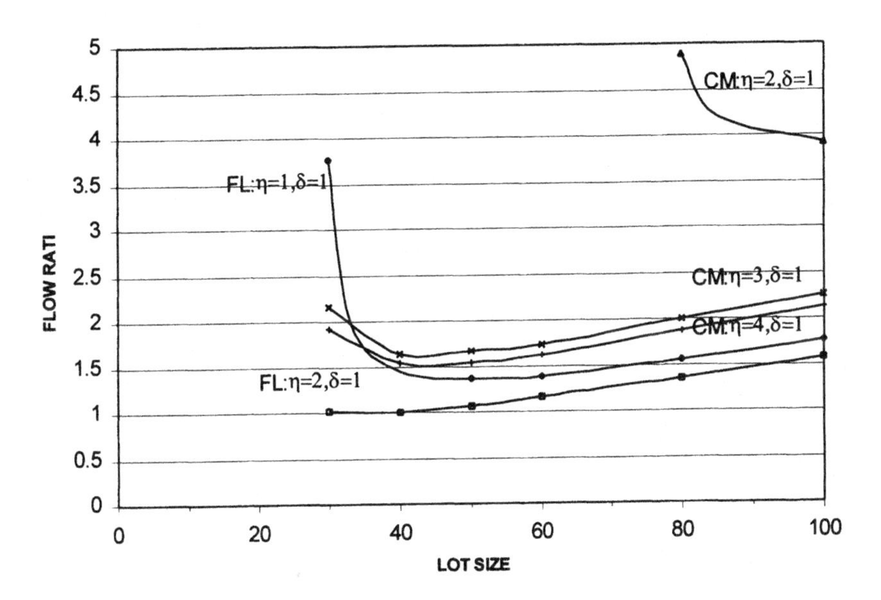

FIGURE 3c. Effects of cross-training.

4.2.2. Labor Assignment And Job Dispatching Rules.

Another aspect of flexibility is the mobility of labor (unlike machines in general), and the ability to dynamically reassign labor to different parts of the system and alleviate congestion. This type of flexibility can be exploited by effective labor organization and job assignment rules.

It may be noted that cross-training is a prerequisite for exploiting this mobility-derived flexibility: greater amount of cross-training provides greater choice for this mobility and reassignment, serving to improve overall system performance.

Limited research to date indicates that the "COMP-LNQ-SPT" labor assignment combination is effective for minimizing flow ratio. This rule permits labor reassignment, to another cell, or else where in the same cell, after completion of every operation ("COMP"), to the longest queue in the system ("LNQ") (within the domain of the worker's skills); and, within this queue, to the job with the shortest setup-plus-processing time ("SPT"). This finding is generally consistent with the results of DRC job shop studies.

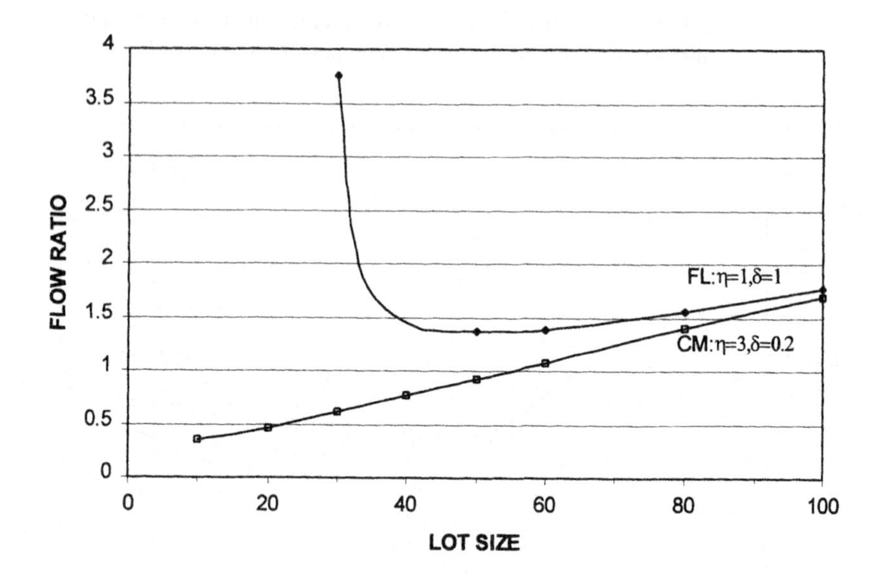

FIGURE 3d. Effects of cross-training and setup reduction.

5. CONCLUSIONS

Much analytical work remains to be done for this problem area, which has long been considered as intractable. The range of parameters for which CL and CM outperform functional organizations needs to be isolated through further development. It is clear that several parameters unique to an environment, such as the setup-to-operation-time ratios, the degree of setup similarities, as well as operating policies like lot sizing, scheduling, inter-cell movements and overlapped operations influence the flow ratio (the relative advantage of CL over efficiently-operated FL system). It is also apparent that the performance of the FL system itself could be improved through a variety of measures, including the use of part-family scheduling, other ways to reduce setup times, cross-training the work force, etc., all within the FL context.

There is now a concerted attempt in industry, to dispense with functional organizations and department-after-department sequence of operations, throughout the firm, through concurrent engineering, business process reengineering, etc., as noted earlier in chapter A. In this process, it is likely that the issue of loss of pooling synergy may be encountered in other ways as well. Viswanadham, Narahari and Raghavan alluded to this earlier, in section §2.1 of chapter E1.

The models developed so far are undoubtedly leading us towards a generalized range of parameters in which CM may be effectively and efficiently utilized. The models have so far addressed problem contexts in which machine and labor pools can be partitioned; in some mid- and low-volume situations, they cannot be satisfactorily partitioned in the first place, precluding the implementation of CM in such cases. Likewise, further analytical development is also necessary to address other, realistic factory configurations where loss of pooling synergy may not be a significant issue. Further empirical research, along the lines of Wemmerlöv and Johnson (1997), is needed to provide realistic inputs for choice of experimental factors, and to validate analytical and simulation models. It must also be recognized that not all real-world factors can be captured fully in analytical and simulation models.

Further empirical research is also needed to assess the real impact of GT in industry. In many of the firms that have implemented CM, the real extent of success with CM is unclear. In some cases, the benefits from CM seem to be marginal, at best. Whether this is due to improper implementation, improper planning systems, or due to underlying problems with CM, etc., is hard to say. But it does leave the impression in such cases that it may be beneficial to merely resort to improvements in the erstwhile FL system through measures such as part-family-oriented scheduling, setup reduction, cross-training and better planning systems.

Empirical research is often constrained by the fact that it is possible to document the performance of CM only with what existed before; and, the *status quo ante* may not correspond to an efficiently-operated FL system. Research has also been confined to the cases where CM has been implemented. It is necessary to include firms from the vast majority of low- and mid-volume situations where CM has not been implemented, and the feasibility of implementing CM in such contexts should be explored, as in chapter E2 earlier, by Johnson and Wemmerlöv.

In firms that have implemented CM, a reduction in lead time is often reported. There is a need to develop reliable metrics for lead time, WIP and other physical measures, and include provisions for cross-checking and validating reported impacts. For instance, if lead times did indeed reduce significantly, we should see commensurate improvements in inventory turnover ratio, across all part types and in overall performance of the factory.

Data gathering is also at times based on information elicited from project champions and/or other change agents within each firm. It is likely that a more realistic assessment of CM may emerge from studies based on a broader respondent base within each firm.

Empirical research is subject to numerous methodological and other difficulties. While further developments in analytical and simulation modeling are clearly necessary, empirical research and case studies can also significantly enhance our understanding at this juncture.

REFERENCES

Agarwal, A., Huq, F., and Sarkis, J., 1993, Partitioning work centers for GT: Insights from a parametric analysis, *Proceedings of Decision Sciences Institute*, Washington, D.C.: Decision Science Institute, 1563-1565.

Ang, C. L. and Willey, P.C.T., 1984, A comparative study of the performance of pure and hybrid group technology manufacturing systems using computer simulation techniques, *International Journal of Production Research*, 22,2, 193-233.

Athersmith, D. and Crookall, J.R., 1974, Some organizational aspects of cellular manufacture based on computer simulation, *Proceedings of the 15th Machine Tool Design Research Conference*, Sept., 18-20, Birmingham, UK.

Baldwin, K.I. and Crookall, J.R., 1972, An Investigation into the application of grouping principles and cellular manufacturing using Monte Carlo simulation, *CIRP Journal on Manufacturing Systems*, 1, 2, 193-208.

Burgess, A. G., 1988, *Evaluation of group technology/cellular manufacturing: A computer simulation*, Unpublished doctoral dissertation, Boston University.

Burgess, A. G., Morgan, I., and Vollmann, T. E., 1993, Cellular Manufacturing: its impact on the total factory, *International Journal of Production Research*, 31, 9, 2059-2077.

Christy, D. P. and Nandkeolyar, U., 1986, A simulation investigation of the design of group technology cells, *Proceedings of the 1986 Annual Meeting of the Decision Sciences Institute*, 1201-1203.

Craven, F. W., 1973, Some constraints, fallacies, and solutions in G.T. applications, *Proceedings of Fourteenth International Machine Tool Design and Research Conference*, 169-175.

Crookall, J. R. and Lee, L. C., 1977, Computer-aided performance analysis and design of cellular manufacturing systems, *CIRP Journal of Manufacturing Systems*, 6, 3, 177-195.

Cummings, G.F., 1980, Simulation model to compare group technology and functional layouts, *Proceedings of Summer Computer Simulation Conference*, LaJolla, CA: Society for Computer Simulation, 626-630.

Flynn, B.B., 1984, *Group technology versus process layout: A comparison using computerized job shop simulation*, doctoral dissertation, Indiana University.

Flynn, B.B., 1987a, Repetitive lots: The use of a sequence-dependent scheduling procedure in group technology and traditional shops, *Journal of Operations Management*, 7, 2, 203-216.

Flynn, B.B., 1987b, The effects of setup time on output capacity in cellular manufacturing, *International Journal of Production Research*, 25, 12, 1761-1772.

Flynn, B.B. and Jacobs, F.R., 1986, A simulation comparison of group technology with traditional job shop manufacturing, *International Journal of Production Research*, 24,5, 1171-1192.

Flynn, B.B. and Jacobs, F.R., 1987, An experimental comparison of cellular (group technology) layout with process layout, *Decision Sciences*, 18, 4, 562-581.

Garza, O. and Smunt, T. L., 1991, Countering the negative impact of intercell flow in cellular manufacturing, *Journal of Operations Management*, 10, 1, 92-118.

Gupta, R.M. and Tompkins, J.A., 1982, An examination of the dynamic behavior of part-families in group technology, *International Journal of Production Research*, 20, 1, 73-86.

Jacobs, F.R. and Bragg, D., 1988, Repetitive lots: job flow considerations in production sequencing and batch sizing, *Decision Sciences*, 19, 2, 281-294.

Jensen, J.B., Malhotra, M.K., and Philipoom, P.R., 1996, Machine dedication and process flexibility in a group technology environment, *Journal of Operation Management*, 14, 1, 19-39.

Johnson, D. and Wemmerlöv, U., 1996, On the relative performance of functional and cellular layouts- An analysis of the model-based comparative studies literature, *Production and Operations Management*, 5, 4, 309-334.

Karmarkar, U.S., 1987, Lot sizes, lead times and in-process inventories, *Management Science*, 33, 3, 409-418.

Karmarkar, U.S., Kekre, S., Kekre, S. and Freeman, S., 1985, Lot-sizing and lead time performance in a manufacturing cell, *Interfaces*, 15, 2, 1-9.

Kekre, S., 1987, Performance of a manufacturing cell with increased product mix, IIE Transactions, 19, 3, 329-339.

Lee, L.C., 1985, A study of system characteristics in a manufacturing cell, *International Journal of Production Research*, 23, 6, 1101-1114.

Leonard, R., and Rathmill, K., 1977a, The group technology myths, *Management Today*, (January), 66-69.

Leonard, R. and Rathmill, K., 1977b, Group technology - A restricted manufacturing philosophy, *Chartered Mechanical Engineer*, 24.

Leu, J.Y., Russell, R.S., and Huang, P.Y., 1993, On the applicability of cellular manufacturing, *Proceedings of the Decision Science Institute*, Washington D.C., Decision Sciences Institute, 1469-1471.

Morris, J. S. and Tersine, R. J., 1989, A comparison of cell loading practices in group technology, *Journal of Manufacturing and Operations Management*, 2, 299-313.

Morris, J. S. and Tersine, R. J., 1990, A simulation analysis of factors influencing the attractiveness of group technology cellular layouts, *Management Science*, 36, 12, 1567-1578.

Morris, J.S. and Tersine, R.J., 1994, A simulation comparison of process and cellular layouts in a dual resource constrained environment, *Computers and Industrial Engineering*, 26, 4, 733-741.

Mosier, C.T., Elvers, D.A. and Kelly, D., 1984, Analysis of group technology scheduling heuristics, *International Journal of Production Research*, 22, 5, 857-875.

Rathmill, K. and Leonard, R., 1977, The fundamental limitations of cellular manufacture when contrasted with efficient functional layout, *Proceedings of Fourth International Conference on Production Research*, 523-546.

Shafer, S.M. and Charnes, J.M., 1993, Cellular versus functional layouts under a variety of shop operating conditions, *Decision Sciences*, 24, 3, 665-681.

Shafer, S.M. and Charnes, J.M., 1995, A simulation analysis of factors influencing loading practices in cellular manufacturing, *International Journal of Production Research*, 33, 1, 279-290.

Shambu, G., Suresh, N.C. and Pegels, C.C., 1996, Performance evaluation of cellular manufacturing systems: A taxonomy and review of research, *International Journal of Operations and Production Management*, 16, 8, 81-103.

Shunk, D.L., 1976, *The measurement of the effects of group technology by simulation*, Unpublished Ph.D. Dissertation, Purdue University, IN.

Suresh, N.C., 1991a, Partitioning work centers for group technology: Insights from an analytical model, *Decision Sciences*, 22, 4, 772-791.

Suresh, N.C., 1991b, The performance of hybrid group technology systems, *Proceedings of the National Decision Sciences Institute*, Miami, FL.

Suresh, N.C., 1992, Partitioning work centers for group technology: Analytical extension and shop-level simulation investigation, *Decision Sciences*, 23, 2, 267-290.

Suresh, N.C., 1993, Job shops and cellular manufacturing: A dual-constrained comparison, *Proceedings of Decision Sciences Institute*, Washington D.C.: Decision Science Institute, 1463-1465.

Suresh, N. C. and Meredith, J. R., 1994, Coping with the loss of pooling synergy in cellular manufacturing systems, *Management Science,* 40, 4, 466-483.

Treleven, M., 1989, A review of the dual resource constrained system research, *IIE Transactions*, 21, 3, 279-287.

Wainwright, C.E.R., Harrison, D.K. and Leonard, R., 1993, Production control strategies within multiproduct batch manufacturing companies, *International Journal of Production Research*, 31, 2, 365-380.

Wainwright, C.E.R., 1996, The application of queuing theory in the analysis of plant layout, *International Journal of Operations and Production Management*, 16, 1, 50-74.

Wemmerlöv, U. and Johnson, D.J., 1997, Cellular manufacturing at 46 user plants: Implementation experiences and performance improvements, *International Journal of Production Research*, 35, 1, 29-49.

Yang, K.K. and Jacobs, F. R., 1992, Comparison of make-to-order job shops with different machine layouts and production control systems, *International Journal of Production Research*, 30, 6, 1269-1283.

AUTHOR'S BIOGRAPHY

(appears at the end of chapter A).

F1

Production Planning and Control Systems For Cellular Manufacturing

J. Riezebos, G. Shambu and N. C. Suresh

1. INTRODUCTION

The performance of a production system depends not only on the quality of the decomposition of the system into cells and departments, but also on the quality of the production planning and control system that is being used. The design of the production planning and control system should meet the requirements of the production system. The goodness of fit between both systems is also of significant importance to take full advantage of the benefits of cellular manufacturing.

Cellular manufacturing (CM) creates coordination needs that cannot be tackled by existing planning systems (Rolstadås 1988). These needs concern both the handling and determination of batches that contain families of parts and the consideration of the cell as one planning unit. Batch sizes cannot be determined in the traditional way, due to setup similarities of various parts within the same family and tooling constraints on the (automated) machines. Considering the cell as a planning unit affects the planning with respect to the cell loading procedure applied and the possibility to control production.

Rolstadås (1988) considered automated flow line cells, but even if other types of layouts within a cell were used, some of the problems unique to CM would remain. Therefore, it is necessary to dwell on the design of production planning and control systems that can, and need to be applied in cellular manufacturing contexts. A number of review articles on production control in CM have appeared, e.g., Sinha and Hollier (1984), and part of the study of Mosier and Taube (1985). In this chapter, we attempt to provide an overview of various systems that are available and to identify their important characteristics if applied to CM.

The paper consists of three main parts. §2 is directed towards production planning systems in CM. §3 gives attention to part family-oriented scheduling. §4 discusses the important problem of lot sizing in CM, in particular, it provides a concise summary of Burbidge's views on lot sizing. This is followed by conclusions and a selection of useful references in the end.

2. PRODUCTION PLANNING AND CONTROL SYSTEMS FOR CM

Petrov (1968) was one of the first to note that a redesign of the production planning and scheduling system is required when applying group technology (GT) principles. He considered various types of flowline cells and determined the planning conditions that are required to improve both the performance of these cells and of the complete system, as this consists of interrelated cells.

Dale and Russell (1983) reported typical production control problems in flow-line CM systems, such as the load balancing problem. The cells consist of various types of machines and operators which often are not equally qualified. In such configurations it can become a problem to maintain a good balance between key machine utilization and operator utilization. Fluctuations in product mix and volume and introduction of new products can aggravate these problems. Redesigning the production system itself to solve these problems is often not possible or acceptable, so the production control system has to deal with them. The same holds for the problems caused by the sharing of key machines between cells. In these cases the realization of the full potential of CM depends mainly on the production planning and control system design. Dale and Russell stated that many problems in firms that reorganized their shop floor layout along GT, have been caused by continuing to apply conventional control thinking, which had worked in a functionally organized production system.

This section is directed towards the design of a production planning and control system for CM. §2.1 presents a framework for production planning in CM. Next, §2.2 describes some existing frameworks and points to their contribution in designing a production planning and control system for CM. §2.3 is directed towards the use of MRP in CM. §2.4 summarizes the views of Burbidge on production planning in CM and the use of Period batch Control. Finally, §2.5 gives attention to other approaches to planning for CM.

2.1. A Framework For Production Planning In CM

There are many differences among firms in the way they plan their production, caused by differences in product characteristics, market position, organization of the production system, capabilities of the planner, available information technology, etc. Therefore, designing a production planning and control (PPC) system is a firm-specific activity. However, some general guidelines and frameworks can be developed. One useful approach to this design process can be found in Banerjee (1997), which is applied for the design of a manufacturing planning and control system to a real-life CM system.

A framework for designing a PPC system for a specific production system should, in our opinion, specify:

1. *The required planning functions*: determine what to produce (orders); determine when to produce (time); determine where to produce (resources)

2. *The direction and contents of the relations between these functions*: specify the following information on the proposed decomposition of the planning process in phases: hierarchical or heterarchical decomposition; aggregation levels per phase with respect to orders, time, and resources; abstraction levels per phase with respect to orders, time, and resources; frequency of (re)planning in the various phases.

Note that a framework does not specify *how* the decisions are taken. Hence, the methods that will be appropriate in a specific production situation to determine what, when, and where to produce are still to be selected.

Aggregation And Abstraction In Planning Decomposition.

Production systems that use CM can often not be planned in the same way as a functionally-organized system. It is therefore important to give attention to the aggregation of resources in layers of the production system when designing a planning system. We identify five layers: *single resource layer*, *shift layer*, *cell or production unit layer*, *cluster layer*, and *system layer*. Some planning functions that are specified may be required only for one layer. For example, loading procedures for a cluster of similar cells. In a functional organization such a layer may be not necessary to take into account. Other planning functions may be required at various layers. For example, material requirements planning (MRP) may be performed both at system layer and within a cell, as described by Love and Barekat (1989).

The *system layer* comprises the total production system and its relation with the environment. The *cluster layer* consists of various clusters of production units within the production system, e.g., assembly cluster, components cluster. The *cell layer* consists of the cells or production units within the cluster. The *shift layer* consists of the shifts within the production units. The *single resource layer* consists of the various resources within a single shift. Types of resources that can be distinguished are machine, operator, tool, buffer place, handling equipment, etc.

This five-layer system can be used to make a more explicit decision on the relations in CM that should be coordinated with the aid of a production control system. It is important to note that the decisions need not to be taken in the sequence of the distinguished layers. In that sense this layer system differs from the NBC-layer system and the architecture for decision making proposed by Jackson and Jones (1987), who apply a hierarchical approach according to this layer system.

The choice for an aggregation level is determined by the required level of information detail for the decisions that have to be taken. The choice for an abstraction level can be based on the cost of timely acquisition of the required information versus the cost of omitting part of the information in the analysis.

Aggregation of *orders* can be done by considering product families, for example, all products of the same model, but with various colors. For the abstraction of *orders* the subset of orders that are placed by a customer can be considered, or the subset of orders that are generated by a reorder point system (forecasted demand), or the subset of orders that are generated to fill capacity, etc.

For the aggregation and abstraction of *resources,* subsets can be constructed using combinations of resource types, such as machines, operators, transportation equipment, storage places, tools, fixtures, information, etc. Within these resource types further aggregation or abstraction is possible, e.g., key machines, tools that are not duplicated, welding operators, etc. The aggregation of time is determined by the length of the time bucket that is considered; the abstraction of *time,* by the length of the planning horizon that is considered.

Characteristics Of A Framework For PPC.

Figure 1 provides a framework for production planning and control for the CM system shown in Figure 2. This framework consists of planning functions, shown as boxes, relations between these functions, planning horizon, period length and replanning frequency applied to these functions, and the layer of the production system at which the function operates. Finally, the framework contains some information on abstraction levels for certain planning functions (e.g. demand management).

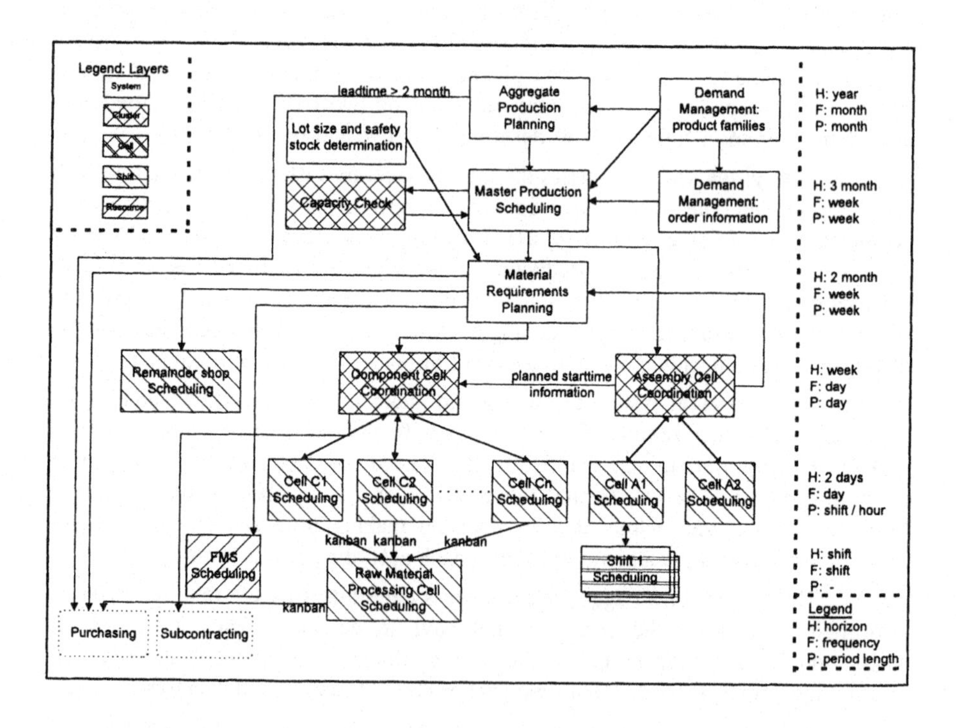

FIGURE 1. Proposed framework for production system of Figure 2.

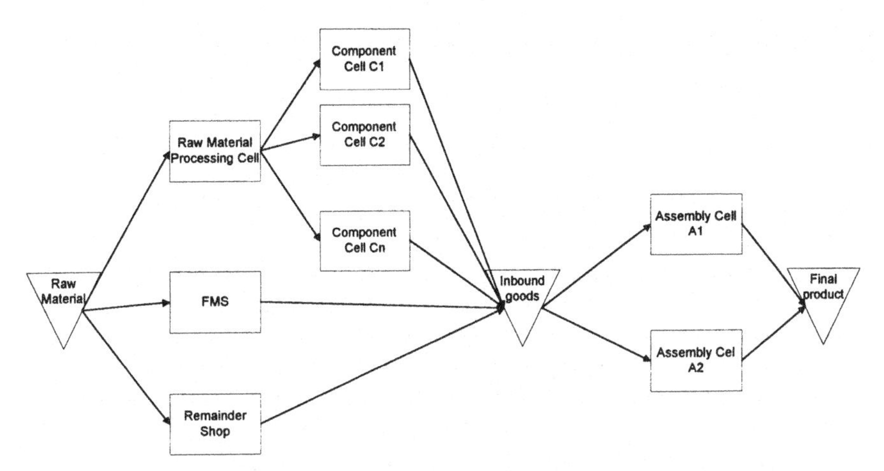

FIGURE 2. Goods flow in a cellular system.

The proposed framework is hierarchical. The highest level contains aggregate production planning and demand management. The next level consists of Master Production Scheduling. Next, MRP uses information on the planned production of the end items (including, for example, spare parts) and the preferred lot sizes and safety stocks to time phase the requirements for the various clusters and production units, using the expected (standard) throughput times. The remainder shop and the FMS construct schedules on the basis of this information. However, the component cells use also information of the assembly cell coordinator on the planned start times of the various assemblies to determine the actual priority of the various released work orders. The component cells obtain orders daily from the cell coordinator. The available capacity in these cells is controlled by the coordinator function, and reallocating work to one of the other cells or an (external) subcontractor is used to solve short term loading problems. Work order release to the cells is performed by the cell coordinator function. The component cells obtain new material from the raw material cell through a kanban system. This cell is therefore not controlled from the MRP planning function. MRP does present information on the expected amount of raw material needed to the suppliers. The flow of this material is also controlled through kanban. The framework also shows that the FMS schedule is being updated far more frequently than the remainder shop schedule.

2.2. Review Of Frameworks For Production Planning In CM

Many authors have proposed an MRPII framework for CM system (e.g., Singh 1996). However, an MRPII framework specifies what planning modules are required and how they are related, but does not give attention to the contents of the relation between the planning modules and the configuration of the production system that has to be controlled. The information content of such a framework is therefore very restricted.

Hyer and Wemmerlöv (1982) proposed a general framework for production planning and control and apply this to cellular manufacture in the component parts manufacturing. Their framework is a hierarchical decision process that consists of three levels: 1) when and in what quantities are final products to be produced; 2) what parts are to be produced during a specific time period in what quantities; 3) when and in what order should jobs be processed at various workstations. At each level capacity checks are required to ensure feasibility of a particular decision. The hierarchical levels specify the sequence of decision making. Feedback loops between the levels are not considered. Note that their framework does not help to determine a suitable time period in level 2 and that the type of coordination between clusters, cells, or shifts is not determined in this framework.

Bauer et al. (1991) developed a manufacturing control systems hierarchy for a batch-oriented discrete parts manufacturing context. Their hierarchical framework represents a hybrid approach to production planning, e.g., it is said to be based on ideas from MRP, optimized production technology (OPT), and just-in-time (JIT) systems. Production activities that require planning can be strategic, tactical, or operational in nature. Strategic activities relate to the products to be manufactured, and the design of the production system. These strategic activities have to result in a realizable master production schedule. The tools that can be used to generate such schedules can as well be obtained from JIT planning techniques as from MPS scheduling techniques (Vollmann, Berry, and Whybark 1997). The specific situation dictates which of these techniques is appropriate and what level of detail in modeling the production system is required.

The tactical planning level consists of a requirements planning function, which is considered to translate the master schedule into weekly or daily requirements of parts and components in the system. The operational planning consists of cell controllers (production activity control) and a factory coordination level, which coordinates activities of the various cells. Factory coordination can be divided into a production environment design task (short term redesign of the production system and the product routings) and an inter-cell goods flow control task. The main contribution of this framework is the recognition that a direct translation from tactical requirements planning, based on planned operation lead times, to operational detail planning of the production process is problematic. The characteristics of the cells can vary, for example with respect to the degree of autonomy, multi-functionality of employees, presence of bottlenecks, shared resources, etc. Therefore, each cell has to be planned and controlled separately (the PAC planning function), while at the same time another planning function is required for coordinating activities between cells (Factory Coordination). If cells are totally independent, both with respect to goods flow and use of resources, this latter function can be omitted.

The approach to consider cells as autonomous organizational units in the design of a production planning system is further elaborated upon in German literature. Rohloff (1993) developed a framework that decentralizes planning to the autonomous units (e.g., cells) as much as possible. The framework places a strong emphasis on the horizontal coordination level, e.g., the direct coordination between various autonomous units. The vertical coordination levels can be considered as an attempt to solve certain remaining planning problems using a hierarchical approach. The planning hierarchy has to take explicit notice of the available capacity in the cell within a certain time frame. This can be accomplished by a load-oriented order release planning function (Bechte 1994). Habich (1989) developed a production planning framework that recognizes the essential planning problem resulting from giving planning autonomy to cells that are interrelated in their primary production process. He views the essential problem of the central planning level to generate an overall optimum from the various local optima that were generated by the decentralized planning of the cells. His approach to this central planning is to consider the set of orders that require subsequent processing in various cells, determine for these orders appropriate sequences between the cells and planned throughput times per cell (e.g., order due dates), such that the cells will be able to finish these orders within their due dates while at the same time enough flexibility is available to optimize the planning within the cell.

2.3. MRP For Cellular Manufacturing

New (1977) was one of the first to consider MRP for CM. He stated that MRP is well equipped to determine parts requirements to meet assembly needs, but it may not be entirely suitable for detailed production control. The MRP model of how the production system operates differed slightly from actual situations in the shop floor. Updated priority lists for already released work orders are often not used at all at the floor, for instance, making the outcome of the system less predictable. Cellular systems can benefit from MRP if one reduces the fixed planned lead times, if the fluctuations in workload of cells can be dampened, and if setup similarities can be exploited. However, this requires fundamental modifications to MRP I approach. Adding a standard capacity requirements planning (CRP) system is also not appropriate for CM. New also attempted to reconcile Burbidge's Period Batch Control (PBC) method with MRP.

Suresh (1979) described an MRP system implemented in an automobile parts manufacturing firm in India, which had cells in operation. The classical MRP system had to be modified to suit differing nature of operations in cells. For one thing, a part-family grouping phase is required after generating net requirements for components using MRP. It was also shown that MRP can be modified to operate like the PBC, by foregoing item-by-item lot-sizing algorithms (e.g., part period balancing), and adopting single-cycle, single-phase operations. Implementing MRP appeared to be simpler in CM due to simplified production control, reduced inventory, better coordination of jobs in the shop floor, etc., all of which obviated the need for a shop floor control system. Inventory accuracy, a known problem in job shops, and a prerequisite for a successful MRP, also seemed to improve in CM.

In the work by Hyer and Wemmerlöv (1982), one of the questions addressed is whether an MRP II system is compatible with the production planning and control requirements of production cells. They explored this question within their framework of the tri-level hierarchical decision-making process (see §2.2). They concluded that MPS generation (level 1) would be unchanged and performing rough cut capacity checks will be easier. They conjectured that impact on the second level would be significant. Lead times would be shorter and more predictable as queue times, setup times and transfer times are smaller due to the proximity of machines in a cell. This would result in modifications of some parameters in the MRP system. The same holds for the lot sizes that are used, as the product families in cellular manufacturing require similar setups of the machines in the cell. Short throughput times in a cell, and the possibility of applying lot streaming, would also make it often not necessary to monitor the status of production orders within the cells.

Wemmerlöv (1988) provided more attention to the choice of the cell as the basic planning unit. He identified a number of factors of relevance to the choice of the basic planning unit: the appropriate level of delegation of planning decisions to cells, the nature of the production process in the cell, the length and variability of throughput times, and the internal flow patterns in the cell. Wemmerlöv (1988) also addressed the problem of how to utilize the advantages of CM in an MRP planning system. The advantage of producing similar parts in one cell should be recognized and handled by the MRP system in order to obtain the benefits of cellular manufacturing. However, the nature of MRP is to convert independent (end item) demand to dependent demand of parts and components. This process does not count for similarities between parts. Lot sizing rules that can be used in MRP try to find a suitable number of subsequent period requirements that can be combined in one order.

Shtub (1990) discussed many of these lot sizing rules and concluded that they do not consider common setups required for a family of components and therefore are not suitable for the MRP/Group Technology lot sizing problem. At the other hand, Wemmerlöv (1988) stated that he does not see family lot-sizing during the MRP explosion process as a realistic approach for most cellular systems, because of the implementation costs and the inflexibility in execution.

Sum and Hill (1993) criticized MRP II framework with respect to the tactical planning level (MRP I level). The basic MRP I does not apply finite scheduling in generating the requirements plan. MRP uses fixed planned order lead times that are based on static planned operation lead times, and these parameters are usually determined independently of order sizes, work center loads, and capacities. In many production situations, including CM, this may not result in realistic plans.

Chamberlain and Thomas (1995) discussed modifications required to MRP. They stressed the importance of building information systems that can be easily modified with respect to the organization of the production system. Flow-line cells are sometimes formed for a period of 3 months, and after this period production will again be performed in other cells. This requires MRP II systems that are flexible in modeling available capacities and their allocation to cells. Restructuring the planning bills of material (BOM) should be made easier. In general, the number of levels in BOM and routings files can be reduced, with less need to control production progress, and with reduced throughput times. The number of parts controlled using MRP can also be reduced, as simple two-bin systems with short cycle shipments often function very well in practice. However, MRP is still considered to be useful as a tactical planning instrument.

To summarize, even though MRP I does provide the necessary basic mechanisms for exploding the Master Production Schedule, using BOM, to generate parts requirements, there is an inherent job-shop orientation in the assumptions and logic of MRP II. Some of the inadequacies of MRP (or MRP II) in a CM context include:

- Each component is treated independently, and this needs to change to a part family orientation, and allocation of production plans to various cells in CM
- Problems such as load balancing among cells need to be addressed
- Actual information on production progress in determining due dates and planned lead times are often not considered
- Capacity requirements planning (CRP) logic also has a job-shop, work-center-orientation, which is altered in CM
- Shop floor control (SFC) modules in MRP II also assume a job-shop mode of operation.

2.4. Period Batch Control For CM

Burbidge's preference for Period Batch Control (PBC) as the appropriate production planning and control system for CM can be traced to some of his earliest works (e.g., Burbidge 1958). Burbidge initially made a distinction between *stock control* methods (like the reorder point system) and *flow control* methods. The term flow control refers to the fact that parts fabrication is in tune with final assembly schedules (as in MRP) instead of being based on independent component lot sizing. PBC is a flow control method based on *single-phase, single-cycle* ordering system, unlike MRP which typically operates as a *multi-cycle* system.

In single-cycle systems like PBC, orders have identical frequency of occurrence, while this may vary in multi-cycle systems such as MRP. The use of single-cycle systems can cause uneconomical usage of setups for the production of some parts, while multi-cycle systems can result in uneconomical usage of system capacity due to fluctuations in the loading of the system. In CM, capacity is decomposed into several independent units (cells), which causes an increased sensitivity for fluctuations in the loading of the system. However, overall production control is simplified. Hence, Burbidge preferred the use of single-cycle flow control ordering methods in combination with cellular manufacturing.

Like MRP, PBC uses time-phased planning of goods flow between stages and applies explosion of the end item demand to determine parts requirements. The essential feature of PBC is the periodicity with which this system operates, causing a synchronization of the goods flow within the production system. All products have equal throughput time T, determined by the product of the number of stages N in the production system and the length of the period P (see Figure 3).

The selection of suitable values for N and P is an important design problem in PBC (Riezebos 1997). If there is little variation in the loading of the cells over time, the dispatching level can accomplish high-quality schedules that make use of similar setups within part families and transfer batches that are smaller than process batches. In that sense, the use of PBC can be easily combined with insights from just-in-time (JIT) (Burbidge 1987; 1989a) and theory of constraints/OPT (Burbidge 1990).

The number of firms using PBC is known to be restricted. Burbidge, Falster and Riis (1991) reported that it would be difficult to find 30 companies in the UK which used PBC. Zelenović and Tesić (1988) reported on several applications in Yugoslavia, Whybark (1984) described an application of a related concept in Finland in the firm Kumera Oy. Recently, a renewed interest in research on the performance, design, and characteristics of PBC systems has evolved (e.g., Yang and Jacobs 1992; Kaku and Krajewski 1995; Steele and Malhotra 1994; Steele, Berry, and Chapman 1995; Rachamadugu and Tu 1997; and, Steele and Malhotra 1997). It is worth mentioning that PBC seems to perform well in the various tests compared with well known systems such as MRP and Kanban. PBC is discussed further in chapter F2 by Steele.

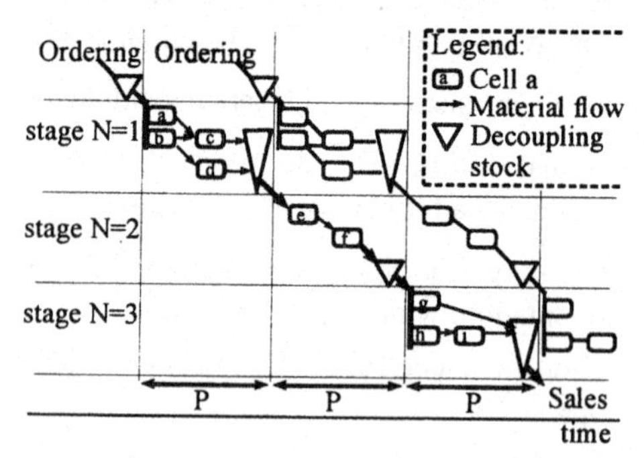

FIGURE 3. Stages and period length in PBC.

2.5. Other Approaches To Planning In CM

Schonberger (1983) pointed to the possibility of combining several elements from JIT into MRP, into a hybrid approach, referred to as the synchro-MRP approach, found at Yamaha. Flapper, Miltenburg and Wijngaard (1991) discussed how to embed JIT into MRP. Kanban is only one among many JIT techniques. Use of MRP for planning raw material and component deliveries, and for looking forward, while kanban is used to control the actual assembly process, is therefore one of the possible ways of embedding JIT into MRP.

Klein (1989) reported on the effect of kanban on the stress of the human system. She concluded that JIT eliminates the ability of workers to control their own work pace, but kanban makes workers to react to each other rather than answering a computer printout or a supervisor. Kanban therefore leads to a *perception* of *increased* control over the flow of production, although the reality may be otherwise.

Buzacott and Shanthikumar (1992) described a general approach for coordinating inter-cell goods flow that can be used for a more systematic comparison of several approaches, such as kanban, CONWIP and MRP. However, their framework focuses only on the sequential coordination between cells.

An interesting approach to planning for cellular manufacturing originated from the early work of Hax and Meal (1975). The hierarchical production planning framework they developed has been applied to group technology manufacturing in Kistner, Schumacher, and Steven (1992). In this approach a strong focus exists on capacity allocation to various layers of production units. Much effort is given to the disaggregation of the complex production planning problem in several less complex subproblems and the description of the interfaces between these subproblems. The type of disaggregation that should be applied strongly depends on the specific characteristics of the cellular manufacturing system, e.g., relations between the cells and flexibility of the system.

Olarunniwo (1996) reported on changes in production planning and control systems when cells are implemented in a firm. Most of the firms he studied were only partly cellularized, e.g., there existed a remaining shop in more than 90% of the firms. A notable result that he found was that almost all firms that used MRP before the implementation of CM continued with this after cellularization took place. However, the number of firms that combined the use of MRP with a kanban system increased from 3.6% to 32.7%. After cellularization, 30.9% of the firms operated MRP alone, while 12.7% only used kanban. The popularity of kanban therefore increases rapidly (up to 50% of the firms) after implementation of cellular manufacturing. His survey makes clear that many firms do not simply choose between various production planning and control systems, but apply a hybrid approach to planning. Wemmerlöv and Johnson (1997) reported in a recent study that 80% of the firms indicated that production planning and control had become simplified with cells.

There does not appear to be one best approach to planning and control of CM systems. Much work remains to be done in this area, after first developing acceptable taxonomies of various CM systems that exist in the first place.

3. PART FAMILY-ORIENTED SCHEDULING

Early studies in cellular manufacturing employed traditional job shop-style (or single-stage) scheduling rules in cells. Subsequent research has attempted to create group scheduling (or two-stage) procedures which exploit part family properties, an aspect ignored by single-stage job shop rules. Mosier, Elvers and Kelly (1984) conducted a thorough examination of several group scheduling rules. This has been expanded in a number of other works. In these studies, a cell is typically dedicated to the production of parts from a single family. Within this family exist subfamilies, created on the basis of setup similarities. The number of subfamilies could be determined, for example, by the number of major setups that are required to be performed in order to manufacture the part family in the cell. There are two stages to any group scheduling procedure. The first stage involves determining the sequence of jobs within a subfamily. This decision is made by employing traditional job-shop rules (job selection heuristics). The second stage involves the selection of the subfamily which will be processed. In addition, it is determined when to switch from the current subfamily to a new subfamily for processing. The second stage uses queue selection heuristics.

Group scheduling rules can be either exhaustive or non-exhaustive. Exhaustive heuristics are such that, once a subfamily has been selected, all the jobs within that subfamily queue, including new arrivals to that queue, are processed before the decision to switch subfamilies is made. When the current subfamily queue is empty, a new subfamily is chosen to be current. While using non-exhaustive heuristics, switching to a new subfamily may be performed even though the current subfamily queue is non-empty. Mahmoodi and Dooley (1991) found that the performance of exhaustive heuristics was superior and more robust than that of non-exhaustive heuristics. Simple heuristics, the study showed, performed better than complex heuristics.

In addition, both static and dynamic scheduling rules have been studied. In the case of the former, a fixed set of jobs is scheduled over a certain horizon, but in the latter, new jobs are continually being generated and added to queues. Dynamic systems have been studied with the help of computer simulation. Finally, both the case of the single-stage system and the multi-stage systems have been examined. For example, Wemmerlöv (1992) found in a single-machine study that part family scheduling rules had a significantly greater impact when there was high variability in arrival and processing times, and that the degree of variability in arrival and service times had a significant impact on flow time. In the case of the multi-stage systems, two types of cells have been considered: job shop cells in which a job can follow almost any routing within the cell; and flow-line cells, in which all jobs follow the same unidirectional routing through the cell. Mosier et al. (1984), Lee (1985), Mahmoodi, Dooley and Starr (1990), Mahmoodi and Dooley (1991), and Ruben, Mosier and Mahmoodi (1993) all used a job-shop cell. Examples of studies that used flow-line cells include Wemmerlöv and Vakharia (1991) and Mahmoodi, Tierney and Mosier (1992) and Russell and Philipoom (1991). See Table 1 for a list of representative studies.

This body of literature is discussed at length in chapter F3 by Mahmoodi and Mosier. A taxonomy of this literature is also provided in Shambu, Suresh and Pegels (1996). Practical issues surrounding cell loading and scheduling are described in a case study of a firm in Puerto Rico by Süer in Chapter F4. A classification of cell operating policies is described by Askin and Iyer in chapter F5.

TABLE 1. Group Scheduling Studies.

	Static Systems	Dynamic Systems
Single-Stage Systems	Hitomi and Ham (1978) Foo and Wager (1983) Ozden, Egbelu and Iyer (1985)	Wemmerlöv (1992) Mahmoodi and Martin (1997)
Multi-Stage Systems	Ham et al. (1979) Vakharia and Chang (1990) Skorin-Kapov and Vakharia (1993)	Vaithianathan and McRoberts (1982) Ang and Willey (1984) Mosier et al. (1984) Mahmoodi et al. (1990) Mahmoodi and Dooley (1991) Wemmerlöv and Vakharia (1991) Russell and Philipoom (1991) Mahmoodi et al. (1992) Ruben et al. (1993) Suresh and Meredith (1994) Kannan and Lyman (1994) Frazier (1996)

4. BURBIDGE'S VIEWS ON ECONOMIC LOT-SIZING

It is generally believed that economic order quantity (EOQ) formulae, or economic batch quantity (EBQ) formulae, generally became discredited with the advent of Japanese manufacturing techniques. Schonberger (1982, p. 19), for instance, pointed out that "it is time to discard some of our EOQ training". He pointed out two reasons for the non-use of EOQ models by the Japanese: 1) Carrying cost and setup costs are only the obvious costs. Quality, scrap, worker motivation and responsibility, productivity are also affected by lot sizes; and, 2) Setup cost is real and significant but not unalterable: with ingenuity and resolve, setup costs can be driven down to permit "stockless production". In a similar vein, Hall (1983, p.14), argued against use of EOQ in the context of zero inventories, and emphasized systematic reduction of setup times, instead.

However, economic lot-sizing techniques came under attack even earlier, by GT pioneers, notably Burbidge, starting with the late 1950s. The objective of this section is to summarize Burbidge's arguments against the use of EOQ models and to focus attention on a debate that took place within the pages of *Operational Research Quarterly* in 1964/65.

With the advent of group technology methods, it became clear that new approaches to production planning and control were required to suit cellular manufacturing. Traditional practices, in particular, the use of economic lot sizing models were seen to be an obstacle in the move towards CM. Burbidge (1958), while describing "a new approach to production control", devoted most of the article for making a case against the economic batch quantity methods. He summarized these arguments more pointedly in a later work (Burbidge 1964).

Around this time, EOQ, or EBQ also began to be criticized for attempting to minimize costs, instead of maximizing returns from inventory investment. The latter objective function was also shown to lead to smaller batch quantities (Eilon 1957; Duckworth 1960). In response to these criticisms, Tate (1964) attempted to show the robustness of EOQ formula using the alternative objectives of maximizing rate of return, and generally concluded that there is no better methodology yet other than EBQ; and so, "the dragons have scorched, but not consumed the methodology of the EBQ".

This prompted a rejoinder by Burbidge (1964), Duckworth (1964), Eilon (1964) and Burbidge (1965). The interested reader is referred to this "dragons in pursuit of EBQ" series of papers for a comprehensive debate on economic lot-sizing formulae. The views of Burbidge (1958; 1964; 1968; 1975) on lot-sizing, that are more closely related to GT/CM context are summarized here. They may be grouped under several broad categories, as described below.

Restrictive Notion Of The Batch Quantity.

First, Burbidge pointed out that there is no single, simple parameter as the "batch quantity", and its use is somewhat restrictive in practice. There are four main ways in which items can be grouped into batches for convenience in production. These include the *order quantity, run quantity, transfer quantity, and setup quantity.*

The order quantity refers to quantity covered by an order. Run quantity is the quantity processed consecutively at a work center in one or more successive working periods. Setup quantity is the number of items, not necessarily the same, processed at a work center between changes of setup. Transfer quantity is the quantity moved together as a batch between work centers. (These may be compared with latter-day definitions of transfer quantities mentioned under OPT).

Each of these four different quantities can be altered independently of the others, within certain limits, and each has a different effect on output variables such as inventory levels, processing costs and purchasing costs. Only run quantity and transfer quantity have an effect on inventory levels; order quantity and setup quantity can be varied, with no direct effect on stock values. All four have some effects on costs, but in different ways. In the case of purchases, order quantity may have en effect on price, for instance, and transfer quantity may affect inventory carrying costs.

More importantly, there is no simple batch quantity in practice, as used in EBQ models. EBQ is a special case in which all four batch quantity variants have the same value and are altered together as one unit.

Varying Batch Quantity Has An Insignificant Effect On Total Costs.

Even with conventional use of EOQ models, it has been known for a long time that the total cost curve is somewhat flat in the vicinity of the optimum point. Smith (1989, p. 129), for instance, illustrates this insensitivity of total costs to changes in batch quantity over a wide range. Given this flexibility (of violating the EBQ), some researchers (e.g., Solomon 1958) proposed the use of an economic lot range, which provides more flexibility in use of lot sizing formulae.

But, besides this intrinsic mathematical property of the EBQ model, Burbidge's contention was that, in practical situations, changes in total costs, with changes in batch quantity, represent only a small percentage change in total cost: real total costs are flat around the optimum point, permitting violation of EOQs in practice.

Costs Are Mainly Fixed: Only A small Part Varies With Batch Quantity.

An investigation of setup (or order preparation) costs and carrying costs in industry reveals that these expenses are mainly fixed and only a small part is variable with batch quantity.

In the case of setup costs, Burbidge argued that doubling the number of batches (setups) will not necessarily double the setup costs. Setup expenses are largely fixed and only a small part is variable with changes in batch size due to many reasons. For one thing, the labor cost component in setup can be often reduced effectively, only in units of one setter's wage. Therefore the labor component in setups is mainly a fixed expense. Also, Burbidge argued that a reduction in batch size may increase setup expense, but may reduce resetting expense: a setup used once every month may take less time than one which is used only once a year.

Setup costs are also affected by sequence of loading. They can be reduced through group technology methods, part-family-oriented scheduling, and many other ways, which have come to be emphasized later, in JIT contexts.

Storage expenses also are mainly fixed and only a small part really varies with batch size. In the case of storage component of carrying costs, storage expenses may also, actually get slightly larger when batch sizes are reduced, to a greater number of transactions.

The Method of Overhead Allocation.

Burbidge (1958) also pointed out that the method of overhead allocation, based on direct labor, *suggests a variability that does not really exist in practice*. A large part of overhead expenses are unaffected by batch quantity.

It must be pointed out that his arguments in this context can be reconciled without difficulty, with contemporary *activity-based costing (ABC)* principles.

Use Of EOQ Constitutes Item-By-Item Suboptimization.
The use of EOQ models involves optimization on an item-by-item basis, a practice imposed by the functional organization, that only results in suboptimization and lack of coordination at the total system level. As Burbidge stated, "the cost reduction possible by varying individual component batch quantities is far less significant, compared with the gains possible through adopting a single-phase system, where order frequency is the same for all components. He stated further that, "Companies which concentrate on improving material flow and high stock turnover are generally more successful than those which accept traditional systems and inflate stocks to minimize costs". The use of EBQ imposes multi-phase ordering, which results in unbalanced inventories, obsolescence, and hidden costs of administration.

Finally, EOQ models also do not consider actual amount of capital available for total investment; instead, inventory levels become the chance result of independent batch quantity calculations.

5. CONCLUSIONS

The choice and design of an effective production planning and control system are important determinants of the performance of cellular manufacturing systems. Many systems and variants have been developed and implemented over the years, but it is still not clear as to which system is best, and under what conditions. Much work remains to be done in these areas using analytical, simulation and empirical research, and to integrate the results usefully for both researchers and practitioners.

REFERENCES

Ang, C. L. and Willey, P. C. T., 1984, A comparative study of the performance of pure and hybrid group technology manufacturing systems using computer simulation, *International Journal of Production Research*, 22, 193-233.

Banerjee, S.K., 1997, Methodology for integrated manufacturing planning and control systems design, 54-88, in *The planning and scheduling of production systems, methodologies and applications*, (Eds. A. Artiba and S.E. Elmaghraby), Chapman & Hall.

Bauer, A., Bowden, R., Browne, J., Duggan, J., and Lyons, G., 1991, *Shop floor control systems, from design to implementation*, Chapman & Hall.

Bechte, W., 1994, Load-oriented manufacturing control, just-in-time production for job shops, *Production planning and control*, 5, 3, 292-307.

Burbidge, J.L., 1958, A new approach to production control, *The Institution of Production Engineers Journal*, May.

Burbidge, J.L., 1962, *The principles of production control*, MacDonald and Evans.

Burbidge, J.L., 1964, Comments on paper: In defence of the economic batch quantity, *Operational Research Quarterly*, 15, 4, 341-343.

Burbidge, J.L., 1964, The case against the economic batch quantity, *The Manager*, January.

Burbidge, J.L., 1965, Batches and dragons (Letters to the Editor), *Operational Research Quarterly*, 16, 2, 248-250.

Burbidge, J.L., 1987, JIT for batch production using period batch control, in *Proceedings of 4th European conference on automated manufacturing*, 12-14 May 1987, IFS (conferences) Ltd., Birmingham, 163-174.

Burbidge, J.L., 1988, Operation scheduling with GT and PBC, *International Journal of Production Research*, 26, 3, 429-442.

Burbidge, J.L., 1989a, A synthesis for success, *Manufacturing engineer*, November, 99-102.

Burbidge, J.L., 1989b, *Production flow analysis for planning group technology*, Clarendon Press, Oxford.

Burbidge, J.L., 1990, Production control: A universal conceptual framework, *Production planning and control*, 1, 1, 3-16.

Burbidge, J.L., Falster, P., and Riis, J.O., 1991, Why is it difficult to sell GT and JIT to industry?, *Production planning and control*, 2, 2, 160-166.

Buzacott, J.A. and Shanthikumar, J.G., 1992, A general approach for coordinating production in multiple-cell manufacturing systems, *Production and operations management*, 1, 1, 34-52.

Chamberlain, W. and Thomas, G., 1995, The future of MRP II: headed for the scrap heap or rising from the ashes, *IIE Solutions*, 27, 7, 32-35.

Dale, B.G. and Russell, D., 1983, Production control systems for small group production, *Omega*, 11, 2, 175-185.

Duckworth, W.E., 1964, Comments on paper: In defence of the economic batch quantity, *Operational Research Quarterly*, 15, 4, 343-345.

Eilon, S., Dragons in pursuit of the EBQ, *Operational Research Quarterly*, 15, 4, 347-354.

Flapper, S.D.P., Miltenburg, G.J., and Wijngaard, J., 1991, Embedding JIT into MRP, *International Journal of Production Research*, 29, 2, 329-341.

Foo, F.C. and Wager, J.G., 1983, Set-up times in cyclic and acyclic group technology scheduling systems, *International Journal of Production Research*, 21, 63-73.

Frazier, G.V., 1996, An evaluation of group scheduling heuristics in a flow-line manufacturing cell, *International Journal of Production Research*, 34, 959-976.

Habich, M., 1989, Koordination autonomer Fertigungsinseln durch ein adaptiertes PPS-Konzept, *Zeitschrift für wirtschaftliche Fertigung*, 84, 2, 74-77.

Hall, R.W., 1983, *Zero Inventories*, Dow Jones-Irwin, Homewood, Ill.

Ham, I., Hitomi, K., Nakamura., and Yoshida, T., 1979, Optimal group scheduling and machining speed decision under due-date constraints, *Journal of Engineering for Industry*, 101, 128-134.

Hax, A.C. and Meal, H.C., 1975, Hierarchical integration of production planning and scheduling, 53-69, in *Studies in management sciences: logistics, vol. 1*, (Ed. M.A. Geisler), Elsevier.

Hitomi, K. and Ham, I., 1977, Group scheduling technique for multiproduct, multistage manufacturing systems, *Journal of Engineering for Industry*, 99, 759-765.

Hitomi, K. and Ham, I., 1978, Machine loading and product-mix analysis for group technology, *Journal of Engineering for Industry*, 100, 370-374.

Hyer, N.L. and Wemmerlöv, U., 1982, MRP/GT: A framework for production planning and control of cellular manufacturing, *Decision Sciences*, 13, 681-701.

Jackson, R.H.F. and Jones, A.W.T., 1987, An Architecture for Decision Making in the Factory of the Future, *Interfaces*, 17, 15-28.

Kaku, B.K. and Krajewski, L.J., 1995, Period Batch Control in group technology, *International Journal of Production Research*, 33, 1, 79-99.

Kannan, V.R. and Lyman, S.B., 1994, Impact of family-based scheduling on transfer batches in a job shop manufacturing cell, *International Journal of Production Research*, 32, 12, 2777-2794.

Kistner, K-P., Schumacher, S., and Steven, M., 1992, Hierarchical production planning in group technologies, 60-74, in *New directions for operations research in manufacturing,* (Eds. G. Fandel, Th. Gulledge, and A. Jones), Springer Verlag.

Klein, J.A., 1989, The human costs of manufacturing reform, greater employee responsibility does not mean greater discretion over time and work, *Harvard Business Review,* 67(March-April), 60-66.

Lee, L.C., 1985, A study of system characteristics in a manufacturing cell, *International Journal of Production Research,* 23, 1101-1114.

Love, D. and Barekat, M.M., 1989, Decentralized, distributed MRP: solving control problems in cellular manufacturing, *Production and inventory management journal,* 30, 3, 78-84.

Mahmoodi, F., Dooley, K. J., and Starr, P. J., 1990, An investigation of dynamic group scheduling heuristics in a job shop manufacturing cell, *International Journal of Production Research,* 28, 1695-1711.

Mahmoodi, F. and Dooley, K. J., 1991, A comparison of exhaustive and non-exhaustive group scheduling heuristics in a manufacturing cell, *International Journal of Production Research,* 29, 9, 1923-1939.

Mahmoodi, F. and Martin, G.E., 1997, A new shop-based and predictive scheduling heuristic for cellular manufacturing, *International Journal of Production Research,* 35, 313-326.

Mahmoodi, F., Tierney, E. J., and Mosier, C. T., 1992, Dynamic group scheduling heuristics in a flow-through cell environment, *Decision Sciences,* 23, 61-85.

Mosier, C. T., Elvers, D. A., and Kelly, D., 1984, Analysis of group technology scheduling heuristics, *International Journal of Production Research,* 22, 857-875.

Mosier, C. and Taube, L., 1985, The facets of group technology and their impacts on implementation - a state of the art survey, *Omega,* 13, 5, 381-391.

New, C.C., 1977, *Managing the manufacture of complex products: coordinating multi-component assembly,* Business Books Communica Europe, London.

Olarunniwo, F.O., 1996, Changes in production planning and control systems with implementation of cellular manufacturing, *Production and inventory management journal,* 37, 1, 65-69.

Ozden, M., Egbelu, P.J., and Iyer, A.V., 1985, Job scheduling in a group technology environment for a single facility, *Computers and Industrial Engineering,* 9, 67-72.

Petrov, V., 1968, *Flow line group production planning,* Business publications, London.

Rachamadugu, R. and Tu, Q., 1997, Period batch control for group technology - an improved procedure, *Computers and industrial engineering,* 32, 1, 1-7.

Riezebos, J., 1997, On the determination of the period length in a period batch control system, *Matador conference proceedings, vol. 32,* MacMillan Press Ltd., 131-136.

Rohloff, M., 1993, Decentralized production planning and design of a production management system based on an object-oriented architecture, *International journal of production economics,* 30-31, 365-383.

Rolstadås, A., 1988, Flexible design of production planning systems, *International Journal of Production Research,* 26, 3, 507-520.

Ruben, R. S., Mosier, C. T., and Mahmoodi, F., 1993, A comprehensive analysis of group scheduling heuristics in a job shop cell, *International Journal of Production Research,* 31, 6, 1343-1370.

Russell, G. R. and Philipoom, P. R., 1991, Sequencing rules and due date setting procedures in flow line cells with family setups, *Journal of Operations Management,* 10, 4, 524-545.

Schonberger, R.J., 1982, *Japanese Manufacturing Techniques: Nine Hidden Lessons in Simplicity,* Free Press, New York.

Schonberger, R. J., 1983, Selecting the right manufacturing inventory system: Western and Japanese approaches, *Production and inventory management journal*, 24, 2, 33-44.

Shambu, G., Suresh, N. C., and Pegels, C. C., 1996, Performance evaluation of cellular manufacturing systems: A taxonomy and review of research, *International Journal of Operations and Production Management*, 16, 8, 81-103.

Shtub, A., 1990, Lot sizing in MRP/GT systems, *Production planning and control*, 1, 1, 40-44.

Singh, N., 1996, Production planning in cellular manufacturing, in *Cellular manufacturing systems, design, planning and control*, (Eds. N. Singh and D. Rajamani), Chapman & Hall, 212-245.

Sinha, R.K. and Hollier, R.H., 1984, A review of production control problems in cellular manufacture, *International Journal of Production Research*, 22, 5, 773-789.

Skorin-Kapov, J., and Vakharia, A.J., 1993, Scheduling a flow line manufacturing cell: A tabu search approach, *International Journal of Production Research*, 31, 1721-1734.

Smith, S. B., 1989, *Computer-Based Production and Inventory Control*, Prentice Hall, Englewood Cliffs, NJ.

Solomon, M.J., 1958, The use of an economic lot range in scheduling production, *Management Science*, 5, 434-442

Steele, D.C., Malhotra and M.K., 1994, Operating characteristics of period batch control, *Proceedings DSI 20-22 November 1994*, 3, 1699-1701.

Steele, D.C., Berry, W.L., and Chapman, S.N., 1995, Planning and control in multi-cell manufacturing, *Decision sciences*, 26, 1, 1-34.

Steele, D.C. and Malhotra, M.K., 1997, Factors affecting performance of period batch control systems in cellular manufacturing, *International Journal of Production Research*, 35, 2, 421-446.

Sum, C-C. and Hill, A.V., 1993, A New Framework for Manufacturing Planning and Control Systems, *Decision sciences*, 24, 4, 739-760.

Suresh, N.C., 1979, Optimizing intermittent production systems through group technology and an MRP system, *Production and Inventory Management Journal*, 20, 4, 77-84.

Suresh, N. C., and Meredith, J. R., 1994, Coping with the loss of pooling synergy in cellular manufacturing systems, *Management Science*, 40, 4, 466-483.

Tate, T.B., 1964, In defence of the economic batch quantity, *Operational Research Quarterly*, 15, 4, 329-345

Vaithianathan, R. and McRoberts, K. L., 1982, On scheduling in a GT environment, *Journal of Manufacturing Systems*, 1, 149-155.

Vakharia, A.J. and Chang, Y., 1990, A simulated annealing approach to scheduling a manufacturing cell, *Naval Research Logistics*, 37, 559-577.

Vollmann, T.E., Berry, W.L., Whybark, D.C., 1997, *Manufacturing planning and control systems*, 4th edition, Irwin/McGraw-Hill, NewYork.

Wemmerlöv, U., 1988, *Production planning and control procedures for cellular manufacturing*, American Production and Inventory Control Society, Falls Church.

Wemmerlöv, U., 1992, Fundamental insights into part family scheduling: the single machine case, *Decision Sciences*, 23, 565-595.

Wemmerlöv, U. and Johnson, D.J., 1997, Cellular manufacturing at 46 user plants: implementation experiences and performance improvements, *International Journal of Production Research*, 35, 1, 29-49.

Wemmerlöv, U., and Vakharia, A. J., 1991, Job and family scheduling of a flow-line manufacturing cell: A simulation study, *IIE Transactions*, 23, 4, 383-393.

Whybark, D.C., 1984, Production planning and control at Kumera Oy, *Production and Inventory Management Journal*, 25, 1, 71-82.

Yang, K.K. and Jacobs, F.R., 1992, Comparison of make-to-order job shops with different machine layouts and production control systems, *International Journal of Production Research*, 30, 6, 1269-1283.

Zelenović, D.M. and Tesić, Z.M., 1988, Period batch control and group technology, *International Journal of Production Research*, 26, 4, 539-552.

AUTHORS' BIOGRAPHY

Jan Riezebos is Assistant Professor of Production Systems Design at the Faculty of Management and Organization, University of Groningen, The Netherlands. His research interests are in flow shop scheduling, planning problems in cellular manufacturing, and Period Batch Control. He has published papers in *Journal of Intelligent Manufacturing*, and *European Journal of Operational Research*, and other journals, besides presentations at several international conferences such as the International Conference on Production Research, Production and Operations Management Society conference, Matador conference, and International Conference on Industrial Engineering and Production Management. He obtained a degree in econometrics from the university of Groningen, specializing in operations research and logistics management based on a research project on the design of an inventory management and control system for a plastics manufacturing firm.

Girish Shambu is Associate Professor of Management at the Wehle School of Business, Canisius College, Buffalo, NY, USA. He received his B.Tech in chemical engineering from Indian Institute of Technology, Kharagpur and his Ph.D. in Production / Operations Management from the State University of New York at Buffalo. His research interests lie in the performance of cellular manufacturing systems. He has published in several journals including *International Journal of Operations and Production Management*. At Canisius College, he has received the Donald Calvert Outstanding MBA Professor Award in 1992 and the Alpha Omega Pi Undergraduate Professor of the Year award in 1997.

Nallan C. Suresh (biograpahy provided at the end of chapter A).

F2

Period Batch Control Systems For Cellular Manufacturing

D. C. Steele

1. INTRODUCTION

Period Batch Control (PBC) is a manufacturing planning and control (MPC) system that has been advocated for cellular manufacturing (CM) by Burbidge (1975). PBC appears attractive for CM many reasons. It matches the stage-like structure of CM, has been supported by a number of authors as appropriate for CM, and has been implemented successfully in some instances. While the use of PBC has not been widespread, this approach has continued to attract interest in recent years.

PBC creates a phased flow of production lots. Each phase is a group of job lots which enter a defined setup loading sequence at each stage or cell in turn. Since lot-for-lot quantities are used, the master production schedule (MPS) represents the exact quantities to be produced at each stage; no residual inventories are produced. The combination of defined setup sequences and lot-for-lot quantities means the MPS directly represents work to be accomplished at each stage. Consequently, PBC provides a global transparency of shop load. Because what is in the MPS is going to be produced at each stage, the MPS can be used to accurately control production demands to match cell capability. Also, since each phase of scheduled work must be completed for each stage and time period in turn, PBC is also transparent to the work force, providing clear knowledge of system-wide schedules. However, it does so through structural rigidity and system-wide lot sizes and lead times that may not fit all environments. Alternatively, conventional systems either have simple requirements (stable demand, few setup restrictions) and tend to be rate-driven (e.g., Kanban), or they resort to decoupled cells where queues provide cells the independence and flexibility necessary for efficient setup loading sequences (e.g., Material Requirements Planning). Kanban and Material Requirements Planning (MRP) are better known and perhaps incrementally easier to implement, but they may not always have the same potential as PBC.

By and large, PBC does not appear to be well understood. In a fundamental sense, the concept has been clearly defined, but the implications of using PBC have received relatively little attention so far from academics and practitioners. Thus the purpose of this chapter is to broadly acquaint the reader with this interesting approach to CM planning and control. In the following pages, we shall first define PBC, review its characteristics, summarize what is known about PBC, and finally highlight absences of key work in understanding PBC as a potential tool in group technology (GT).

2. PBC DEFINED

Period Batch Control is based on the notion of a production cycle recurring at a periodic fixed interval with all needed products or parts scheduled each cycle. This is accomplished by PBC as a *single-cycle, single-phase planning system that uses single offset planning.*

Single cycle means that the order frequency (period order quantity) of all orders for all stages is exactly the same, namely, one-period. In the language of MRP, orders are lot for lot. *Single phasing* means that all orders are released simultaneously at a single time, the beginning of each period. Orders are not released continually. *Single offset timing*, as an extension of single phasing, means that lead time for each stage is planned as one period. Thus equal, one-period planned lead times are allocated to each production stage.

This phased-flow of lot-for-lot orders creates a pre-planned schedule where each production stage produces in turn the exact equivalent of what the preceding stage produced in the preceding period. Thus the master production schedule (MPS) or final assembly schedule is represented at each prior stage by a corresponding schedule, each one back-scheduled or offset one period earlier than the following stage. In effect, a simple bill-of-material explosion of each period in the master schedule and a planned lead time of one period provides the corresponding schedule of requirements for each stage of production.

Production Stage	Period 1	Period 2	Period 3	Period 4	Period 5
Stage 1 (Procurement)	**Cycle #4 Material**	Cycle #5 Material			
Stage 2 Cells (Fabrication)	Cycle #3 Material	**Cycle #4 Material**	Cycle #5 Material		
Stage 3 Cells (Assembly)		Cycle #3 Material	**Cycle #4 Material**	Cycle #5 Material	
Stage 4 Cells (Test and Ship)			Cycle #3 Material	**Cycle #4 Material**	Cycle #5 Material

FIGURE 1. Period Batch Control illustration.

Figure 1 provides a depiction of the schedule flow. In the figure, the material representing single-cycle orders is shown progressing through manufacturing one stage each period, converging on the last stage in the last period. Each group of material is identified as a cycle number corresponding to the shipping period. Thus, product orders scheduled for shipment in period 4 are identified as cycle 4 material, beginning as purchased components in period 1 and finishing as product in period 4.

Within CM, each stage in production can be represented by one or more cells. A cell may contain multiple operations, but since only stages are scheduled, external planning complexity is reduced. And with the potentially low setup time of a cell, small lot production in short cycle times is practical. Further, producing in balanced part sets with no remaining inter-stage inventory simplifies control. Thus, with only a few stages to be planned, no inventory for netting, and short cycle times, the resulting tight lot-for-lot flow creates a very simple, highly transparent schedule at each stage.

3. ORIGINS OF PBC AND RELATIONSHIP TO CM

Burbidge (1975) proposed PBC as a necessary part of the successful implementation of group technology. According to Burbidge (1975), PBC was devised by R. J. Gigli around 1926. Describing it as "short-cycle flow control", he contrasted it with reorder point systems and later (Burbidge 1988), with MRP. Both other systems tended to use economic order quantities (EOQs) which provided random order timing and loading of cells. In contrast, PBC cycles orders at a fixed one-period frequency and all part requirements are coordinated through an "explosion." When combined with group layout and planned loading sequences, PBC was seen as an effective enabler. Group layout essentially permits small lot sizes since handling, setup, and administrative costs can be very low. Handling is low because of machine proximity, setup is low because of limited parts range and family setup sequences, and administrative costs are low because of the resulting simplicity. Further, planned loading sequences within a cell can implement the most efficient family setup sequences, further supporting small lot production. Through PBC, the fixed, short one-period ordering cycle permitted system-wide control of the order timing and quantity being placed on all cells in turn, namely a one-period batch each period. Thus, the name "Period Batch" Control is apt and its function, valuable.

Burbidge (1975) noted that shorter period in PBC permitted shorter order cycles and increased flexibility, but saw two problems in implementation. First, the period size cannot be less than the throughput time of any component at any cell. Since the small transfer lots and overlapping production shorten throughput time in CM, this was not seen as a major problem. However, Burbidge (1988) acknowledged the importance of close scheduling long lead time parts to accommodate the period size. The second problem is that the period cannot be so short that setup time uses up needed capacity. However, the use of loading sequences possible with PBC and CM reduces setup time and minimizes this problem.

New (1977) proposed MRP as an appropriate planning system for use with CM, but, like Burbidge, noted that random timing of economic order quantities would disrupt the needed load control of cells. Without level loads, planned lead times would be extended and a major benefit of CM lost. New proposed that PBC could be implemented through the mechanics of MRP, taking advantage of single cycles and single phasing.

Suresh (1979) in fact reported the successful use of an MRP system modified to provide PBC-type controls for a cellular manufacturing system in India. He saw short lead times in CM as making unnecessary the need for shop floor priority control systems, but requiring single-cycle PBC form of ordering to maintain balanced sets of parts and shop loading. The reduced shop control and inventories and easier monitoring also contributed to a successful PBC-type of MRP system.

Hyer and Wemmerlöv (1982) provided further insight into the use of MRP in CM, but suggested that PBC, though possible within an MRP context, may be unsophisticated and may have restrictive assumptions. They viewed PBC as unsophisticated because of minimal use of computers. They also saw PBC as assuming that demand for a full planning sequence would exist each period (predictable sequence) and that demand fluctuation would not exceed capacity limitations (over and under loading). Still, they compared a small-cycle PBC to successful just-in-time production.

In an alternative view, Burbidge (1988) dealt further with the realities of PBC shop scheduling by noting that PBC at a floor level so simplifies the scheduling problem that any knowledgeable cell foreman can resolve it reliably and cheaply within the cell without resorting to computers. In particular, by careful use of appropriate sequencing and overlapping within the cell, problems of bottleneck machines and excessive throughput time can be surmounted. Only four pieces of information are required: the list of orders, period load, optimum loading sequence, and parts in throughput time sequence. With such information, a well trained foreman can determine appropriate action.

4. OTHER RELATED WORK

Several other works have been reported which support the notion of PBC. One is Whybark's (1984) description of a cyclical ordering system at Kumera Oy. The ordering system was similar to PBC in that fixed single cycles were used. A period size of 5 weeks was used, over which a fixed set of product groups was released. However, it was also multi-phase in that the product groups were released sequentially in phases over the 5-week period. Significant performance improvements were reported. Another example of a periodic system was reported by Vollmann, Berry, and Whybark (1992) as a "weekly wash" control of inventory. Here the focus is on what quantity is required in a given period, not on when a product was to be produced. Both are similar to PBC in that a particular sequence of products each period is assumed.

Rees, Huang, and Taylor (1989) offered another insight into PBC when they compared an MRP lot-for-lot system to Kanban. While this was not a group technology experiment, MRP using short cycles and reduced setup time was shown to be comparable to Kanban under several conditions. That is, though Kanban is usually thought to be a superior system (particularly true in group technology), MRP lot-for-lot under these conditions was often equal or better than Kanban. The experiments suggest a major value of PBC may simply be the small lot sizing, although PBC was not tested in this work. All of this work does suggest that systems similar to period-oriented, short-cycle PBC have shown performance advantages.

5. PBC CHARACTERISTICS AND VALUE

Each one of the various elements of PBC contributes to its unique characteristics. Single cycling, single phasing, and single offset timing are most predominant, but resulting system-wide constraints and control are also important characteristics.

Single cycle is the most visible element of PBC, controlling the size and frequency of all orders very rigidly. Single-cycle orders mean that system-wide all order quantities correspond exactly to a single period's requirements. With no independent lot sizing, no inter-stage inventory exists; component or sub-assembly inventory exists only as in-process material. Consequently, planning is simple and schedule requirements very clear. However, if the period of an economic order quantity (EOQ) is not similar between stages, then there is potential for excessive capacity or inventory costs. EOQs are often viewed as inappropriate today; Burbidge (1975) in fact has been a most vocal opponent of EOQ. Certainly, if incremental capacity is not expensive, then for stages with higher setup times, capacity can be acquired and smaller lot sizes produced that better match other stages and the system overall. However, if stages with high setup time have high incremental capacity costs, then extra setups are costly and the overall system would tend to be forced to larger lot sizes to avoid those setups. Consequently, PBC is a system with planning and control simplicity, but lot-size constraints.

Single phasing creates a unique flow of work. At the beginning of each period, each stage has a full queue of the period's work. The advantage is that at each individual stage one can review the entire queue for sequencing (a planned loading sequence) before starting. Thus work for the entire period can be run in the most advantageous sequence. The disadvantage is that the rigid period phasing means that no work from adjacent periods can be combined for family scheduling purposes. Only the window of work established by the period schedule is available to each stage in turn. Therefore, single-phasing facilitates family setup sequences, but only if a period of work provides an efficient sequence for all cells where cost is high or capacity limited.

Single offset timing means a common planned lead time is required of all production stages. This creates a lock-step production schedule that is highly visible. However, similar to the lot size constraint, stages with longer flow times will tend to force the entire system to a longer lead time offset than might otherwise be used. If incremental capacity costs are high for longer flow-time stages, it may not be practical to reduce the flow time. Indeed, some process characteristics may prohibit it. Since the period-size lead time must accommodate all stages, then the period lead will be set longer than otherwise needed by the other stages. So again, the clarity of a system-wide time line can be offset by excess capacity, flow time, or both.

A system-wide constraint occurs not only in that period lot sizes and planned lead times must be the same for all stages, but also in that *both* must be equal to the period size. The longest requirement for either constrains both. Therefore, the period size serves as a multi-purpose parameter for the entire system. Once set, the period size must be honored by all stages of production, producing to the same lot size and lead time.

System-wide control is a final characteristic and advantage. Once PBC is in place, the master production schedule (MPS) controls the composition of each phase of work arriving at cells. No lot sizing or netting creates loads in any cell different than that represented by the MPS. Thus, highly accurate global, system-wide control of cell load is possible. Highly transparent load information is possible using only a simple rough-cut capacity planner such as a capacity bill or resource profile (see Vollmann, Berry, Whybark 1997). A further advantage is the family scheduling it permits. Any setup or family groupings established at the MPS will be replicated through all stages. Since cells are often dissimilar, the MPS control can be directed to only one or a few cells. However, for a critical cell, control over the family composition of a period schedule presented to the cell for further sequencing could enhance cell performance and thus that of the entire production system. In this respect, it provides control similar to Drum-Buffer-Rope (DBR), a system developed from a Theory of Constraints perspective.

PBC can thus be seen as a rigid, though highly transparent system. Both the rigid structure and the transparency seem to match what is often found in CM. The rigid structure and single-cycle, single-phase production matches the stages and short-cycle manufacturing often found in CM. And the transparency facilitates control of family scheduling load sequences, both in controlling the content of a period's work globally from the MPS and in presenting a full phase of work to each cell for optimum local sequencing.

Thus PBC is conceptually attractive. Zelenović and Tesić (1988) also reported it attractive in practice. Reporting on the application of GT and PBC in 32 plants in Yugoslavia, the improvements provided by GT in setup reduction was not matched by appropriate inventory reductions. Also, converting from reorder point and MRP systems to PBC did. Inventory and throughput time were significantly improved. Further, production managers were able to better focus their cell capacities on customer demand rather than just internal utilization.

6. CURRENT RESEARCH

Several recent research studies have investigated some aspects of PBC more closely. One study compared PBC, Kanban, and MRP performance in a simulation, representative of an actual CM shop (Steele, Berry, and Chapman 1995). The objective of the work was to determine which system performed better and how environment conditions affected the results. The specific shop simulated involved multi-cell manufacturing using flow cells and assembly. Setup sequencing was not tested, only how conditions affected lead times and inventories. Environmental conditions tested included MPS volume variation, MPS mix variation, material commonality, rate similarity, setup time (and lot size), and parts range (number of parts). Relative system performance was tested over three cases: (1) JIT conditions with low MPS mix variation, (2) JIT conditions with high MPS mix variation, and (3) non-JIT conditions. Table 1 shows the variable settings used for the three cases. Further sensitivity analysis was conducted for the individual variable settings.

TABLE 1. Base Case Variable Settings (from Steele et al. 1995)

Environmental Variable	Case 1	Case 2	Case 3
MPS volume variation	low	low	**high**
MPS mix variation	low	**high**	**high**
Material commonality	high	**low**	**low**
Rate similarity	high	high	**low**
Setup time	low	low	**high**
Part range	low	low	**high**

(Underscored factors are changed from Base Case 1)

The results showed that the production environment was a major factor in relative system performance. Three of the environmental variables were found to be important: MPS volume variation, MPS mix variation, and setup time (lot size). Further, resolution was found to be influential. Resolution was defined as the incremental adjustment in quantity possible compared to full MPS daily quantities. Very small lot sizes of less than one-period usage (possible with Kanban) provide a fine resolution; minimum period-size quantities (required in PBC and MRP) provide a more coarse resolution.

Figure 2 summarizes the comparative results. PBC was found to be the most effective system at handling high MPS variation when setup time could be made reasonably small (which, of course, may be facilitated by PBC). MRP was better at absorbing high setup times: large lot sizes could be tailored to the needs of each cell. Kanban was better when lot sizes were *very* small and resolution fine: lot sizes could be much less than MPS order sizes, better matching individual requirements variations. All systems performed about the same under "JIT conditions": low MPS variation and small lot sizes, but coarse resolution. These experiments suggest not only which environmental factors are important, but that there is a range of conditions where one system may be more appropriate. For PBC, setup time is consistent with short period sizes, PBC may be better able to handle an MPS that needs to contain high variation, particularly if it is a mix variation and not an overall volume variation.

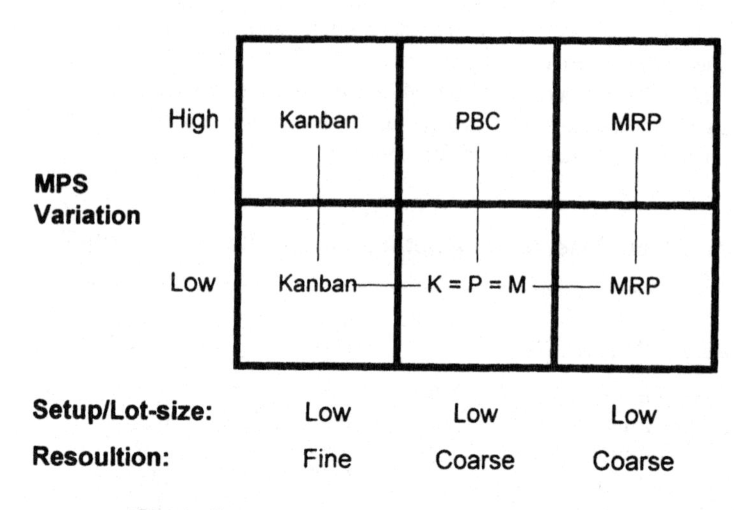

FIGURE 2. Comparative results (from Steele et al. 1995).

In some follow-on experiments, Steele and Malhotra (1997) tested PBC more critically. The purpose was to identify the key environmental factors affecting PBC performance. The study was focused on the same shop used earlier (Steele et al. 1995), simulating a variety of conditions. Period size, cell capacity imbalance, and MPS load variation were key variables tested.

Predictably, period size was found to be a key predictor of PBC performance. Performance improved as the period size was reduced. However, a minimum period size of PBC was shown to be associated with the capacity utilization and with the cell exhibiting the shortest make-span. Simply put, if the manager can identify the weakest cell (longest make-span and highest utilization), then he or she can determine the both minimum period size necessary for successful PBC (the make-span) and the amount of overall inventory required (one period of requirements per cell). Also, MPS load variability, the amount of capacity imbalance between cells, and transfer batch size further affected performance.

Kaku and Krajewski (1995) also evaluated PBC period size, but using a cost model. They showed that a minimum cost period size could be evaluated through use of such a model, minimizing inventory for a given capacity investment. Using four different cases, their model demonstrated that the cell structure design directly influences inventory and overtime costs and thus the best choice of period length. Further, demand increases and demand uncertainty were confirmed as key factors. Thus, selecting the appropriate amount of capacity slack is an important cell design consideration. Their study also showed that safety stock could reduce cost and cycle length.

Riezebos (1997) proposed an alternative perspective in that if external conditions determine the overall throughput time (T) required, then there are still tradeoffs between period size (P) and the number of production stages (N) used in PBC. That is, since T = N*P, one could choose to include more or fewer cells in a single stage, thus decomposing the production chain into more or fewer stages. Doing so means that P can increase as N decreases. The objective of his paper was to identify factors important in determining N and P. Cell start/finish idle times, setup time, and the need for overlapping production were all identified as potentially important.

7. ABSENCE OF SYSTEMATIC WORK

While PBC is well described, systematic research in almost all major aspects has been missing. The application of PBC is not well understood and research so far has been erratic. This absence of work can be seen in the four following dimensions.

7.1. Design

The design of PBC actually parallels three aspects that Burbidge (1975) proposed as essential to successful GT implementation in general. These were group layout, flow control, and planned loading sequences. From a PBC design standpoint, these correspond to stage definition, period size, and family sequencing.

Stage definition is not well understood. PBC is designed for the management of stages of production corresponding to periods of time. However, each stage need not be a single cell; a combinations of cells with shorter throughput times may balance well with one cell with a longer throughput time. Each stage need only be reasonably balanced in throughput time with the others *and* be able to accommodate the planned single-phase arrival of orders at the beginning of each period. Although some factors have been proposed, relationships between PBC and the variety of balance levels, external flow patterns, and internal flow patterns are not known.

Period size also remains a central issue in PBC design. While some work has established maximum makespan and cost minimization as important elements, the understanding of the role of period size is still not fully developed. Further, the value of the linking of both lead time and lot size to a single period has been untested.

Family sequencing is also poorly developed. A unique character of PBC is in the possibility of preset setup sequences (part families) at either MPS or stage levels. The replication of the MPS sequence into each stage provides an almost unparalleled ability for the MPS to control the work content in a given period at the cell, both in setup and load content. When and how this MPS control is useful has not been investigated. Indeed, the possible need to meet constraints of several cells simultaneously could raise interesting questions. In addition to the MPS, the control of the setup sequence *within* the cell and period further determines cell load and performance. In some respects this may parallel the single-machine scheduling problem. However, the possible variety of internal cell routings and process times and the potential for overlapping, compressed part schedules complicate the problem.

7.2. Environment

The shop environment can create quite different constraints upon a system. For PBC, a variety of dimensions could be considered: material flow pattern, lead time variety, product structure, MPS uniformity and stability, MPS period size (resolution), setup time (lot size), and stage setup or make-span variety.

Wemmerlöv (1985) observed that schedule stability and flow pattern could be expected to influence performance, though PBC was generally lumped with MRP. Steele, Berry, and Chapman (1995) experiments showed that some environmental factors could affect the performance of PBC compared with other systems. Small lot, high MPS variation, coarse resolution environments favored PBC.

Still, the work is insufficient for understanding how the environment affects the desirability or the design of PBC.

7.3. Relationship To Other Systems

The linkage to other MPC systems is undeveloped in two respects. First, synergy with other types of systems is important, both in how PBC might link to other systems in the non-CM part of the shop and in how the transition from another system to PBC might take place. With respect to non-CM sections of the shop, PBC could be used only where stages of production consistent with period flow can be identified. Consequently, it is likely many non-CM parts of the shop would retain other types of systems. One system, MRP, is a push system that matches PBC in character and information. Another, Kanban, seems to be inconsistent with PBC. Thus the effectiveness of this linkage to other systems may determine the practicality of using PBC in many firms.

Also, given that most firms will have a system in place when PBC is considered, the transition difficulty will also need to be known. Again, with its similarity to PBC, MRP may be a most logical change (Steele and Malhotra 1995). Indeed, several have viewed PBC as a special case of MRP (New 1975)(Hyer and Wemmerlöv 1982). However, little has been developed beyond a conceptual level with either non-CM area or transition linkage.

A second linkage relates to system choice. Three alternatives to PBC would probably be most widely considered for CM: MRP, Kanban, and Drum-Buffer-Rope (DBR). MRP is probably used in more CM facilities (Wemmerlöv and Hyer 1989). Kanban has also been widely used with CM. DBR is relatively new, but continues to gain interest. Some initial work compared systems (Steele et al. 1995), but did not include DBR. Very little is known about how system choice should be made within a CM environment, or even which CM environments are complex enough to gain advantage from an MPC system.

7.4. Application

Ultimately it is the application of a concept that determines its utility. There is no doubt a number of PBC or PBC-like systems in practice. However, there are very few reports. An exception is the Zelenović and Tesić report (1985). Either environmental conditions are inappropriate for PBC, performance characteristics of PBC are unacceptable, or the value of PBC is unknown. Which of these is true needs resolution.

Burbidge (1985) suggested that systems similar to PBC almost necessarily would come into play. Otherwise, the full value of GT and CM would never be realized. The continued implementation of CM suggests the potential is important to investigate. Indeed, the absence of work in all of the preceding areas suggest the is fertile ground for continued thought into the value of cyclical systems such as PBC.

8. CONCLUSION

PBC has been demonstrated to be a unique planning and control system. Single-cycle orders constrained to low quantities and single-phase, one-period lead times create a system-wide planning consistency. And while the constraint to one-period creates rigidity, it also creates transparency and good potential for setup family control. Current research partially supports PBC utility, but the picture is incomplete. PBC's unique character remains to be fully tested. However that occurs, it is clearly a system with value that should not be ignored for CM.

REFERENCES

Burbidge, J.L., 1975, *The Introduction of Group Technology*, John Wiley and Sons, New York.

Burbidge, J. L., 1979, *Group Technology in the Engineering Industry*, Mechanical Engineering Publications, Ltd., London.

Burbidge, J. L., 1988, Operation scheduling with GT and PBC, *International Journal of Production Research*, 26, 3, 429-442.

Burbidge, J. L., 1996, *Period Batch Control*, Clarendon Press, Oxford.

Campbell, G.M. and Mabert, V.A., 1991, Cyclical scheduling for capacitated lot sizing with dynamic demands, *Management Science*, 37, 4, 409-427.

Dale, B.G. and Russell, D., 1982, Production control systems for small group production, *Omega*, 11, 2, 175-185.

Hall, R. W., 1988, Cyclic scheduling for improvement, *International Journal of Production Research*, 26, 3, 457-472.

Hyer, N. and Wemmerlöv, U., 1982, MRP/GT: A framework for production planning and control of cellular manufacturing, *Decision Sciences*, 13, 681-701.

Kaku, B.K. and Krajewski, L.J., 1995, Period batch control in group technology, *International Journal of Production Research*, 33, 1, 79-99.

New, C. C., 1977, MRP and GT, A new strategy for component production, *Production and Inventory Management Journal*, 3rd Quarter.

Rees, L.P., Huang, P. and Taylor, B.W. III., 1989, A comparative analysis of an MRP lot-for-lot system and a kanban system for a multistage production operation, *International Journal of Production Research*, 27, 8, 1427-1443.

Riezebos, J., 1997, On the determination of the period length in a period batch control system, *Proceedings of the Thirty-Second International Matador Conference (A.K.Kochhar, Ed.)*, 131-136, University of Manchester Institute of Science and Technology and MacMillan Press Ltd.

Steele, D.C., Berry, W.L., and Chapman, S.N., 1995, Planning and control in multi-cell manufacturing, *Decision Sciences*, 26, 1, 1-34.

Steele, D.C. and Malhotra, M.K., 1997, Factors affecting performance of period batch control systems in cellular manufacturing, *International Journal of Production Research*, 35, 12, 421-446.

Suresh, N. C., 1979, Optimizing intermittent production systems through group technology and an MRP system, *Production and Inventory Management Journal*, 4th quarter.

Vollmann, T., Berry, W.L., and Whybark, D.C., 1997, *Manufacturing Planning and Control Systems*, Irwin, Homewood, IL.

Wemmerlöv, U., 1988, *Production Planning and Control Procedures for Cellular Manufacturing Systems*, American Production and Inventory Control Society, Falls Church, VA.

Wemmerlöv, U. and Hyer, N., 1987, Research issues in cellular manufacturing, *International Journal of Production Research*, 25, 3, 413-431.

Wemmerlöv, U. and Hyer, N., 1989, Cellular manufacturing in the U.S. industry, *International Journal of Production Research*, 27, 9, 1511-1530.

Whybark, C., 1984, Production planning and control at Kumera Oy, *Production and Inventory Management Journal*, 1st quarter.

Zelenović, D. and Tesić, Z., 1988, Period batch control and group technology, *International Journal of Production Research*, 26, 3, 539-552.

AUTHOR'S BIOGRAPHY

Daniel C. Steele is an Associate Professor of Management Science at the College of Business Administration at the University of South Carolina. Coming from an industrial background, his activities center around manufacturing planning and control systems, Just-in-Time, cellular manufacturing, and manufacturing strategy. Professor Steele has Electrical Engineering and Industrial Management degrees from Purdue University and a Ph.D. from the University of Iowa. His industrial experience spans 15 years in both staff and line positions. Professor Steele has published in *Decision Sciences, Production and Inventory Management Journal, International Journal of Operations and Production Management, International Journal of Production Economics, International Journal of Production Research, and International Journal of Logistics Management.* He received the 1996 Romeyn Everdell award for the best article in the *PIM Journal.*

F3

Scheduling Rules For Cellular Manufacturing

F. Mahmoodi and C. T. Mosier

1. INTRODUCTION

Cellular manufacturing systems should be effectively managed to augment their advantages while some of their disadvantages minimized. Given an appropriate classification of parts into families and machines into cells, many researchers believe that utilizing scheduling procedures which capitalize on the unique features of cellular manufacturing is essential to counter its disadvantages and enhance the likelihood of successful implementation (e.g., Flynn 1987; Mahmoodi, Dooley and Starr 1990a).

Such scheduling procedures, often referred to as *group scheduling* heuristics, can reduce overall machine setup time by taking advantage of setup similarities among part families. Furthermore, they potentially increase the flexibility and utilization of cells by adapting them to a more diverse range of part families (Lee 1985), thus reducing a major disadvantage associated with cellular systems (Greene and Sadowski 1984; Suresh and Meredith 1985; Suresh 1992; Suresh and Meredith 1994). It is this area that practitioners most often cite a need for further study (Mosier and Taube 1985; Wemmerlöv and Hyer 1989).

Group scheduling procedures exploit similarities in setups that exist among the individual parts within a given part family; something that the traditional job shop scheduling procedures by themselves do not explicitly consider. To implement group scheduling procedures we first partition the parts within the part family into *subfamilies* based on setup similarities. This is accomplished by determining the number of different *major* setups (long machine changeovers required between the production of two dissimilar parts or batches/lots of many parts) that need to be performed at each workcenter (Vaithianathan and McRoberts 1982). For example, if three different major setups are required on a workcenter (within a cell) to manufacture the entire part family in that cell, then the part family can be decomposed into three subfamilies. Different subfamily designations may exist within different workcenters. Thus, jobs within the same subfamily require the same major setups but may require different minor setups. For example, jobs within the same subfamily could require the same tool turret but require different NC codes, or use the same raw materials but in different mixtures (Mahmoodi and Dooley 1992). Procedures designed to minimize the number of major setups can be executed manually based on actual knowledge of the production process, or by utilizing coding schemes (if available).

There are a variety of manufacturing environments other than cellular manufacturing in which group scheduling procedures can be applied. Let us consider two examples (from Mahmoodi and Dooley 1992). A particular automobile model may be designated for body painting at a workcenter in a cell. The cellular arrangement can take advantage of the similarities of automobiles coming through this workcenter and automate the paint dispersion process. However, within that model, color requirements will differ and thus will require different setups. Here, color could be used to designate subfamilies.

Now consider a plastic part produced in an injection molding process. Injection molding takes raw materials in the form of pellets and heats them into liquid, injecting them into the mold cavity. Whenever the "pellet recipe" is altered a new setup is incurred. Two parts in the same part family may have the same design (i.e., require the same mold) but use different pellet mixes because of strength requirements. The pellet mixture could thus distinguish subfamilies. Many other examples in processing, chemical, and assembly industries can be found that exhibit similar characteristics.

It is worth noting that the need for group scheduling procedures increases as the demand for flexibility increases (Mahmoodi and Dooley 1992). This is because flexibility necessitates the use of versatile machining centers that perform many functions at a single location. As such flexible machines become more utilized within cells, the variety of parts that the cell can handle increases, and thus the variety of setup requirements increases as well.

In summary, the group scheduling approach typically involves two stages. The first stage involves sequencing jobs within each subfamily. These decisions are made by utilizing traditional job shop scheduling rules. The second stage consists of determining which subfamily queue to select, and when to switch to another subfamily queue at each workcenter. These decisions are addressed by queue selection heuristics.

A hint of the inherent complexity of seemingly simple scheduling problems is illustrated in the next section. Following, a summary of the results of group scheduling research findings and their managerial implications is presented. This is followed by a presentation of a number of effective group scheduling heuristics and their application to a small problem. Finally, conclusions relevant to the practitioners are outlined.

2. PROBLEM COMPLEXITY

In today's world of burgeoning computer power it seems that scheduling problems should be easily solved, if not by efficient algorithms, then by complete enumeration. We put the notion of complete enumeration to rest with a simple demonstration. Consider a simple static sequencing problem with n jobs.

Furthermore, assume that our computer is relatively fast, capable of generating all possible sequences and evaluating them at a rate of 1 millionth of a second per job within each feasible sequence. For example, if there are 3 jobs (say A, B, and C), then we have 6 (i.e., $n!$) possible sequences of 3 jobs each:

$$A \rightarrow B \rightarrow C \quad A \rightarrow C \rightarrow B$$
$$B \rightarrow A \rightarrow C \quad B \rightarrow C \rightarrow A$$
$$C \rightarrow A \rightarrow B \quad C \rightarrow B \rightarrow A$$

Thus, the total time required to solve this problem by complete enumeration is:

$$\left(\frac{1}{1{,}000{,}000}\right) \times 6 \times 3 = 0.000018 \text{ second}$$

The logical expansion of this to larger values of n is presented in Table 1. Clearly, even for reasonably sized problems, complete enumeration is not a tractable approach to solving the problem. Furthermore, faster computers will not help, at least in the reasonably near future. If we speed up our computer a billion-fold, it will still take 4.82 x 10^{43} years to determine an optimal sequence of 50 jobs using complete enumeration. Thus, researchers have sought alternative strategies such as heuristic approaches for determining reasonable, if not optimal, solutions.

TABLE 1. Solution Times For Complete Enumeration.

Problem Size (n -jobs)	Time for Complete Enumeration
1	0.000001 second
2	0.000004 second
10	36.288 seconds
20	1.54x10^6 years
50	4.82x10^{52} years

Two main types of heuristic have been developed:

1. Heuristics that work so well in terms of their targeted performance measure(s) that the positive results of their operation override any negative aspects. Or equivalently, heuristics which schedule successfully in terms of all possible performance criteria.

2. Heuristics that are a combination of two or more single-focus heuristics, thus garnering the benefits of each. This requires the specification of acceptable ways of combining heuristics in logical and successful ways. In a heuristic made up of a composite of other heuristics, we would expect that the impact of individual heuristics on their particular targeted performance criteria would be reduced.

The first strategy has driven most previous job shop scheduling research; unfortunately, it has not proven successful. The "holy grail" of heuristics, an all powerful "super" heuristic that solves all sequencing problems according to all criteria seems impossible to achieve. The second strategy currently drives group scheduling research. Of course, the strategy of combining heuristics to achieve a desired result is not new and has been utilized in the job shop scheduling for many years.

Note that the problem context considered in the previous discussion is almost the simplest possible. The number of jobs is small, the number of machines is trivial, and the problem context is static. This set of conditions is almost never true in industry. The number of jobs and machines is always large, and the problem context is always dynamic. Under these more complex conditions, the evaluation of the performance of alternative rules is almost always performed using computer simulation. Although analytic models provide many insights, they are unable to capture the complex dynamic nature of the problem.

3. GROUP SCHEDULING LITERATURE

Group scheduling rules can be classified as either exhaustive or non-exhaustive. Exhaustive heuristics are such that once a subfamily is chosen, all the jobs within that subfamily, including new arrivals, are processed before another subfamily can be considered. Only when there are no jobs left in that subfamily are jobs in other subfamilies processed. Conversely, non-exhaustive heuristics allow switching to another subfamily even when the current subfamily queue is not empty. Thus, non-exhaustive heuristics consider processing jobs from other subfamily queues upon the completion of each job, while exhaustive rules avoid processing jobs from other subfamilies until all jobs in the current subfamily are processed. While exhaustive rules tend to minimize major setups and are simpler to implement, some non-exhaustive rules may enhance due date performance (Mahmoodi and Dooley 1991).

A significant body of research exists on scheduling in cellular manufacturing environment (e.g., Mosier and Taube 1985; Mahmoodi and Dooley 1992; Shambu, Suresh and Pegels 1996). While some studies have examined only traditional single-stage rules, most have focused on two-stage, group scheduling heuristics. Some studies have considered a static set of jobs to schedule over a fixed horizon, while others have focused on a dynamic set of jobs. Furthermore, a variety of environmental factors, shop structures, performance measures, and assumptions have been included in these experiments. For further discussion, we classify the group scheduling research into two major categories: scheduling of single and multi-stage systems.

3.1. Single-Stage Group Scheduling

A number of researchers have addressed the group scheduling problem for a single facility, i.e., a single machine system. These studies have used a variety of analytical techniques to develop optimizing algorithms, mainly under static conditions. The performance criteria used for evaluation ranges from minimizing the setup time or cost (Foo and Wager 1983; Ozden, Egbelu and Iyer 1985; Irani, Gunasena, Davachi and Enscore 1988) to maximizing the production rate (Hitomi and Ham 1978), and minimizing the total job tardiness (Nakamura, Yoshida and Hitomi 1978).

More recently, a few studies have examined the problem under more realistic dynamic conditions using computer simulation. Wemmerlöv (1992) tested two group scheduling heuristics and two dispatching rules under a variety of experimental conditions. It was concluded that the advantages of group scheduling are accentuated in cells with high utilization, a high degree of instability with respect to job arrivals and run times, a high setup time to processing time ratio, and a low number of part families. Also, it was concluded that group scheduling heuristics generally tend to produce lower flow time and lateness averages and variation than single-stage rules for any given environment. Finally, it was concluded that the performance of group scheduling procedures tend to be very robust with respect to a number of environmental factors. More recently, Mahmoodi and Martin (1997) developed an efficiency oriented queue selection heuristic with the goal of minimizing aggregate setup times. This heuristic included a feature for dynamically assessing variations in a subfamily's arrival rate, enhancing its suitability for highly transient-state environments. Results indicated good flow time and due date performance by this heuristic in a variety of experimental conditions.

In conclusion, although these studies reveal many insights to the group scheduling problem, their results are limited to single-stage production facilities (as opposed to a cell that consists of multi-stages). Another important aspect of these studies is in their operationalization of setups. Specifically, some studies have assumed sequence-dependent setup times (i.e., current setup times are dependent on current machine settings), other studies have presumed major setup times are independent of subfamily sequence. Further, analytical studies usually focus on optimizing a single criterion (as specified by their objective functions) without considering others. Finally, the computation time required to solve realistically sized problems via the analytical approaches is often excessive.

3.2 Multi-Stage Group Scheduling

The research on multi-stage cellular manufacturing systems can be classified into two categories: scheduling of flow-through cells and of job shop cells. The difference between flow-through and job shop cells is the cell's internal flow pattern. In a flow-through cell (in its pure form) all parts have identical routes; in a job shop cell, parts may arrive and depart at different workcenters and have different routings.

Flow-through cells are popular since they are usually used in conjunction with JIT programs. In flow-through cells, scheduling of a job takes place only once at the first workcenter, and that sequence is maintained throughout the system. Most of the group scheduling research addressing flow-through cells has focused on either analytical techniques or heuristics. On the other hand, since job shop cells manufacture a variety of parts with different routings, loading and dispatching these parts takes place at every workcenter. Thus, simple analytical solution algorithms do not exist and heuristic approaches have been employed most often.

Scheduling Flow-Through Cells.

A number of researchers have addressed the group scheduling problem in a flow-through cell. Many analytical studies have applied a branch-and-bound approach (e.g., Hitomi and Ham 1977; Ham, Hitomi, Nakamura and Yoshida 1979). Solution procedures determine both optimal subfamily and job sequences such that makespan (total flow time of all the jobs) is minimized. Vakharia and Chang (1990) and Skorin-Kapov and Vakharia (1993) have developed new scheduling heuristics based on simulated annealing and tabu search techniques, respectively. They assumed static conditions and their procedures are extremely complex to implement in practice.

A number of studies have examined the dynamic dispatching problem under a broad variety of experimental conditions. They showed that the group scheduling procedures almost always perform better than single-stage job shop scheduling procedures, with respect to both average and variation of a variety of performance measures (Wemmerlöv and Vakharia 1991; Mahmoodi, Tierney and Mosier 1992). The lower the utilization, the less advantageous are group scheduling heuristics. However, the ability of group scheduling heuristics to avoid setups during peak loads makes their performance relatively insensitive to changing experimental conditions (Wemmerlöv and Vakharia 1991; Mahmoodi et al. 1992). Such robust performance, in itself, may be a good reason for considering group scheduling heuristics.

Results also indicated that unlike the single stage rules, the performance of the group scheduling heuristics is tightly clustered (i.e., selecting the "wrong" heuristic is less serious when using group scheduling heuristics than when single stage rules are used). This is especially true in stable, under-utilized manufacturing environments. Thus, in the presence of such environments, the most easily implemented group scheduling heuristics should be the choice (Mahmoodi et al. 1992). In fact, Frazier (1996) indicated that more complex heuristics showed poorer performance than simpler heuristics, under a variety of experimental conditions.

It was also concluded that the selection of the group scheduling heuristic should be based on the choice of performance measures. For example, if flow time is the primary consideration, MSSPT (described in §4.2) should be selected. However, if customer service (due date adherence) is top priority, DDSI (described in §4.2) should be selected (Mahmoodi et al. 1992). In Frazier's (1996) study the scheduling rule that selected the subfamily with the most waiting jobs and sequenced them according to SPT (i.e., WOSPT) generally performed the best.

Finally, Russell and Philipoom (1991) showed that as setup times increases, the queue selection decision becomes more critical than due date setting procedures in terms of job due date adherence. They also concluded that the choice of due date procedure has a significant impact on due date and flow time performance as well as on the appropriate choice of group scheduling heuristic.

Scheduling Job Shop Cells.

Many group scheduling heuristics have been examined in job shop cells. Numerous simulation studies have compared the performance of a wide variety of heuristics under different experimental conditions (e.g., Vaithianathan and McRoberts 1982; Ang and Willey 1984; Mosier et al. 1984; Flynn 1987; Mahmoodi et al. 1990a; Mahmoodi, Dooley and Starr 1990b; Mahmoodi and Dooley 1991; Ruben, Mosier and Mahmoodi 1993; Wirth, Mahmoodi and Mosier 1993; Suresh and Meredith 1994; Kannan and Lyman 1994).

As in the flow through cell environment, it has been shown that group scheduling heuristics result in significant improvements in performance over the single stage rules, under a variety of experimental conditions (e.g., Mosier et al. 1984; Flynn 1987; Mahmoodi et al. 1990a; Suresh and Meredith 1994). More specifically, Kannan and Lyman (1994) concluded that utilizing group scheduling heuristics always resulted in better flow time performance regardless of transfer batch sizes. Furthermore, Mahmoodi et al. (1990a) showed that the performance of several group scheduling heuristics are robust to a variety of experimental factors.

Results of a later study (Ruben et al. (1993)) showed that performance comparable to that of group scheduling heuristics can be obtained with the single stage rules when factors that lessen the impact of setup times are in place. In particular, they showed that the presence of low setup-to-run-time ratio, shop load, less variable interarrivals, and subfamily dominance substantially reduces the advantages of group scheduling heuristics.

The performance of exhaustive heuristics is generally superior and more robust than those of non-exhaustive heuristics. In fact, overall simpler heuristics performed better than complex heuristics (Mahmoodi and Dooley 1991). Non-exhaustive heuristics only dominated exhaustive heuristics on average tardiness in lightly loaded manufacturing cells where a large due date allowance existed. This may be because such environments allow the more complex logic of the non-exhaustive heuristics to be executed effectively (Mahmoodi and Dooley 1991).

Including labor as an additional constraint in the cell environment had a significant impact on the performance of several group scheduling heuristics. For example, MSSPT, which had exhibited poor average tardiness performance in previous machine-constrained studies demonstrated excellent performance across all performance measures, including average tardiness, in a study where both machines and labor were considered as constraints (Wirth et al. 1993). Furthermore, the performance of non-exhaustive heuristics deteriorated with the inclusion of labor as an additional constraint.

The performance of group scheduling heuristics also depended on the dispatching rule utilized. Generally, heuristics that employed FCFS performed poorly on average flow time and percentage tardy, while heuristics that utilized SPT demonstrated very good performance on those measures (Mahmoodi and Dooley 1991, Ruben et al. 1993, Kannan and Lyman 1994). However, the average tardiness performance of the heuristics that employed SPT was poor. Finally, group scheduling heuristics that utilized SI^x were among the best performing heuristics. This can be attributed to the fact that SI^x retains the superior flow time performance of SPT while minimizing the extreme completion delay of a few jobs (Mahmoodi and Dooley 1991). Overall, the MSSPT heuristic (described in detail in §4.2) displayed the best average time in system and percentage tardy performance while DDSI (also described in §4.2) performed best on average tardiness measure (Mahmoodi et al. 1990; Mahmoodi and Dooley 1991; Ruben et al. 1993; Wirth et al. 1993).

In most studies the performance of manufacturing cells deteriorated as the cell load and the variability in job interarrivals increased (e.g., Mosier et al. 1984; Mahmoodi et al. 1990a; Mahmoodi and Dooley 1991; Ruben et al. 1993; Wirth et al. 1993). This indicates that management can substantially improve the cell's performance by focusing on production planning activities to reduce the variation in job interarrival times and leveling the cell load.

4. APPLICATION OF EFFECTIVE GROUP SCHEDULING HEURISTICS

The group scheduling procedures discussed in this section have generally shown very favorable results in their respective studies. A small example is utilized to illustrate the application of a limited number of non-exhaustive and exhaustive group scheduling procedures. For the sake of simplicity, we assume a single machine, static scheduling problem, thus our results are merely indicative of relative performance. Clearly, the exhaustive evaluation of group scheduling heuristics in a multi-stage dynamic environment requires complex simulation models and in depth experimentation.

4.1. Non-Exhaustive Heuristics

Economic (ECON).

ECON (Mosier, Elvers and Kelly 1984) can be classified as a *non-exhaustive queue selection* heuristic (the second stage of group scheduling heuristic). The primary focus of ECON was to attempt to balance the time-based "costs" due to setups between subfamilies of jobs sequentially processed on a workstation. Operationally ECON determines: (a) when to perform a setup, and (b) if required, which setup to perform. ECON dynamically addresses the question "*when to perform a setup.*"

For implementation, information is required concerning the time required for performing a setup (or changeover) from one subfamily of parts to another. Furthermore, since ECON is typically used in conjunction with a specific queue discipline, whatever information needed for implementation of the queue discipline is also required.

Initially, a priority value for each setup or changeover (or equivalently, each sub-family) is computed. ECON uses a priority that measures the potential setup savings to be derived by changing to (or remaining with) a setup, minus the potential setup savings lost by "leaving" the current setup. Thus, for the setup that is currently in place, the priority, i.e., the potential setup savings to be derived by remaining with that setup, is the number of jobs in the queue that require that setup, multiplied by the expected time required to perform the setup.

For those setups that are not currently in place, the potential gain to be derived by changing to the new setup is the number of jobs requiring that setup (minus one - the cost of doing the setup) multiplied by the expected time required to perform the setup. The priority of setups not currently in place is the potential gain described above minus the priority of the setup currently in place. This can be represented as:

$$P = (1-\delta)\left[E(S_N) \times (N_N - 1) - E(S_C) \times N_C\right] + \delta\left[E(S_C) \times N_C\right] \quad [1]$$

where:

$$\delta = \begin{cases} 1 & \text{if we are evaluating the queue currently being serviced} \\ 0 & \text{else} \end{cases}$$

$E(S_N)$ = the expected number of setup time of jobs in a new queue,

N_N = the number of jobs in a new queue,

$E(S_C)$ = the expected number of setup time of jobs in the queue currently being serviced, and

N_C = the number of jobs in the queue currently being serviced.

Initially, ECON operated by performing the setup with the highest priority. Thus, the job for next processing is the job within the selected setup subfamily with the highest priority defined by any number of scheduling heuristics. Later, ECON was modified for better operational control. Instead of simply choosing the setup with the highest priority for next process, the ratio priority of the current setup to the other setup priorities was more carefully considered. An example is provided to illustrate ECON's implementation.

Example: Consider the following simple problem with a job queue as specified in Table 2. Assuming that the workstation is currently set up for job type A, the priority of setup A is 15, the priority of setup B is 60, and the priority of setup C is 90. Thus, in the original consideration of the heuristic, the machine setup would be changed to C.

Recently, Mosier and Collins (1992) introduced an option for providing more control over the number of setups. Designating the priorities (of setups A, B, and C as defined in the example) as P_A, P_B, and P_C, respectively, we can specify (and evaluate) the maximum of *ratio* of priorities of current setup to competing setups in this case as:

$$\frac{P_C}{P_A} = \frac{90}{15} = 6$$

TABLE 2. ECON Example Problem.

Job	Setup Family	Expected Setup Time	Expected Processing Time	Job	Setup Family	Expected Setup Time	Expected Processing Time
401	A	15 min	7 hr	406	C	35 min	2 hr
402	B	25 min	3 hr	407	C	35 min	4 hr
403	C	35 min	6 hr	408	B	25 min	1.5 hr
404	B	25 min	3 hr	409	B	25 min	1 hr
405	B	25 min	2 hr				

The initial implementation of ECON was to change the setup if this ratio was greater than one. The extension proposed by Mosier and Collins (1992) (and subsequently incorporated in the study by Ruben et al. (1993)) was to restrict changeovers to cases where this ratio was greater than a specified value (defined as k, thus the new name ECON-k). Thus in this example if $k = 5$ a changeover to setup C would be prescribed, if $k = 7$, no changeover would occur. The specification of the best value of k for a specific environment is determined experimentally.

Caveats: The studies of this heuristic (e.g., Wirth et al. 1993; Ruben et al. 1993) were conducted in an environment where the notion of sequence dependent setups was somewhat restrictive. That is, it was assumed that only a limited number of distinct setups were required. If a job followed a job with the same setup, no setup was required. Thus, in truth, ECON addresses the reduction of the activity that might be better termed "major" setups. Further, the rule was studied in relatively simplified job shops, prohibiting lot splitting, and using fairly random processing, setup, and interarrival times, and, in some cases, explicitly considering sequence dependent setups.

4.2. Exhaustive Heuristics

Table 3 presents an example problem, i.e., a list of jobs and their corresponding processing times, setup family, and due dates, labeled sequentially as they have arrived to the system. Table 4 specifies the major setup matrix, i.e., the setup time for changing the machine from one setup family to another. The symbol Ø implies no setup is present. Note that the setup times are sequence-dependent, i.e., the matrix is not symmetric. For example, to change from setup family 1 to setup family 2 takes 1 time unit, while to change from setup family 2 to setup family 1 takes 2 time units.

Alternative scheduling strategies are demonstrated by applying various scheduling procedures to this problem. For each heuristic we calculate average flow time, average lateness, average tardiness and percentage of jobs tardy. Note that the first two characters in each group scheduling procedures' acronym refers to the queue selection heuristic while the remaining characters indicate the imbedded dispatching rule.

TABLE 3. Example Problem.

Job	Processing Time	Setup Family	Due Date	Job	Processing Time	Setup Family	Due Date
A	7	2	38	E	4	3	6
B	9	1	15	F	2	1	5
C	1	3	10	G	5	2	17
D	3	2	37				

TABLE 4. Major Setup Matrix.

	1	2	3
Ø	3	2	1
1	0	1	1
2	2	0	2
3	3	1	0

First-Come-First-Come-First-Served (FCFCFS).
This heuristic, which is based on the "repetitive lots" rule (Jacobs and Bragg 1988), utilizes FCFS to sequence jobs within each subfamily and selects for next processing the subfamily whose first job arrived first. FCFCFS attempts to minimize the number of setups by not processing from another subfamily until all the jobs in the current subfamily are processed (i.e., it is exhaustive). The subfamily of which the last processed job is a member is referred to as the current subfamily. FCFCFS is dynamic, i.e., it considers all jobs in the subfamily when selecting the next job for processing, but does not consider job due dates. FCFCFS was the best performing heuristic in Flynn's (1987) and Wemmerlöv and Vakharia's (1991) studies, but not in the Mahmoodi et al. (1990a) study.

TABLE 5. Results of FCFCFS Applied to the Example Problem.

Job	Processing Time	Setup Family	Start Time	Finish Time	Lateness	Tardiness	% Tardy
A	7	2	2	9	-29	0	
D	3	2	9	12	-25	0	
G	5	2	12	17	0	0	
B	9	1	19	28	13	13	
F	2	1	28	30	25	25	
C	1	3	31	32	22	22	
E	4	3	32	36	30	30	
Totals				164	36	90	
Averages				23.43	5.14	12.86	57.14%

Table 5 presents the schedule and corresponding shop performance levels for FCFCFS applied to the example problem described in Tables 3 and 4. Initially each job is placed in its appropriate subfamily, i.e., jobs B and F in subfamily 1, jobs A, D and G in subfamily 2, and jobs C and E in subfamily 3. Then, jobs are ordered within each subfamily according to FCFS (e.g., in subfamily 1 job F is placed after job B). Subfamily 2 is processed first since its first job (i.e., job A) arrived earlier than the other subfamilies' first jobs (i.e., jobs B and C). After all the jobs in subfamily 2 are processed, subfamily 1 is processed since job B arrived earlier than job C.

Due-Date-SI^x (DDSI).

This procedure utilizes the SI^x heuristic to sequence jobs within subfamilies. SI^x is a two-class truncated shortest processing time (SPT) heuristic that dynamically assigns priority to jobs with zero or negative slack times and orders by SPT (Oral and Malouin 1973). Note that a job's slack time is calculated by subtracting its processing time from the difference between its due date and the current time. DDSI attempts to minimize average number of setups while maintaining due date performance by selecting the subfamily queue whose first job has closest due date. This procedure is dynamic and exhaustive and attempts to minimize number of very late jobs. DDSI displayed the best average tardiness performance in the Mahmoodi et al. (1990a) job shop study and Mahmoodi et al. (1992) flow-through cell study.

Table 6 presents the schedule and corresponding shop performance for the DDSI procedure. Similar to FCFCFS, initially each job is placed in its appropriate subfamily. Then jobs are ordered within each subfamily according to the SI^x heuristic; note that in subfamily 2 job G is placed in the priority queue and is processed before job D due to its negative slack time. Subfamily 1 is processed first since its first job (i.e., job F) has a more imminent due date than the first jobs in the other subfamilies (i.e., jobs C and G). After all the jobs in subfamily 1 are processed, subfamily 3 is processed since job C has an earlier due date than job G.

TABLE 6. Results of DDSI^x Applied to the Example Problem.

Job	Processing Time	Setup Family	Start Time	Finish Time	Lateness	Tardiness	% Tardy	
F	2	1	3	5	0	0	0	
B	9	1	5	14	1	-1	0	
C	1	3	15	16	-6	6	6	
E	4	3	16	20	-14	14	14	
G	5	2	21	26	-9	9	9	
D	3	2	26	29	8	-8	0	
A	7	2	29	36	2	-2	0	
Totals				146		18	29	
Averages				20.86		2.57	4.14	42.86%

Minimum-Setup-Shortest-Processing-Time (MSSPT).
This procedure utilizes the SPT heuristic to sequence jobs within subfamilies. MSSPT attempts to minimize both the number of setups and the total setup time by exploiting the sequence-dependent nature of the setup times. The subfamily queue that requires the least amount of setup is selected. This procedure is dynamic and exhaustive and has also shown very good average flow time and percentage tardy performance in the Mahmoodi et al. (1990a) job shop cell study as well as in the Mahmoodi et al. (1992) flow-through cell study. Furthermore, this heuristic performed even better on all performance measures (including average tardiness) in the Wirth et al. (1993) dual constrained manufacturing cell study.

Table 7 presents the schedule and corresponding shop performances for the MSSPT procedure. After each job is placed in its appropriate subfamily, jobs are ordered according to SPT. Subfamily 3 is processed first since it incurs minimum setup time (i.e., 1 time unit). After all the jobs in subfamily 3 are processed, subfamily 2 is processed since changing from setup 3 to setup 2 takes less time than changing from setup 3 to setup 1 (see Table 4).

TABLE 7. Results of MSSPT Applied to the Example Problem.

Job	Processing Time	Setup Family	Start Time	Finish Time	Lateness	Tardiness	% Tardy
C	1	3	1	2	-8	0	
E	4	3	2	6	0	0	
D	3	2	7	10	-27	0	
G	5	2	10	15	-2	0	
A	7	2	15	22	-16	0	
F	2	1	24	26	21	21	
B	9	1	26	35	20	20	
Totals				116	-12	41	
Averages				16.57	-1.71	5.86	28.57%

WOrk-Shortest-Processing-Time (WOSPT).
This procedure utilizes the SPT heuristic to sequence jobs within subfamilies. Queue selection is based on the WORK procedure that selects the subfamily queue with the largest total work content across all of its jobs. A queue choice is made only when the queue is emptied (exhaustive). The performance of WOSPT has been previously reported by Mosier et al. (1984) and Frazier (1996). In fact, overall, this heuristic performed the best among all the heuristics examined by Frazier (1996).

Table 8 presents the schedule and corresponding shop performances for the WOSPT procedure. Similar to MSSPT procedure, each job is placed in its appropriate subfamily and within each subfamily ordered according to the SPT heuristic. Subfamily 2 is processed first since its jobs require the most work (i.e.,

$3+5+7=15$ time units). After all the jobs in subfamily 2 are processed, subfamily 1 is processed since it contains a larger work content than subfamily 3.

TABLE 8. Results of WOSPT Applied to the Example Problem.

Job	Processing Time	Setup Class	Start Time	Finish Time	Lateness	Tardiness	% Tardy
D	3	2	2	5	-32	0	
G	5	2	5	10	-7	0	
A	7	2	10	17	-21	0	
F	2	1	19	21	16	16	
B	9	1	21	30	15	15	
C	1	3	31	32	22	22	
E	4	3	32	36	30	30	
Totals				151	23	83	
Averages				21.57	3.29	11.86	57.14%

5. CONCLUSIONS

A large number of research studies have addressed scheduling of different cellular manufacturing systems including a broad variety of environmental factors, shop structures, performance measures, and operational assumptions. In this chapter we have categorized the relevant research and discussed the research findings in each category. We now summarize these findings in an outline of general conclusions relevant to practitioners. Note that these conclusions generally apply to both single-stage and multiple stage cellular systems.

1. Group scheduling heuristics almost always perform better than single-stage scheduling rules with respect to both average performance levels and performance variation for a wide variety of flow time and due date oriented performance measures.

2. The performance of group scheduling heuristics is much more robust with respect to a wide variety of experimental factors than those of single-stage rules. This characteristic, in itself, may be a good reason for utilizing group scheduling heuristics.

3. The performance advantage of group scheduling heuristics is enhanced, and the performance of manufacturing cells deteriorates as cell utilization, setup times, and variation in job interarrival and processing times increase, number of part families decrease, and part families are partitioned equally. This result has important implications for the practicing manager:

   - Management can improve cell performance significantly by concentrating on production planning activities directed at reducing variation in job interarrival times and leveling the cell load.

- The results of the part family formation procedure are crucial to the success of the group scheduling approach. For example, when part families do not partition equally, it is likely significant expediting will be required to maintain an effective cell. In such cases, manager might spend their time more effectively balancing the cell load.

4. Differences in performance among the group scheduling heuristics are much narrower than those of single-stage rules. Thus, choosing an inappropriate group scheduling heuristic is less serious than selecting an inappropriate single-stage rule.

5. Unfortunately, research to date has not produced a universally successful, all encompassing heuristic; the "holy grail" of heuristics that performs the best on all performance measures and under all environments. Thus, the selection of the group scheduling heuristic should be based on the choice of performance measures. For example, MSSPT is clearly the choice for firms that place a high emphasis on flow time performance. However, due to its rather poor performance with respect to average tardiness measure in some environments, it is more appropriate for make-to-stock firms as a high level of safety stock will be necessary with its use in the make-to-order environment. On the other hand, DDSI, while not yielding the best throughput performance, demonstrates extremely good average tardiness performance and is more appropriate for make-to-order firms where adherence to due dates is more critical. Note that both MSSPT and DDSI heuristics are fairly easy to implement.

6. Generally, the simpler exhaustive group scheduling heuristics perform better than their more complex non-exhaustive counterparts. Furthermore, the performance of exhaustive heuristics are more robust with respect to a wide variety of experimental factors than those of non-exhaustive heuristics.

The operational characterization of group scheduling approaches is still incomplete. The development of a scientifically derived knowledge base of heuristics and their performance profiles is a necessary prerequisite for the meaningful evolution of "automated" rule-based scheduling systems. A published report (Fordyce, Dunki-Jacobs, Gerard, Sell, and Sullivan 1992) described the use of such a system, the so-called "*Logistics Management System*," at the IBM Burlington Semiconductor development laboratory and manufacturing site. As manufacturing systems become more complex and as the influx of automated systems increases, the "smart" scheduling systems will become more commonplace.

Further, there is a need for research focusing on the development and testing of hybrid methods, combining analytic and heuristic methods to contribute to the knowledge base underlying "smart" scheduling systems. There is a need for primary research characterizing performance, but the problem is operationally more complex. A diversity of issues need to be considered including:

- problems intrinsic to the gathering and organizing of data to be used as input to these scheduling systems,

- the performance evaluation structure - necessarily some aggregate and/or compromise of often operationally conflicting performance measures, and
- the inherent sensitivity of heuristics (or analytical approaches) to the demand demographics and shop environmental factors.

We conclude with a future view of cellular manufacturing, possibly of all manufacturing. With the influx of "smart" machinery and the incredible increases in available computing power, the ability to gather data and to make system adjustment "on-the-fly" should be greatly enhanced. Some modifications may be operational and reactive, e.g., when a job or part is considered for processing, all cost and logistical consequences will be immediately evaluated. Further, the system will be capable of action beyond reactive, to proactive, e.g., able to search for the "next best" job or part to be completed using criteria gleaned from user input and/or pre-specified system goals.

Given the availability of yet unheard of power and flexibility in our material transport systems, the system layout may be altered dynamically. *Virtual cells*, dynamically determined using the results of ongoing analyses of the "demographics" of the production demand stream and associated production requirements, may become the dominant manufacturing system design paradigm.

REFERENCES

Ang, C.L. and P.C.T. Willey, 1984, A comparative study of the performance of pure and hybrid group technology manufacturing systems using computer simulation techniques, *International Journal of Production Research*, 22, 193-233.

Flynn, B.B., 1987, Repetitive lots: The use of a sequence-dependent set-up time scheduling procedure in group technology and traditional shops, *Journal of Operations Management*, 7, 203-216.

Foo, F.C. and J.G. Wager, 1983, Set-up times in cyclic and acyclic group technology scheduling system, *International Journal of Production Research*, 21, 63-73.

Fordyce, K., R. Dunki-Jacobs, B. Gerard, R. Sell, and G. Sullivan, 1992, Logistics management system: An advanced decision support system for the fourth decision tier dispatch or short-interval scheduling, *Production and Operations Management*, 1, 70-86.

Frazier, G.V., 1996, An evaluation of group scheduling heuristics in a flow-line manufacturing cell, *International Journal of Production Research*, 34, 959-976.

Greene, T.J. and P.R. Sadowski, 1984, A review of cellular manufacturing assumptions, advantages and design techniques, *Journal of Operations Management*, 4, 85-97.

Ham, I., K. Hitomi, N. Nakamura, and T. Yoshida, 1979, Optimal group scheduling and machining-speed decision under due-date constraints, *Transactions of the ASME: Journal of Engineering for Industry*, 101, 123-134.

Hitomi, K. and I. Ham, 1977, Group scheduling technique for multiproduct, multistage manufacturing systems, *Transactions of the ASME: Journal of Engineering for Industry*, 99, 759-765.

Hitomi, K. and I. Ham, 1978, Machine loading and product-mix analysis for group technology, *Transactions of the ASME: Journal of Engineering for Industry*, 100, 370-374.

Irani, S.A., U. Gunasena, A. Davachi, and E.E. Enscore, 1988, Single machine setup-dependent sequencing using a setup complexity ranking scheme, *Journal of Manufacturing Systems*, 7, 11-23.

Jacobs, F.R. and D.J. Bragg, 1988, Repetitive lots: Flow-time reductions through sequencing and dynamic batch sizing, *Decision Sciences*, 19, 281-294.

Kannan, V.R. and S.B. Lyman, 1994, Impact of family-based scheduling on transfer batches in a job shop manufacturing cell, *International Journal of Production Research*, 32, 2777-2794.

Kelly, F.H., C.T. Mosier, and L.R. Taube, 1986, Impact of cost structures on group technology implementation, *Proceedings of New Developments and Integrative Opportunities in Information Systems and Operations Management*, Greensboro, NC.

Lee, L.C., 1985, A study of system characteristics in a manufacturing cell, *International Journal of Production Research*, 23, 1101-1114.

Mahmoodi, F. and K.J. Dooley, 1991, A comparison of exhaustive and non-exhaustive group scheduling heuristics in a manufacturing cell, *International Journal of Production Research*, 29, 1923-1939.

Mahmoodi, F. and K.J. Dooley, 1992, Group scheduling and order releasing: Review and foundations for research, *International Journal of Production Planning and Control*, 3, 70-80.

Mahmoodi, F., E.J. Tierney, and C.T. Mosier, 1992, Dynamic group scheduling heuristics in a flow-through cell environment, *Decision Sciences*, 23, 61-85.

Mahmoodi, F., K.J. Dooley, and P.J. Starr, 1990a, An investigation of dynamic group scheduling heuristics in a job shop manufacturing cell, *International Journal of Production Research*, 28, 1695-1711.

Mahmoodi, F., K.J. Dooley, and P.J. Starr, 1990b, An evaluation of order releasing and due date assignment heuristics in a cellular manufacturing system, *Journal of Operations Management*, 9, 548-573.

Mahmoodi, F. and G.E. Martin, 1997, A new shop-based and predictive scheduling heuristic for cellular manufacturing, *International Journal of Production Research*, 35, 313-326.

Mosier, C.T. and N.S. Collins, 1992, The development and preliminary analysis of an adaptive group scheduling heuristic, *Proceedings of the National DSI San Francisco Meeting*.

Mosier, C.T., D.A. Elvers, and D. Kelly, 1984, Analysis of group technology scheduling heuristics, *International Journal of Production Research*, 22, 857-875.

Mosier, C.T. and L. Taube, 1985, The facets of group technology and their impacts on implementation - A state-of-the-art survey, *OMEGA-International Journal of Management Sciences*, 13, 381-391.

Nakamura, N., T. Yoshida, and K. Hitomi, 1978, Group production scheduling for minimum total tardiness, *AIIE Transactions*, 10, 157-162.

Ozden, M., P.J. Egbelu, and A.V. Iyer, 1985, Job scheduling in a group technology environment for a single facility, *Computers and Industrial Engineering*, 9, 67-72.

Ruben, R.A., C.T. Mosier, and F. Mahmoodi, 1993, A comprehensive analysis of group scheduling heuristics in a job shop cell, *International Journal of Production Research*, 31, 1343-1369.

Russell, G.R. and P.R. Philipoom, 1991, Sequencing rules and due date setting procedures in flow line cells with family setups, *Journal of Operations Management*, 10, 524-545.

Shambu, G., N.C. Suresh, and C.C. Pegels, 1996, Performance evaluation of cellular manufacturing systems: A taxonomy and review of research, *International Journal of Operations and Production Management*, 16, 81-103.

Skorin-Kapov, J. and A.J. Vakharia, 1993, Scheduling a flow line manufacturing cell: A tabu search approach, *International Journal of Production Research*, 31, 1721-1734.

Suresh, N.C., 1992, Partitioning work centers for group technology: Analytical extensions and shop-level simulation investigation, *Decision Sciences*, 23, 267-290.

Suresh, N.C. and J.R. Meredith, 1985, Achieving factory automation through group technology principles, *Journal of Operations Management*, 5, 151-167.

Suresh, N.C. and J.R. Meredith, 1994, Coping with loss of pooling synergy in cellular manufacturing systems, *Management Science*, 40, 466-483.

Vaithianathan, R. and K.L. McRoberts, 1982, On scheduling in a GT environment, *Journal of Manufacturing Systems*, 1, 149-155.

Vakharia, A.J. and Y. Chang, 1990, A simulated annealing approach to scheduling a manufacturing cell, *Naval Research Logistics*, 37, 559-577.

Wemmerlöv, U. 1992, Fundamental insights into part family scheduling: The single machine case, *Decision Sciences*, 23, 565-595.

Wemmerlöv, U. and N.L. Hyer, 1989, Cellular manufacturing in the U.S. industry: A survey of users, *International Journal of Production Research*, 27, 1511-1530.

Wemmerlöv, U. and A.J. Vakharia, 1991, Job and family scheduling of a flow-line manufacturing cell: A simulation study, *IIE Trans*actions, 23, 383-393.

Wirth, G.T., F. Mahmoodi, and C.T. Mosier, 1993, An investigation of scheduling policies in a dual-constrained manufacturing cell, *Decision Sciences*, 24, 761-788.

AUTHORS' BIOGRAPHY

Farzad Mahmoodi is Associate Professor of Operations & Production Management, and Director of Undergraduate Business Programs at Clarkson University. He is also an adjunct Associate Professor of Mechanical and Aeronautical Engineering at Clarkson. He received his B.S. in 1982, M.S. in 1983, and Ph.D. in Industrial Engineering in 1989 from the University of Minnesota. He has published many articles in journals such as *Decision Sciences, International Journal of Production Research, Journal of Operations Management, Computers and Industrial Engineering, International Journal of Computer Integrated Manufacturing, International Journal of Production Planning and Control, The Logistics and Transportation Review, Quality Progress,* and in a variety of books and conference proceedings. He consulted for firms in U.S., Canada, and the Middle East. His research interests are in scheduling of cellular and flexible manufacturing systems, shop order release and due-date assignment, systems modeling and simulation, and quality management. He was the recipient of the 1995 Clarkson University John W. Graham, Jr. Faculty Research Award, the 1995 School of Business Faculty Leadership Award, the 1994 Tau Delta Kappa Teacher's Excellence Award, and 1990 Clarkson University Computer Curriculum Award.

Charles T. Mosier (please see biography provided at the end of chapter B2).

F4

Operation and Control of Cellular Systems at Avon Lomalinda, Puerto Rico

G. A. Süer

1. INTRODUCTION

This chapter briefly discusses various projects undertaken in Avon Lomalinda Inc., a jewelry manufacturing company located in San Sebastian, Puerto Rico from March 1990 to December 1995. In particular, we focus on cell scheduling, loading, and knowledge-based loading projects. During these six years, the company went through several changes in administration, products, processes, and manufacturing facilities. Often, various projects were carried out under different conditions in terms of number of products, product types, number of manufacturing cells, automation level and the emphasis by the top management. Despite the continuous changes in the company priorities, the top management always supported our efforts and allowed us to push the limits. The issues, concerns and the proposed solutions are described under the original circumstances when they were addressed.

The projects involved several people from the company such as product development engineers, industrial engineers, planners, material analysts, master schedulers, cell supervisors, business unit managers and the plant managers. In some projects, weekly training sessions were held where the real cases from the company were studied. The effect of the training sessions was evaluated by giving regular quizzes and exams. On the other hand, some of the complex problems were assigned to graduate students as master thesis topics, and five master theses were completed.

The work in this chapter is limited to the labor-intensive manufacturing cells. Labor-intensive cells are characterized by the presence of light weight, inexpensive and usually small machines and equipment, and often simple tools. In such a case, machine utilization and investment costs are of less significance compared to machine-intensive cells, in which usually expensive, big and heavy machines and equipment are used. Machines cannot be easily relocated to accommodate variations in product mix and volumes, and the output rate is determined by the bottleneck operations. But in labor-intensive cells, it is feasible to duplicate machines to maximize the production rates. The key issues in this case become the availability and the allocation of manpower among different cells and products. As a result, the output rate in labor-intensive manufacturing cells is usually determined by the available manpower.

2. DESCRIPTION OF THE COMPANY

A brief explanation is given about the company and its operations. However, the reader is reminded that some of the information provided here may not reflect the current situation in the company today.

2.1. Manufacturing Plants.

The general manager of Avon Puerto Rico Operations is in charge of three plants; Lomalinda (Puerto Rico), Mirabella (Puerto Rico) and Marbella (Dominican Republic). Each plant has a plant manager and they all report to the general manager. The Mirabella plant is the first Avon plant opened in the Island (1976). The manufacturing takes place in a facility rented from the government. At times this brings limitations to the efficient use of space and other resources. Having experienced difficulties with Mirabella, the newer plant Lomalinda has been designed by Avon to meet their specific needs. It opened its doors in 1982. The Marbella plant is the most recent one added to the group to take the advantage of much lower labor rates in the neighboring Dominican Republic. Usually, it works as a subcontractor for both Lomalinda and Mirabella plants in Puerto Rico. The general manager reports to Avon USA VP Manufacturing who is also in charge of other Avon products besides jewelry. The product design, marketing and purchasing functions are handled by the jewelry group located in the Home Office in New York City. The work discussed in this chapter is about the projects undertaken in Avon Lomalinda plant.

2.2. Products.

Avon Lomalinda produces fashion jewelry products. The products are either gold or palladium plated. Recently, silver products were also introduced. The products that are usually assigned to Avon Lomalinda are earrings (casting, stamping, plastics and porcelain), rings, bracelets and chains. Due to the nature of the fashion business, the life cycle of the products is very short (3-6 months), product diversity is considerably high (over 100 products/week) and the demand for many products is highly uncertain.

2.3. Manufacturing Strategy.

The company went through conversion from process layout to cellular layout in 1988 through the help of a New York-based consulting firm. Before the conversion, the company was operating truly in a process layout environment. Similar operations were grouped into departments and parts had to be handled among departments leading to high inventory levels and long lead times. Switching to cellular layout improved the results as the lead times were reduced and the inventory levels dropped. Only the plating operation could not be integrated into manufacturing cells. Four basic product families were identified and the number of cells needed for each family were roughly estimated. As a result, the company is organized around nearly self-sufficient four business units (focused units). Each business unit has a manager who is in charge of various functions such as product development, material

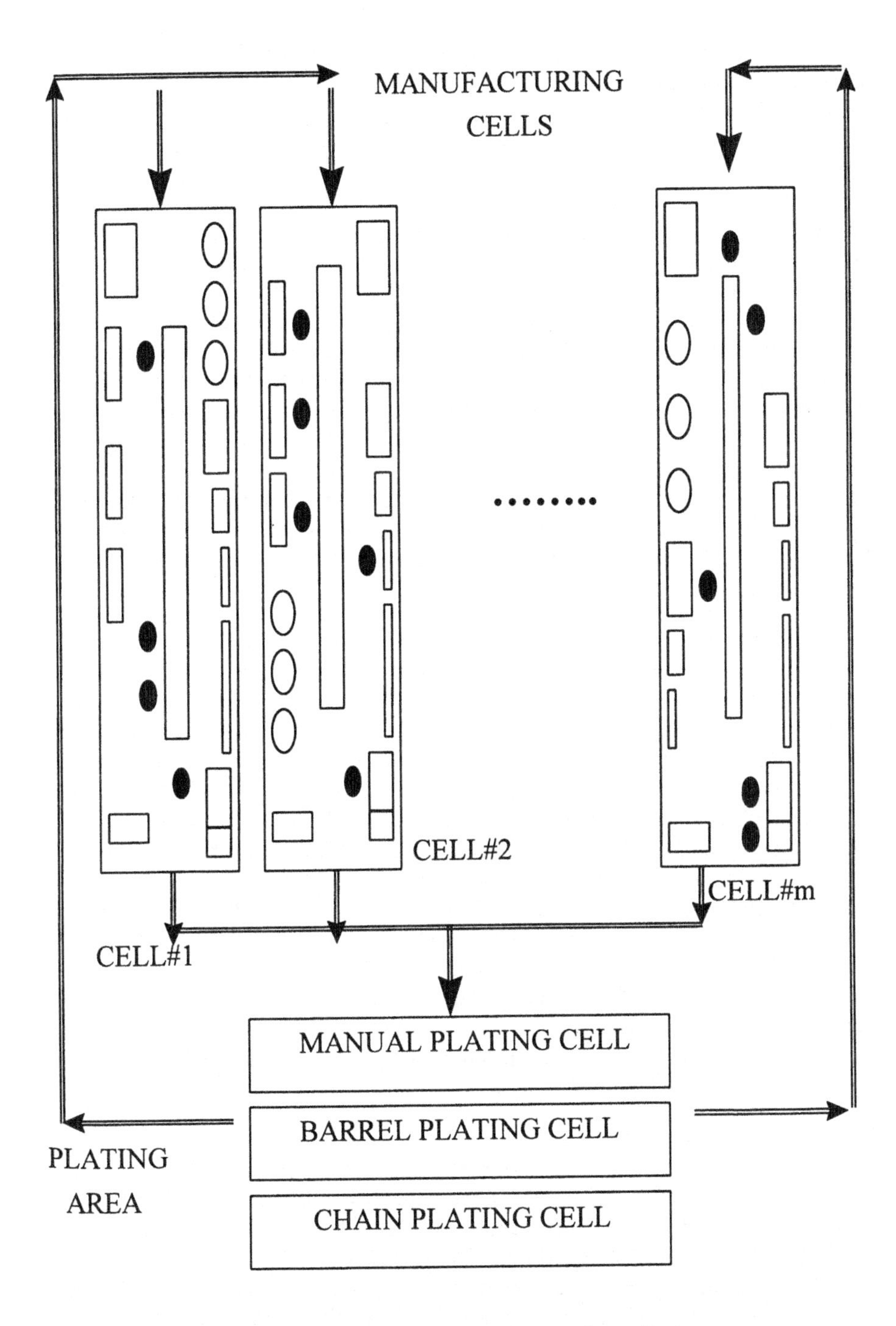

FIGURE 1. Cellular Structure at Avon Lomalinda.

analysis, manufacturing, quality, maintenance, master scheduling and detailed scheduling. Some functions such as human resources, accounting and finance remain centralized.

2.4. Product Development.

New products are conceptually developed by the designers in Home Office. They interact with the marketing people to learn the most recent trends in the fashion jewelry market. The manufacturability of the new product is evaluated by the product engineers in the jewelry group in Home Office. Often, there is communication between product engineers in Home Office and Puerto Rico to finalize the product development. Once the design is finalized, product costing is performed by the engineers in the plant. Home Office evaluates the cost estimate and final decision is made as to whether the product will be manufactured or not.

2.5. Marketing.

Puerto Rico operations is the only source for all the jewelry marketed in the United States by Avon. The company uses famous "Avon Ladies" as its major sales force. Marketing people continuously gather sales data to revise projections as often as 3 times a week.

2.6. Operations.

The operations are mostly labor intensive, and the company is choosing low cost, light weight equipment and machinery to increase the flexibility of the manufacturing system. The crew size in a manufacturing cell varies between 20 and 30 employees depending on the desired production rate. Only a very small percentage of products are completely processed within a cell (5-10% of the products). For the majority of products, there is inter-cell transfer involved. Some of the operations are performed in the manufacturing cell. Then, the parts leave the cell to be plated in one of the plating cells. Later, the parts return to the same manufacturing cell for the remaining operations to be completed. Each product in this category requires at least one of the three types of plating; manual, barrel or chain, depending on the processing requirements.

The worker mobility within and between the cells is feasible due to simplicity of the operations. The number of operations for the majority of the products varies from 6 to 10. The operation times vary from 5 to 20 seconds for almost 90% of the operations.

2.7. Manufacturing Cells.

A cell consists of several workbenches that can accommodate up to 30 employees, simple machines and equipment. The total number of cells varied between 15-17 during the course of these projects. The flexibility within a business unit was very high since the machines could easily be relocated from one cell to another. Indeed, the machines could even be moved to other business units on a temporary basis. The pictorial representation of the manufacturing system is shown in Figure 1.

3. OVERVIEW OF THE PROJECTS UNDERTAKEN

A number of different projects were undertaken during our involvement with the company:

- Cell Scheduling
- Cell Loading
- Knowledge-Based Cell Loading
- Cellular Design

Since schedules for each day are quite likely to be different, the execution heavily depends on how closely they are implemented every day. Undoubtedly, this daily challenge is a very demanding task for planners, schedulers, supervisors and employees. To develop a simple scheduling system requires that jobs start and finish at regular intervals on a periodic basis. This periodic behavior eliminates the need to prepare a different schedule everyday. Therefore, a fixed-time interval based synchronization approach has been proposed to tackle the cell scheduling task in the company.

The objective of the cell loading project was to develop finite-capacity cell loads and use it as the Master Production Schedule (MPS) to be fed into the MRP process. This approach can be defined as Schedule-Based Materials Requirements Planning (SB-MRP) and is substantially different than the classical MRP approach. What was discovered during the cell scheduling project was that the success of detailed scheduling heavily relied on the Master Production Schedule prepared by the planners. As part of this project, several cell loading rules and algorithms were proposed and a customized software was developed to run any desired combination.

Having a customized software which is capable of generating 192 MPS(s) increased the possibility of finding a better MPS. However, this also created another problem; selection of the best MPS(s) among all alternatives. The need to select the best MPS led to the third project which involved the development of a knowledge-based system.

Finally, the last important project was to analyze the design of manufacturing cells. It became very clear that Master Production Scheduling is limited to the success of cellular design. If focusing has not been done correctly, capacity analysis has not been done right and the number of cells have not been determined accurately then the software and/or planners will have their hands tied in terms of preparing successful Master Production Schedules.

The chapter focuses on cell scheduling, loading and knowledge-based cell loading projects. The cellular design is briefly mentioned at the very end and finally general remarks are made in terms of successful implementation.

4. THE CELL SCHEDULING PROJECT

As mentioned before, the first task tackled was cell scheduling. Cell scheduling deals with the scheduling of operations in a cell once a product has been assigned. Worker mobility, cell size, automation level, intra-cell and inter-cell material transfers, lot-size, product mix, number of machines and equipment, number of shifts and overtime considerations impact the design and the effectiveness of a cell scheduling system. Since the objective was to design a simple and effective scheduling system, a synchronization approach would be ideal to address the problem. Synchronization can be defined as a systematic way that provides a perfect flow of material through the production system by making it available for the right resource at the right time using a periodic approach with the objective of simplifying the scheduling task and minimizing the work-in-process inventory and the flow time.

There are several methods that can be included in the category of efforts towards achieving synchronous manufacturing; assembly line balancing, Period Batch Control (PBC) technique and the variant in Kumera Oy, the Finnish firm, as discussed in chapter F2, kanban system, Optimized Production Timetable, etc. The Uniform Time Bucket Approach (see related references in the end) may be included in this category. This method aims at synchronization in multi-stage flow shop systems. This approach has its origins in the assembly line balancing concept but it is extended beyond the assembly lines towards the inclusion of the entire manufacturing system.

4.1. Motivation for Synchronization

This section discusses the need for implementing a synchronization method in the company. The reasons can be grouped in three categories:

Inter-cell Material Transfers: All of the operations of a product cannot be completed within a single manufacturing cell. Therefore, the parts leave the cell to continue further processing in the plating area and then return to the same manufacturing cell as shown in Figure 1. The synchronization of the material flow from a cell to the plating area and also from plating back to the original cell becomes very critical for maximizing the utilization of resources. Furthermore, the plating area is common to all of the manufacturing cells. An unsynchronized material flow even from a single manufacturing cell can easily disrupt the synchronization between other cells and the plating area. Therefore, a true synchronization can only be achieved by controlling the material transfer to and from each cell simultaneously.

Intra-cell Material Transfers: The flow of parts is not continuous within a manufacturing cell. Both successive transfer (in batches) and overlapping transfer (unit transfer size) methods are used between operations based on the characteristics of the operation. This causes an interruption in the flow of the materials within a cell itself. This is another reason why synchronization is a valid approach.

Nonlinearity in Production: In each manufacturing cell, most of the fabrication operations are performed early in the week, usually on Mondays, Tuesdays and Wednesdays as shown in Figure 2. The plating area receives and returns most of the parts between Wednesday noon to Friday noon. The finishing and packing operations are usually performed on Thursdays and Fridays. This creates an excess amount of work-in-process inventory in the manufacturing cells and also adversely affects the load distribution in the plating area, which is underutilized early in the week with shorter delivery times and then overused during the rest of the week with increased chaos, delivery times, overtime hours and shifts. This made it necessary to design a synchronization method that will force the system to produce in a periodic manner so that linearity (uniform production) is achieved throughout the week, plating capacity is uniformly used, overtime and additional shift work is avoided and work-in-process inventory is reduced.

Operation	Week Days				
	Monday	Tuesday	Wednes.	Thursday	Friday
Fabrication					
Plating					
Finishing and Packing					

FIGURE 2. Typical work schedule in the company.

4.2. The Methodology Used

In this project, a modified version of the Uniform Time Bucket Approach was used. The Time Bucket Approach is a means of achieving a synchronization between the finished products and their lower level subassemblies and parts. It uses the "right part", the "right quantity" and the "right time" principle. Operations are grouped into (s) stages and planning horizon is divided into buckets. If a batch is scheduled to be completed at stage (s) at the end of bucket (t), it is processed at stage (s-1) in bucket (t-1), at stage (s-2) in bucket (t-2) and so on. The approach can be summarized as a variable quantity but fixed time interval-based synchronization. Another way of defining this technique is "controlled input/output" system. This technique helps to convert a dynamic and stochastic scheduling problem into a static and deterministic one which makes it easier to implement in a real setting.

The processing times of all operations have been considered and the maximum one (3.5 hours) established the lower bound for the length of the bucket.

Available hours in a day had to be considered in making the final decision so that the number of buckets in a day would be a rounded number. As a result, it was determined that it is the best to set the length of Time Bucket to 4 hours and have two buckets in a day. The morning bucket (AM) was between 7:00 am-11:00 am and the second bucket (PM) was between 11:30 am-3:30 pm.

The number of stages has been determined as five for most of the products considering both transfer sizes between the operations and whether the operations have been performed in or outside the cell. There are some operations with unit transfer size while others require batches. The use of the batches is required because some machines/equipment can process the parts economically only by batches. For this reason, it is necessary to accumulate the parts and form the batches before they are processed on these machines and equipment. The consecutive operations that have a unit transfer size are grouped into the same stage and the operations that require batches form separate stages. The processing requirements for a representative product and the definition of stages are summarized in Table 1.

A schedule prepared based on the synchronization approach includes 1.5 week period as shown in Figure 3. This covers the current period and the later half of the previous week. To finish a batch on Monday Morning (bucket #1) of the current week, the operations need to start on Thursday Morning of the previous week. A batch is completed at a stage in 4 hours and then transferred to the next stage. The last batch of the week, batch 10, needs to be released to production at bucket #6 (Wednesday Afternoon). As one can easily observe, the leadtime for any batch is 5 buckets or 2.5 days. This schedule is prepared for each cell independently. However, as claimed earlier, the start and the completion times of the batches remain the same for all cells and weeks thus simplifying the scheduling task greatly. What varies in the schedules from one week to another and from one cell to the next is what products those batches represent.

TABLE 1. Processing Requirements and Stages for a Sample Product.

Operation Number	Opn. Description	Transfer Size	Processing Time (min)	In/Out Cell	Stage
1	Casting	1	0.25	In	1
2	Degating	1	0.10	In	1
3	Cleaning	1	0.30	In	1
4	Deburring	1	0.25	In	1
5	Putting Sleeve	1	0.15	In	1
6	Tumbling	Batch	210.00	In	2
7	Racking	1	0.40	In	3
8	Plating	Batch	180.00	Out	4
9	Unracking	1	0.25	In	5
10	Carding	1	0.30	In	5
11	Packing	1	0.15	In	5

Stage	Previous Week				Current Week									
	Days													
	Thurs.		Friday		Mon.		Tues.		Wed.		Thurs.		Friday	
	AM	PM	AM	PM	AM	PM	AM	PM	AM	PM	AM	PM	AM	PM
	Bucket Number													
	7	8	9	10	1	2	3	4	5	6	7	8	9	10
	Batch Number													
1	1	2	3	4	5	6	7	8	9	10				
2		1	2	3	4	5	6	7	8	9	10			
3			1	2	3	4	5	6	7	8	9	10		
4				1	2	3	4	5	6	7	8	9	10	
5					1	2	3	4	5	6	7	8	9	10

FIGURE 3. Synchronized manufacturing in the company.

4.3. Manpower Assignment

Another variation of this problem from a typical cellular manufacturing system is that the mode of the scheduling in each cell is in the form of line balancing by manpower assignment rather than flow shop or job shop. The number of workers in a cell is greater than the number of operations. Therefore, it is very natural to expect parallel stations (operators) for some operations. A mathematical programming model and an algorithm to determine the optimal assignment of available manpower to different operations were developed. The mathematical model maximizes the production rate as given in equation (1). Equation (2) guarantees that there will be enough operators assigned to each operation to achieve the production rate. The limitations in terms of number of resources available for each operation (machines, equipment, skilled operators, etc.) are enforced by equation (3). Finally, the total number of operators available in a cell cannot be exceeded as given by equation (4).

Objective Function: $\text{Max } Z = P$ (1)

Subject to:

$$[\,(X_i) * (1/ST_i)\,] - P \geq 0 \qquad i=1,2,..,n \qquad (2)$$

$$X_i \leq U_i \qquad i=1,2,..,n \qquad (3)$$

$$\sum_{i=1}^{n} (X_i) \leq M \qquad (4)$$

X_i integer and positive for all i; P, positive

where, P = production rate; X_i = number of operators assigned to operation I; ST_i = standard time for operation I; U_i = upper bound on the number of operators for operation i ; M= total number of operators available in the cell; n = number of operations.

4.4. Implementation Results

The implementation efforts of the synchronization project can be grouped in three phases:

Pilot runs controlled by industrial engineers: In this phase of the pilot, a product was selected and a weekly detailed schedule was prepared by the industrial engineers along with the required number of operators per operation. A short seminar was given to the employees as to how this system works and what they are expected to do. At end of the week, the information gathered was analyzed and it was determined that the timing accuracy was 76% (percent of times parts were transferred by the end of time buckets) and the production schedule performance was 90%. Having encouraged by these results, a second pilot was run by using another product. In this case, the timing performance improved to 92% and schedule performance to 96%.

Pilot runs controlled by supervisors: The next trial was to see if the cell supervisors could run the pilot themselves. The cell supervisors and other related staff of the selected business unit such as material analysts, master scheduler and the unit manager went through the training sessions prepared using the actual product, cell and demand information. The synchronization project was simulated using lego-type toys in the workshops organized. Having demonstrated that everybody is ready for the second round of pilot runs, this phase of the implementation started and lasted approximately for a period of three months. Although there were moments when part delivery was not perfect, the company internally evaluated the results and decided to go ahead with the factory-wide implementation.

Factory-wide implementation: The implementation responsibility was transferred to the company staff. The training sessions and workshops were repeated for each business unit. A 15-month plan was devised to implement synchronization in a new cell every three weeks. However, the factory-wide implementation required more organizational changes to be made. The responsibilities of the involved parties had to be studied and re-defined as proposed in Figure 4. The objective was to make sure that everybody has the same understanding and the objectives. Most of the cells eventually moved to synchronization or at least experimented with it. However, it took much longer than 15 months as originally planned.

4.5. Comments

The synchronization project was not as simple as suggested in this chapter. Some products had subassemblies that required processing in different manufacturing and plating cells. Furthermore, some complex products (gift sets) indeed consisted of quite distinct products that needed to be processed in different business units. These products required extra effort in planning and execution not to disrupt the synchronization.

The number of different products that a cell can run in a week is limited to the number of buckets. Therefore, in this case it was limited to 10. However, there were times when more than 10 products had to be assigned to a cell due to Master Schedule requirements. In that case, a cell was partitioned into two or more mini-cells and each mini-cell was assigned some of those products and expected to follow the synchronization approach. One of the supervisors also tried 2-hour time bucket as well. During the peak demand seasons, this approach was adapted to two or even three-shift work schedule with minor modifications.

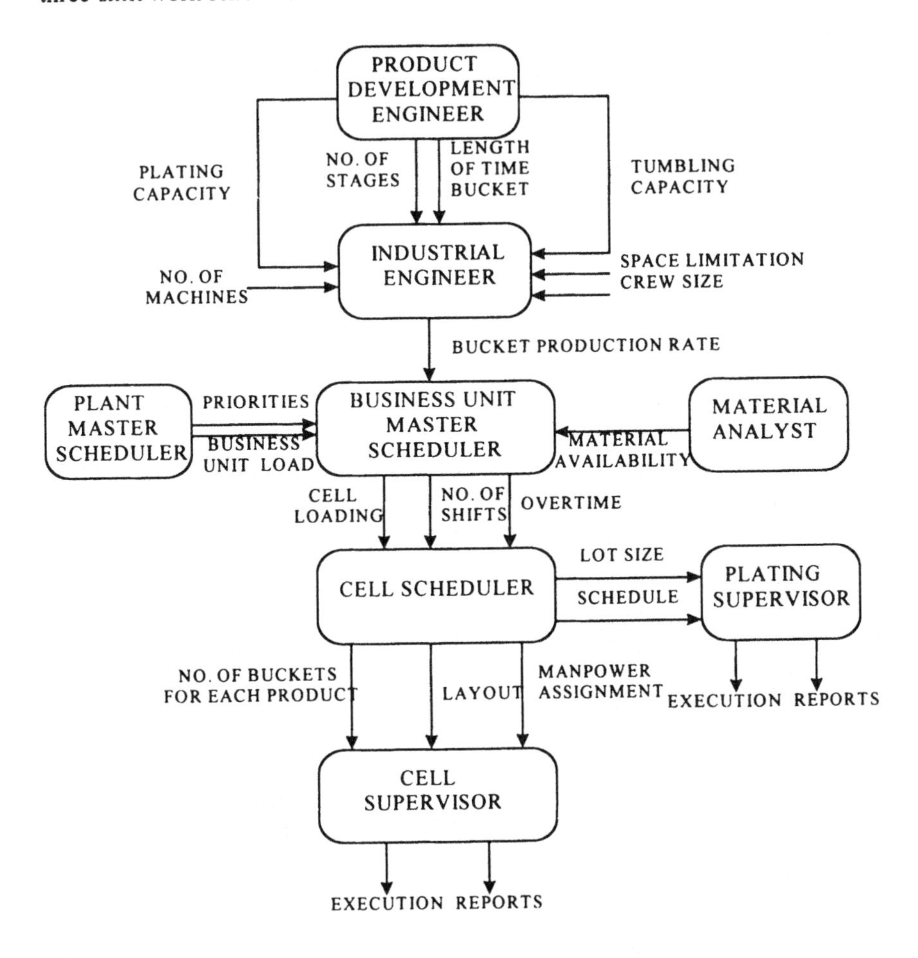

FIGURE 4. A Framework for information flow.

Although simple, synchronization requires a lot of discipline. This was a much tighter system than what many employees and supervisors were used to. I believe that for that reason some employees and supervisors were never fully committed to the project. Each supervisor was after improving their own performance by pushing the products through limited resources in the plating area regardless whether that was the best for the company at that point. Unfortunately, planning office failed to communicate the priorities effectively and plating continued to be the competing grounds. Supervisors were too nervous to really comprehend that if uniform production could be achieved, this would comfort plating cells (by generating uniform loads) and in return, their lives would have been much smoother as well.

The most important conclusion reached was that all other supporting units (purchasing, receiving, inspection, warehouse, engineering, etc.) needed to be involved in this project as well. As once one supervisor said, the manufacturing area should be the last to implement the synchronization approach. Despite the difficulties, the majority of the supervisors adapted synchronization with varying degrees and obtained satisfactory results. (See related references for a discussion of other relevant factors for successful implementation of synchronization approach).

5. THE CELL LOADING PROJECT

In this section, the cell loading project carried out in the same company is discussed. Despite the fact that numerous articles have been published in cellular manufacturing, cell loading has been rarely addressed in the literature. The cell loading problem exists in any multi-cell environment regardless whether it is machine-intensive or labor-intensive cellular environment. Cell loading involves determining to which cell(s), among the feasible ones, products should be assigned, how many units should be produced and in what order they should be produced.

5.1. Motivation For The Study

During the synchronization project it became very clear that the success of cell scheduling very much depended on what products have been assigned to each cell, i.e. cell loading. We realized that some of the difficulties with synchronization occurred because the planners simply ignored that the cells were running products in a synchronized fashion. There was no tool available in the company to do cell loading efficiently. It heavily depended on the experience of the planners. The decision rules used by the planners were not clear and moreover not consistent with each others. The products were assigned to cells based on the sales requirements and cell feasibility restrictions. Even though there were attempts to consider the impact of the load on some key machines by performing capacity planning, when the resources were scarce and the job had to be finished, it was usually the cell supervisor's responsibility to create the "miracles". The irony is that even though there were employees with educational background in planning and scheduling, the supervisors who lacked formal training would end up preparing schedules. The MRP software in use was not flexible enough to handle the task efficiently either. The Master Production Scheduling module limited the assignment of a product to only a single cell and the cell would remain the same unless database changes are made. In other words, it didn't allow a product to be assigned to different cells in different periods based on the load. Furthermore, you couldn't assign a product to more than one cell either (i.e., no lot-splitting was allowed). These significant limitations crippled the use of MPS module for any serious purpose. The biggest deficiency in the current system was that the impact of cell loads on the plating cells was never considered during the planning process. This quite often led to unbalances in the loads of plating cells. While one plating cell was quite underutilized, the others might have been overloaded sometimes even beyond third shift. This adversely affected the synchronization efforts in the manufacturing cells as well since plating cells have failed to deliver the parts back to manufacturing cells in four hours. Therefore, there was a need to develop a cell loading software which will have look-ahead feature and generate finite-capacity Master Schedules taking synchronization into account.

5.2. The Methodology Used

The methodology used in the project consisted of selection rules and loading algorithms. No attempt was made to develop mathematical models at the time due to obvious computational requirements. Developing fast procedures was especially significant considering that this was an industry-cosponsored project and real problems had to be addressed and solved without making any simplifying assumptions. However, the author has been focusing in mathematical modeling in recent years and has done some preliminary work in this area.

The cells in the company fall into the category of "connected" cells. The explanation to this lies in the description of the processes in the company. As mentioned before, loading the manufacturing cells is critical since most of the products visit the plating area (one of three plating cells) and then return to the original cell. There are two possible approaches to the loading problem; (1) ignore plating area and treat the manufacturing cells as "independent cells" and (2) consider the impact of cell loads on the plating area during the process of planning. The former approach is simple but may not be efficient and adversely affect cell synchronization as previously discussed. On the other hand, the latter approach is more complex in terms of planning and developing software but more accurate and should be pursued. Another factor to decide is whether to allow lot-splitting or not. Eventually four different scenarios were created by; 1) independent cells-no lot-splitting, 2) independent cells-lot splitting allowed, 3) connected cells-no lot-splitting and 4) connected cells-lot splitting allowed.

TABLE 2. Product And Cell Priority Rule Combinations.

Product Priority Rules	Cell Priority Rules
1. PP/EDD/NFC/Min/CL/Min	25. CP/CL/Min/EDD/NFC/Min
2. PP/EDD/NFC/Min/CL/Max	26. CP/CL/Max/EDD/NFC/Min
3. PP/EDD/NFC/Min/NFP/Min	27. CP/NFP/Min/EDD/NFC/Min
4. PP/EDD/NFC/Min/NFP/Max	28. CP/NFP/Max/EDD/NFC/Min
5. PP/EDD/NFC/Min/PM/Min	29. CP/PM/Min/EDD/NFC/Min
6. PP/EDD/NFC/Min/PM/Max	30. CP/PM/Max/EDD/NFC/Min
7. PP/EDD/NFC/Max/CL/Min	31. CP/CL/Min/EDD/NFC/Max
8. PP/EDD/NFC/Max/CL/Max	32. CP/CL/Max/EDD/NFC/Max
9. PP/EDD/NFC/Max/NFP/Min	33. CP/NFP/Min/EDD/NFC/Max
10. PP/EDD/NFC/Max/NFP/Max	34. CP/NFP/Max/EDD/NFC/Max
11. PP/EDD/NFC/Max/PM/Min	35. CP/PM/Min/EDD/NFC/Max
12. PP/EDD/NFC/Max/PM/Max	36. CP/PM/Max/EDD/NFC/Max
13. PP/EDD/NCR/Min/CL/Min	37. CP/CL/Min/EDD/NCR/Min
14. PP/EDD/NCR/Min/CL/Max	38. CP/CL/Max/EDD/NCR/Min
15. PP/EDD/NCR/Min/NFP/Min	39. CP/NFP/Min/EDD/NCR/Min
16. PP/EDD/NCR/Min/NFP/Max	40. CP/NFP/Max/EDD/NCR/Min
17. PP/EDD/NCR/Min/PM/Min	41. CP/PM/Min/EDD/NCR/Min
18. PP/EDD/NCR/Min/PM/Max	42. CP/PM/Max/EDD/NCR/Min
19. PP/EDD/NCR/Max/CL/Min	43. CP/CL/Min/EDD/NCR/Max
20. PP/EDD/NCR/Max/CL/Max	44. CP/CL/Max/EDD/NCR/Max
21. PP/EDD/NCR/Max/NFP/Min	45. CP/NFP/Min/EDD/NCR/Max
22. PP/EDD/NCR/Max/NFP/Max	46. CP/NFP/Max/EDD/NCR/Max
23. PP/EDD/NCR/Max/PM/Min	47. CP/PM/Min/EDD/NCR/Max
24. PP/EDD/NCR/Max/PM/Max	48. CP/PM/Max/EDD/NCR/Max

Two search strategies were used, namely product priority (PP) and cell priority (CP). If a product is selected first and then a cell is chosen, that strategy is called product priority. As one can expect, the cell priority uses exactly the reverse search strategy. The primary product rule is Earliest Due Date (EDD). In case there are ties in terms of due dates, secondary product rules were proposed; Number of Feasible Cells (NFC), Number of Cells Required (NCR). As to the cell rules, three were proposed; cell load (CL), number of feasible products (NFP) and product mix (PM). The selection of the product or the cell could be based on the one with maximum value (MAX) or minimum value (MIN). The rules suggested were combined in all possible ways thus producing 48 combinations of which 24 were product priority and the remaining 24 were cell priority as listed in Table 2. When the "connected cells" approach was used, the rule to select the plating cell was Minimum Load (ML). These rules have been used in a dynamic fashion where the values have been updated as soon an assignment was made.

For example, the combination PP/EDD/NFC/Max/NFP/Min means that this application is a product priority case where first a product is selected based on the EDD rule. If there are several products with the same due date, then the product with maximum number of feasible cells is given the highest priority. The next step is to select the feasible cell to which this product will be assigned. In this case, the product is assigned to the feasible cell that can process the minimum number of products at the time of selection.

5.3. Algorithms Developed

The algorithms for four different scenarios can roughly be described as follows:

Algorithm 1: Independent Cells- No Lot-Splitting: 1) Sort jobs by EDD; 2) Select a job (If there is a tie, use the selected secondary product rule); 3) Select a feasible cell using the selected cell rule; 4) Assign the job to the cell; 5) Go to step 2 (until all jobs are assigned).

Algorithm 2: Independent Cells- Lot-Splitting Allowed: 1) Sort jobs by EDD; 2) Select a job (If there is a tie, use the selected secondary product rule); 3) Select a feasible cell using the selected cell rule; 4) Assign the job to the cell. If the job is completed by its due date go to step 6; Otherwise continue with the next step; 5) Search for another feasible cell for the same job so that due date can be met by lot-splitting as needed; 6) Go to step 2 (Until all jobs are completed).

Algorithm 3: Connected Cells- No Lot-Splitting: 1) Sort jobs by EDD; 2) Divide the sorted jobs into three sets based on their plating requirements (manual, barrel, chain); 3) Determine the plating cell with minimum load; 4) Select the set of jobs that needs to be processed on the plating cell with minimum load; 5) Select a job from that set (If there is a tie, use the selected secondary product rule); 6) Select a feasible manufacturing cell to run the job by using the selected cell rule; 7) Assign the job to the cell; 8) Go to step 2 (until all jobs are assigned).

Algorithm 4: Connected Cells- Lot-Splitting Allowed: 1) Sort jobs by EDD; 2) Divide the sorted jobs into three sets based on their plating requirements (manual, barrel, chain); 3) Determine the plating cell with minimum load; 4) Select the set of jobs that needs to be processed on the plating cell with minimum load; 5) Select a job from that set (If there is a tie, use the selected secondary product rule); 6) Select a feasible manufacturing cell to run the job by using the selected cell rule; 7) Assign the job to the cell. If the job is completed by its due date go to step 9. Otherwise continue with the next step; 8) Search for another feasible cell for the same job so that due date can be met by lot-splitting as needed; 9) Go to step 2 (until all jobs are assigned).

The algorithms discussed above consider the product priority approach. The same algorithms had to be modified to consider the cell priority approach as well. However, they are not included in this chapter considering that modifications to the procedures are minor.

The 48 rule combinations can be run under 4 different scenarios thus generating 192 possible results (Master Production Schedules) as summarized in Figure 5.

The inputs to the loading system were demand forecast, product-cell feasibility matrix, work hours/week, number of shifts and finally 4-hour production rates. The four-hour production rates were determined by using the linear programming model described in section 5.3.

An extensive experimentation was carried out by using the data from one business unit where there were 5 subfamilies and 5 cells. Demand forecast for a period of two years was divided into four six-month segments and various comparisons were made. See related references in the end for a discussion of some of these results.

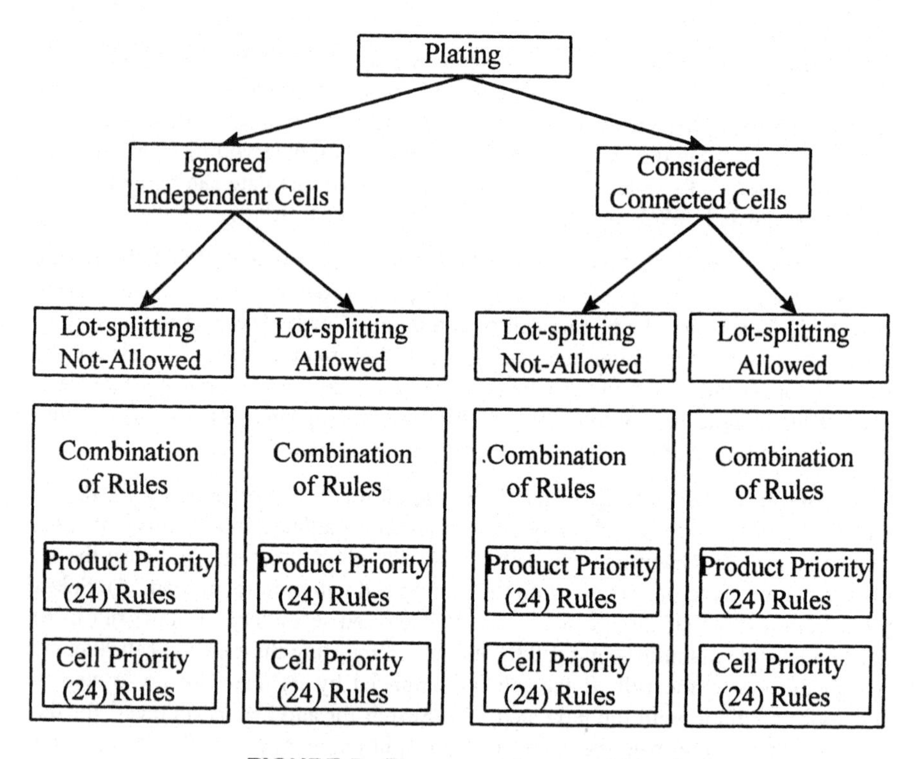

FIGURE 5. Four scenarios considered.

5.4. Implementation Efforts

The programs developed in Turbo Pascal 6.0 were used along with Topaz version 3.0 database manager and objects from Object Professional version 1.1. The programs were run on IBM-compatible machines. It would take approximately 10 minutes to generate summary statistics for all alternative Master Production Schedules. The planners were expected to review the summary statistics and select the best Master Schedule(s). Later, the same programs were run to generate the detailed reports for the selected Master Schedule(s). The cell loads were generated for several weeks based on the 4-hour time buckets. The Master Production Schedule thus obtained indeed established the packing stage schedule (Stage 5) for the synchronized cells. Once the schedule for stage 5 has been determined, schedules for stages 4,3,2 and 1 could be easily determined by offsetting stage (5) schedule with 4-hour time buckets. This helped us to link cell loading and cell scheduling consistently. In other words, integrated, simple, fast and yet efficient loading and scheduling systems were developed.

There are two general ways to execute the Cell Loading Program. The first one is to generate 192 summarized MPS(s) and the second one is to generate the detailed output of a specified combination. The performance measures considered are the number of critical tardy jobs, number of non-critical tardy jobs, total tardiness, maximum tardiness and average cell utilization. A critical job is the one that is due within the next three weeks, whereas a non-critical job is the one that can be late for a reasonable period of time without causing any significant consequence. The cell loading program can handle up to 7 cells and the number of shifts can be defined as 1, 2 or 3. The length of planning horizon can be set to eight, ten or thirteen periods.

5.5. Comments

The company was interested in using the cell loading software for generating an acceptable Master Production Schedule and then input to the MRP process as mentioned earlier. However, there was a concern by the planners in terms of the number of alternatives generated. If the planner wanted to take the full advantage of the software, he needed to run the summary statistics option of the software and then evaluate the results and eliminate the poor Master Schedules. Having reduced the alternatives to a handful good ones, the planner was expected to run the software with a request for a detailed report for the selected MPS(s). A question has been raised if there was a way to further improve the selection of the MPS(s). This led us to the next project discussed in the next section.

6. THE KNOWLEDGE-BASED CELL LOADING PROJECT

The interest in Artificial Intelligence continues to grow among academicians. In recent years, many research results have been published that includes the use of expert systems, neural networks, fuzzy logic and genetic algorithms in manufacturing as well. Unfortunately, the use of these techniques in a real environment has been limited to a few cases only.

6.1. Motivation for the Study

As a result of the concerns raised in terms of the complexity of evaluating alternative MPS(s), the knowledge-based cell loading project was launched. The overall objective was to present only a handful acceptable Master Schedules to the planners for a final selection.

6.2. The Methodology Used

This application was suitable for a Knowledge-Based System (KBS) since the cell loading software could generate 192 alternative MPS(s) and also there were 64 possible permutations of performance measures. Two sources of information have been used in building the knowledge base, the expert knowledge and the experimentation results. The knowledge-base has been developed in M.1 Version 3.0. It provides an interface with dBASE III to access data that has been compiled by other conventional programs and stored as a database file. The input data required for the knowledge-based system are: (1) The MPS set to be evaluated, (2) specification of scenarios to be considered (if not all), 3) the number and the ranking of performance measures, and 4) acceptable values for the performance measures considered. The KBS searches for the MPS(s) that meets the user specified values with respect to the first performance measure. Among the selected MPS(s), another search starts with respect to the second performance measure and so on. At the end, the knowledge-based system presents all master schedules that meet the all of the user requirements. The next step is to quit the KBS and go back to cell loading program again to generate the detailed versions of the selected MPS(s).

If the knowledge-based system cannot find any MPS meeting all of the user requirements, it lowers the acceptable limits to certain default values and continues the search process after informing the user. If it can find master schedules that satisfy the modified values of performance measures and the user accepts the results, the search ends. Otherwise, the KBS recommends to execute the Cell Loading Program by increasing the system size (number of cells, shifts or both) based on the number of tardy jobs and cell utilization figures.

The KBS developed for the selection of the MPS was planned to be incorporated into Schedule-Based MRP Approach, as shown in the Figure 6.

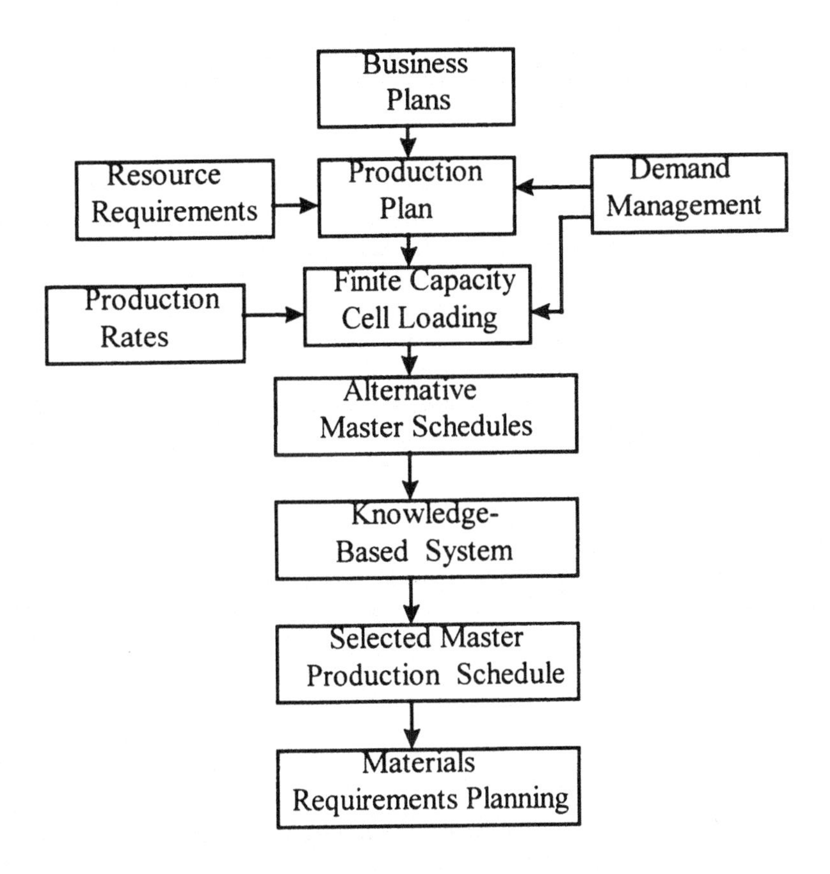

FIGURE 6. The use of KBS in schedule-based MRP.

6.3. Comments

The performance measures number of critical tardy jobs, number of non-critical tardy jobs, maximum tardiness and total tardiness are related to each other. The experimentation carried out showed that the behavior of these performance measures depend on the system size. Toward the completion of this project, the expert interviewed for this project was promoted to the supervisory position in the planning department and turned over master scheduling task to another employee. However, his knowledge about MPS selection process, acceptable ranges and other valuable information has been retained in a knowledge-based system. This justifies one more time that the KBS is a valid approach for manufacturing problems as well.

This project helped us to demonstrate the feasibility of Knowledge-Based Finite Capacity MRP approach. This is just the reverse of classical MRP approach in the sense that a finite capacity master schedule is generated first and then fed into MRP process which is now expected to plan the purchased items to support the MPS. The role of knowledge base was to evaluate the alternative master schedules and select the better ones.

Regrettably, despite the hours of demonstrations and training by our team, and the support of master scheduler and top management, other schedulers remained distant to our efforts and actually never used the software for real scheduling purposes. One of the main reasons for this could have been the insufficient preparation and academic background of schedulers, the unfortunate assignment of master scheduler to a supervisory position and the intimidation due to not knowing much about computers. Apparently, management also failed to make the necessary adjustments in the organization despite their desire to improve operations, willingness to support our efforts by both allocating resources (engineers, planners, etc.) and financial support.

7. THE CELL DESIGN PROJECT

While working on both cell loading and knowledge-based cell loading projects, it became very obvious that cell loading couldn't be considered independent of the cellular configuration. The design of cellular system directly impacts its performance. Therefore, both design and control aspects had to be successfully addressed. Cell scheduling cannot alone solve all problems if cell loading has not been designed well. Similarly, neither cell loading nor cell scheduling can create miracles, if the cellular system has not been designed correctly.

Therefore, there was a need to integrate all these efforts towards an integrated and meaningful solution for the company. An itegrated hierarchy of these of decisions is shown in Figure 7.

8. CONCLUSIONS

Our efforts appear to have actually contributed positively to the company operations. Most managers praised our efforts and encouraged us to continue working in the same direction. Some top managers were claiming that we had already achieved more than what they expected from our projects. It created an opportunity for many issues to be brought up and discussed in different levels of the organization. Some key staff members were motivated, excited and actively participated and contributed to the projects significantly. Unfortunately, the company started having difficulties in terms of new product design, marketing and hence sales which eventually forced us out.

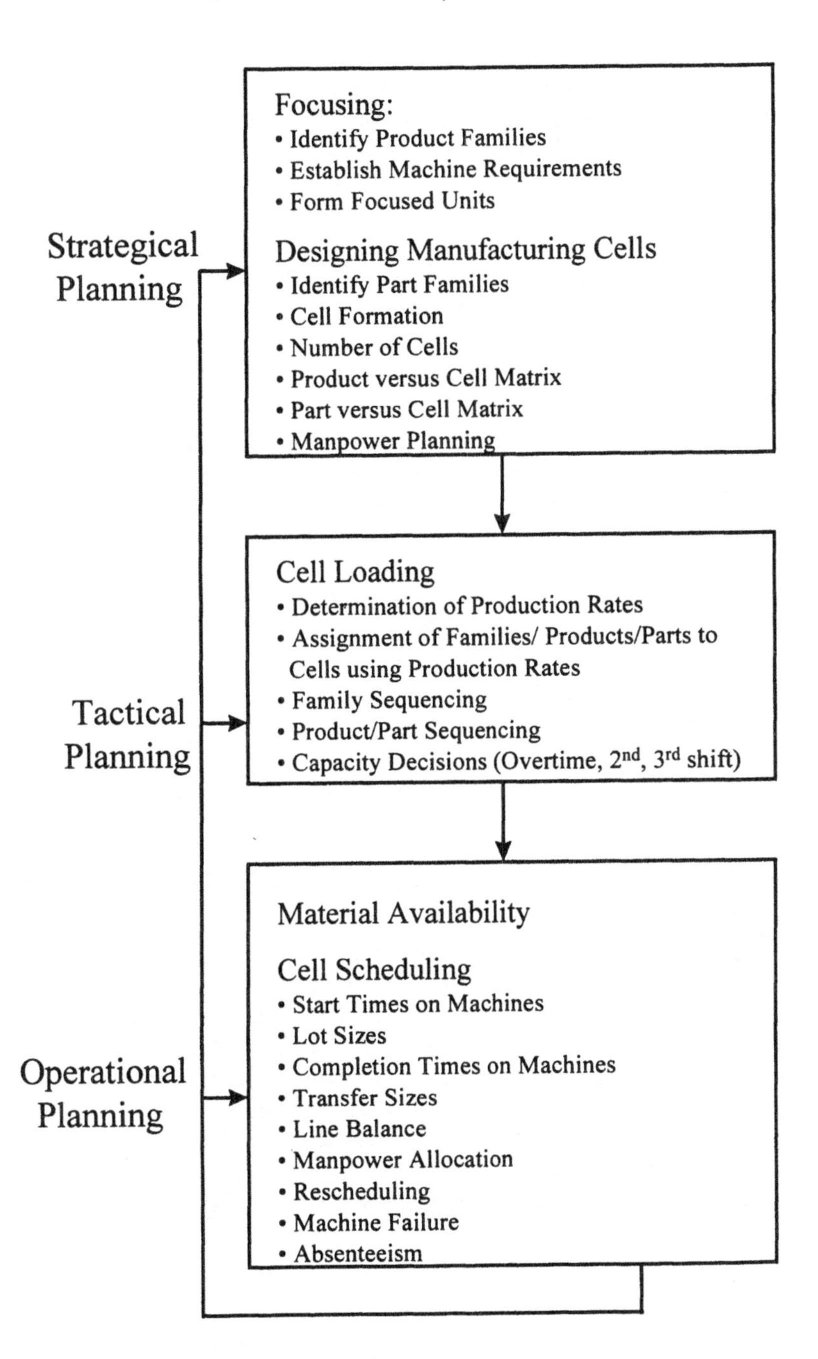

FIGURE 7. Hierarchy of cellular manufacturing decisions.

Despite much progress, the difficulties that the company was experiencing put an end to all our efforts. We got very close but couldn't reach to the ultimate target and satisfaction of linking and implementing all of these projects. On the other hand, this was a great opportunity for us to make observations about both design and control of CM. The company was a "learning factory" and gave us a chance to criticize the work in the literature from implementation feasibility perspective. Some of these thoughts are summarized in the next few paragraphs.

What we have learned is that design and control should go hand-to hand and in fact it is an iterative process. Unfortunately the majority of the literature focuses on the design aspects of cellular manufacturing. Minimizing inter-cell and intra-cell materials transfers are very limited objectives. Without a doubt, machine sharing among cells results in lower investment costs. One can also reduce the handling times and relevant costs by minimizing moves. On the other hand what about the costs associated with complicated planning and scheduling tasks since some machines are now shared? You need more planners, schedulers, and more complicated software (if available) to deal with connected cells as a result of machine sharing. This reality is ignored in total cost calculation by many researchers. I strongly believe that the simplicity of planning and scheduling should be a cellular design target as well.

Another observation is that it is not sufficient to form families based on the processing similarity only. Parts that are likely to form families due to processing similarity need to be further analyzed by considering the similarity of demand characteristics as well (stable, unstable, etc.). It may be wise to keep parts with predictable demand patterns apart from the parts with high uncertainty to reduce the overall system nervousness and avoid rescheduling as much as possible. In other words, an unpredictable-demand part in a cell may ruin the entire schedule for other parts with predictable demand patterns.

It may be almost impossible to design a complete cellular system out of a manufacturing system. There is a need to develop methodologies that will help to design hybrid manufacturing systems with dedicated cells, manufacturing cells and remainder job shop. Forcing every part to belong to a family might lower the quality of the cellular system designed. Furthermore, operation sequences are too important to ignore in family formation. Only a few procedures consider its impact. Finally, demand variability is the case for many products and therefore flexibility in designing manufacturing cells cannot be ignored. The flexibility can be economically achieved by creating overlapping cells among closely related families.

RELATED REFERENCES

Süer, G. A., 1989, *SB-MRP: Integration of Production Scheduling Techniques with Material Requirements Planning Systems*, Ph.D. Dissertation, Wichita State University, Wichita.

Süer, G.A., 1992, An algorithm to find number of parallel stations for optimal cell scheduling, *Proceedings of the 14th Conference for Computers and Industrial Engineering*, Cocoa Beach, Florida, March 9-11.

Süer, G.A. and Gonzalez, W., 1993, Synchronization in manufacturing cells: A case study, *International Journal of Management and Systems*, 9, 3, Sept-Dec.

Süer, G.A. and Dagli, C., Product sequencing in labor-intensive manufacturing cells, (under revision, *Computers and Industrial Engineering*)

Süer, G.A., 1996, Optimal manpower assignment in labor-intensive manufacturing cells, *Proceedings of the 19th International Conference on Computers and Industrial Engineering*, Miami 4-6, 1996.

Süer, G.A., 1996, Minimizing the number of tardy jobs in multi-period cell loading problems, *Proceedings of the 20th International Conference on Computers and Industrial Engineering*, Kyongju, Korea, October 6-9.

Süer, G.A. and Sanchez, I., 1997, Common cell size determination and cell loading in labor-intensive manufacturing cells, *Proceedings of the 21st International Conference on Computers and Industrial Engineering*, San Juan, Puerto Rico, March 10-12.

Süer, G.A. and Gonzalez, W., 1992, Evaluation of cell loading rules to incorporate with schedule-based Material Requirements Planning application, *Proceedings of the National Science Foundation Grantees Conference*, Atlanta, Georgia, January 7-10.

Süer, G.A., Saiz, M. and Gonzalez, W., 1993, Evaluation of cell loading rules to incorporate with schedule-based Material Requirements Planning application: Results, *Proceedings of the National Science Foundation Grantees Conference*, Charlotte, North Carolina, January.

Süer, G.A., Saiz, M., Gonzalez, M. and Dagli, C., 1995, Manufacturing cell loading rules and algorithms for connected cells, *Planning, Design and Analysis of Cellular Manufacturing Systems*, (Eds. H. Parsaei, A. Kamrani and D. Liles), Elsevier Science.

Süer, G.A., Saiz, M. and Gonzalez, W., Evaluation of manufacturing cell loading rules for independent cells, (under revision, *International Journal of Production Research*).

Süer, G. A. and Ortega, M., 1997, Flexibility considerations in design of manufacturing cells, *Group Technology and Cellular Manufacturing: Methodologies and Applications*, (Eds. A. Kamrani and L. Logendran), Gordon and Breach.

AUTHOR'S BIOGRAPHY

Gürsel A. Süer is Professor in Industrial Engineering Department at the University of Puerto Rico-Mayaguez. He received BSIE and MSIE degrees from Middle East Technical University, Ankara, Turkey and Ph.D. in IE from the Wichita State University, Kansas. He is actively involved with the industry and focuses on developing and implementing integrated solution approaches for the complex production problems. His interest areas are hierarchical planning, manufacturing scheduling, design and control of manufacturing cells, genetic algorithms and expert systems in production management. He made several presentations, participated in national/international conferences, contributed to books and published in some journals. He also organized several sessions, workshops, panels in various meetings and conferences and was the co-chair of the 21st International Conference on Computers and Industrial Engineering held in San Juan, Puerto Rico this year. In his spare time, he enjoys playing classical guitar and soccer.

F5

Classification and Analysis of Operating Policies for Manufacturing Cells

R. G. Askin and A. Iyer

1. INTRODUCTION

While the problem of forming manufacturing cells has received considerable attention in the literature (see chapter D1), the development of guidelines for operating cells once they are constructed has not received as much attention. Nevertheless, some research has appeared on the impact of worker cross-training, cell loading rules, the use of sublot transfers, queue dispatching rules, and the assignment of workers to tasks. In this chapter we summarize the findings to date and propose important directions for future research.

In practice, manufacturing cells consist of an empowered team of workers and the resources required to produce a family of related products. Cells cells require significant changes in organizational policies for personnel, wage administration, accounting and scheduling. Among the demands placed on workers as a result of this transition is the expansion of their skill bases. Typical cell workers are cross-trained to perform many jobs within their cell. It also not atypical for the machines to outnumber workers. Unfortunately, little is known about the best operating policy under these circumstances. By operating policy, we mean a protocol for setting lot sizes, transfer batch sizes, cell work-in-process limits and machine queue dispatching as well as worker assignment rules. Some components of operating policies have been examined in isolation previously. However, no attempt has been made to examine the interaction between these components in the context of their being part of an operating protocol. Thus, in this study we seek to characterize operating policies by understanding the nature of the interactions between their components.

Even with the limited literature on scheduling in cellular manufacturing systems, there is a wide array of terminologies and test configurations. In order to place the various results resulting from these studies in perspective as well as compare and contrast these results, a basic unifying framework is required. Once a framework has been established, it is our belief that more focused efforts can be directed towards answering the question of which approach is best for which type of system. Casting the various policies studied so far in the mold of a unifying framework also has another advantage. Not only do we recognize the explicit assumptions made in those models, but we are also forced to identify the implicit assumptions that lie in the background. By bringing all the facets of different policies to the fore, we are able to identify otherwise ignored or hidden flaws and opportunities. In a sense, the minutiae are assembled to reveal the big picture.

An eight-parameter general framework for cell operation has been developed, from which specific policies can be recovered as special cases by assigning appropriate values to each of the parameters. The utility of the framework representation is illustrated by applying it to four families of policies, each embodying a different operating philosophy (and thus different choices of parameter values). Aside from the pedagogical motivation mentioned earlier, the framework view of policies also has practical implications. The general framework representation of policies can be used in conjunction with traditional modeling constructs such as event graphs to develop a general purpose simulator. Due to the generic nature of the representation, such a simulator has significant value, since simulation models for a wide variety of configurations and policies could be developed easily without major programming effort. The value of such a tool is enhanced by virtue of the fact that developing analytical models of operating policies is very difficult due to the presence of multiple resource constraints, general distributions and state-dependent control systems. As a result, simulation is the primary method of studying operating policies. A prototype of this framework-based simulator, termed **COPS** (Cell Operating Policy Simulator), has been developed. Implementation details are available in (Iyer and Askin-3 1997).

2. LITERATURE REVIEW

Burbidge (1968, 1971, 1975) was the first to work on developing production planning and control polices for manufacturing cells. One outgrowth of this work is the Period Batch Control (PBC) approach to cell scheduling, described in chapter F2. In PBC, traditionally, periods were typically between one week to one month in length, with Burbidge noting that the cycle length is bounded above by allowable lead time and bounded below by the minimum time need to produce average period demand and setup each part type. Setup time reduction programs can lead to short periods of one day or less, effectively blending PBC and just-in-time (kanban) policies.

Some of the earliest work on dynamic scheduling for a GT shop is presented in (Ham et al. 1985). Sequencing procedures that explicitly considered family setup times were investigated. Also presented in Ham et al. (1985) are scheduling rules based on permutation schedules for flow shops. Other group scheduling research has, in the past, concentrated on developing *exhaustive and non-exhaustive heuristics* for managing flow through a manufacturing cell. Exhaustive heuristics are such that once a sub-family is chosen, all the jobs in that sub-family including new arrivals are processed first. Only when there are no jobs left in that sub-family does processing switch to another sub-family. Vaithianathan and McRoberts (1982) performed, perhaps the first study in the area by decomposing part family groups based on setup familiarities into subfamilies, such that each of these subfamilies could be treated like a flow shop problem. Once the subfamilies were formed, five queue selection heuristics were proposed to sequence the subfamilies.

In general, empirical scheduling research has shown that due date-based dispatching rules are important when due dates are moderately tight, but when due dates are loose, then all rules tend to perform well on due date-based criteria and when due dates are very tight, no rule performs particularly well. In studying a printed circuit board assembly cell, Askin et al.. (1994) found that assigning tasks to machines in order to balance workload followed by shortest-total-processing-time sequencing proved more effective at minimizing makespan than forming part families and using exhaustive heuristics. Sundaram (1983) proposed two static heuristics to find near-optimal sequences using only one performance criterion, namely, minimization of makespan. In a later study, two cost oriented and exhaustive queue selection rules were proposed by Mosier et al. (1984). This study was motivated by the heuristic scheduling studies which focus on the cost attributes of a job. Mahmoodi and Dooley (1991) presented a comparison between exhaustive and non-exhaustive heuristics for group scheduling. Their simulation studies indicate that non-exhaustive heuristics show promise in lightly loaded cells where due dates are not set tightly.

Most of the scheduling rules in the literature discussed until this point assume that the production system is machine-limited. As early as 1967, Nelson (1967, 1970) pointed out that research on machine-and-labor-limited systems must be emphasized. This conclusion was based on studying operating data from job shops over a seven month period. A labor-limited system is a system in which labor is also a constrained resource. This means that the conventional assumption of machine capacity being the only limiting factor in a production system is no longer valid. Subsequently, several simulation studies of dual-resource-constrained systems appeared in the literature (Hogg et al.-1 1975, Hogg et al.-2 1975, Huang et al. 1984, Park and Bobrowski 1989). Most of these studies focused on the interaction between job structure and labor assignment rules as well as the efficacy of various assignment rules. Nelson (1970) concluded that the choice of queue disciplines at each workstation produced a large effect on the mean and variance in the system for jobs. However, there was a negative interaction between the mean and variance in that an improvement in one measure was obtained at the expense of the other. The labor assignment rule, on the other hand, had a smaller effect on the performance measures. However, in this case there were no tradeoffs between the mean and variance of the time in system. He also concluded that the mean and variance of the time in system for a job shop with a highly flexible workforce was much lower than the corresponding measures for a job shop with an inflexible workforce. Weeks and Fryer (1976) also confirmed Nelson's conclusion that making a worker available for reassignment after every operation reduced the mean flow time of jobs through the system. The alternative reassignment policy (called decentralized control) was to allow workers to switch only when they became idle i. e., when the queue ahead of them emptied. A study by Malhotra and Kher (1994) concluded that when transfer delays (i.e. delays incurred when workers are switched between machines) are significant, centralized control caused the performance of the job shop to deteriorate. They also concluded that with a heterogenous workforce (not all workers are equally efficient at each machine), the timing of the workers assignment (WHEN) was not as important as the the machine the worker is sent to (WHERE). Specific components of operating policies have been examined in isolation as well.

There has been relatively little work on analytical models for describing or optimizing cell operations. Askin and Iyer (1993, 1994) compared traditional batch processing rules with multiproduct dedicated cells and a worker batch assignment strategy where each batch is assigned a champion to perform all operations and push the job through the cell for machine-limited systems as well as machine and labor-limited systems. The multiproduct approach was shown to perform best for low to moderate utilization. The worker batch assignment strategy was also shown to perform well in labor-limited systems. This study assumed complete labor flexibility, *i.e.*, all workers were completely cross trained. Also, this study assumed that all part types had the same processing times at all machines. Iyer and Askin-2 (1997) use a two stage queueing model to evaluate the performance of dedicated cell loading in flow shops with dual resource constraints. The stages correspond to pure dedication of the cell to a single job and the period of overlap when one job is nearing completion and the next has entered the cell. Bartholdi and Eisenstein (1993) describe a labor assignment policy called TSS (Toyota Sewn Products System) and show that for an unbalanced, identical item TSS line, the line is stable only if workers are sequenced slowest to fastest along the line. Using simulation, Bischak (1993) showed that reduction in processing time variability between machines and the presence of buffers are beneficial for TSS lines. Zavadlav et al.. (1994) compared the efficiency of dual-constrained production lines for various levels of cross-training. They found that free floating assignments (where workers were fully cross-trained and when blocked returned back upstream to the first machine that was not blocked or starved) outproduced fixed path and limited overlap orbits for the same level of WIP. It was assumed that workers could change machines with no time loss. Complete cross-training allows workers to leapfrog when encountering a busy worker on their upstream search instead of interrupting a job/worker in progress. Vembu and Srinivasan (1997) studied the joint problem of allocating a limited number of workers to a serial set of cells and sequencing part families through the cells. Within cells workers push one part at a time, but flow between cells is by full batches. A hierarchy of increasingly complex heuristics was proposed based on problem size and closeness to optimality desired.

While the use of transfer batches (also referred to as lot splitting/streaming) appears to be widely practised, there is relatively little discussion in the academic literature to provide further insights. One of the earliest references to this can be found in (Szendrovits 1975) where an economic lot sizing model is considered for a single product flow shop under conditions of continuous, deterministic demand, linear holding costs and deterministic unit processing times. It is assumed that each process batch will be split into a few predetermined equal sized transfer batches. The objective is to minimize the sum of fixed costs and inventory costs (both in-process inventory and finished goods inventory). Baker and Pyke (1989) discuss the determination of transfer batch size with a view to minimizing makespan for single and multi-product flow shops.

3. GENERAL FRAMEWORK

As we have seen, a variety of issues must be considered when describing the operation of a cell. In this section we describe a general framework for describing the operating policy of a cell.

There is no basic framework yet under which past studies can be examined systematically so as to compare their assumptions and results on an equal footing. By defining a framework within which various operating policies can be compared, we hope to show the germane literature in a unified light. The framework provides a taxonomy for full description of cell operations including control of workers, machines and material flows.

As results are accumulated and placed in this framework, operations managers may be able to select the policy most appropriate for their environment. Researchers will be able to characterize the conditions under which methods have been tested and proven to work well. The framework description allows us to represent different policies in a manner suitable for direct translation into computer simulation models of policies. This fact has been exploited to develop a fast general purpose simulator. This simulator is used to compare various operating policies.

In understanding the utility of the framework, it is helpful to make a distinction between the generic parameters which make up our representation and the information required to assign *values* to these parameters. Different operating environments and different policies can be represented by assigning values to these parameters.

The remainder of this chapter is organized as follows. In the next section we describe a prototypical cell and the notation used to describe the configuration and operating environment of the cell. We then describe the generic parameters used to describe operating policies. This is followed by an illustration of the use of the parameters for describing specific operating policies. Finally, we summarize the results of a comprehensive simulation experiment based on this framework. Further details on the framework and demonstration of its completeness are provided in (Iyer and Askin-1 1997; Iyer and Askin-3 1997).

4. CELL OPERATION

We restrict ourselves to a single cell and describe the various components which make up the cell. To view the entire manufacturing facility and describe the coordination between various cells, we make note of the fact that at that level of description, the PAC (Production Authorization Card) framework (Buzacott and Shanthikumar 1992) is applicable by considering cells in their entirety and disregarding the internal policies of the cell. The PAC framework describes the flow of information and materials between cells. We, on the other hand, are only concerned with making decisions that affect the cell internally.

Two sets of parameters are required to describe cell operation. The first set describes the configuration of the cell *vis-a-vis* the relatively static attributes of the physical entities in the cell and the dynamic state of the cell. The static configuration of the cell can be described in terms of the three basic entities namely, machine resources (which includes tools and material handling equipment), parts, and workers. This includes, but is not limited to, describing: 1) the number of machines by type; 2) the tools available on machines; 3) setup matrices for machines; 4) the processing rate of jobs on machines; 5) the routing structure for jobs; 6) number of workers in the cell; 7) the cross-training level of each worker; 8) efficiencies of individual workers on specific machines; and, 9) transportation times and load sizes between machines. The static configuration parameters are obtained from the solution to the cell design problem.

The second set of parameters describe the dynamic state of the cell. These parameters affect, and are in turn affected by the operating policies used. This usually involves providing information on: 1) the jobs currently loaded on a machine; 2) the amount of processing time that remains for a job on a machine; 3) the composition of the input and output buffers of a machine; 4) the current position of each worker; and, 5) the jobs that a worker is currently responsible for. These are usually obtained from the shop floor reporting and control system.

The parameters which comprise the general framework description (namely, the parameters whose values we can set to obtain various policies and/or different levels of performance) are:

- **Admittance Condition for the Cell.** This parameter describes the condition which must be satisfied for a batch to be admitted to the cell. At any given time, there are batches of jobs waiting in the dispatch queue to the cell which enter the cell when certain conditions are met For example, in a CONWIP system (Spearman et al. 1990), the AC for a batch would be met when the WIP in the cell falls below a predetermined limit. Several other types of conditions have been proposed in the literature as well. For example, Fredendall et al. (1996) describe some admittance conditions for DRC cells such as checking for the presence of free worker. The admittance condition is used in conjunction with the **Admittance Rule**.
- **Admittance Rule for the Cell:** When multiple batches are eligible to enter the cell, the admittance rule is used to determine which one(s) will be admitted. One of the simplest rules used is the FCFS discipline at the dispatch queue to the cell.
- **Switching Instants for Units:** Switching Instants are used to specify the material movement policy. Switching instants are defined for every unit of a batch at every machine and represent the conditions that must be satisfied in

order for the routing policy for that unit to be invoked. Thus, process batch movement from a machine would be defined by specifying the switching instant for every unit of a batch at a machine to be the end of batch processing at that machine. However, if the SI for every unit is defined as the end of unit processing, transfer batch movement in sizes of one is implied.

- **Switching Rules for Units**: Switching Instants define when the units move from machine to machine but do not specify the actual routing decision to be made. Switching Rules define the dynamic routing decision that is made when the process or transfer batches are to be transported. Thus, if multiple machines are eligible to perform the next operation, the switching rule would define which one would be used. When no strict precedence constraints exist, the switching rule would specify both the next operation to be performed as well the machine to perform it on. Commonly used rules are to perform the next operation on the machine with the shortest queue or on the machine with the lowest sequence dependent setup cost.
- **Part Switching Instants for Workers**: Part Switching Instants (PSI) are defined for every worker at all part-operation combinations. The PSI for a worker defines the condition under which he or she can switch to a new part. Typically, the PSIs in non preemptive sequencing policies are defined to be the end of batch processing at a machine. Thus, the worker can potentially switch to a new batch only after he or she completes the current batch at the current machine. However, defining the PSI to be the end of unit processing allows us to define preemptive policies in conjunction with the Part Switching Rule (PSR) for the worker-part-operation combination.
- **Part Switching Rules for Workers**: PSRs specify the next batch a worker must switch to given that the conditions of the PSI are met. Thus, the PSR for FIFO dispatching at a machine with a dedicated server would specify the worker should switch to the part with the earliest arrival time to the queue when the conditions specified by the PSI (completion of current batch processing) are met. When there are fewer workers than machines, specifying the PSI and PSR may not completely define the labor assignment policy since the part (re)assignment may also necessitate a machine reassignment. In order to completely specify the labor utilization policy, it will be necessary to define two additional parameters which are invoked for DRC cells only.
- **Machine Switching Instants for Workers**: Machine Switching Instants (MSI) are defined to specify the conditions which must be satisfied for a worker to be reassigned to another machine. In polling systems with exhaustive service for example (Srinivasan 1988; Takagi 1990), the MSI for the lone worker would be when the queue at the current machine is cleared.
- **Machine Switching Rules for Workers**: Machine Switching Rules (MSR) represent the decision rule used to dispatch workers to machines when the conditions of the MSI are met. In cyclical service systems with **M** machines, the MSR at machine **m** is to switch to machine **m + 1**. In DRC cells, it is possible that conditions for both the MSI and PSI are met simultaneously. It is therefore also necessary to specify the precedence relationship between the PSR

and MSR. Consider the case when both the MSI and the PSI are defined to be the end of batch processing at a machine and the PSR takes precedence over the MSR. Under these circumstances, the worker would switch to the next batch based on the PSR and the MSR would be interpreted in this context with the current batch for the worker being the new batch he or she was assigned.

To describe the operating policies of the cell, it is also necessary to include the material handling system within our framework. Recent work (Malhotra and Kher 1994) has shown that transfer times strongly interact with other components of an operating policy. In the framework, material handling equipment can be treated as a machine which performs the transfer operation on a part. Assignment of workers who are designated material handlers can then be treated in a manner consistent with the description of cell operation. For convenience of description we will make the assumption that material handling is not a limiting constraint and will therefore not deal with it explicitly. The next section describes four families of operating policies and their framework-based descriptions.

5. DESCRIBING POLICIES

The first family of policies we consider is based on traditional job shop operation and is termed Machine-based Batch Loading (MBL). In the MBL policy, batches of parts enter the cell at the first machine in their routing sequence. There is no limit on WIP. The queue discipline at each machine is first come, first served. Batches are moved in their entirety. When there are fewer workers than machines, they must be shared by all the machines. A machine can begin processing parts from the queue only when the worker arrives from the assignment station. Workers remain at machines until the input queue is empty .When no work remains at the machine, the worker moves to a central assignment station from where he/she is dispatched to the unattended machine with the longest input queue. When workers are not cross-trained to work at all machines, the PSR and MSR for the worker are modified to apply only within their orbits (defined as the set of machines the worker is cross-trained to work on). The choice of parameters which corresponds to MBL policy is :

Admittance Condition	None
Admittance Rule	None
Switching Condition	End of batch processing for all units.
Switching Rule	Next machine required in the routing.
Part Switching Instant	End of Batch Processing at current machine.
Part Switching Rule	Switch to Batch at the head of the queue.
Machine Switching Instant	End of Batch Processing for last job in queue.
Machine Switching Rule	Machine with the longest input queue.
Precedence	MSI >> PSI

The second type of policy we consider is called Dedicated Cell Loading (DCL). The DCL policy uses a temporary dedication strategy wherein workers are assigned to one batch of parts at a time and released only when it is determined that the current batch no longer requires them. The effect of dedication is maximized by transporting units of the batch in transfer batches. In this paper, we assume transfer batches of size one although any other appropriately determines sizes may be used. Although batches are moved in transfer batches, workers move from a machine only after they have finished processing the entire batch at a machine. Once a worker finishes processing a batch at a machine, he/she either moves to the next idle unstaffed machine in the orbit set required by the current batch. If no longer required by the current batch, he/she then switches to a new batch. The WIP in the cell is no longer a predetermined constant but a function of the processing requirements of successive batches *vis-a-vis* machines and workers. However, the upper bound on the WIP is *min*(Number of workers, Number of machines). For a flowshop operated under the DCL policy, the framework parameters can be written as follows.

Admittance Condition	None
Admittance Rule	None
Switching Condition	End of Unit Processing for all units.
Switching Rule	Next machine required in the routing.
Part Switching Instant	End of Batch Processing at machine m such that number of remaining operations <= K - 1.
Part Switching Rule	Switch to Oldest Batch in the cell.
Machine Switching Instant	End of Batch Processing for current job.
Machine Switching Rule	If MSI is also PSI, move to the first unstaffed machine in the next batch's routing. Else, move to the first unstaffed machine in the current batch's routing.
Precedence	PSI >> MSI

Note that the framework description appears to suggest that the DCL policy for a flowshop can be described as though there is no limit on the WIP in the cell. This insight has been used to develop a fast analytical approximation method to estimate flowtimes of batches (Iyer and Askin-2 1997).

The third policy we consider, called Worker Batch Assignment (WBA) is motivated by the Japanese practice of using cross-trained workers and excess capacity. Under WBA, workers are assigned to process the entire batch at all machines within their orbit. If the next operation requires a machine outside the worker's orbit, the batch is handed off to a worker (determined by some decision rule) whose orbit the machine is in. Workers may be responsible for more than one batch at a time. Thus, when a worker is blocked waiting for a machine, he/she may switch to one of the other batches they are responsible for using some pre-determined criteria. The maximum number of batches that could be in process within the cell is thus the sum of the maximum number each worker may be responsible for.

Admittance Condition	Number of batches in the cell <= Number of workers times maximum number of batches they can be assigned to.
Admittance Rule	FCFS
Switching Condition	End of Batch Processing for all units.
Switching Condition	Next machine required in the routing.
Part Switching Instant	End of Batch Processing at current machine.
Part Switching Rule	If next machine for current batch is available, do not switch parts. Else, switch to oldest batch at an available machine
Machine Switching Instant	End of Batch Processing for current batch.
Machine Switching Rule	Next machine required by current batch.
Precedence	PSI >> MSI

The final policy we consider is the Toyota Sewn Products System (TSS) (Bartholdi and Eisenstein 1993; Bischak 1993). This policy is well defined only for a flow shop environment. Workers are positioned along the line and do not pass each other. Each worker moves down the line with a batch until they are either blocked, relieved by their successor or are required to move outside their orbit. Within the orbit when the last worker in the line finishes a batch, he/she moves upstream and relieves the next worker of their batch even if the batch is being processed. If the preceding worker is blocked at the last machine, the last worker does not switch since he/she takes over from the waiting worker. On being relieved, a worker will move upstream and relieve his/her predecessor. The new batch this worker is assigned to depends on the circumstances under which this worker was relieved. If this worker was processing a batch and there are other jobs (with workers) in queue, the worker takes over the first batch waiting in queue thus relieving the worker who was responsible for that batch. On the other hand, if there is no queue at the machine, this worker moves upstream to relieve their predecessor.

Admittance Condition	Number of batches in the cell <= Number of workers.
Admittance Rule	FCFS
Switching Condition	End of Batch Processing for all units.
Switching Condition	Next machine required in the routing.
Part Switching Instant	End of Batch Processing at last machine.
Part Switching Rule	If a queue exists at current machine, switch to the batch at the head of the queue. Else, switch to the first batch encountered upstream.
Machine Switching Instant	End of Batch Processing at last machine, End of batch Processing at current machine
Machine Switching Rule	Switch to the machine required by the current batch.
Precedence	PSI >> MSI

When workers reach the end of their orbit, the batch is dropped off where it waits for a downstream worker to start working on it. Several variants of this policy are possible. In the first variant, workers are not blocked waiting for the next machine within their orbit when it is occupied. Instead, the batch is dropped off at the next machine for the downstream worker to work on. This variant is referred to as BSS. Yet another variant uses the basic TSS policy with transfer batches. A third variant combines the first two wherein workers move with transfer batches but are never blocked waiting for a downstream machine. This variant is referred to as DSS. The framework-based description that follows is for the basic TSS policy assuming completely cross-trained workers.

6. EXPERIMENTAL RESULTS

A simulation experiment was conducted to compare the operating policies described previously under flowshop conditions. The experiment was designed to address the issues of cross training policy, admittance control, and choice of strategy. The factors studied included Number of Machines, Number of Workers, Utilization, Degree of Cross-Training, Interarrival Time Variability, Processing Time Variability and Processing Time Patterns. Two systems of five and nine machines were studied with the number of workers being set to two, four and five for the five machine system and four, seven and nine for the nine machine system respectively. The utilization levels studied were 60% and 80%. Interarrival and processing time variability were measured by the squared coefficient of variation of the interarrival time and processing time distribution respectively which were set to values of 0.3 and 1.0. Degree of cross-training is measured by the degree of overlap between the orbit sets of workers. Three levels of overlap were studied - complete, low and high. Pattern of processing times refers to the pattern in which mean batch processing times vary along the flow line. For the various systems studied, the processing time pattern was set at Increasing, Decreasing and Random. A complete factorial simulation experiment, resulting in 384 cases, was run using the framework-based simulator for 7500 hours of shop time with the initial 2500 hours being discarded to remove initialization bias. Three replicates were run for every combination of factor levels. Common random numbers were used across all cases to increase the precision of comparison.

A variety of policies were tested. We report the results for the best overall performer in each class. By class we mean a policy such as DCL. The choice of policy in each class was based on the best overall performer or that which performed among the best in average cycle time with the lowest level of average work-in-process. Several modified policies also proved worth considering. In the sections that follow, the numerical descriptor n at the end of each policy indicates the maximum level of WIP allowed by the admittance policy. The admittance control policy is to admit jobs until the number of batches is equal to nK where K is some integer. A suffix of O indicates the original policy without any modifications made to the admittance rule. Batch sizes were fixed at ten units.

In order to allow meaningful comparisons, the observed average batch flow times were normalized by the ideal batch cycle times. The ideal cycle time is defined as the sum of mean batch times at all machines and represents the total cycle time for a batch when it never has to vie for limited resources with other batches. Thus, a normalized value of 1.0 would mean that the batch had no waiting time in the system. Values less than 1.0 are also possible when transfer batches are used.

Simulation Insights

Initially, eleven policies were studied. These were MBLO, MBL1, MBL4, WBA2, WBA4, TSSO, BSS1, BSS4, DCLO, DCL4 and DSS1. Preliminary analysis using SAS revealed that although the number of machines and the pattern of processing times were not very significant, there were a large number of significant higher order interactions. In order to derive some insights into the performance of policies, the initial list of policies was pared down to six by eliminating policies which showed poor performance. The final policies selected were DCLI, DSS1, MBL4, WBA4, BSS4 and TSSO. A comparison of these six policies for various conditions appears in Figures 1 to 4. In the figures, the conditions on the x-axis represent the simulated systems. For example, 5d2cllh represents a 5 machine cell with 2 workers with mean batch processing times decreasing along the line. Workers are completely cross-trained as represented by the *c*. The last three letters represent utilization level, processing time variability and arrival time variability respectively with the *l* representing the low factor setting and the *h* representing the high factor setting.

One of the strongest interactions observed was between the cross-training level and the admittance control rule particularly at high levels of utilization and processing time variability. At low levels of processing time variability and utilization, different variants of the same class of policies were almost identical. For example, at low levels of utilization and variability, different variants of the DCL policy do not show any significant differences. At higher levels of utilization and variability, however, the original DCL policy which in previous studies (Askin and Iyer 1993; Iyer and Askin-2 1997) was one of the best policies performed very poorly in a majority of the simulations in this study. DCLI and DCL4 , by contrast, were among the best policies across most conditions. The reason for this is the assumption made in the other studies that all workers were completely cross-trained. Under these conditions, a new batch was admitted whenever the current batch released a worker assigned to it. With limited cross-training, the flow line has in fact been partitioned into sub-lines staffed according to the degree of cross-training. The benefits of transfer batches are lost since workers may not be available to work on single units as they are moved to the next machine in their sequence. In addition, by adhering to the original admittance policy, workers are being kept idle even when they cannot be switched to machines required by the current batch due to limited cross-training. If the cell were viewed as a single server machine with the batch flow times being the service times at this machine, the queue at the machine becomes unstable since the arrival rate exceeds the service rate. One way of recovering lost capacity is to change the admittance policy. Thus, when the allowable WIP limits are increased, workers are no longer idle when they reach the limits of their cross-training orbits but can instead begin working on the oldest batch in their section of the cell. From the framework point of view, we have redefined the PSI for each worker. In the case of WBA, the number of workers is an equally significant factor. With tight admittance control as with WBA1, restricted orbits result in bad performance since workers get blocked but there are no avenues for recourse. With few workers, complete cross-training is preferred since workers rarely block each other but must prevent the accumulation of work at the hand-off points which is a factor only with limited cross-training. When the number of workers increases, they get blocked often with a larger degree of overlap between the orbit sets whereas with little or no overlap this is prevented. However, if the admittance policy allows for it, even with some overlap workers can be prevented from being blocked provided the admittance control policy allows for it by allowing workers to switch to other batches. A similar effect is observed with the TSS family as well. The original TSS policy with complete cross-training results in workers being blocked frequently. Allowing workers to place batches in buffers at the hand-off zones and start moving backwards when coupled with an appropriate admittance control policy (BSS4) greatly mitigates this effect. When transfer batch movement is added to this policy (DSS1), the performance greatly improves.

Another clear trend observed was the dominance of transfer batch-based policies. Comparing the average flow times for all the policies across all conditions clearly showed that the policies could be classified on the basis of batch transfer policy. The transfer batch-based policies such as DCL and DSS usually outperformed the other process-based policies. DSS1 and DCLI were, in fact, the best policies.

Careful analysis of the figures reveal several interactions which might account for these differences. It would appear that the relative superiority of transfer batch policies is magnified as the utilization and processing time variability increases. However, the magnitude of this difference is also a function of the number of machines and workers in the system. The disparity between transfer and process batch-based(TB and PB respectively) policies is greater when there are more machines and workers in the system. With few machines or few workers, the magnitude of the disparity decreases.

This phenomenon is explained by considering the effect transfer batches create. With transfer batches, several units of the same batch can be under production at different machines. Thus, processing (and setup) times at one machine are nested within the processing times at a previous machine. The actual number of units that can be under parallel production. is given by *min*(Batch Size, Number of Workers, Number of Machines). When the number of workers is low, the dominance of TB-based policies is muted due to the fact that workers must be shared between many machines. Cross-training has the effect of partitioning lines into smaller lines, the size of the line being determined by number of workers and degree of cross-training.

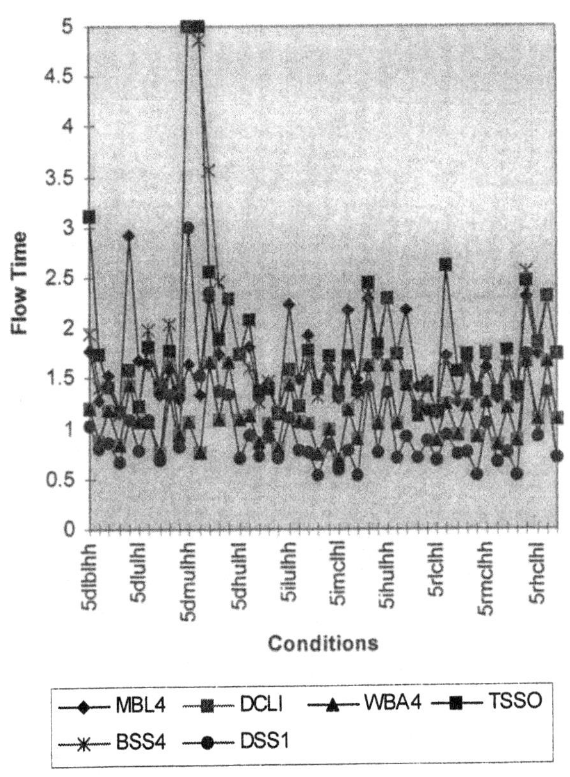

FIGURE 1. M=5; rho=low; Var(Proc.Time)=high.

In summary, we have the following major effects. There are several significant interactions to be considered in the choice of a good operating policy. Isolated optimization of parameters is likely to lead to bad performance. However, some effects appear to be prevalent across most conditions. Transfer batches provide an opportunity for significant reduction in cycle time relative to batch processing in manufacturing cells. Second, worker cross-training can be important but its impact is highly correlated with admittance control. Required WIP levels can be reduced by increasing cross-training.

7. FUTURE RESEARCH DIRECTIONS

We now have a basic understanding of how several factors affect cell operations. There remains a need for the development of analytical models that closely approximate the behavior of cells for various strategies. Much of the research on cell scheduling has assumed flow line movement or specific shop configurations. It would be useful to propose methods to transport flow shop strategies to cells with general material flow patterns and model the performance of such systems.

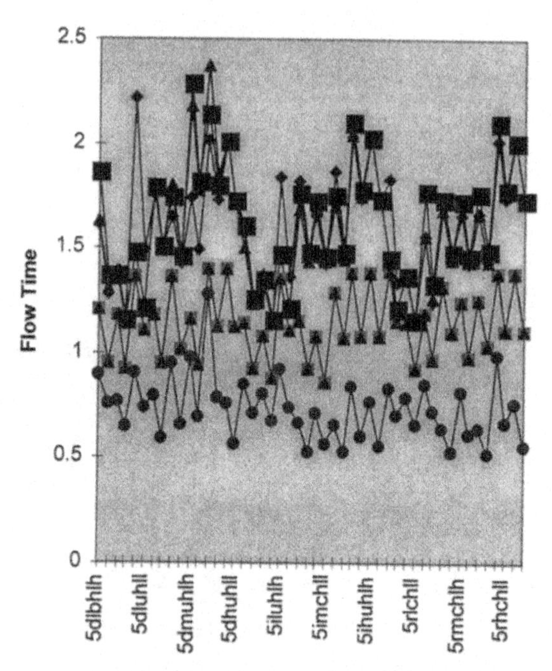

FIGURE 2. M=5; rho=high; Var(Proc.Time)=high (Legend as in Figure 1).

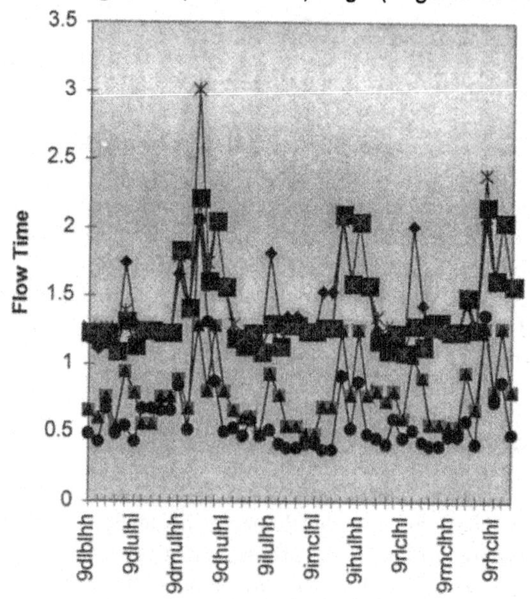

FIGURE 3. M=9; rho=low; Var(Proc.Time)=high; (Legend as in Figure 1).

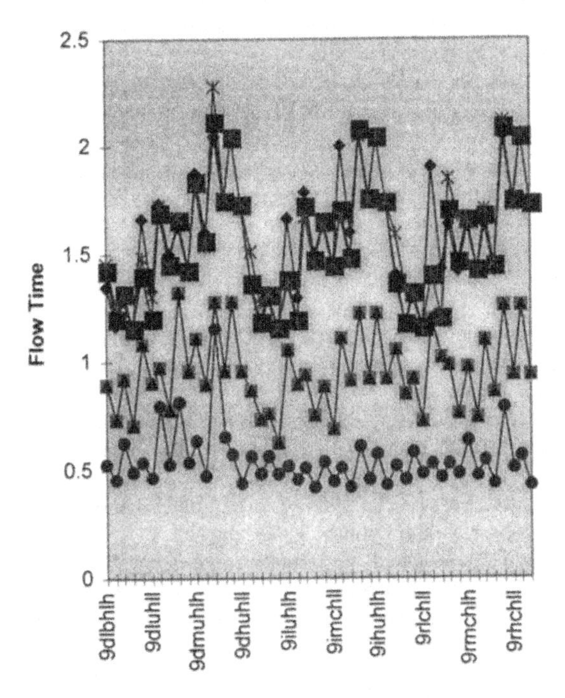

FIGURE 4. M = 9; rho=high; Var(Proc.Time) = Low; (Legend as in Figure 1).

The results to date on the effect of cross-training are complex with significant interactions with other factors. A more complete model of the value of cross-training would be useful for planning cell operations. It would be interesting to survey industries to ascertain the operating policies they are currently using and correlate their experiences to the results of previous simulation and analytical models. This would test the value of the current research models and possibly indicate policies that should be studied further. The framework has been presented but not optimized. We should be able to use the framework to show that certain policies are or are not optimal from a technical perspective. For instance, does a threshold exist for when cross-training is valuable? Can we build an economic model that determines the optimal level of cross-training given training costs?

REFERENCES

Askin, R. G. and Iyer, A., 1993, A comparison of scheduling philosophies for manufacturing cells, *European Journal of Operational Research*, 69, 438-449.

Askin, R. G. and Iyer, A., 1994, *Strategies for Controlling the Flow of Material Through Manufacturing Cells*, Technical report, Department of Systems and Industrial Engineering, University of Arizona, Tucson, AZ.

Baker, K. R. and Pyke, D. F., 1989, Solution procedures for the lot streaming problem, *Decision Sciences*, 21, 475-491.

Bartholdi III, J. J. and Eisenstein, D. D., 1993, *A Production Line That Balances Itself*, Technical report, Georgia Institute of Technology, School of Industrial and Systems Engineering.

Bischak, D., 1993, *Throughput of a Manufacturing Module with Moving Workers,* Technical report, Department of Business Administration, University of Alaska, Fairbanks.

Burbidge, J. L., 1968, *Principles of Production Control*, MacDonald & Evans, Ltd., London.

Burbidge, J. L., 1971, *Production Planning*, Heinemann Press, London.

Burbidge, J. L., 1975, *The Introduction of Group Technology*, John Wiley and Sons, New York.

Buzacott, J. A. and Shanthikumar, J. G., 1992, A general approach for coordinating production in multiple-cell manufacturing systems. *Journal of Production and Operations Management*, 1, 1, 34-52.

Fredendall, L. D., Melnyk, S. A. and Ragatz, G., 1996, Information and scheduling in a dual resource constrained job shop, *International Journal of Production Research.*, 34, 10, 2783-2802.

Ham, I., Hitomi, K. and Yoshida, T., 1985, *Group Technology Applications to Production Management,* Kluwer-Nijhoff Publishing, Boston, MA.

Hogg, G. L , Phillips, D. T., Maggard, M. J.; and W. G. Lesso, 1975, GERTS QR: a model for multi-resource constrained queueing systems: Part I, *AIIE Transactions*, 7, 2, 89-99.

Hogg, G. L, Phillips, D. T., Maggard, M. J., and Lesso, W. G., 1975, GERTS QR: A model for multi-resource constrained queueing systems: Part II: An analysis of parallel-channel dual-resource constrained queueing systems with homogenous resources, *AIIE Transactions*, 7, 2, 100-109.

Huang, P. Y., Moore, L. J., and Russell, R. S., 1984, Workload versus scheduling policies in a dual resource constrained job shop, *Computers and Operations Research*, 11, 1, 37-48.

Iyer, A., and Askin, R. G., 1997, A general framework for comparing operating policies in manufacturing cells, *Annals of Operations Research*, to appear.

Iyer, A., and Askin, R. G., 1997, *Operating Manufacturing Cells with Dedicated Cell Loading*, Working Paper, Department of Systems & Industrial Engineering, The University of Arizona.

Iyer, A., and Askin, R. G., 1997, Modeling and simulating operating policies for manufacturing cells, *IIE Transactions,* to appear.

Mahmoodi, F. and Dooley, K. J., 1991, A comparison of exhaustive and non-exhaustive group scheduling heuristics in a manufacturing cell, *International Journal of Production Research*, 29, 9, 1923-1939.

Malhotra, M. K. and Kher, H. V., 1994, A comparison of worker assignment policies for dual resource constrained shops, *International Journal of Production Research*, 32, 5, 1087-1103.

Mosier, C.T., Elvers, D. A., and Kelly, D., 1984, Analysis of group technology scheduling heuristics, *International Journal of Production Research*, 22, 857-875.

Nelson, R. T., 1967, Labor and machine limited production systems, *Management Science*, 13, 9, 648-671.

Nelson, R. T., 1970, A simulation of labor efficiency and centralized assignment in a production model, *Management Science*, 17, 2, B97-B106.

Park, P. S., and Bobrowski, P. M., 1989, Job release and labor flexibility in a dual resource constrained job shop, *Journal of Operations Management*, 8, 3, 230-249.

Spearman, M.L., Woodruff, D. L., and Hopp, W. J., 1990, CONWIP: A pull alternative to kanban, *International Journal of Production Research*, 28, 879-894.

Srinivasan, M. M., 1988, An approximation for mean waiting times in cyclic server systems with non-exhaustive service, *Performance Evaluation*, 9, 17-33.

Sundaram, R.M., 1983, Some scheduling rules for a group technology manufacturing system, in *Computer Applications in Production and Engineering*, (Ed. Ackerman), 765-772.

Szendrovits, A. Z., 1975, Manufacturing cycle time determination for a multi-stage economic production quantity model, *Management Science*, 22, 293-308.

Takagi, H., 1990, Queuing analysis of polling models : An update, from *Stochastic Analysis of Computer and Communication Systems*, (Ed. H. Takagi), Elsevier Science Publishers B.V., North Holland, 267-318.

Vaithianathan, R., and McRoberts, K. L., 1982, On scheduling in a GT environment, *Journal of Manufacturing Systems*, 1, 149-155.

Vembu, S., and Srinivasan G., 1997, Heuristics for operator allocation and sequencing in product line cells with manually operated machines, *Computers and Industrial Engineering*, 32, 2, 265-279.

Weeks, J. K. and Fryer, J. S., 1976, A simulation study of operating policies in a hypothetical dual-constrained job shop, *Management Science*, 22, 12, 1362-1371.

Zavadlav, E., McClain, J. and Thomas, L. J., 1994, *Self-buffering, Self-balancing, Self-flushing Production Lines*, Working Paper Series 94-02, Johnson Graduate School of Management, Cornell University, Ithaca, NY.

AUTHORS' BIOGRAPHY

Dr. Ronald G. Askin is Professor and Associate Department Head of Systems and Industrial Engineering at the University of Arizona. He is a Fellow of the IIE, and an active member of INFORMS, ASQC, and SME. Currently he serves on the editorial boards of the *IIE Transactions on Design and Manufacturing* and the *Journal of Manufacturing Systems*. He has consulted widely, and published in many professional journals, predominantly in the area of design and analysis of manufacturing systems. Dr. Askin has co-authored the text *Modeling and Analysis of Manufacturing Systems* which was awarded the 1994 IIE Joint Publishers Book of theYear Award. Other awards he has received include the Shingo Award for Excellence in Manufacturing Research, *IIE Transactions* Development and Applications Award (coauthor), the ASEE/IIE Eugene L. Grant Award (coauthor), and a National Science Foundation Presidential Young Investigator Award.

Dr. Anand Iyer is a member of the Advanced Product Development group at Irving-based i2 Technologies where he is involved in the development of models for integrated supply chain optimization. He received his Ph. D. in Systems and Industrial Engineering from the University of Arizona. His professional and research interests include the development and analysis of models for supply chain optimization and manufacturing systems management.

G

Cells and Flexible Automation: A History and Synergistic Application

K. E. Stecke and R. P. Parker

1. INTRODUCTION

Flexible manufacturing equipment is the technological union of computer integration and machine tools. Flexible manufacturing is commonly understood as offering certain types of 'flexibility' for manufacturers to mitigate or manage the significant uncertainty they may encounter. For example, with this investment they are better able to respond to customers' needs and desires in product design, product availability, and adaptation for successive design of part types. Such promises have resulted in widespread investment in these technologies in industry and significant investment in research in academia. The competitive pressures in global manufacturing markets, primarily during the 1980's, drove companies to improve product quality and 'flexibility' without sacrificing operational cost reductions already achieved in many industries. The use of flexible automation was suggested as one way to answer these pressures, even though these new technologies were more expensive. However, it soon became clear that mere capital investment was insufficient to satisfy these challenges: significant technical and managerial issues needed to be resolved before the full benefits of these technologies could be attained.

There is a continuum of what is considered flexible automation. Although there has been great progress in the area of flexible assembly, the majority of installations and research has focused on flexible machining. The basic building block of such flexible automation is the machine tool which could be a lathe, drill press, milling machine, or other material (usually metal) removal device.

A description of the technologies and their historical development is discussed in §2. This discussion leads to an outline of the various research streams associated with flexible automation in §3. The literature suggests that these technologies must necessarily be managed differently, compared to conventional technologies, to realize the promised benefits. In order to understand these differences, the 'flexibility' exhibited by flexible automation needs to be understood. The various types of flexibility are examined in §4. §5 contains a discussion of the synergistic relationship of flexible automation with group technology and cellular manufacturing. Issues for future research are highlighted in §6, and the concluding remarks are in §7.

2. FLEXIBLE AUTOMATION: DESCRIPTION AND HISTORY

Flexible automation has many accepted definitions and meanings. For example, a small example of flexible automation is a stand-alone CNC machine tool, whereas a totally automated system contains multiple CNC machine tools, possibly with robotic arms loading and unloading parts, automated guided vehicles transporting parts between machines, inventories sometimes accessed by automated storage and retrieval systems, and a coordinating computer overseeing the scheduling and production planning of the system. Installations of flexible automation can be found throughout the world that lie on the automation continuum anywhere between the above examples. The level of human intervention varies inversely with the level of automation, with the appropriate mix determined by the particular application.

Flexible automation may be found in both fabrication and assembly, and the various types of manufacturing included in these categories. However, the focus in this chapter is on flexible automation in fabrication and machining. This is because of both historical circumstances and the particular focus here on cellular manufacturing. That said, it would be remiss to not highlight the rapid advances of flexible automation in assembly operations. Automated assembly is extremely complex because of the necessary planning for all contingencies that can occur during operation. There is also the complicated nature of the process itself: the combination of multiple physical parts brought together into an assembly. This aspect can require an enormous number of monitoring devices to ensure a successful operation, an expensive proposition. Since automated assembly is currently used mostly in simple tasks, and flexible assembly automation, where multiple part types may be assembled at a single station, is still in the development phase, we do not discuss it further. Interested readers are referred to Boothroyd (1992).

Various texts provide extended histories of process automation technologies (Cohen and Apte, 1997; Buzacott and Shanthikumar, 1993) including the evolution of flexible automation. A shorter description is now provided. Flexible automation has its source in fixed automation which began to appear in the early 20th century, particularly in the Ford Motor Company in the United States. Fixed automation is a group of special purpose technologies and was developed to attend to a regimen of relatively simple tasks repeated many times. The automation fit perfectly with Henry Ford's philosophy of automobiles with few variants or options but at high volumes to spread the fixed costs over many units. This type of automation evolved during the first half of the 20th century and was used in conjunction with human operators.

In the wake of the second World War, development on machining centers (usually for prismatic rather than rotational workpieces) took place with the arrival of the first numerical control machine tools. These technologies delivered commands to the machine tool, controlling the spindle speed, axes location and movements, and tool movements and speeds. In general purpose machine tools such as lathes, milling machines, or drill presses, the operator would be responsible for these actions, as can be seen schematically in part (a) of Figure 1a. Initially these commands were hardwired into the first numerical control machine tools, typically milling machines and drill presses. The next development was to have instructions delivered via punched paper tape, allowing some greater flexibility in the variety of programs that could be executed on a machine, as seen in part (b) of Figure 1a. The responsibility of inserting the tools into the machines was next given to automated tool changeovers (part (c) of Figure 1a).

The rapid advances in computer technology subsequently permitted direct numerical control (DNC) of machine tools, where a centralized computer issued commands to numerous numerical control machine tools (part (e) of Figure 1b). This was appropriate when computer technologies were still somewhat expensive and physically bulky. However, the advent of microprocessors allowed the computer control of the machine tools to become localized, with one computer dedicated to a single machine tool (part (d) of Figure 1a). This computer numerical control (CNC) lessened the impact of computer downtime, a mishap which disabled all the machine tools under DNC but affected only the local machine tool under CNC. As with all computer technologies, the reliability of numerical control of CNC machine tools today is not questioned; the cause of unplanned downtime on CNC machines today is more likely to be a broken tool or another unexpected contingency. Part (f) of Figure 1b shows Cook's (1975) view of the ultimate "computer-managed parts manufacturing" system. It shows a single centralized computer controlling pallet movement, workpiece loading and unloading at several machines, and the delivery of individual commands to the machines directly. A more common reality is that a centralized computer does control the material handling system but delivers part programs to computers located at the individual CNC machine tools, which then execute the programs (i.e., control the operations) locally.

The CNC machine tool is the basic building block of flexible automation in machining. This is because it is easy to download part programs consisting of code describing machine operations using a reasonably high-level language for a variety of part types. These part programs can be easily adapted for engineering design changes or generational changes in the part design. Consequently, access to a library of 'standard' program routines, which may be executed from larger programs, enables far greater versatility than equivalent paper-tape machine tools. Interface with a variety of ancillary machine tool equipment such as robot arms, tool exchangers, tool magazines, automated pallet changers, inspection equipment, part conveyors, CAD/CAM, and automated guided vehicles provide greater flexibility capability.

The typical operation of a flexible machine is as follows. A part is delivered to the machine either by a human operator, an automated guided vehicle, a gravity feed system, or a conveyor. Parts are attached to fixtures designed to hold the parts during machining. The fixtures are attached to pallets which are held on a table inside the CNC machine. The part may be put onto a fixture by operators at the machine or it may be manually put on the same fixture and pallet at a load/unload station and then travels through the system. Sometimes the table inside the machine can rotate so that the spindle holding the tool may cut multiple sides of a part, or multiple parts may be attached to different sides of a fixture. The pallet itself is sometimes fixed semi-permanently to the machine table or sometimes it is automatically exchanged with another pallet containing a new part. The part program is downloaded into current memory and the sequence of operations described by the code commences, controlling the axial movement, spindle speed, tool exchanges, table rotations, coolant delivery, and many other tasks. Once the task sequence has finished, the part/fixture/pallet is changed for another part either automatically (pallet exchanger, robotic arm) or by an operator. The new part may be the same type or different from the newly completed part. The entire sequence occurs again without delay because a different part program may be downloaded almost instantaneously and a different task sequence can commence, requesting different tools and tasks.

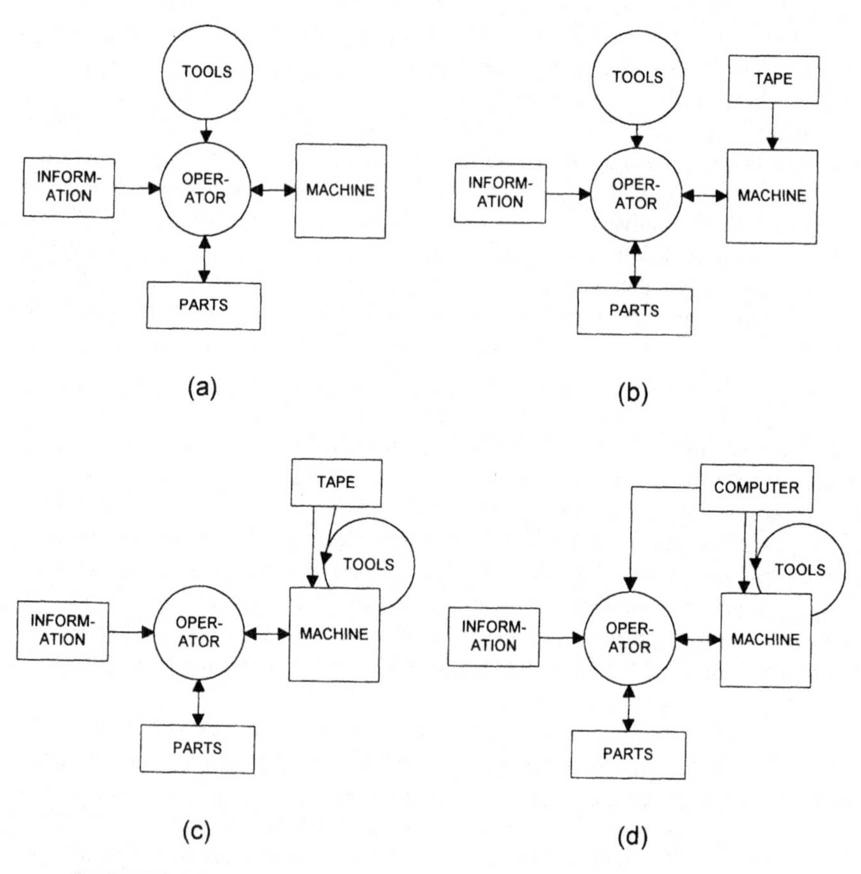

FIGURE 1a. Evolution of flexible automation (from Cook 1975).

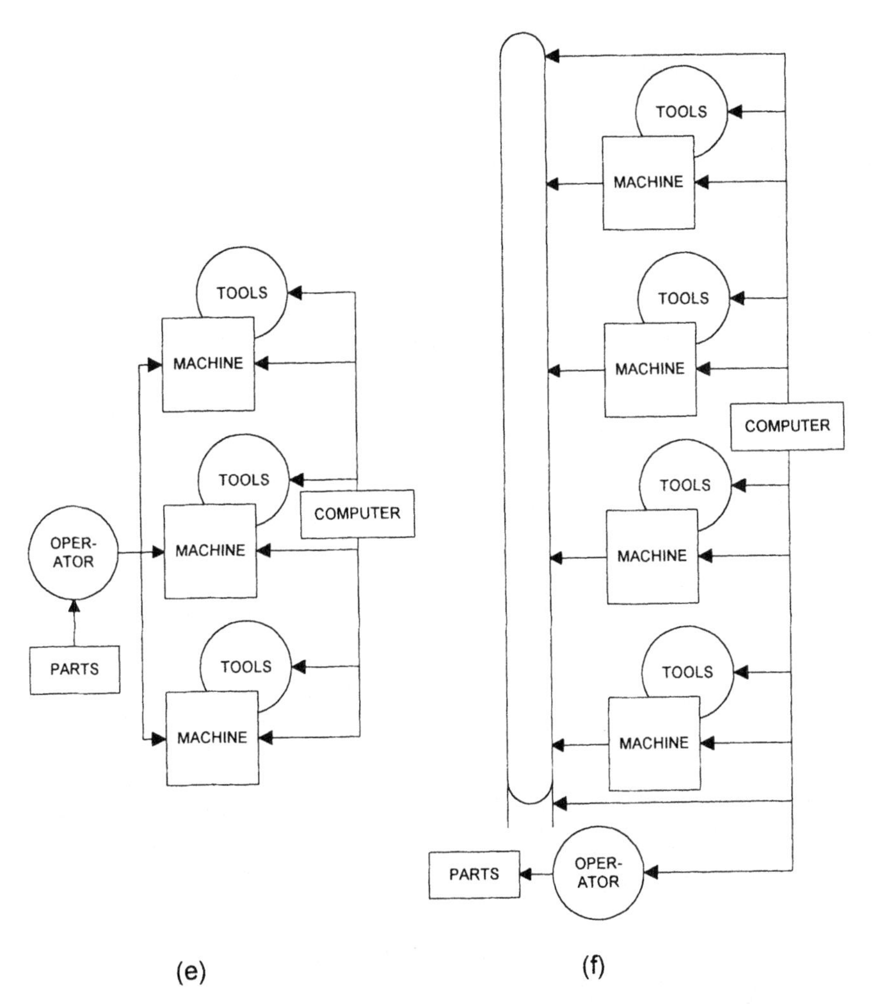

FIGURE 1b. Evolution of flexible automation (from Cook 1975).

Due to a finite tool magazine capacity, care must be given to the decisions concerning which tools are loaded. This is one production planning problem (the FMS loading problem) that must be planned for in order to successfully operate a flexible manufacturing system. This and other production planning problems are discussed in §3.

It may seem that a system having all of the ancillary hardware listed above may be the most flexible of flexible automation. However, it is generally considered that the hierarchy of flexibility follows that in Figure 2. The placement of flexible manufacturing systems, a common form of flexible automation, is best positioned to make a medium variety of part types and a medium volume of parts of each type.

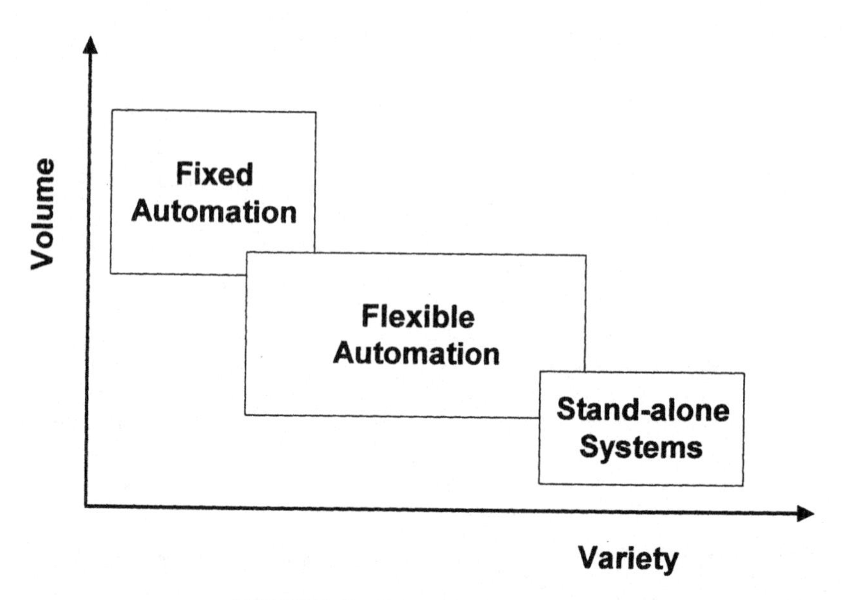

FIGURE 2. Domain of flexible automation.

A dedicated manufacturing technology makes a limited variety of parts (typically one) but at high volumes. A stand-alone machine tool, perhaps operated by a skilled worker, is considered more flexible in that many more part types can be made but with a correspondingly smaller volume. It is commonly said that flexible automation attempts to gain the efficiencies of large batch processing while retaining the flexibility of a job shop. Flexibility here means an ease of changing between manufacture of different part types; other types of flexibility are presented in §4.

3. FLEXIBLE AUTOMATION: LITERATURE AND RESEARCH

The research described as flexible automation research using management science and operations research techniques began to appear regularly in academic journals in the early 1980's. Here, we highlight the various research topics, survey articles, and pieces which contributed significantly to the field. Survey articles include Gunasekaran, Martikainen, and Yli-Olli (1993), which surveys many articles related to flexible manufacturing, Buzacott and Yao (1986), which reviews analytical models, and Sethi and Sethi (1990), which reviews various aspects of manufacturing flexibility.

In a study of early flexible manufacturing practice, Jaikumar (1986) described differences in FMS usage in Japan and the United States. The study found that the more of the capabilities of FMSs were being utilized in Japan, where greater numbers of part types were being made concurrently, and more part designs were being introduced on the machines every year. The American counterparts were using the systems to make high volumes of a few part types, turning over the designs of parts infrequently. These observations should be regarded with caution when considering the situation since then. In a more recent study, Mansfield (1993) ascertained that the rate of acquiring flexible manufacturing systems was no slower in the United States than in Japan but that the first installations in particular industries occured later in the U.S., and the expected rate of return in the U.S. was lower than in Japan. While not exploring the differences in usage as seen in Jaikumar (1986), Mansfield suggests that there could simply be a lag in experience in the usage of such systems between companies in the two countries. This would imply that this difference in usages of system potential could merely be a symptom of this lag, and anecdotal evidence and factory visits strongly suggest this. The difference in usage might also be explained merely by different needs from the system.

As Gunasekaran et al. (1993) notes, management decisions regarding flexible manufacturing systems can be divided into pre-release and post-release decisions. The former deal with decisions that pertain to the planning of parts, machines, and tools before production begins, whereas the latter deals with the decisions once production has begun. Stecke (1983) provides a framework for the production planning (i.e., pre-release) decisions relating to FMSs. This framework suggests the following five problems, with some representative articles:

1. Part type selection problem: From a set of part types that have production requirements, determine a subset for immediate and simultaneous processing (Stecke and Kim 1988; Hwang and Shogan 1989).

2. Machine grouping problem: Partition the machines into machine groups in such a way that each machine in a particular group is able to perform the same set of operations (Purcheck 1985; Kumar, Tewari, and Singh 1990).

3. Production ratio problem: Determine the relative ratios at which the part types selected in problem 1 are to be produced (Stecke 1992).

4. Resource allocation problem: Allocate the limited number of pallets and fixtures of each fixture type among the selected part types (Montazeri, Gelders, and Van Wassenhove 1988; Widmer and Solot 1990)

5. Loading problem: Allocate the operations and required tools of the selected part types among the machine groups subject to technological and capacity constraints of the FMS (Kumar, Singh, and Tewari 1987; Sarin and Chen 1987).

This framework has formed the basis for much research in FMS production planning. These five problems are interrelated and ideally should be solved simultaneously. Indeed many researchers have solved two or more of these problems simultaneously, although it is more common to observe single problem models. Most of these problems are solved subject to the constraints of the other problems, resulting in feasible solutions. Although not all these problems are unique to FMSs, some are very specific, such as the loading problem. The loading problem arises since a machine in an FMS has automated tool exchanging from a tool magazine which typically hold 60 tools, although some have up to 160. Determining which tools should be allocated to a magazine is a challenging management problem that can affect various other functional areas (Gray, Seidmann, and Stecke 1993). FMS planning problems are seen as more complicated than the comparable dedicated system decisions because the system needs to adapt to a variety of contingencies that may arise during automated production. Examples are the completion of the operations for a given part type which must be followed by a different setup, or a tool breaking on one machine and a reallocation of work to the remaining similarly equipped machines in the group. An FMS is typically known for its multiple and frequent setups whereas a more conventional technology has few and infrequent setups. Other pre-release decisions would be the machine and equipment selection (Whitney and Suri 1985; Fine and Li 1988; Monahan and Smunt 1989), design of the FMS layout (Heragu and Kusiak 1988; Kouvelis and Kiran 1989; Afentakis, Millen, and Solomon 1990), and economic justification of the investment (Meredith and Suresh 1986; Wallace and Thuesen 1987; Fine and Freund 1990; Son 1992; Rajagopalan 1994).

The post-release decisions are scheduling problems. Some representative articles are Shalev-Oren, Seidmann, and Schweitzer (1985); Doulgeri, Hibberd, and Husband (1987); Garetti, Pozzetti, and Bareggi (1990); Montazeri and Van Wassenhove (1990); and Sawik (1990).

4. MANAGING FLEXIBILITY

There are many meanings of flexibility in manufacturing. As a result of recent developments in manufacturing technology, there has been much work on defining workable definitions. Browne, Dubois, Rathmill, Sethi, and Stecke (1984) were the first to formally present a taxonomy of eight manufacturing flexibility types, with Sethi and Sethi (1990) reviewing the large literature on the field that grew in the intervening years. Sethi and Sethi (1990) identified about fifty terms relating to eleven flexibility types. Recent activity has focused on developing valid measures for these flexibility types. To a lesser extent, there has been empirical work attempting to determine the 'true' nature of manufacturing flexibility. Some work in this area is Dixon (1992) in the textile industry, Upton (1994, 1995) in the paper industry, and Maffei and Meredith (1995) in various industries including machine tools, medical equipment, engines, valves, and injection molding machines. Gupta and Somers (1992) and Slack (1987) gathered the opinions of executives to investigate manufacturing flexibility. They found that managers' knowledge of flexibility is not consistent, providing impetus for further study of the topic.

Flexibility has been examined before in the context of economics (e.g., Kreps 1979) and mathematics but we focus on manufacturing applications. The commonly accepted flexibility types (Browne et al. 1984; Sethi and Sethi 1990) are:

- *Machine flexibility* (of a machine) refers to the various types of operations that the machine can perform without requiring a prohibitive effort in switching from one operation to another.
- *Material handling flexibility* is the ability to move different part types efficiently for proper positioning and processing through the manufacturing facility it serves.
- *Operation flexibility* of a part refers to its ability to be produced in different ways, commonly by changing the sequence of operations.
- *Process flexibility* of a manufacturing system relates to the set of part types that the system can produce without major setups.
- *Product flexibility* is the ease with which new part types can be added or substituted for existing part types.
- *Routing flexibility* of a manufacturing system is the ability to produce a part by alternate routes through the system.
- *Volume flexibility* of a manufacturing system is its ability to be operated profitably at different overall output levels.
- *Expansion flexibility* of a manufacturing system is the ease with which its capacity and capability can be increased when needed.
- *Program flexibility* is the ability of the system to run virtually untended for a long period.
- *Production flexibility* is the universe of part types that the manufacturing system can produce without adding major capital equipment.
- *Market flexibility* is the ease with which the manufacturing system can adapt to a changing market environment.

This piecewise segmenting approach to manufacturing flexibility implicitly assumes that there are various types of flexibility which are more effectively captured by individually considering them. Indeed, much work has focused on developing measures for individual flèxibility types (for example, Brill and Mandelbaum 1990). Other researchers (Chung and Chen 1989) took a more 'holistic' approach. The variety of approaches is a strength in this challenging field and is not considered contradictory.

While acknowledging multiple types of flexibility, other aspects of flexibilities are also to be addressed For example, Slack (1983) considers flexibility to be a measure of potential rather than performance, whereas Browne et al. (1984) see both potential and actual aspects of flexibility. This distinction between the potential and actual dimensions of flexibility is also discussed by Dixon (1992). Furthermore, other 'dimensions' of flexibility have been used to develop measures. These include:

Action vs. State (Mandelbaum 1978). Action flexibility is considered present when active decisions are needed to counter new circumstances after some change. State flexibility is considered present when the system can cope with changes without intervention.

System vs. Machine level (Buzacott 1982) refers to the source of the capability for the flexibility type. For example, machine flexibility is a machine level flexibility since it is self contained, and independent of the other system elements. Routing flexibility, on the other hand, is a system level flexibility since it requires system resources (machines tools, materials handling system, coordinating software) to be used.

Range vs. Response (Slack 1987) refers to the distinction between the extent to which a system can adapt to a change, and how quickly it adapts to the change. Slack (1987) and Upton (1994) observed that managers thought about flexibility more constructively when aligning their thoughts along these dimensions.

Static vs. Dynamic (Carlsson 1992) refers to the scope of uncertainty surrounding the changes to the system. If a system is static flexible, it is capable of handling foreseeable changes such as fluctuations in demand. Dynamic flexibility refers to the system's ability to cope with changes that were largely unpredictable such as sudden regulatory or environmental changes.

Potential vs. Actual (Browne et al. 1984) draws the distinction between possessing a flexibility to use when a circumstance arises, and utilizing a flexibility consistently in operations regardless of contingencies. For example, a potential use of routing flexibility would be having the AGV routes programmed and laid out between various machine tools and load/unload stations but using only a fixed route until a machine breakdown or some other event occurs. The active use of routing flexibility would be sending parts through the system independent of changes.

Short, Medium vs. Long term (Carter 1986) refers to the time horizon over which the flexibility type influences the system or the system's environment.

Range, Mobility vs. Uniformity (Upton 1994) refers to three aspects of length of measure, ease of movement, and similar performance at different operating positions.

Now, how can manufacturing flexibility be managed? As can be seen from the above taxonomies, flexibility is a complicated concept. However, it may be more beneficial for managing flexibility by thinking along these dimensions and types rather than attempting to manage the rather nebulous 'manufacturing flexibility' as a whole. Upton (1995) suggests that flexibility can be managed but that it is folly to assume that automation alone will provide this flexibility; the involvement of human operators is more important for effective utilization of manufacturing flexibility than computer integration. Upton (1995) found that inexperienced workers may provide more operational mobility flexibility than experienced workers; this was attributed to not being as conscious of equipment limits. Maffei and Meredith (1995) investigated the importance of operational infrastructure to achieving the benefits of flexible manufacturing technologies. The three aspects of operational integration considered were: operator's role, the production planning systems, and the integration of technology, and specific relationships between these infrastructure and system benefits were proposed.

5. FLEXIBLE AUTOMATION, GT AND CELLULAR MANUFACTURING

Flexible automation, group technology, and cellular manufacturing are three distinct concepts that can be applied separately, with positive results. However, applied concurrently, and sensibly, they may result in a very synergistic combination.

We first establish some common definitions for sake of clarity. Group technology (GT) is a technique which groups part types into families based on some common factor or factors, such as manufacturing process or functionality. Cellular manufacturing is a layout strategy based on grouping manufacturing equipment into cells to produce a family of part types. These terms are sometimes used interchangably in the literature. As discussed earlier, flexible automation is a technology which permits the concurrent manufacture of multiple part types. These are, of course, simplified definitions for the sake of exposition.

Group technology literature offers many techniques and criteria for grouping of part types. The primary criteria for categorizing part types are some form of commonality in part geometry or processing requirements. The operating objective of classifying the part types in this way is to subsequently manufacture more efficiently, in some sense, than could otherwise be done with all of the part types in one large family, by exploiting some aspect of commonality in the manufacture while still maintaining a portfolio of heterogeneous part types. For example, if all the cylindrical part types requiring a specific customized tool were grouped into the same family, they could be manufactured on a single lathe machine, in a cell with other machines for the remaining operations. However, the criteria by which parts are grouped into families can vary markedly, depending on objective, which is a primary challenge for GT researchers.

Cellular manufacturing (CM) is often regarded as directly related to group technology because the techniques can naturally be applied together. It is, however, a separate concept that may be applied independently. Simply stated, CM is the grouping of several machines into a configuration that specializes in the production of a part type family. The suggested benefits of cellular manufacturing are reduced material handling, less work in process inventory, fewer quality problems through more direct feedback, greater operator commitment, reduced production leadtimes, and fewer machine setups. Manually operated cells are often U-shaped for ease of movement for the operator, flexibility and balance for operators, more immediate rework of mistakes, team-oriented, and easier linking with other U-shaped lines. More automated cells can be any shape.

Radharamanan (1994) suggests that flexible automation should be considered differently when analyzed in group technology. He suggests that machine tools can be grouped logically rather than physically since they have automated material handling available. He then suggests methods for grouping part types that are particularly suited for flexible manufacturing systems, which overcomes limitations of usual methods, and a group scheduling heuristic for batch production processes.

More pertinent than the automated material handling equipment is the availability of automated tool handling and tool magazines, which enable a variety of operations to be performed on a single machine tool. For an FMS, the set of similar part types produced simultaneously can easily and quickly change over time, just by changing cutting tools. A logical machine tool can be changed continuously by exchanging cutting tools resident in the automated tool magazine. The distinction between physical and logical machines is an interesting one and potentially very useful; it should be explored further in future research.

Burbidge (1985) states that GT is a useful precursor to flexible automation. He suggests that GT permits suitable tooling families to be found for automation investment in the future. Specifically, he details how a group technology technique, Production Flow Analysis (PFA), can be used as a tool to plan for flexible automation. PFA is a technique used to plan a change from a process-focused plant to a product-focused plant. PFA divides part types into "families" and machines into associated "groups" to define the part type families. This simultaneous finding of families and groups highlights the difference of PFA from the usual classification and coding (C&C) systems such as OPITZ, KC-1, KK-1, MICLASS, TELKA, CODE-MDSI, DCLASS, and BRISCH-BIRN. These C&C schemes are used to decompose part types into descriptions of geometric shape, material, dimensional accuracy, function, and so forth. PFA incorporates C&C but goes the additional step of examining the flow of material at the company, factory, department, and machine group level.

PFA involves five progressive techniques: company flow analysis, factory flow analysis, group analysis, line analysis, and tool analysis. Burbidge suggests that the sequential solution of these PFA phases up to and including line analysis results in a useful plan for the design of FMSs. Line analysis is based on examining the flow of material between machines in a group. Given a large proportion of "simple" part types (i.e., part types that could be made on a combination of lathes, drills, and mills) in a family, the selection of flexible automation could greatly simplify the flow. The final PFA phase is tool analysis, which considers each machine in a group and determines the list of tooling required, tooling families of part types made with the same tools, and information needed to reduce tooling variety and standardize the tooling used in a group.

The successive nature of PFA suggests that the decisions taken at one stage are fixed for subsequent stages. However, the nature of flexible automation suggests that PFA is an inappropriate tool for planning either the design of FMSs, the part type assignments to FMSs, or the tool loading of FMSs. Since many tools can be allocated to a single tool magazine on a flexible machine, it seems inappropriate to allocate machines, part types, and tools at different stages. Also, tools are changed over time, to allow the production of different part types on different days, for example.

The notion that GT can form a basis to design FMSs contrasts with the experiences of Slomp, Molleman, and Gaalman (1993). They studied seven Dutch manufacturers who found that acquiring flexible automation first, served as a useful precursor to the development of cellular manufacturing in their firms. Several of the companies that were studied, had acquired flexible machine tools first, which then acted as a catalyst for the transition from a functional layout to a CM product layout.

A good use of flexible automation requires an examination of how the part types are allocated to the equipment, so the installation of the FMSs prompted the companies to undergo a GT analysis and to eventually implement a CM structure. The transitions took from 3 months to 2 years. For these companies, the CM structure was not fixed. The companies fully expected, and experienced, that any current CM structure (people and machines) needed to be reallocated within 1 to 3 years because of changes in the part type mix and/or new technology. These expectations suggest that there are benefits to incorporating flexible automation with cellular manufacturing to accommodate such changes in the future. Gunasekaran et al. (1993) suggests that for the FMS grouping problem, "the GT approach needs to code the characteristics of the parts and their associated components such as tools, pallets, and fixtures" and consider the "limits on the number of part types and associated components in each group". That is, the concepts are not always incompatible but need to be adapted for successful implementation.

Slomp (1997) relates another industrial experience. SVEDEX, a manufacturer of doors, found that the installation of two CNC machines greatly simplified the material flow on their shop floor clearing the way for production expansion. The CNC equipment integrated three different existing manufacturing processes, and the company expects further manufacturing lines to be added. Indeed, a couple of assembly processes will immediately follow the CNC operations, incorporating teams. In essence, the company believes that flexible automation was a prerequisite for CM.

Further, Slomp (1997) suggests that there is an inherent contradiction between CM and flexible automation. With CM there is a tendency to decentralize several tasks, such as maintenance, quality control, planning and control, etc. In flexible automation there is a tendency to centralize more: maintenance is complex and usually dedicated to an FMS; also, parts of the planning and control are more complex. These opposing tendencies create an interesting problem for those intending to implement a blend of CM and flexible automation. The experiences of Slomp et al. (1995) demonstrate that the introduction of flexible manufacturing is often a catalyst for a firm to reorganize its production using GT and CM.

Some researchers claim that GT and/or CM can be used to help design FMSs. See, for example, Mitchell (1991) and Luggen (1991). It is easy to see why. GT and CM group like part types and form small cells of the few machines necessary to produce families of similar part types. However, we believe that using only GT and CM are insufficient to design FMSs for many reasons.

One reason is because of the flexibility and the automation. Because of the versatility of the machine tools to perform many (sometimes all) types of operations required and the cutting tool magazine quick interchange capability, often a part type only requires one machine visit for its entire processing. Also, for the same FMS, a wide variety of part types can sometimes be produced, with different mixes of different part types produced over time.

Then using only GT and CM to design an FMS appears to be too limiting. GT/CM selects only a minimum number of machines of each type that are necessary to produce a family of similar part types. Such GT/CM techniques are incapable of designing an FMS consisting of a number of identical CNCs that are to be used to produce changing mixes of part types over time.

Another reason is that the modes of operation are very different for GT cells and FMSs. A GT cell tends to be a fixed route, unbalanced flow system, often with workers helping one another, producing the same family of part types. An FMS tends to have variable routing, continuously changing part types over time, and the capability of balancing the workloads for various mixes of part types. Indeed, the mode of operation can change from day to day. As an example, consider an FMS consisting of eight identical CNCs. If the FMS is "sufficiently" flexible and reliable, it can operate in many ways. For example, if there is a rush for an order (and there are sufficient cutting tools and all machines are equally capable), it is possible to tool all machines to produce only that rush part type. Over time, these same machines can be tooled in many ways. Sometimes a large number of part types need to be processed and each operation is assigned to only one machine. At other times, fewer part types can be selected and two or more machines can be tooled for each operation. There could be a wide variety of ways to operate an FMS.

Indeed, current books such as Askin and Standridge (1993) and Chang, Wysk, and Wang (1998) treat these two topics independently in separate chapters. Often, more capacity-oriented methods are more appropriate to use to design FMSs; GT/CM techniques are too limited.

An interesting question is how flexible automation can operate harmoniously and profitably with group technology and cellular manufacturing. An answer may lie in which of the flexibility types listed in §4 are used to advantage. Cohen and Apte (1997) describe an interesting application of the three concepts at Cummins Engine Company, where an innovative mixture of cellular manufacturing and flexible automation has been incorporated for the manufacture of diesel engines.

After enjoying a stable, profitable market during the 1970's, Cummins experienced competitive and regulatory pressures during the 1980's forcing them to increase product variety, reduce costs, and decrease delivery leadtimes. The company adopted cellular manufacturing to replace their existing process-focused production. But as the variety of part types proliferated, their cells became congested with the changeovers needed. The answer lay with innovatively investing in flexible and dedicated equipment arranged in cellular layouts. They chose to allocate their high volume part types to traditional layouts and cells with dedicated technologies, and the large number of remaining part types with smaller volumes to the cells with flexible automation. Part types with unstable designs were also allocated to the flexible systems, which could handle the design changes. The company experienced substantial improvements. This is an example of Burbidge's hypothesis that cellular manufacturing can establish the foundation for flexible technologies.

With conflicting theories and experiences regarding the relationship between GT/CM and flexible automation, investigation of how these concepts and technologies can operate harmoniously may be a fruitful research topic. It should also prove to be useful for firms intending to implement cellular manufacturing with flexible automation but are apprehensive about the risks inherent in investing in expensive technologies; guidelines resulting from academic research as to how to best invest could be invaluable for such companies.

6. SOME ISSUES FOR FUTURE RESEARCH

There are a variety of possible directions for combining group technology, cellular manufacturing, and flexible automation. First, there is the question of whether it is appropriate to invest in flexible automation before or after reorganizing for cellular manufacturing. The current evidence from practice for determining this is contradictory and further rigorous empirical studies could suggest appropriate implementation strategies. Resolution of this sequencing issue would be of significant interest to practitioners. It is highly likely, as in other fields of study, that the solutions to answer this problem may lie in industrial practice. As with any empirical study, however, caution must be exercised so that field observations do not mislead the development of theory. Care should be taken to ensure that observations are common enough to be generalized, not aberrant exceptions resulting from inefficient industry practice. The next step is to find theoretical foundations for these decisions. The most appropriate domain for this research is academia, building mathematical models of cellular manufacturing organization and the investment of flexible automation.

The integration of group technology with CAD/CAM activities is another issue. It is apparent that the incorporation of aspects of GT with CAD/CAM will produce superadditive benefits. For example, using a CAD/CAM system, firms can determine quantitative methods to find similarities in part types in both manufacturing process and physical configuration (size, material, etc.). Consequently, a more effective grouping of machines can be found. In addition, the design of various part types could be adapted to accommodate some better grouping of part types.

7. SUMMARY

Technological developments in manufacturing present challenges and opportunities for both practitioners and observers (academics, consultants, etc.). Flexible manufacturing is one of these developments. It has presented challenges to overcome in integrating it into existing concepts, practices, and theory. The opportunities it presents focus on integrating the technology in a way to produce a greater effect, that is a synergistic effect, than the current practices or the new technology could by themselves.

In this chapter we have introduced and described the technologies collectively known as flexible automation. We summarized a history of its development which draws primarily from the combination of standard machine tool technology and the advent of the computer integrated circuit, which permitted the automation of various tasks. These tasks involve programmed operation commands such as the transit of the spindle, authorizing spindle speed, using lubricant and coolant, exchanging tools from a magazine, input and output of a part using automated material handling equipment and pallet exchangers, and rotating the table inside the workspace to gain access to various sides of a part.

Then, various research streams involving flexible automation were described, citing relevant research articles. Five production planning problems were reviewed and other problems involving acquisition, design, layout, and scheduling were listed. A discussion of various flexibilities that have been observed in practice and studied in theory, was provided, along with several conceptual dimensions to highlight the difficult problems posed by attempting to measure and capture these flexibilities in models, as a step towards attempting to manage them effectively.

This was followed by a discussion of how flexible automation may mesh with group technology and cellular manufacturing. There is not much theoretical grounding for recommending a particular acquisition sequence of these technologies and the evidence from practice is somewhat conflicting, suggesting an important research topic for the future.

REFERENCES

Afentakis, P., Millen, R.A., and Solomon, M.M., 1990, Dynamic layout strategies for flexible manufacturing systems, *International Journal of Production Research*, 28, 311-323.

Askin, R.G. and Standridge, C.R., 1993, *Modeling and Analysis of Manufacturing Systems*, John Wiley and Sons, New York, NY.

Berrada, M. and Stecke, K.E., 1986, A branch and bound approach for machine load balancing in flexible manufacturing systems, *Management Science*, 32, 10, 1316-1335.

Boothroyd, G., 1992, *Assembly Automation and Product Design*, Marcel Dekker, New York, NY.

Brill, P.H. and Mandelbaum, M., 1989, On measures of flexibility in manufacturing systems, *International Journal of Production Research*, 27, 747-756.

Brill, P.H. and Mandelbaum, M., 1990, Measurement of adaptivity and flexibility in production systems, *European Journal of Operational Research*, 49, 325-332.

Browne, J., Dubois, D., Rathmill, K., Sethi, S.P., and Stecke, K.E., 1984, Classification of flexible manufacturing systems, *The FMS Magazine*, 4, 114-117.

Burbidge, J.L., 1985, Production flow analysis and the design of FMSs, in *Advances in Production Management Systems 85: Proceedings of the Second IFIP WG 5.7 Working Conference on Advances in Production Management Systems - APMS 85*, Elizabeth Szelke and Jimmie Browne (Eds.), Elsevier Science Publishers B.V., Amsterdam, The Netherlands, 43-55.

Burbidge, J.L., 1992, Change to group technology: Process organization is obsolete, *International Journal of Production Research*, 30, 5, 1209-1219.

Buzacott, J.A., 1982, The fundamental principles of flexibility in manufacturing systems, *Proceedings of the First International Congress on Flexible Manufacturing Systems*, Bedford, UK, 13-22.

Buzacott, J.A. and Shanthikumar, J.G., 1993, *Stochastic Models of Manufacturing Systems*, Prentice Hall, Englewood Cliffs, NJ.

Buzacott, J.A. and Yao, D.D., 1986, Flexible manufacturing systems: A review of analytical models, *Management Science*, 31, 890-905.

Carlsson, B., 1992, Management of flexible manufacturing: An international comparison, *OMEGA*, 20, 1, 11-22.

Carter, M.F., 1986, Designing flexibility into automated manufacturing systems, in *Proceedings of the Second ORSA/TIMS Conference on Flexible Manufacturing Systems: Operations Research Models and Applications*, (Eds. K.E. Stecke and R. Suri), Elsevier Science Publishers B.V., Amsterdam, The Netherlands, 107-118.

Chang, T.-C., Wysk, R.A., and Wang, H.-P., 1998, *Computer-Aided Manufacturing*, 2nd Edition, Prentice Hall, Englewood Cliffs, NJ.

Chung, C.H. and Chen, I.J., 1989, A systematic assessment of the value of flexibility from FMS, in *Proceedings of the Second ORSA/TIMS Conference of Flexible Manufacturing Systems: Operations Research Models and Applications*, (Eds. K.E. Stecke and R. Suri), Elsevier Science Publishers B.V., Amsterdam, The Netherlands, 27-34.

Cohen, M.A. and Apte, U.M., 1997, *Manufacturing Automation*, Irwin, Chicago, IL.

Cook, N.H., 1975, Computer managed parts manufacture, *Scientific American*, February, 22-29.

Dixon, J.R., 1992, Measuring manufacturing flexibility: An empirical investigation, *European Journal of Operational Research*, 60, 131-143.

Doulgeri, Z., Hibberd, R.D., and Husband, T.M., 1987, The scheduling of flexible manufacturing systems, *Annals of the CIRP*, 36, 1-14.

Fine, C.H., 1993, Developments in manufacturing technology and economic evaluation models, in *Logistics of Production and Inventory*, (Eds. S.C. Graves, A.H.G. Rinnooy Kan, and P.H. Zipkin), North Holland, Elsevier Science Publishers B.V., Amsterdam, The Netherlands, 711-750.

Fine, C.H. and Freund, R.M., 1990, Optimal investment in product-flexible manufacturing capacity, *Management Science*, 36, 4, 449-466.

Fine, C.H. and Li, L., 1988, Technology choice, product life cycles, and flexible automation, *Journal of Manufacturing and Operations Management*, 1, 372-399.

Garetti, M., Pozzetti, A., and Bareggi, A., 1990, On-line loading and dispatching in flexible manufacturing systems, *International Journal of Production Research*, 28, 1271-1292.

Gerwin, D, 1993, Manufacturing flexibility: A strategic perspective, *Management Science*, 39, 4, 395-410.

Gray, A.E., Seidmann, A., and Stecke, K.E., 1993, A synthesis of decision models for tool management in automated manufacturing, *Management Science*, 39, 5, 549-567.

Gunasekaran, A., Martikainen, T., and Yli-Olli, P., 1993, Flexible manufacturing systems: An investigation for research and applications, *European Journal of Operational Research*, 66, 1-26.

Gupta, Y.P. and Somers, T.M., 1992, The measurement of manufacturing flexibility, *European Journal of Operational Research*, 60, 166-182.

Heragu, S.S. and Kusiak, A., 1988, Machine layout problem in flexible manufacturing systems, *Operations Research*, 32, 258-268.

Hwang, S.S. and Shogan, A.W., 1989, Modelling and solving an FMS part selection problem, *International Journal of Production Research*, 27, 1349-1366.

Jaikumar, R., 1986, Postindustrial manufacturing, *Harvard Business Review*, 64, 6, 69-76.

Kouvelis, P. and Kiran, A.S., 1989, Layout problem in flexible manufacturing systems: Recent research results and further research directions, in *Proceedings of the Third ORSA/TIMS Conference of Flexible Manufacturing Systems: Operations Research Models and Applications*, (Eds. K.E. Stecke and R. Suri), Elsevier Science Publishers B.V., Amsterdam, The Netherlands, 147-152.

Kreps, D.M., 1979, A representation theorem for "Preference for Flexibility", *Econometrica*, 47, 3, 565-578.

Kumar, P., Singh, N., and Tewari, N.K., 1987, A nonlinear goal programming model for the loading problem in a flexible manufacturing system, *International Journal of Production Research*, 25, 13-20.

Kumar, P., Tewari, N.K., and Singh, N., 1990, Joint consideration of grouping and loading problems in a flexible manufacturing system, *International Journal of Production Research*, 29, 1345-1356.

Kusiak, A., 1986, Applications of operational research models and techniques in flexible manufacturing systems, *European Journal of Operational Research*, 24, 336-345.

Luggen, W.W., 1991, *Flexible Manufacturing Cells and Systems*, Prentice-Hall International, Englewood Cliffs, NJ.

Maffei, M.J. and Meredith, J., 1995, Infrastructure and flexible manufacturing technology: theory development, *Journal of Operations Management*, 13, 4, 273-298.

Mandelbaum, M, 1978, Flexibility in Decision Making: An Exploration and Unification, Ph.D. Thesis, Department of Industrial Engineering, University of Toronto, Canada.

Mansfield, E., 1993, The diffusion of flexible manufacturing systems in Japan, Europe, and the United States, *Management Science*, 39, 2, 149-159.

Meredith, J.R. and Suresh, N.C., 1986, Justification techniques for advanced manufacturing technologies, *International Journal of Production Research*, 24, 1042-1057.

Mitchell, F.H., 1991, *CIM Systems: An Introduction to Computer-Integrated Manufacturing*, Prentice-Hall, Englewood Cliffs, NJ.

Monahan, G.E. and Smunt, T.L., 1989, Optimal investment of flexible manufacturing systems, *Operations Research*, 37, 2, 288-300.

Montazeri, M., Gelders, L.F., and Van Wassenhove, L.N., 1988, A modular simulator for design, planning, and control of flexible manufacturing systems, *International Journal of Advanced Manufacturing Technology*, 3, 15-32.

Montazeri, M. and Van Wassenhove, L.N., 1990, Analysis of scheduling rules for an FMS, *International Journal of Production Research*, 28, 785-802.

Purcheck, G., 1985, Machine-component group formation: An heuristic method for flexible production cells and flexible manufacturing systems, *International Journal of Production Research*, 23, 911-943.

Radharamanan, R., 1994, Group technology concepts as applied to flexible manufacturing systems, *International Journal of Production Economics*, 33, 133-142.

Rajagopalan, S., 1994, Capacity expansion with alternative technology choices, *European Journal of Operational Research*, 77, 3, 392-403.

Sarin, S.C. and Chen, S.C., 1987, The machine loading and tool allocation problem in a flexible manufacturing system, *International Journal of Production Research*, 25, 1081-1094.

Sawik, T., 1990, Modelling and scheduling of a flexible manufacturing system, *European Journal of Operational Research*, 45, 177-190.

Sethi, A.K. and Sethi, S.P., 1990, Flexibility in manufacturing: A survey, *International Journal of Flexible Manufacturing Systems*, 2, 4, 289-328.

Shalev-Oren, S., Seidmann, A., and Schweitzer, P.J., 1985, Analysis of flexible manufacturing systems with priority scheduling: PMVA, *Annals of Operations Research*, 3, 115-139.

Slack, N., 1983, Flexibility as a management objective, *International Journal of Operations and Production Management*, 3, 3, 4-13.

Slack, N., 1987, The flexibility of manufacturing systems, *International Journal of Operations and Production Management*, 7, 4, 35-45.

Slomp, J., 1997, Personal communication.

Slomp, J., Molleman, E., and Gaalman, G., 1993, Production and operations management aspects of cellular manufacturing - a survey of users, in *Advances in Production Management Systems*, (Eds. I.A. Pappas and I.P. Tatsiopoulos), Elsevier Science Publishers B.V. (North Holland), Amsterdam, 553-560.

Slomp, J., Molleman, E., and Suresh, N.C., 1995, Empirical investigation of the performance of cellular manufacturing systems, in *Management and New Production Systems*, (Eds. D. Draaijer, H. Boer, and Koos Krabbendam), University of Twente, Enschede, The Netherlands, 432-441.

Son, Y.K., 1992, A comprehensive bibliography of justification of advanced manufacturing technologies, *The Engineering Economist*, 38, 1, 59-71.

Stecke, K.E., 1983, Formulation and solution of nonlinear integer production planning problems for flexible manufacturing systems, *Management Science*, 29, 3, 273-288.

Stecke, K.E., 1992, Procedures to determine part mix ratios for independent demands in flexible manufacturing systems, *IEEE Transactions on Engineering Management*, 39, 4, 359-369.

Stecke, K.E. and Kim, I., 1988, A study of FMS part type selection approaches for short-term production planning, *International Journal of Flexible Manufacturing Systems*, 1, 1, 7-29.

Stecke, K.E. and Suri, R. (Eds.), 1985, Flexible Manufacturing Systems: Operations Research Models and Applications, *Annals of Operations Research*, Baltzer, Basel, 3.

Stecke, K.E. and Suri, R. (Eds.), 1988, Flexible Manufacturing Systems: Operations Research Models and Applications II, *Annals of Operations Research*, Baltzer, Basel, 15.

Upton, D.M., 1994, The management of manufacturing flexibility, *California Management Review*, 36, 2, 72-89.

Upton, D.M. 1995, Flexibility as process mobility: The management of plant capabilities for quick response manufacturing, *Journal of Operations Management*, 12, 205-224.

Upton, D.M., 1997, Process range in manufacturing: An empirical study of flexibility, *Management Science*, 43, 8.

Wallace, W.J. and Thuesen, G.I., 1987, Annotated bibliography of investing in flexible automation, *The Engineering Economist*, 32, 247-257.

Whitney, C.K. and Suri, R., 1985, Algorithms for part and machine selection in flexible manufacturing systems, *Annals of Operations Research*, 3, 34-45.

Widmer, M. and Solot, P., 1990, Do not forget the breakdowns and the maintenance operations in FMS design problems, *International Journal of Production Research*, 28, 421-430.

AUTHORS' BIOGRAPHY

Kathryn E. Stecke is Jack D. Sparks/Whirlpool Corporation Research Professor in Business Administration at The University of Michigan, Ann Arbor, MI. She received an M.S. in Applied Mathematics, and an M.S. and Ph.D. in Industrial Engineering from Purdue University. She has authored numerous papers on various aspects of FMS production planning and scheduling in numerous journals including *Management Science*, *Operations Research*, *International Journal of Production Research*, *European Journal of Operational Research*, *IIE Transactions*, and *IEEE Transactions on Engineering Management*. She is the Editor-in-Chief of the *International Journal of Flexible Manufacturing Systems*. She has delivered seminars at educational institutions and companies around the world, and has consulted for and visited many corporations in the area of flexible manufacturing. She is a member of INFORMS, SME, and IFIP Working Group 5.7.

Rodney P. Parker is a Ph.D. candidate at the School of Business Administration, The University of Michigan, Ann Arbor, MI. He has an M.S.E. in Industrial and Operations Engineering from the University of Michigan and a Bachelor's degree in Industrial Engineering and a Master's degree in Manufacturing Management from the University of Melbourne in Melbourne, Australia. He has worked in the steel, automotive, and aerospace industries. His current research focuses on capacity planning and evaluation and technology selection in manufacturing. He is a member of INFORMS.

H1

Human Resource Management and Cellular Manufacturing

R. Badham, I. P. McLoughlin and D. A. Buchanan

1. INTRODUCTION

In recent years, cellular manufacturing (CM) has come to be linked with a variety of different manufacturing philosophies and human resource management (HRM) strategies such as Japanese 'lean production' methods (Schonberger 1992), European socio-technical 'production island' philosophies (Brodner 1989) as well as with other new manufacturing strategies such as the 'fractal' factories (Warnecke 1993).

There is a similar degree of variation on the side of HRM. As Legge (1995) observes, there are both 'soft' and 'hard' forms of HRM, the former associating it with the promotion of a particular approach to HRM involving high levels of teamwork and cooperation, devolution of responsibilities, the existence of 'strong' cultures etc., and a 'hard' form that is more concerned with aligning human resource practices with the strategic needs of the organization. Consequently, for some research on HRM and CM, the focus is strongly on the benefits of high levels of semi-autonomous group work (Ingersoll 1990), whereas for others the issue is one of discerning the real benefits, or costs, of HRM activities and rhetoric. (Buchanan 1994). This ambiguity is matched by the promotion of very different models of teamwork as the 'group' work associated with CM, some models drawing strongly on 'Japanese' forms of teamwork and other on 'Western' models of 'self-managing' work teams. (Buchanan and Preston 1994; Badham, Couchman and Selden 1995). As if this was not confusing enough, there are some definitions of CM that refer only to the technical operations management dimensions (Wemmerlöv and Hyer 1982), while others include new forms of group work within the definition (Burbidge 1979; Ingersoll 1996).

This chapter attempts to steer a mid-course between overviewing these debates and providing a clearer, yet narrower definition of CM and its human resource dimensions. The focus of the chapter will be a critical appraisal of previous human resource debates on CM, focusing in particular on the linkages drawn between CM and teamwork. While risking the dangers of over-stereotyping positions, the chapter outlines a traditional 'one-dimensional' model of the relationship between CM, work redesign, teamwork and performance, and contrasts this with the more recent 'two-dimensional' approaches that use bi-polar models of types of effective teamwork, and the desirability or undesirability of the new forms of teamwork. These two models are then contrasted with an emerging 'multi-dimensional' approach to CM and teamwork. This is aimed at providing a more sophisticated framework for exploring the technical and human resource character and dynamics of CM. In conclusion, it is argued that not only is additional empirical work needed, but also, importantly, further theoretical development is required.

2. THE TRADITIONAL VIEW: ONE DIMENSIONAL ANALYSIS

The explicit concern with the human aspects of CM rose to prominence in the 1970s and early 80s. During this period, however, it tended to be dominated by what an earlier French philosopher rather clumsily called a 'humanist-economist' problematic' (Althusser 1978). This approach regards technology as driven by natural scientific laws and engineering knowledge and techniques. Following this approach, CM was, in much of the early engineering and organizational literature, defined as a clear and relatively uncontested technical methodology. At the same time, the 'human' consequences of this technology were strongly emphasized but treated in terms of individual psychology, universal human needs and the full development of human potential or realization of human rights. The 'people' issues associated with CM were primarily seen as created by the 'impact' of the technology, and as the most crucial factors in obtaining the promised results of CM.

This 'traditional' approach has four main components:

1. a defined technical characterization of CM;
2. an evaluation of the benefits of CM as clearly outweighing its costs
3. a view that CM enables and makes necessary or desirable enriched jobs and group work; and;
4. a focus on overcoming implementation problems

This approach is defined here as 'one-dimensional' because of its overarching view that CM is fundamentally *one* type of production system, and that there is, in general, one type of work organization 'required' or 'best suited' to its effective operation.

2.1. The Technical Dimension: Group Technology To Assembly Cells

CM is presented as a system of production that stands between the functional layout traditionally adopted in batch manufacturing and line-based mass production, combining the flexibility of batch production with the efficiencies of flow production while avoiding the costs and inefficiencies of both (Badham and Benders 1997 forthcoming). A dominant focus of the earlier CM research was its use in batch fabrication environments (see definition of CM by Wemmerlöv and Hyer (1989, p. 1511). An additional focus, also recognized by Wemmerlöv and

Hyer (1987, p. 414), is the application of CM principles to assembly. Unlike the assembly line, however, cells retain a greater degree of flexibility to make or assemble a range of similar products. In mass assembly this has meant the creation of a series of shorter, parallel and 'dock' assembly systems to create the degree of cell autonomy needed to deal with variations in product and demand (Aguren and Edgren 1980) This broader focus on fabrication and assembly has also been adopted by Ingersoll Engineers in their survey of 300 UK engineering companies.

When addressing the HRM aspects of CM, the traditional one-dimensional approach often defines CM in this broad sense. It is important to note, however, that while this general definition is adopted as the focus of study, both Wemmerlöv & Hyer and Ingersoll Engineers make the point that there are different types of cells. Wemmerlöv and Hyer (1987, p. 415) point out the weaknesses of adopting such a broad definition for research on HRM issues, and highlight the need for development of a cell taxonomy. However, neither survey goes on to explore human resource aspects of different cell types, preferring to concentrate on CM in general.

2.2. Costs And Benefits

The identification of the clear benefits of CM is a key factor amongst researchers and consultants committed to CM. (Ingersoll 1994; Brodner 1989; Badham et.al. 1995) Both productivity and work satisfaction are identified as clear overall benefits, despite evidence of early costs, lack of suitability to particular industrial environments, and trade-offs between the benefits of traditional batch and line production and cellular production. This is an important element reinforcing the one-dimension focus on *the* benefits of CM as *one* particular type of production system.

2.3. The Social Dimension: Job Redesign And Group Work

In the early discussions of group technology (GT) and CM, in the post-war period and prior to the 1970s, there was hardly any reference to human resource factors (Badham and Benders 1997 forthcoming). However, during the 1970s this changed. GT became more closely linked to group 'work'- influenced in the UK by Burbidge (1979) and the socio-technical research of the Tavistock Institute (Buchanan 1977), in Sweden by the Volvo experiments, and in Europe generally by a rising interest in work humanization and the quality of working life (Burbidge 1976). Since that time, CM techniques have been seen as enabling, facilitating and generally having an 'elective affinity' with group work or, in stronger terms, either requiring job redesign and group work or being explicitly defined as incorporating such elements.

Consequently, Burbidge defines 'group technology' as: *"an approach to the organization of work in which the organizational units are relatively independent groups, each responsible for the production of a given family of products. The smallest organization unit is the group, but the same principle of organization is used when forming larger organizational units such as departments...A group is a combination of a set of workers and a set of machines and/or other facilities laid out in one reserved area, which is designed to complete a specified set of products. The workers in a group share a series of common output targets in terms of lists of products to be completed by a series of common due-dates. The number of workers in a group is limited by the need to obtain social cohesion."*

Despite later refinements, Burbidge's analysis has been highly influential in defining the framework of the one-dimensional approach and remains one of the most comprehensive and clearly expressed. As outlined in Table 1, Burbidge defines the 6 key elements of GT as having a clear social benefit. The overall result of group technology is not only one of increased visibility and improved productivity but also a much improved 'climate' for job satisfaction.

TABLE 1. Desirable Characteristics of Groups (Burbidge 1979).

	Desirable characteristics	**Social Factor**
1. Team	Set of workers special to group	Feeling of belonging
2. Facilities	Set of machines and equipment special to group	Association with the means of production
3. Group Layout	One area special to the group	Association with a territory. (Territorial imperative)
4. Products	Set of products made only in the group	Association with the completion of a significant product.
5. Target	Output targets set for group as a whole	Common objective and simple feedback necessary for achievement motivation.
6. Size	Small number of workers	Group stability and cohesion

Establishing a theme that has reverberated through the later literature, he accompanies this analysis of social desirability with a deterministic argument i.e. that "The change to Group Technology will <u>inevitably</u> affect working conditions in the factory" and, consequently, that "the most difficult job in introducing Group Technology is to persuade people to accept new ways of working which are different to those to which they have been accustomed throughout their working lives." (Burbidge 1979, p.49). In particular, he asserts that there are a number of immediate impacts on direct workers i.e. a movement from a concentration on individual tasks on one type of machine and little social cooperation to flexible transfer between machines and extensive cooperation between cell members in balancing capacity and load. He also emphasizes that groups can now coordinate their own work as simplified processes makes it possible to devolve this responsibility. Moreover, by enabling groups to carry out all operations on 'their' family of parts, and being able to clearly trace the origins of parts to the work of particular cell groups, it is possible to allocate responsibility for quality control and improvement, as well as meeting orders and due-dates.

These themes have continued up to the present as surveys by Ingersoll Engineers (Ingersoll 1990) and case studies (Buchanan and Preston 1992) continue to define CM as an integrated technical and organizational phenomenon and part of several new 'philosophies' of manufacturing. As Dawson argues in his case study of CM, teamwork can be seen as 'complementing' the creation of group technology operations. (Dawson 1994, p.106) Most simply, Proctor, Hassard and Rowlinson (1995, p.48) observe, "when machines are brought together so too are the workers who operate them." In both the latter instances, detailed case studies are drawn upon that continue a similar focus on CM and the new 'flexible' work arrangements for operators and the changing role of supervisors, line management and support staff.

2.4. Implementation As The Key Problem

In 1990, 1994 and 1997, Ingersoll Engineers confirmed the importance of the organizational aspects of implementation in two ways. Firstly, it was illustrated by the fact that the most successful implementers had invested a higher proportion of their expenditure on 'human factors' in addressing the issues outlined above. In the 1990 survey this was indicated by the top 10% of performers having spent in excess of 40% of their investment on people, and 90% of the overall sample identified training and flexibility in the teams and communication as key success factors (Ingersoll 1990, p.39). This was further confirmed in both the 1994 survey in the UK and the 1997 survey in Australia. Secondly, the most widely perceived factors leading to success were commitment at all levels; training/flexibility of the team; and clearly defined responsibilities/ accountabilities (Ingersoll 1990, p.34).

In terms of problems that need to be overcome, Wemmerlöv and Hyer's (1982) survey continued Burbidge's emphasis on addressing the 'people' factors as a key implementation issue. They discovered in their survey that organizational change and associated human resistance was the first of the three sets of most common problems in implementation.

In recent years, Huber and Brown (1991) have provided the most useful extension of Burbidge's analysis of organizational issues and the implementation problems that they raise. Their analysis focused on the direct links between CM, its conformance to principles of socio-technical design, and the human resource problems that are created by its introduction. Drawing on the design principles of Albert Cherns and other socio-technical writers, Huber and Brown focus on issues of necessary skills and training for the new participatory and multi-skilled environment, the stress experienced by new forms of accountability and peer pressure, and motivational problems due to the loss of traditional task stability, actual reductions in autonomy, perceived failure of the change to realize promised higher levels of autonomy, or difficulties in creating new equitable and accepted team based pay systems.

The overarching framework within which they interpret these results remains, however, within the one-dimensional approach. As they assert, "if an organization is to attain the full potential of CM, technical changes must be accompanied by parallel changes in the organization's social system". The focus of research is clear: to investigate the 'social' in order to help "realize high levels of success" and help "avoid implementation problems" (Wemmerlöv and Hyer 1987, p.156).

This traditional 'one dimensional' view has played an important role in industry and academia in promoting consideration of human resource issues in the introduction of CM. It has also acted to promote the general concept of CM as a combined socio-technical innovation. Within this traditional view, however, these are commonly perceived as management's HR 'implementation problems' that are an inherent or possible consequence of CM, and that management must address these in order to realize the overall clear productivity and QWL benefits of CM.

3. THE CRITICAL APPROACH: TWO-DIMENSIONAL VIEW

The 'two-dimensional' approach recognizes two fundamentally different types of CM and teamwork, and evaluates the resulting forms of work organization as either leading to autonomy (good) or increasing management control (bad) .

Rather than assuming the character of CM, assessing its overall costs and benefits, and arguing for greater attention to overcoming HR problems - the focus is on the existence of different technical options and the choice between two radically different types of CM: 'lean production' and 'socio-technical'. The former is identified with the Japanese, and particularly 'Toyotist', model of more 'tightly coupled' workstations and cells and limited forms of teamwork, and the latter with European, 'Volvoist', model of 'uncoupled' autonomous cells and semi-autonomous work groups. In addition, HR issues associated with CM are not simply identified as 'problems' to be addressed in implementing an inherently advantageous socio-technical system. Rather, the ambiguities and potential contradictions in concepts of HRM are recognized and explored.

Thus, whilst one 'autonomy' tradition of thought on teamwork celebrates the potential for enhancement of worker autonomy through enriched jobs and increasingly autonomous forms of group work, another 'control' tradition focuses on the new, and more or less subtle and insidious forms of management control embedded within increasingly visible production systems, new forms of electronic and group surveillance, intrusive normative controls, and an increasing ability of management to monitor and set group goals (Badham and Jurgens 1997).

Table 2 outlines the different types of approaches to CM that emerge as a result of making these distinctions. It is important to note that this categorization is for classification and orientation purposes only, it should not be taken to reflect overly rigidly defined positions nor to encompass the whole range of thought on the issue by the authors whose general orientation only is pointed to by their position in particular 'boxes'. Similarly, the contrast between 'one-dimensional' and 'two-dimensional' perspectives helps to illustrate the major themes and issues, as we have seen individual researchers may have ideas or make comments that straddle these divides, despite their bias towards one approach or another.

3.1. Autonomy In Lean Production

In the work of Womack et.al. (1990), Wickens (1994) and Adler (1993a; 1993b), a strong contrast has been drawn between the utopian 'craft romanticism' of attempts by companies such as Volvo to cellularize mass production car assembly lines into relatively autonomous cell teams and the 'lean production' model of tightly coupled 'U' cells introduced by Japanese automobile manufacturers such as Toyota. As outlined in Table 3's contrast between the GM-Toyota NUMMI (lean production model) and the Volvo Uddevalla (socio-technical model) of cellular assembly, there are clear differences in technical configuration, job design and teamwork. As we have discussed elsewhere, this difference reflects long-standing and well established contrasts between 'Japanese' and 'Western' models of teamwork and associated technical design philosophies and configurations, with the Japanese models appearing to have more of an influence on S.E.Asian countries initiatives in this area (Badham, Couchman and Selden 1995; Benders et.al. 1995; Dankbaar 1995). Advocates of lean production models such as Adler and Cole (1993) criticize as unproductive and, in many cases, less satisfying to the workers, 'Volvoist' attempts to enrich individual work by creating technical systems that enable long work cycles and establish semi-autonomous work group operations 'uncoupled' from line pacing and upstream and downstream customers and suppliers.

In Adler and Borys (1996), this argument is extended into a more general critique of the anti-bureaucratic 'metaphysical pathos' embedded in the socio-technical models. Autonomy, they argue, is more realistically and appropriately seen as participation by the workforce in the *design* of the production systems within which they work *not* as an escape from standardized work procedures and tightly coupled production processes in their execution of work tasks. What this means for CM is support for a more tightly integrated 'flow line' model of cell layout and organization, with highly standardized tasks and work procedures, tightly coupled work cells, yet a commitment to strongly involving the workforce in the design and improvement of cell and work operations. Given the constraints of mass production, this is seen as the only realistic area of autonomy for production workers. Moreover, substantial evidence is brought forward to argue that workers may prefer working in such systems rather than dealing with the stress and uncertainty of determining how to perform more loosely defined work tasks.

TABLE 2. Typology of Approaches to Teamwork

	Lean production	Sociotechnical production
Control	Sewell and Wilkinson, 1992; Parker and Slaughter, 1988	Barker, 1993
Autonomy	Adler and Cole, 1993; Wickens, 1994; Womack, Jones and Roos 1990	Jurgens, Malsch and Dohse, 1993; Buchanan, 1994; Berggren, 1993

3.2. Autonomy In Socio-Technical Production

In the work of critics of lean production such as Berggren (1993a; 1993b; 1994) and Jurgens, Malsch and Dohse (1994), there is a similar contrast between lean production and socio-technical models of CM. They are, however, strongly supportive of both the QWL aspects of the socio-technical model and its productive potential. Berggren (1993b), in particular, strongly defends such a view, disagreeing strongly with the data on both the productivity failures of Uddevalla and workers attitudes to working in tightly coupled assembly systems. This approach has been most strongly taken up by European supporters of the 'production island' (Fertigungsinseln) model of CM as highly autonomous cellular production units. (Hartmann 1989; Brodner 1989). Whereas the debates between protagonists such as Adler & Cole and Berggren have been focused on automobile assembly, the advocates of production islands are more concerned with the traditional province of GT: batch production and fabrication in the machine tool and mechanical engineering sectors. It is in these areas that strong arguments have been made for the appropriateness of uncoupled work cells and associated semi-autonomous work teams for 'diversified quality production' (Sorge and Streeck 1988) i.e. a production concept arguably better suited to mid-European conditions of highly skilled labor and production of lower volume, high quality products. While supporters of production islands cover a wide range of political positions, in general they are more concerned than lean production advocates with democratic corporate governance and the contribution of 'direct' autonomy in working practices to overall industrial democracy.

TABLE 3. NUMMI vs. Uddevalla (Adler & Cole 1993)

Feature	NUMMI (USA) Toyota-GM	UDDEVALLA (Sweden) Volvo
Assembly layout	Fordist assembly line	parallel "dock" assembly
Job design	highly standardized, but team can re-define standards subject to management approval ("Democratic Taylorism")	much less standardized, focus on balance of tasks within assembly cycle
Work cycle time	1 minute	2 hours
Process coupling	tight: internally via machine-paced line, externally via JIT delivery	much looser: no paced line and buffer stocks
Team size	4 -5	10
Team leader	selected by union reps. and management	teams select own leader and may rotate role
Team responsibilities	assembly, quality control, preventative maintenance, team job rotation schedules, improvement of work process	assembly, balance work tasks, quality control, preventative maintenance, job rotation schedules, set overtime schedules, select own members
Training	high emphasis	high emphasis
Pay for skills	No	yes

3.3. Management Control In Lean Production

Advocates of either lean production or socio-technical forms of CM have tended to concentrate on the benefits that flow from the forms of worker autonomy embodied within the different models. The optimism of these approaches is strongly denounced by a number of critics of working conditions and new forms of management control embedded within such systems (Badham and Jurgens 1997). Lean production or total quality management cells and work teams have been a particular focus of criticism as a form of 'management by stress' (Parker and Slaughter 1988) and surveillance by an 'electronic panopticon' (Sewell and Wilkinson 1992a; EGOS 1997). In direct contrast to the autonomy approach, these approaches are highly critical of the mystifying and sometimes consciously ideological character of human relations rhetoric and 'team ideology' (Sinclair 1992; Jenkins 1994) that surrounds the teamwork aspects of CM. Challenging the focus of much of the teamwork literature on the self-management of production tasks, these alienation critics have re-emphasized the political 'win-lose' dimension of team based control systems and broader industrial relations issues (Knights, Willmott and Collinson 1985). This approach has drawn on both labor process and radical organizational theories, as well as more mainstream sociological and neo-Weberian perspectives on organizations (Thompson and McHugh 1995). Teamwork, it is argued, necessarily embodies new forms of employee control. It is not seen as a removal, or even, necessarily, a relaxation of management control but, rather, as embodying an often insidious and intrusive control mechanism. It is argued that the 'developmental humanist' rhetoric of the organic doctrine often highlights the 'obvious' increases in employee control over their immediate work environment (Perrow 1986; Storey 1987; Hendry and Pettigrew 1990), while systematically obscuring the forms of control it introduces. These include such elements as:

1. continued or increased control over outputs (within cells 'responsible' for their output and sharply constrained by 'tight' just-in-time systems),
2. new forms of electronic surveillance and individual and group visibility (as operations are simplified, work-in-progress is removed, performance goals and attainments are clearly laid out in the 'visible factory', and computer based performance monitoring systems are introduced)
3. tight restrictions on 'task autonomy' (Klein 1989) through pre-determined work procedures and the rigors of just-in-time supply systems
4. peer pressure (team members , normative indoctrination, and undermining of the collective bargaining strength of the workforce).

For the more pessimistic, the workforce's autonomy and discretion is, overall, decreased as a result of such changes. As Sewell and Wilkinson (1992b) argue, the view is one of "empowerment and trust as rhetoric and the centralization of power and control as the reality" (Sewell and Wilkinson 1992b, p.102). As noted by Oliver (1995) and Oswald (1997), this critique has been strongly influential in the UK literature on total quality management (EGOS 1997)

3.4. Management Control In Socio-Technical Production

Many of these criticisms of lean production forms of CM would be shared by the alternative socio-technical model. For some critics, however, the problem lies deeper than the particular weaknesses of the lean production model. More insidious and obtrusive forms of control are deeply embedded in both models of teamwork. Thus, Barker (1993), for example, argues that rather than the 'iron cage' of bureaucracy being loosened by delegated responsibility to semi-autonomous work teams, the ironic paradox is that a *more* effective disciplinary system is created. The 'eye of the norm' becomes more powerful as teams take on company values and create a tightly prescribed set of self-imposed rules and regulations to interpret and implement these values. The emphasis of such critics is on the establishment of new forms of peer pressure, normative indoctrination by management supported by increased monitoring of individual and team performance, and undermining of the collective bargaining strength of the workforce by splitting up into teams. As identified by Badham, Benders and Sewell (EGOS 1997), support for such control mechanisms is as apparent in the literature promoting self-managing work teams as it is the object of criticism in the control literature. As management relies more strongly on the 'responsible autonomy' of team based cells, particularly within highly interdependent systems, a barrage of new HR evaluation and performance control systems are created to ensure that this independence is used in a 'responsible' manner.

These two-dimensional approaches have greatly improved our understanding of both the technical and organizational options available in the general area of CM, and have provoked an important and highly emotive debate over the economic and social benefits of the different models. In doing so, they have extended and systematized many of the more loosely addressed issues raised by the one dimensional approaches. As a result, however, they have imposed a rather rigid theoretical straight-jacket on our conceptualization of CM options and associated forms of teamwork. The key choices are presented in the form of strongly worded 'either-or' options. The choice is often presented in overly stark forms between lean production or socio-technical, autonomy of the workforce or increased control by management. Despite the usefulness of inevitably simplified models, there is a danger that these models will become equated with reality, and these alternatives presented as 'the' real choices available. The point made earlier must be emphasized again here - the overall corpus of work of the researchers labeled here as falling within the different 'boxes' cannot be adequately captured by such stereotyped positions. While their overall emphasis in particular texts *lends* itself to such an interpretation, this pigeon holing should be recognized for what it is, a means for clarifying types of perspectives, not adequately representing the real opinions or whole corpus of the researchers in question.

4. THE MULTI-DIMENSIONAL VIEW

Prompted by such concerns, a more multi-dimensional approach is emerging, informed by detailed action research and longitudinal case studies. These go beyond concentrating on rectifying the neglect of the social aspects of CM, or arguing for the costs and benefits of one or other socio-technical options. The focus of this new research is on the complex processual nature of the change process, the opening up and closing off of a variety of technical options with more or less direct organizational consequences, and the many different forms of teamwork that have emerged - influenced by a variety of economic, organizational and contextual factors as well as CM technology. The picture that emerges from this research is one of a far more complex, messy and socially constructed socio-technical system than in the past.

4.1. A Processual Case Study - The SMART Project

In the early 90s, a series of case study investigations have been produced that provide a more in-depth examination of the implementation of CM and associated work teams (see Dawson (1994), Dawson (1996), Buchanan and Preston (1992), Buchanan (1994) , Hassard and Procter (1991), Procter, Hassard and Rowlinson (1995)). These point to the multi-dimensional and open-ended character of socio-technical change processes in general and those concerning cellular manufacture in particular. For example, in case studies by Dawson (1984) in Australia, and Procter, Hassard and Rowlinson (1995) in the UK, the transition to CM is identified as part of ongoing changes or 'waves of change' within the respective firms studied. Dawson's study within a General Motors-Holden plant especially shows how a complex and shifting pattern of contextual and internal political processes served to shape this ongoing process, for example, in relation to the manner in which shopfloor and unions participated in the change process.

Such processual case studies can, therefore, offer insights into the mediated and negotiated character of change within the different organizational and societal settings. In the main, such studies have been executed through relatively conventional qualitative research methodologies. That is, the researchers have played the role of 'participant observers' seeking to understand change from the perspectives of the participants but, whilst observing the process, at the same time preserve as far as possible the boundary between themselves and the researched. However, recently completed research by the present authors has involved a further development of this approach which promises even greater insight into the detail of socio-technical change (see Badham, Couchman and McLoughlin 1996, Badham, Couchman and McLoughlin 1997). This research - known as the SMART project - drew on both action research traditions and ethnographic methods employed in the sociology of technology to engage much more directly with participants in change programs to introduce cellular manufacture into three Australian companies. Boundaries between researcher and researched were deliberately blurred. This permitted the researchers to play the role of one agency seeking to direct and shape the process of change itself and to engage with other stakeholders directly in seeking to negotiate and influence desired change outcomes. The upshot was a much richer and, arguably, authentic comprehension of change as a process - in particular its micro-political dimensions - set within a much more subtle appreciation of the socially constructed nature of the socio-technical factors which 'enable' and 'constrain' change processes within their adopting context.

For example, whilst other process-orientated case studies have focused strongly on variations in teamwork and the influence of contextual organizational factors on specific developments, little detailed attention has been paid to the social shaping of 'technical factors' in such decisions as the initial grouping of parts, the choice of machine layout, selection from amongst a range of production scheduling and control options etc. The SMART project research, on the other hand, allowed the interplay and interaction of the social and the technical to be more adequately understood. For example, in the design of a cellular assembly cell for car instrument panels in on of the case study firms - Wombat Plastics -, the project case studies show how key decisions were made about the length of work cycles and the size of buffers and the uncoupling of the cell from the downstream assembly line. These 'technical' decisions were strongly contested by different actors, and the compromise that ensued was the result of unintended effects of other projects, changes in plant manager, the engineering culture of the plant, as well as the power resources available to, and the psychological disposition of, key individuals. The consequences of the different choices had immense implications for the degree to which individual operators and work teams were able to participate in production planning, continuous improvement and so on.

Similarly, in the second firm - Kangaroo White Goods - decisions were made on 'technical' grounds by engineers about the compromises that had to be made with the ideal cell design e.g. the need to share machines such as a common degreaser of sheet metal rather than providing individual machines for each cell or introducing a new greasing technique that could retain degreasing within the cells. Moreover, layout decisions were made without considering their organizational implications that meant the cells consequently needed to share specialized transport workers - because it was now 'uneconomic' to take operators off machines to retrieve and use transport equipment located in other parts of the shop. All of these decisions had direct implications for the number of tasks that were devolved to operators (e.g. including transport into and out of the cell) and the autonomy of cell teams (e.g. independence, accountability and range of machine operation skills was restricted by the need to share a common degreasing machine).

Such 'technical' decisions and trade-offs are not restricted to trivial 'details' of CM but are rooted in basic engineering concepts. As outlined in Badham (1995) and Badham and Benders (1997 forthcoming), CM is an engineering concept with a long and complex history. Ambiguities and tensions within the concept facilitate negotiations and trade-offs in site specific design and implementation processes. Key decisions have to be made about: the grouping of parts (and consequently the nature of the 'whole' products that workers are responsible for); the co-location of machinery (and thus the area of responsibility and range of production skills to be exercised by the team; and a general tension between the 'engineering flow line' feature of CM (that aims to reduce variation and uncertainty, and thus ultimately standardizes production operations and worker tasks) and the 'socio-technical autonomy' feature of CM (that has as its object the control of inevitable variability and uncertainty within production, with consequent devolution of significant responsibility to individuals and the work team).

The SMART case studies revealed how these ambiguities are played out differently in different contexts in the complex process of socio-technical negotiations and decision making in the design and implementation process. Thus in the case of Wombat Plastics a strong 'Fordist' engineering culture which had shaped previous attempts at manufacturing innovation ultimately proved the decisive factor in determining the nature of socio-technical arrangements for cellular manufacture and team-based working. Most notably in restricting the nature and extent of the team autonomy that could be developed to a 'lean production' model whilst presenting a micro-political context which ultimately defeated attempts by SMART researchers to promote alternative models of team-based working influenced by European practices. At Kangaroo Whitegoods a traditional manufacturing culture also existed but in this case micro-political conditions permitted more leeway for the promotion of team concepts along European lines, although these conditions also constrained and limited the pace of such developments to a significant degree. Finally, by way of contrast, it is worth noting the case of the third firm - Wombat Plastics. Here socio-technical ideas were already well embedded in manufacturing culture as a result of earlier and ongoing change programs. Here the problem was one of fine tuning a change program which was suffering from the effects of being over-ambitious in its objectives, having to deal with the inevitable discontinuities in change management that occur in long term programs, and struggling with a changing corporate context which introduced resource constraints, unintended deviations and slow downs in the change program and generally failed over time to offer full and ongoing support.

4.2. Multi-Dimensional Theoretical Frameworks

Despite this more complex and multi-dimensional approach to investigating case studies of HRM and CM, processual case studies have not been informed by a clear and systematic framework for either (a) explaining the source of variation in the structure and performance of team based cells or (b) conceptualizing this variety with the use of a more sophisticated conceptual model. In this regard, the experience of the SMART project goes some way towards suggesting such a framework.

The Process of Change.

Dawson effectively deploys a 'processual' change framework that distinguishes between the substance, context and politics of change at various stages in the change process. While not offering any particular explanatory hypotheses, this does provide a basic conceptual schema for addressing the nature of the changes involved and the influences upon those changes. Badham (1995) extends further the conceptualization of the substance and analysis of change by providing a 'socio-technical configuration' model for conceptualizing the nature of team based cells and their internal and external influences. This model views CM systems as complex production processes. These incorporate site specific technological configurations, specific configurations of operator skills, attitudes and roles, and, most importantly, a specific set of supervisory, engineering and HRM 'intrapreneur' personnel that maintain and develop the overall system. These factors are all included *within* the CM socio-technical 'system'. They are the production system elements that are being drawn upon and changed in the transition to CM. These detailed 'local' configurational factors are, however, operating within broader contextual conditions that influence, constrain and are influenced by these detailed activities. An adequate understanding of the technical and human resource issues involved in CM requires an analysis of both this local configuration process and the broader contextual influences. Without the former, human resource analyses are in danger of ignoring human resource influences on and consequences of detailed technical designs. Moreover, the local negotiations over the form taken by teamwork and the influences on the outcome of these negotiations are unlikely to be adequately captured. At the same time, the influence of more general factors and conditions that allow generalizations to be made should not be ignored. The socio-technical configuration model addressing these issues, and drawn upon in the case studies, is outlined in Figure 1. Both processual and socio-technical configuration models are frameworks that can help further HRM research in the appropriate direction.

FIGURE 1. Configurational Model.

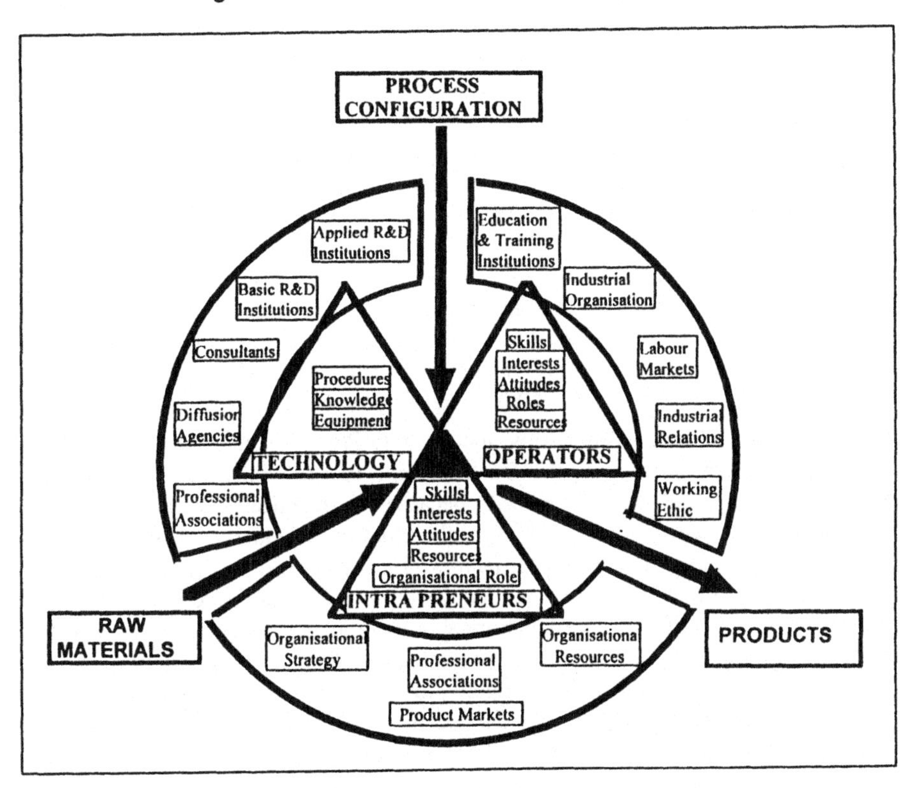

What such models assume but only conceptualize at a rudimentary level, however, is the complex interdependence of technical and organizational factors in the structuring of CM and associated forms of teamwork. As we have shown above, the one dimensional views of teamwork tended to adopt a more deterministic view of CM i.e. that the technology required changes in organization or that it had a clear impact on organizational structures. The two-dimensional views provided us with a view of technical options and teamwork alternatives, and hence contributed to an awareness of the existence of choice and negotiation between different designs. There has, however, only been a general analysis of the factors affecting such choices, normally addressing such macro factors as product markets (degree of mass production, diversity and quality of the goods produced etc.), labor markets (levels of skills, protection of labor) and societal culture (complex blend of the above with patterns of work organization, employee and management attitudes etc.).

The multi-dimensional views begin to explore some of the more complex interactions involved. Some of these features have long been recognized but inadequately explored. While overall working within a one-dimensional model, Burbidge, for example, clearly outlines that there is a greater need for cooperation amongst workers in cellular systems where in the old batch production system there tended to be more machines than men, and in the new cellular environment balancing capacity and load between different machines becomes more critical with less surplus machine capacity. He thus recognized the impact of another intervening variable - number of machines - on the 'deterministic' impact of CM. Similarly, he argues that sometimes there is scope for some workers to remain specialized workers on particular machines while others become more multi-skilled between machines. This is only the case, however, where some machines are more heavily loaded than others and specialists can be effectively allocated to them. Again, this brings in the dimension of the degree of loading of individual machines as an additional technical-economic factor influencing work organization choices. There are, therefore, a complex set of product and technology factors that act as intervening or influencing factors on the deterministic impact of CM: it is crucially the *type* of technical and production environment that exercises and influences work organization not just *one element (CM)* of that environment.

Similarly, from Wemmerlöv and Hyer to Procter et.al., there are more or less frequent comments about what Buchanan and Preston talked of as the 'radical potential' of CM not being 'realized' because of organizational and industrial relations factors. The 'impact' of CM is, therefore, clearly mediated or influenced by organizational as well as technical factors. Burbidge 's work also illustrates this point. He uses phrases such as the 'tendency' for the delegation of responsibilities to the work team, stating that production and coordination responsibility 'can be delegated' to the team because of the increasing autonomy and simplicity of group coordination, CM creates a 'climate' where job satisfaction can grow, and there is a 'desirability' for workforce participation. (Burbidge 1979, p.40-50). There is a clear shift at times in the work of Burbidge and later analysts between talk about 'effects', 'impacts' and 'requirements' to 'implies', 'conducive to', 'compatible with', 'enables' etc.

The HRM consequences of CM are, therefore, clearly a complex, interactive and contextual phenomena that need to be grasped in their complexity by future models. These can only be adequately addressed if an understanding of this complexity is fully integrated into our conceptual and explanatory models rather than loosely expressed comments that deviate from a more simplistic one-dimensional framework. This involves drawing upon contemporary sophisticated models of technological change in organizations (McLoughlin and Clark 1994; McLoughlin and Harris 1996), and clearly going beyond the often simplistic technicism or human relations advocacy of the one dimensional model or the often stereotyped divisions apparent in the two-dimensional models.

Conceptualizing Teamwork in Manufacturing Cells.

The nature and form of 'group' or 'team' work is one of the central human resource issues involved in the design and implementation of CM. Yet, somewhat surprisingly, even the more detailed case study investigations have failed to use or develop anything other than rudimentary models of teamwork. As Badham and Jurgens (1997) have observed, there has been a general tendency for teamwork researchers to adopt simple models of levels of teamwork, and then evaluate the effectiveness and desirability of these levels. In the case of CM, from Burbidge to Dawson there has been a common concern with changes to the jobs of operators, foremen and support staff, and much of this has been interpreted in terms of the desirability and appropriateness of different levels of teamwork. However, current comparative international research on work teams has revealed the absence of clearly identifiable levels of teamwork in actual case study firms. Teamwork varies considerably in the form that it takes in different contexts, in one context there may be a greater development of scheduling skills and tasks, in another context greater control over performance monitoring and assessment, and in another of continuous improvement activities and decisions, and so on. As illustrated in Figure 2, outlining the well known 'spider' diagrams used by German work psychologists, the pattern of teamwork may take very different forms.

It is still the case, however, as Buchanan (1994, p.218) observes, that the 'concept of autonomy is still poorly conceptualized and is in operation applied in a wide variety of ways.". Teams take very different forms in different CM environments, and the costs and benefits of the different forms have not been adequately explored. At one level this is a basic problem of conceptualization. Because of the emotive conflict between supporters of either 'lean' or 'socio-technical' models of teamwork, there has been a tendency to focus on stereotyped contrasts of different levels of teamwork, rather than addressing the crucially important task of conceptualizing and analyzing the very different forms and patterns of teamwork that emerge. "Teamwork", as Thompson and Wallace (1997, p.1) observe, "is not a fixed package." When the complex reality of teams is addressed, we have consequently been left either with piecemeal anecdotal descriptions rather than a systematic conceptualization and analysis or an attempt to fit this complexity into simplistic bipolar contrasts. At another level, the problem is one of accurately assessing the costs and benefits of the different forms of teamwork. As Buchanan (1994, p.219) comments, 'the basis of the potential benefits that are supposed to flow from autonomous group working are not well understood.' Because of the very different forms taken by teamwork, and the difficulty of separating out the effects of teamwork from the multitude of other factors involved in the change process (both technical and organizational), there has been little in the change process (both technical and organizational), there has been little systematic analysis of the benefits of different teamwork options within CM arrangements. Despite the degree of consensus on the productivity benefits of lean production type teams, this weakness applies to these type of teamwork arrangements as well as variants of the more socio-technical model.

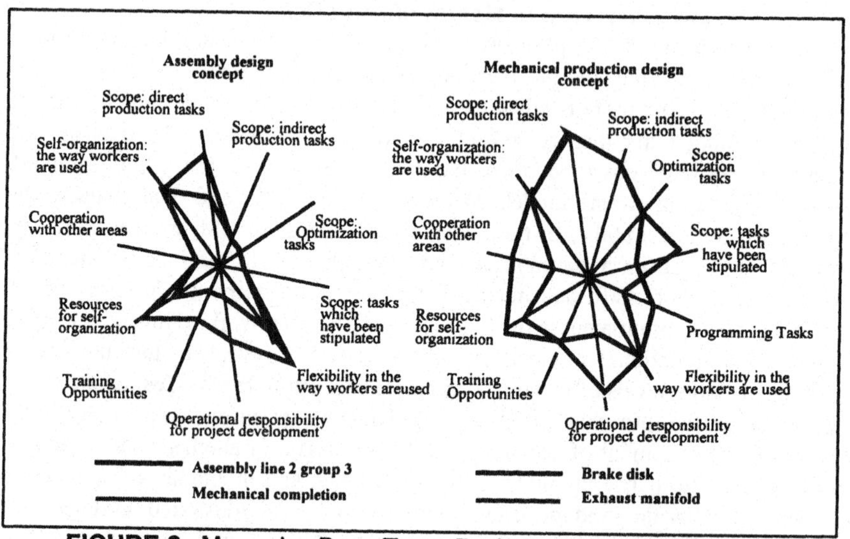

FIGURE 2. Mercedes Benz Team Profiles: SOFI (Springer 1996).

In addressing the creation of new conceptualizations, the work of Badham and Jurgens (1997) provides a useful overview of current debates, and extends the 'spider' diagram models into a three dimensional model that includes patterns of team control over *operational controls* in finance and production, the *design* of these operational controls, and *governance* in the same areas (i.e. ability to choose between designs and their role in the design process). The first area of operational control is that frequently focused on by traditional supporters of team autonomy, whereas the latter area of governance is the one that has more concerned the critics of teamwork as a new form of management control. In the work of Weber (1995) and his colleagues, this focus on the 'bucket' of tasks and responsibilities controlled by the team is added to by a simultaneous focus on the distribution of responsibilities within the team and, thereby, the extent to which they can be considered as a full 'team'. This involves, in particular, examining the extent to which teams are *operator controlled* or *team leader* controlled. All of these models are presently at an early stage of development, despite drawing on well established frameworks for conceptualizing work design and teamwork. They do, however, provide new and important models for further characterizing teamwork associated with CM.

5. CONCLUSION

In 1994, Buchanan (1994, p.222) commented of the CM and work teams literature that, "The key questions for the future, therefore, concern how the thinly substantiated credibility of the autonomous group approach and its socio-technical systems underpinning can survive the continuing lack of conceptual clarity, the demonstrable lack of rigorous and strong supporting empirical evidence, and the invidious comparisons with Japanese manufacturing methods, the demonstrable lack of rigorous and strong supporting empirical evidence, and the invidious comparisons with Japanese manufacturing methods. These questions imply an interesting research agenda which will require organizational as well as academic commitment if they are to be addressed effectively."

Many of the traditional assumptions of the one-dimensional and two-dimensional approaches need to be discarded. Firstly, the issue of the degree of autonomy is not simply a question of examining the degree 'required' or 'determined' by CM - it is also a question of determining the degree that is desired or selected by actors operating within and constrained by given cultural and organizational settings. Secondly, there are no simple measures of 'degree' of autonomy, there are a multiplicity of different *forms* of autonomy, and these need to be more precisely defined and evaluated if we are to consciously and systematically guide the selection of work forms. Thirdly, there is no *a priori* reason for either accepting or rejecting new forms of teamwork as a desirable and productive method of work in CM systems, further research is required to improve both the criteria used to assess desirability and performance and provide a more in depth analysis of actually existing conditions and developments.

There has been calls for more detailed conceptualizations of CM, the use of more multi-plant studies, and the conduct of a greater amount of longitudinal analyses. As summarized in Alford (1994), there has been a greater degree of research in these areas but, as indicated in the research evidence briefly reviewed above, this has not led to or been informed by sufficiently precise categories or insightful theoretical frameworks. This has been most evident in lack of attention paid to the detailed interactions between technological and organizational factors even by researchers who are aware of the importance of researching such issues. Lack of theoretical development in the area of CM research means that even such serious commitments to qualitative research need better theoretical support.

REFERENCES

Adler, P.S. and Borys, B., 1996, Tools or weapons, *Administrative Science Quarterly.*

Adler, P.S., 1993a, The learning bureaucracy: New United Motor Manufacturing, Inc., in *Research in Organizational Behaviour*, 15, 111-194, JAI Press, New York.

Adler, P.S., 1993b, Time and motion regained, *Harvard Business Review*, Jan-Feb, 97-108.

Aguren, S., and Edgren, J., 1980, New factories: Job design through factory planning in Sweden, Swedish Employers Confederation, SAF Editorial Section, Stockholm.

Althusser, L., 1978, *Essays in self-criticism*, New Left Books, London.

Badham, R., 1995, Managing socio-technical change: A configuration approach to technology implementation, in J.Benders, J de Haan and D.Bennett (eds.), op.cit.

Badham, R. and Benders, J., 1997 (forthcoming), History of cell-based manufacturing, in *History of Work Teams*, (Ed. H. Beyerlaan).

Badham, R., Benders, J. and Sewell, G., 1997, From cop to coach (man): A critical analysis of self-managing work teams, EGOS Conference, Budapest, 3-7 July.

Badham, R., Couchman, P., and McLoughlin, I., 1996, Implementing vulnerable socio-technical change projects, in *New Perspectives on Technology, Organization and Innovation,* (Eds. I. McLoughlin and M.Harris), International Thomson, London.

Badham, R., Couchman, P., and McLoughlin, I., 1997 (forthcoming), Kangaroo white goods, in *Strategic Management,* (Ed. M.Browne).

Badham, R., Couchman, P., Kaebernick, H., Moriarty, D., Santiago, B., and Wells, A., 1995, Team based cellular manufacturing series, Booklets 1 to 5., MITOC, University of Wollongong.

Badham, R., Couchman, P., and Selden, D., 1995, Alternative socio-technical systems in the Asia-Pacific region: An international survey of team-based cellular manufacturing, 6th APROS International Colloquium on Organizations: Debating New Agendas, Issues and Perspectives, Guernaveca, Mexico, 11-14 December; forthcoming in *Global Management: Universal Theories and Local Realities*, (Eds. S. Clegg, E. Ibarro and L. Bueno), 1997, Sage, New York.

Badham, R. and Jurgens, U., 1997 (forthcoming), Images of Good Work and the Politics of Teamwork, *Economic and Industrial Demoracy*.

Barker, J. R., 1993, Tightening the iron cage: Concertive control in self-managing teams, *Administrative Science Quarterly*, 38, 408-437.

Benders, J., de Hann, J. and Bennett, D., (Eds.), *The symbiosis of work and technology*, Taylor & Francis, London.

Berggren, C., 1993a, *Alternatives to lean production: Work organization in the Swedish auto industry,* Cornell ILR Press, Ithaca, NY.

Berggren, C., 1993b, The Volvo Uddevalla plant: Why the decision to close it is mistaken, *Journal of Industry Studies*, 1, 1, October, 75-87.

Berggren, C., 1994, NUMMI vs. Uddevalla, *Sloan Management Review*, Winter, 1994, 37-49.

Bršdner, P., 1988, *The Shape of Future Technology: The Anthropocentric Alternative*, Springer-Verlag, Berlin.

Buchanan, D., 1979, *The Development of Job Design Theories and Techniques*, Saxon House, Aldershot.

Buchanan, D., 1994, Cellular manufacturing and the role of teams, in *New wave manufacturing strategies: Organizational and human resource management dimensions*, (Ed. J. Storey).

Buchanan, D., and Preston, D., 1992, Life in the cell: Supervision and teamwork in a manufacturing systems engineering environment, *Human Resource Management Journal*, 2 , 4, 55-76.

Burbidge, J.L., 1975, *The Introduction of Group Technology*, Heinemann, London.

Burbidge, J.L., 1976, *Group production methods and humanization of work: The evidence in industrialised countries*, International Institute for Labor Studies, Geneva.

Burbidge, J.L., 1979, *Group Technology in the Engineering Industry*, Mechanical Engineering Publications Ltd, London, 36.

Dawson, P., 1994, *Organizational Change: A Processual Approach*, Paul Chapman, London.

Dawson, P., 1996, New technology and the development of cellular forms of work organization, in I.McLoughlin and M.Harris op.cit.

Dohse, K., Jurgens, U., and Malsch, T., 1985, From Fordism to Toyotism? The social organization of the labor process in the Japanese automobile industry, *Politics and Society*, 4, 3, 22-4.

Drucker, P.F., 1990, The emerging theory of manufacturing, *Harvard Business Review*, 68, 3, May-June, 94-102.

EGOS, 1997, *Proceedings of the EGOS Conference*, Budapest, 3-7 July.

Hartmann, M., 1989, A West German lesson for increased flexibility and innovation in manufacturing: Autonome fertigungsinseln, Program in Science Technology paper presentation at MIT, May.

Hassard, J. S., and Procter, S. J., 1991, Manufacturing change: Introducing cellular production in two British factories, *Personnel Review*, 20, 4, 15-24.

Ingersoll Engineers, 1990, *Competitive Manufacturing: The Quiet Revolution*, Ingersoll Engineers, Rugby.

Ingersoll Engineers, 1994, *The Quiet Revolution Continues*, Ingersoll Engineers, Rugby.

Jenkins, A., 1994, Teams: From 'ideology' to analysis, *Organization Studies*, 15, 849-860.

Procter, S. J., Hassard, J., and Rowlinson, M., 1995, Introducing Cellular Manufacturing: Operations, Human Resources and High-Trust Dynamics, *Human Resource Management Journal*, 5, 2, 46-64.

Jurgens, U., Malsch, T., and Dohse, K., 1993, *Breaking from Taylorism: Changing forms of work in the automobile industry*, Cambridge University Press, Cambridge.

Klein, J. A., 1989, The human costs of manufacturing reform, *Harvard Business Review*, March-April, 60- 66.

Knights, D., Willmott, H., and Collinson, D., (Eds.), 1985, *Job Redesign: Critical Perspectives on the Labor Process*, Macmillan, London.

Legge, K., 1995, *Human Resource Management: Rhetorics and Realities*, Macmillan, London.

Mathews, J., 1995, *Catching the Wave*, Oxford University Press, Sydney.

McLoughlin, I., and Clark, J., 1994, *Technological Change at Work*, Open University Press, Buckingham.

Parker, M. and Slaughter, J., 1988, *Choosing Sides: Unions and the Team Concept*, Labor Notes, Detroit.

Perrow, C., 1986, *Complex organizations: A critical essay*, (3rd ed.), McGraw-Hill, New York.

Storey, J., 1987, Developments in the management of human resources: An interim report, Warwick papers in industrial relations, IRRU, School of Industrial and Business Studies, University of Warwick.

Hendry, C. and Pettigrew, A., 1990, Human resource management: An agenda for the 1990s, *International Journal of Human Resource Management*, 1, 1, 17-44.

Schonberger, R., 1982, *Japanese Manufacturing Techniques*, Free Press, New York.

Sewell, G., and Wilkinson, B., 1992a, Someone to watch over me: Surveillance, discipline and the just-in-time labor process, *Sociology*, 26, 2, 271-289.

Sewell, G., and Wilkinson, B., 1992b, Empowerment or emasculation? Shopfloor surveillance in a total quality organization, in *Reassessing Human Resource Management*, (Eds. P. Blyton and P. Turnball), Sage, London, 97-115.

Sinclair, A., 1992, The tyranny of a team ideology, Organization Studies, 13, 4, 611-626.

Sorge, A., and Streeck, W., 1988, Industrial relations and technical change: The case for an extended perspective, in *New Technology and Industrial Relations*, (Eds. R. Hyman and W. Streeck), Blackwell, Oxford.

Procter, S. J., Hassard, J., and Rowlinson, M., 1995, Introducing cellular manufacturing: Operations, human resources and high-trust dynamics, *Human Resource Management Journal*, 5, 2, 46-64.

Thompson, P., and McHugh, D., 1995, *Work organizations: A critical introduction*, Second Edition, Macmillan, London.

Warnecke, H., 1993, *The Fractal Company*, Springer Verlag, Berlin.

Weber, W., 1995, Action regulation, co-operative structures and teamwork in manufacturing systems, 5th IFAC Symposium on Automated Systems Based on Human Skill - Joint Design of Technology and Organization, Fraunhofer IPK, Berlin, September 25-28.

Wemmerlöv, U., and Hyer, N.L., 1987, Research issues in cellular manufacturing, *International Journal of Production Research*, 25, 3, 413-431.

Wemmerlöv, U., and Hyer, N.L., 1989, Cellular manufacturing in the U.S. industry: a survey of users, *International Journal of Production Research*, 27, 9, 1511-1530.

Wickens, P., 1994, *The Ascendant Organization*, Heinemann, London.

Womack, J.P., Jones, D.T., and Roos, D., 1990, *The Machine that Changed the World*, Rawson Associates/Macmillan, New York.

AUTHORS' BIOGRAPHY

Dr. Richard Badham is Professor in Management in Department of Management and the BHP Institute for Steel Processing and Products at the University of Wollongong, New South Wales, Australia. He is the author of *Theories of Industrial Society* (Crrom Helm, 1986) and has published over one hundred papers on technology, work and organization. He is currently carrying out research projects on concurrent engineering, expert systems and artificial intelligence, socio-technical methods, and the micropolitics of technical and organizational change.

Dr. Ian McLoughlin is Reader in Management Studies and Head of the Division of Management Studies in the School of Business and Management at Brunel University, West London, UK. He is co-author of *The Process of Technological Change* (Cambridge University Press, 1988); *Technological Change at Work* (Open University Press, 1988; 2nd edition, 1994); and co-editor of *Innovation Organisational Change and Technology* (ITP Business Press, 1997).

Dr. David Buchanan is Professor of Organizational Behaviour at the School of Business, De Montfort University Leicester (UK). He is author/co-author of several books, including the best-selling Organizational Behaviour: An Introductory Text (with Andrzej Huczynski: third edition, 1997), The Expertise of the Change Agent (with David Boddy, 1992), and numerous papers and book chapters. He is Director of the Organization Development and Change Research Group at Leicester, and his main research interests lie in change management and innovative work design.

H2

Teams and Cellular Manufacturing: The Role of The Team Leader

P. D. Carr and G. Groves

1. INTRODUCTION

This chapter discusses the role of the team leader within group technology, and describes an action research study which tested a new approach to making the transition from scientific management to lean production Group Technology (GT) within a UK manufacturing facility. The approach focused on a balance between social and technical change, and involved a strong role for the team leader.

Historically, manufacturing in most of the industrialized world has taken the form of scientific management (Taylor 1911). This form of manufacturing organization is based, however, on a specific view of people. It assumes a relatively low level of intelligence and a strong aversion to work. Work is therefore organized in a process form with minimum skill requirements for direct operators, a plethora of expert functions and a strong policing mechanism to control the people who would otherwise do no work. While this form of work organization creates jobs that are unpleasant, the nature of people determines that it is the best (or so it was argued).

Group Technology proposes an alternative form of work which, it is suggested, will result in better productivity, quality, production lead times, process improvements and greater job satisfaction for the people who do the work. Individual workers will be more highly skilled and most expert functions will be carried out by the production team. Higher levels of motivation will remove the need for policing. Instead, managers will increase their leadership activity, providing vision, direction and support. It implies a view of people which is radically different from that in scientific management. It assumes that people are capable of higher levels of intelligence and self motivation. This form of work organization - linking the technical organization of work with the nature of people - was known as socio- technical systems (Trist and Bamforth 1951).

In the 1960s and 70s this approach had many supporters. Factories were designed to create groups which it was thought would be more profitable. Interest waned when at best the case was not proven and at worst GT failed to deliver the results promised. In recent years interest in GT has re-emerged. Advocates of lean production have argued for the adoption of cellular manufacturing as a more appropriate form of organization to reduce waste.

Within lean production GT there has been growing attention on the team leader and it has been argued that their role is critical in the successful implementation of lean production. The differences between the socio- technical systems approach and the lean production approach may explain this difference.

2. THE NEW WORK ORGANIZATION

A gap has been identified between expectations of lean production (LP) performance and its practice in Western companies. While expectations were high, and improvements have been experienced in many areas, these have not measured up to the improvements experienced by practitioners in Japan, for example (Womack, Jones and Roos 1990; Delbridge and Oliver 1991). At the same time concern is emerging of the people aspects of lean production (Klein 1989: Fucini and Fucini 1990). Substantial changes in the activity of employees is expected in a lean production environment which enable its benefits to be achieved (Monden 1983). While hopes had been expressed of a motivating manufacturing environment, substantial evidence exists to suggest that this has not transpired in a significant number of implementations (Parker and Slaughter 1988; Inman and Brandon 1992). Further, the people aspects of lean production were being perceived as a problem by those implementing them (Mayes and Ogiwara 1992).

This problem appears very similar to that which was experienced in the early days of Taylorism and which spawned the work of the human relations and socio-technical schools. Perhaps a solution might be found here.

A human relations approach to lean production implementation and operation would focus on the people aspects of work, while assuming that the imperatives of the production systems would remain the same. Initial understanding of lean production systems in Japan appears to have reflected the human relations approach. It focused on employee conditions outside the immediate work environment and argued that it was practices such as lifetime employment, company unions, etc. which determined the different employee behavior (Pang and Oliver 1988). Some examples of a human relations approach to lean production may be discerned in the literature, even if they are not explicitly expressed as such (Wood 1993).

A socio-technical approach to lean production is not evident in the literature. One example of integration of socio-technical systems with lean production appears from Berggren (1993). He argues that the benefits of lean production can be combined with socio-technical systems, while the aspects of lean production which cause difficulty are not applied. Setup time reduction, integrated quality control and the application of statistical analysis are argued to constitute good practices which the West should adopt. However, he argues that other aspects, such as unconditional lean production control, create undesirable working conditions and should be rejected.

One aspect of lean production which is likely to be problematic is the decoupling, or weakening of the links between the stages of production. Monden (1983) describes the main objective of the Toyota Production System as the elimination of waste. Decoupling the production system could only be achieved at the expense of waste creation as buffer stock is created to ease the pressure of 'coupling'. While this may be desirable in terms of its human benefits there is an undeniable cost.

The synchronization of production activities requires a daily socio-technical compromise between the demands of the wider organization and the needs of the autonomous work group. This role would appear to be critical in integrating just-in-time (JIT) system and a more human-centered approach. Further, the success of this integration can be measured in the success that is achieved in waste reduction, while maintaining human motivation. Lowe (1993) argued that the development of supervision under a Taylorist regime was not remotely appropriate to this enhanced role. Instead of being the 'middle man' the supervisor would become a critical factor in lean production. With human-centered lean production, this becomes more critical.

The implementation of a socio-technical systems approach to LP within an existing, Taylorist, manufacturing company implies a substantial transfer of power, from service departments to production areas and from managers to shopfloor teams. At the same time, there is an increased need for system integration to minimize waste. A very delicate balance of power must be maintained. Not enough power on the shopfloor will result in a failure to achieve the responsiveness and flexibility that LP requires, while the benefits of socio-technical systems will not transpire. Too much, and integration will be poor and waste will result

3. TEAM LEADER'S ROLE IN CELLULAR MANUFACTURING

Prior to the adoption of Taylorism, supervisors had wide-ranging responsibilities for managing production. The use of increasing levels of technology and the increasing scale of production combined with the separation of task conception from execution, led to a high degree of specialization and division of labor. This in turn led to an increase in the number of specialist departments which took responsibility away from the supervisor for both technical and labor activities. For example, technical departments would specify working practices to be followed, maintenance departments took on responsibility for maintaining equipment, toolmakers and setters took over machine setting, personnel departments took on increasing aspects of labor relations. Child and Partridge (1982) assert that this led to the role of the supervisor being undermined. Pay and status declined (Pollard 1968). This situation was exacerbated as graduates were recruited into these specialist departments who became candidates for management promotion. A social and educational gap opened between supervisors, who were increasingly marginalized in the decision making process, and managers. Wray (1949) describes the resulting supervisor as 'the marginal man', who transmitted management decisions but no longer had a part in making them.

The role played by supervisors in Taylorist industry today continues to reflect this development (British Institute of Management 1976). Instead of exercising extensive control over production activity, supervisors can be seen as reconciling the demands of managers and specialist departments and the needs of employees. While possessing a limited authority, they are responsible for deciding on which, often conflicting, demands to implement (Child and Partridge 1982).

The possibility of leaderless teams has been raised as one form of autonomous team development (for example, Donovan 1987, Semler 1989). It is argued that the removal of an authority figure prompts the group to take responsibility itself for necessary decisions on what work is to be done, who will do it, how to do it, who to recruit, maintenance, training etc. Difficulties have been experienced with this type of work group, though (Bailey 1983). He notes that research in Denmark (Larsen 1979), on autonomous work groups, showed improved internal group cohesion led to improved group cooperation and group productivity, but inter-group problems emerged. Groups isolated themselves, became aggressive towards other groups, resisted change in group membership and became inflexible.

Research in recent years has looked at the effective operation of group working systems, and evidence has continued to emerge of attitudinal and performance benefits (for example Wall et al. 1986). However, it is argued that empirical research remains weak. More fundamentally, there is a criticism of concentration on social aspects of socio-technical systems and neglect of technical aspects (Wall and Martin 1994). Research has found both a weakening and a strengthening of the team leader's role today (Buchanan 1994).

Partridge (1989) argues that supervision may be becoming more important. The nature of this increased importance is explored by Thorsrud (1984), who lists the roles that supervisors and foremen would play in an autonomous group working environment. While teams would be expected to conduct their own internal co-ordination, under normal conditions, supervisors and foremen would increasingly concentrate on the co-ordination between departments.

The socio-technical approach to supervision is still being developed. Two positions are clear. First, that supervision will decline and disappear as autonomous group working is developed. The removal of hierarchical authority and the incorporation of previous service functions will liberate the team to organize itself, motivated by the additional responsibility to produce more effectively. The second approach suggest that far from being removed, the role of supervision will remain but change to support the operation of the autonomous work group through undertaking a wider range of planning, integration and development tasks. The evidence lies in favor of this second role at present as being the appropriate complement to socio-technical systems.

Little work has been undertaken on the role played by supervisors in a Lean Production environment. One study (Lowe 1993) concludes that Supervision has a crucial role in the successful implementation of Lean Production. This crucial role stems from the efforts within Lean production implementation to devolve responsibility to production teams (Groebner and Merz 1994). It is argued that production supervisors develop a new range of activities as facilitators, coordinators and trainers (Forza 1994). However, the studies of supervision in a Lean production environment focus on greenfield sites. Another study is included in Fucini and Fucini (1990) at Mazda's US Flat Rock plant.

Lowe (1993) studies the development of supervision in the introduction of Lean production in a Taylorist UK automotive manufacturer. He notes that there is a substantial gap between the tasks performed in Nissan and those in the plant studied. A substantial change in the tasks undertaken was envisaged by the company and supervisors were trained in what this would entail. However, Lowe found that in eighty percent of cases, job content was little changed. Ninety percent felt relationships with management had not changed and that they continued to lack management support. Seventy-five percent felt inadequate or incapable of fulfilling the requirements of their new role.

Lowe's (1993) work highlighted a critical problem. The role of supervision had failed to keep pace with the other changes made towards lean production in the company studied. There was thus the cost of failing to maximize the possible benefits of lean production and the danger that supervision will be a substantial barrier to implementation success. This conclusion is also drawn in a study by Andersen Consulting (1993).

4. CASE STUDY: ALCAN EXTRUSIONS AND TUBES

A case study to develop and test a model for implementing a new approach to group technology-based lean production implementation was undertaken. In recent years, there has been an ever increasing need in operations management for case study research, action research, etc. In light of various limitations posed by traditional, scientific methods of inquiry. For a comprehensive rationale and justification for such approaches, the reader is referred to Hill (1987), Raimond (1993), and McCutcheon and Meredith (1993).

4.1. The Context

The case study pertains to Alcan Aluminum's Extrusions and Tubes division in the United Kingdom. This division comprised four companies at the time the research was conducted, producing extrusions for a wide range of markets, ranging from aerospace and defense, to more commercial applications like double glazing and caravans, to the high volume automotive sector. The division employed about 1000 people with sites at Banbury, Redditch, St. Helens and Warrington. The main testing activity took place at the company's Banbury site, which then employed about 500 people. Redditch had also focused on Lean Production/Group Technology implementation and was used to compare results

The company's decision to implement lean production-based group technology was based on a number of factors. Alcan had a close relationship with Nippon Light Metal, a Japanese aluminum company and visits to Japan by senior managers had led to a high degree of interest in new production systems. This was combined with growing corporate pressure to improve upon sluggish company performance. The researcher had spent the year prior to the research working with the company and this increased the company's confidence in proceeding with radical change.

The aluminum market is a turbulent one. Aluminum prices fluctuate wildly and, as a consequence, so do the fortunes of the companies operating within this sector. This creates a difficult environment in which to conduct research and, indeed, the effects of this were apparent as the research progressed Aluminum extrusion is the process of pressing hot aluminum billet or log through a die to create a desired shape. This shape is cut to length and then proceeds through a variety of downstream operations, depending on customer requirements. This can include heat and or solution treatment to achieve particular metallurgical characteristics, rolling or drawing to achieve shape or particular diameters, further sawing to achieve tighter tolerances than can be achieved at the initial saw stage, packing and shipment. The normal range of support services that would be found in a western manufacturing environment exist (Maintenance, Sales, Purchasing, Finance etc.). Each site produces specific types of aluminum extrusion for particular markets, although some overlap exists between sites. At each site the shopfloor was formally divided into teams which completed all or a substantial part of a product. A product-focused organization structure had been established but a scientific management approach was used to manage it.

4.2. Team Leader Training: Program Development And Delivery

An initial plan based on the previous discussion was developed by the researcher (Peter Carr). This was presented to the production management team at Banbury. Agreement was reached on the design of the program which would be used. The use of a modular system, focusing on each implementation area in turn was agreed with the production management team, to allow the implementation to be effectively managed. The widespread nature of the changes envisaged meant that everything could not be done at once, and the large number of people involved in the program meant that a clear structure of implementation, understood by all would be necessary.

The application of each module would be developed as the program progressed. The researcher would prepare detailed proposals after discussion with managers and team leaders. Applying the action research approach, the researcher would seek to ensure that an overall lean production direction was maintained, while working with managers with expertise and knowledge in each module subject area. For example, the maintenance module involved the researcher working with the maintenance manager, while the extrusion process module involved work with the technical manager and members of his department. Human elements of change were encompassed in both the content of the modules and the module process. Content was designed to include the development of skills and change of activities, for example, employee participation in working method development in the standard operating methods module.

A delivery system was developed which was intended to achieve the desired focus on manufacturing team leaders. Managers involved in the development of module content would also be involved in managing the implementation in that module area. A standard module format was developed which would be used in each module. First, a seminar would be conducted to introduce the subject area concerned to the team leaders and their managers. All team leaders, covering production areas, were expected to attend the seminar at the start of each module. This would last 1-2 days, depending on the subject matter. The seminar would provide any necessary skills and tools to assist in the planning of the change desired. Team leaders would then have one week to prepare a plan, jointly between shift based team leaders covering the same production areas. The plan would be presented to the researcher and the manager who the team leaders reported to. The plan would be discussed and agreed, detailing time scales and resources required. The plan would then be implemented by the team leaders, utilizing internal or external services where appropriate. Once implementation was completed, (on average 6 weeks later), the results would be presented to members of the production teams plus members of the Banbury production management team.

This delivery system was intended to ensure that the team leaders became the focus of the program. They would be responsible for initiating change in their own production area and implementing the change desired. It was hoped that this would lead to higher levels of understanding and ownership of the changes implemented. The implementation would be supported by managers and the researcher, once the plans were agreed. Where team leaders experienced difficulties, efforts would be made to assist, especially where access to resources outside the control of the team was required.

A paper would be produced describing proposed module content and circulated to all managers and team leaders for comment, prior to the conduct of each module. The proposal would be amended where suggestions were made which were thought to enhance the objectives of the module concerned.

Thus, a program to trial the implementation of the approach was established. While maintaining the emphasis on lean production the program sought to maximize the involvement of manufacturing team leaders, through instituting a leading role in implementation design and management. The modules and their objectives are listed in Table 1.

4.3. Program Evaluation

A series of activities were designed to evaluate the program. Prior to the conduct of each module, clear objectives were identified which would be used to evaluate the success of that individual module. After module conduct, an understanding of the extent of the success achieved could be gained from comparison with actual results. At the end of each module seminar, a questionnaire was completed by all participants, gathering data on the perceptions of module seminar conduct and any suggestions for improvement. This provided valuable information for use in evaluating the seminars and also in identifying areas for attention in implementation. At the end of each module, team leaders presented the results achieved to their peers, managers and team members. The researcher also maintained a diary of module implementation events. It was thus possible to develop a descriptive account of the research.

In addition to describing the practical activity which formed each module, survey activity was undertaken to enable a more objective picture to be established and to enable comparison to be made between Alcan sites at Redditch and Banbury. The characteristics of the different sites have previously been outlined. Surveying was required to understand the development of the sites during the test period and to enable comparisons to be made. Three surveys were developed. The first was an interview survey, which was conducted with all members of the production management above team or shift leader level. This would allow an outline of activity to be developed, to aid in the interpretation of the second survey. This was followed by a survey of all employees, which was intended to identify differences in perception of the activities on each site, by employees at all levels of the company. This would aid the evaluation of the impact of the trial activity, and was conducted three months after the trial was completed. Finally, a survey was undertaken three years after the trial completion, of team leaders at Banbury, to assist the assessment of the continuing impact of the trial. The surveys were intended to both provide comparative data for the researcher and to provide information to the company which would aid the development of future company activity.

4.4. Module Examples

The modules are listed in Table 1. Here are two examples:

4.4.1. Housekeeping.

Through attention to housekeeping activity, it was intended that the working environment be improved. Further, it was intended that control over the nature of the working environment would be moved from management to the team leader, who, following module completion, would be responsible for maintaining the working environment at a high standard. From a lean production perspective, housekeeping is intended to enable greater visibility of the production process, to assist the reduction of waste through its easier identification.

Housekeeping activities were intended to make the workplace safer, be easier for employees to work in, be more efficient when tools were more easily found and faults were more easily identified, less tiring and aggravating and generally a more pleasant place to work. The objective of this module was to develop, implement and maintain a company standard for housekeeping. The module seminar was presented and the projects were coordinated by the Employee Relations manager and the researcher.

The impact of housekeeping activity was visually dramatic. Whole departments were transformed over a weekend. Team members established the new standards themselves and were involved in planning the change. The standard was, by and large, maintained. Visible systems were established to monitor housekeeping standard maintenance duties. Shadow boards were established for tools to create greater visibility. Departments supported each other with provision of housekeeping supplies, labor etc.

FIGURE 1. A housekeeping shadow board developed during the program.

4.4.2. Standard Operating Methods (SOMs).

The standardization of working methods is often noted as a critical component of lean production (e.g., Monden 1983). It is argued that standardization allows LP to operate effectively but can restrict employee influence over working methods (Berggren 1993). In the module, it was intended that working methods be standardized, but that content of the standardized methods would be influenced by the team and controlled by the team leader. Thus, while the benefits of standardization could be gained, it was hoped that this would not be at the expense of the benefits of employee participation in work method content. Rather, a design was adopted which was intended to give employees more control over working methods, to vary the work cycle and undertake more preparatory tasks, as advocated in socio-technical systems theory.

TABLE 1. A Summary of Module Content.

Module	Objectives
1. Introduction	To brief participants on the program objectives, content and structure and to introduce visible systems and project planning.
2. Continuous Improvement	To enable continual improvement skills to be learned and practiced for application throughout the program.
3. Housekeeping	To develop, implement and maintain a company standard for housekeeping.
4. Performance Measurement	To establish performance measurement which results in improvement with particular emphasis on On Time Delivery In Full.
5. Standard Operating Methods	The implementation of visible Standard Operating Methods.
6. Training	The establishment of training responsibilities and practices within the team using visible systems and trainers.
7. Safety	The improvement of current safety practices by increasing awareness of safety responsibilities through implementing the Safety Tag System.
8. The Extrusion Process	To provide a full understanding of the extrusion process in order that team leaders may maximize the contribution of the resources at their disposal.
9. Maintenance	To develop team equipment condition monitoring so that a fuller part may be played in maintenance.
10. Cost Management	To improve control of metal costs through improving measurement and introducing a pull system of work flow.
11. Final Presentation	To allow team leaders to show other team leaders, team members, managers and senior managers progress they have achieved through the program.

A standard operating method was defined by Alcan as "the best way currently known to perform each operation in the production process which achieves requirements for safety, quality and productivity". By creating a common working method, activity is possible to study how work is carried out and identify where improvements might be made.

A format for the establishment of standard operating methods was developed by the team leaders in the seminar. Module projects involved members of their team in establishing SOMs. Commencing on a Monday, the steps culminate in the display and implementation of the new standard operating method one week later (day 6). The development and implementation team is composed of the team leader (or supervisor or foreman) plus one representative per shift, of operators who perform the operation being studied. This team is responsible for drafting the standard operating method and consulting other team members about its content.

4.4.3. Summary.

The module content was developed jointly between the researcher and the company. It reflected the wishes of both in conducting the research for this work and in making a real improvement in company operation of lean production. Each module involved the implementation of change intended to move the company towards lean production and applied socio-technical systems theory. The positive spirit in which the program was conducted was reflected in the following poem which emerged from the maintenance department when a tagging system for safety repairs was developed:

"You don't have to nag,
to get work on a tag,
Just write it out clear,
and we will appear,
Nothing too large and nothing too small,
we will try to please you all,
Endeavor we will to do your task,
Good manners and politeness is all we ask."

5. EVALUATION AND DISCUSSION

A series of surveys were carried out at both Banbury and Redditch to evaluate the program. An interview survey of manufacturing managers shortly after program completion, was used to establish the managerial understanding of lean production activity which had taken place on each site. This was intended to aid the interpretation of the employee survey, a survey based on scored responses, which was completed by over 800 employees throughout the company, achieving a response rate of about 80%. Finally, three years after program completion, the Banbury team leaders were surveyed to analyze the continuing impact of the program and enable triangulation of the team leader responses with the employee survey.

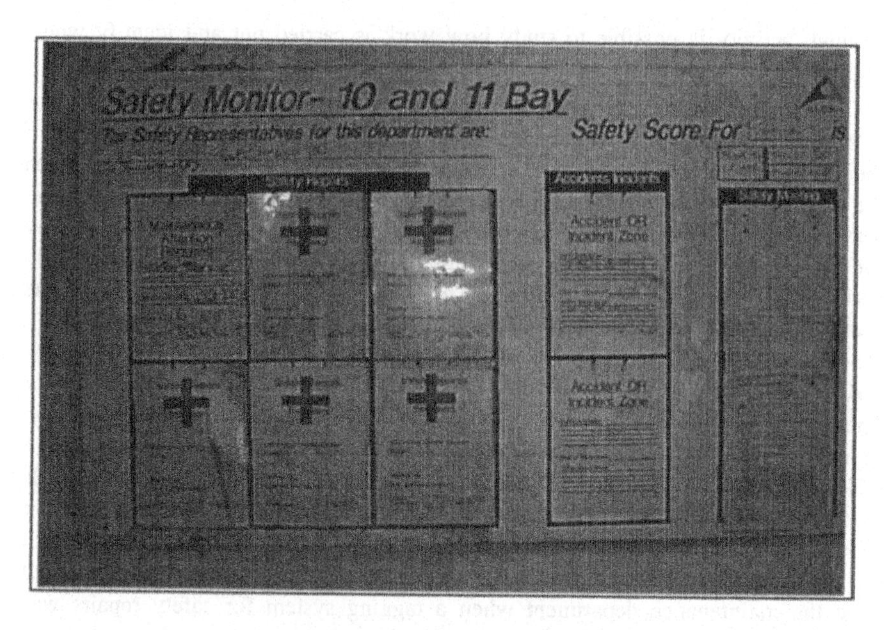

FIGURE 2. The safety tag board.

The manager survey found that at Banbury the main lean production activity had been the team leader program while at Redditch it was found that substantial effort had been apparent, dominated by a team of experts, including a secondment from Japan, who were driving the lean production activity, leading projects and designing change. This change included the establishment of a cellular manufacturing structure at the site. It was reasonable to conclude that Redditch represented a centrally driven approach to lean production oriented cellular manufacturing while Banbury adopted the team leader approach.

The survey results are first discussed in relation to the success of different programs in overall lean production implementation. Next their impact on the role of the team leader is analyzed. The effect on the company's technical performance is discussed and the human aspects of the different programs are considered.

5.1. Overall Lean Production Implementation

The employee survey asked respondents for their views on lean production implementation in their company. It asked whether they thought that it had resulted in improvement or deterioration in a range of areas including flexibility, set up times, quality etc. and whether they thought that the implementation was likely to improvements in these areas in the future. Overall the Banbury senior and middle managers had more positive views of implementation than those at Redditch. Employees at other levels in the company recorded no difference in perception between the sites. The team leader survey at Banbury three years after program completion found that the team leaders felt that the program itself had achieved substantial change, but that its completion had been followed by a slowing in the rate of lean production improvement. The team leader survey also revealed that the systems which had been established by the team leaders during the program were largely still in place e.g. the control of workflow, housekeeping, standard operating methods etc.

It is difficult to establish the precise impact of the program on company performance. Market volatility during the program and differences in the markets served by Banbury and Redditch make direct relationships between cause and effect unclear. To try to overcome this the employee survey identified perceptions within the company of performance change in a range of performance factors including productivity, stock levels, lead times etc. At both sites there was a perception that performance had improved. Banbury scores for this question ("How do you feel that JIT influenced the performance of the company in the past five years? Scale: 1 = Made Worse, 3 = No Influence, 5 = Improved) were on average slightly higher than those at Redditch, as shown in Table 2.

TABLE 2. JIT Influence on Firm Performance In Last Five Years.

	Banbury	Redditch
Senior Managers	3.57	3.09
Middle Managers	3.6	3.21
Team Leaders	3.27	3.46
Office Employees	3.09	3.00
Shopfloor Employees	2.83	2.74

It is concluded that there is a general perception throughout the company that performance improved during the program activity.

There is evidence, then, that the program successfully established key elements of lean production-oriented group technology at Banbury and that these continued to be operated three years after program completion. Further, managerial perceptions suggest that the method used was thought to be more successful than that applied at Redditch, where a more conventional approach was used. It can therefore be argued that the new approach developed can be used successfully to implement lean production oriented group technology.

5.2. Development of the Team Leaders' Role

Previous discussion argued that successful implementation of lean production oriented group technology required a radical change in the role of the team leader. No previous work was found on how to achieve that change. The program at Banbury trialled a new method for doing this. Had it been successful?

The implementation at Banbury attempted to shift power within the organization towards team leaders, from managers and specialist departments, while at Redditch change was managed by a small central team with team leaders being involved in projects where the team thought this was appropriate. Evidence that the power shift occurred at Banbury is found in the team leader responses to the survey completed by all employees. Results of responses to the question ("In the past five years, how do you feel responsibility in your present job has changed? Scale: 1 = Much Decreased, 3 = No Change, 5 = Much Increased) are shown in Table 3.

TABLE 3. Change In Job Responsibility In Last Five Years.

	Banbury	Redditch
Senior Managers	4.29	4.25
Middle Managers	4.00	4.38
Team Leaders	4.43	4.18
Office Employees	3.76	4.04
Shopfloor Employees	3.73	4.05

Here it is found that they feel that responsibility in their jobs has increased substantially more than team leaders at Redditch feel. In addition middle managers at Banbury felt that responsibility in their jobs had increased substantially less than those at Redditch and that this trend would continue in the future. It is therefore concluded that a power shift away from middle managers occurred at Banbury but did not occur at Redditch as a result of the team leader oriented program.

5.3. Human Aspects

The human side of lean production has two key aspects. First, the effects on employee welfare (Fucini and Fucini 1990; Parker and Slaughter 1988) and second the weakness in Western workforce characteristics compared to those in Japanese companies (Mayes and Ogiwara 1992). It has been argued that lean production leads to deterioration in employee working conditions and that Western employees cannot achieve the performance of those in Japan. A method which was able to improve on existing Western applications in these areas would appear to be valuable.

The results of the program on the human aspects of lean production display both positive and negative attributes. Job satisfaction scores were slightly more negative at Banbury on the shopfloor, in spite of effort at Banbury to incorporate aspects of socio-technical systems in the module design. It is difficult to attribute reasons for this, but it may be due to the time at which the survey was undertaken. A new management team had been installed and much company reorganization had taken place, which may have influenced this score. It would be wrong to be complacent though.

There was a stronger perception of the contribution of employees by management at Banbury than at Redditch, although the indications are mild.

Banbury senior managers and team leaders perceive a stronger employee contribution through acquisition of new skills, flexibility and team working, than perceived at Redditch. Table 4 summarizes the response to the question ("How do you feel that Flexible Working influenced performance of the company in the past five years? Scale: 1 = Made Worse, 3 = No Influence, 5 = Improved).

The perceptions of Banbury team leaders suggested that employee contribution improved strongly during the program but deteriorated afterwards, due to the organizational changes the company made. On a series of factors (Job Satisfaction, Employee Involvement, Employee Responsibility, Work Discipline, Employee Motivation, Employee Problem Solving, Employee Job Security, Employee Skills) there was a perception that employee activity improved markedly.

TABLE 4. Flexible Working and Firm Performance In Last Five Years.

	Banbury	Redditch
Senior Managers	4.29	3.67
Middle Managers	3.79	4.2
Team Leaders	4.13	3.69
Office Employees	3.73	3.78
Shopfloor Employees	3.43	3.45

It may therefore be argued that the program has influenced the contribution made by employees, that evidence exists to support the proposition that the contribution of employees increased during the program. However, the effects of the program on employee job satisfaction are unclear. It is therefore cautiously argued that the program achieved some success in improving the human aspects of lean production group technology implementation - but that there is much work still to be done on this.

5.4. The New Approach

There is then evidence that the new approach adopted at Banbury, focusing on a central role for team leaders can provide an answer to some of the current difficulties associated with lean production group technology implementation. The company management perceived that lean production activity had led to substantial benefits, more so than in the site that had adopted a conventional approach. The shift of power to the team leaders was successfully achieved, the first evidence of this having been successfully done in the published literature, providing a method to achieve that which existed in successful greenfield lean production companies but which had not been recorded on a brownfield site. On the shopfloor, evidence existed to show that the approach adopted had resulted in some improvement in the human aspects of implementation, although work remained to be done.

The approach used at Banbury is a significantly different approach from that used in other documented implementation of lean production group technology.

It addressed problems which had been previously highlighted and provided evidence and hope for a way to overcome them.

6. CONCLUSIONS

Three years on, the participants in the changes undertaken at the Alcan site in Banbury retained positive views of what was achieved. Changes made continued to give the company benefits The role of the first line manager was changed in this company research and a method for achieving this appears to have been found.. While shopfloor employees, middle and senior managers were involved in the program implementation, the main focus was on team leaders.. It would appear valuable to consider the additional activity support which might be desirable at these levels to assist the change to the new operating environment.. However, it is likely that caution would need to be exercised here. A strength of the existing approach is the power given to the team leader and the responsibility that they have for leading change. Activity in other areas should complement this.

The technical performance improvement has not yet been clearly established. While the approach has resulted in similar performance improvement at Banbury as occurred at Redditch, it had been hoped that this approach would result in a way of closing the gap between the Japanese levels of manufacturing performance, from lean production, and those in the West. The time scale under which the application was undertaken and the limited research base, were insufficient to allow definite conclusions on the performance outcomes, although there was scope for optimism.

Inconclusive results on the effects of the implementation on employee welfare could be resolved with a wider research base and with the application of a proven survey of job satisfaction, compared with that achieved in a Taylorist environment.

REFERENCES

Andersen Consulting, 1993, *The Lean Enterprise Benchmarking Project Report*, Andersen Consulting.

Bailey, J., 1983, *Job Design and Work Organization*, Prentice-Hall International Inc.

Berggren, C., 1993, *The Volvo Experience: Alternatives to Lean Production*, The Macmillan Press Ltd.

British Institute of Management, 1976, *Front Line Management*, British Institute of Management.

Buchanan, D., 1994, Cellular manufacture and the role of teams, in *New Wave Manufacturing Strategies: Organizational and Human Resource Management Dimensions*, (ed.) J. Storey, Paul Chapman Publishing Ltd, UK.

Child, J. and Partridge, B., 1982, *Lost Managers*, Cambridge University Press.

Delbridge, R. and Oliver, N., 1991, Narrowing the gap? Stock turns in the Japanese and Western car industries, *International Journal of Production Research*, 29, 10, 2083-2095.

Donovan, M., 1987, Employees who manage themselves, *Journal For Quality and Participation*, 12, 1, 58-61.

Forza, C, 1994, Work organization in lean production and traditional plants: What are the differences?, *International Journal of Operations and Production Management*, 16 (2), 42-62.

Fucini, J.J. and Fucini, S., 1990, *Working For The Japanese - Inside Mazda's American Auto Plant*, Free Press.

Groebner, D.F. and Merz, C.M., 1994, The impact of implementing J.I.T. on employees' job attitudes, *International Journal of Operations and Production Management*, 14, 1, 26-37.

Hill, T.J., 1987, Teaching and research directions in Production/Operations Management: The manufacturing sector, *International Journal of Operations and Production Management*, 7, 4, 5-12.

Inman, R.A. and Brandon, L.D., 1992, An undesirable effect of J.I.T., *Production and Inventory Management Journal*, First quarter, 55-58.

Klein, J., 1989, The human costs of manufacturing reform, *Harvard Business Review*, March - April, 60-66.

Larsen, H.M., 1979, Humanisation of the work environment in Denmark, in *The Quality of Working Life in Western and Eastern Europe*, (eds.) Cooper, C.L. and Mumford, E., Associated Business Press.

Lowe, J., 1993, The changing role of the production supervisor, *Journal of Management Studies*, 30, 5, 739-758.

Mayes, D. and Ogiwara, Y., 1992, Transplanting Japanese success in the U.K., *National Institute Economic Review*, November, 99-105.

McCutcheon, D.M. and Meredith, J.R., 1993, Conducting case study research in operations management, *Journal of Operations Management*, 11, 239-256.

Monden, Y., 1983, *Toyota Production System, An Integrated Approach To Just In Time*, Industrial Engineering and Management Press.

Pang, K.K. and Oliver, N., 1988, Personnel strategy in eleven Japanese manufacturing companies in the U.K., *Personnel Review*, 17, 3, 16-21.

Parker, M. and Slaughter, J., 1988, Choosing sides: Unions and the team concept, Labor Notes.

Partridge, B., 1989, The problem of supervision, in *Personnel Management in Britain*, (ed. Sisson, K.), Basil Blackwell.

Pollard, S., 1968, *The Genesis of Modern Management*, Penguin.

Raimond, P., 1993, *Management Projects, Design Research and Presentation*, Chapman and Hall.

Semler, R., 1989, Managing without managers, *Harvard Business Review*, 58, 1, 76-84.

Taylor, F.W., 1911, *The Principles of Scientific Management*, Harper and Row.

Thorsrud, E., 1984, The changing structure of work organization, in *Managing and Developing New Forms of Work Organization*, (ed.) Kanawaty, G., International Labor Organization.

Trist, E.L. and Bamforth K.W., 1951, Some social and psychological consequences of the Longwall method of coal-getting, *Human Relations*, 1, 3-38.

Wall, T.D. and Martin, R., 1994, Job and Work Design, in eds. Cooper, C.L. and Robertson, I.T., Key Reviews in *Managerial Psychology*, John Wiley and Sons.

Wall, T.D., Kemp, N.J., Clegg, C.W., and Jackson, P.R., 1986, An outcome evaluation of autonomous work groups: A long term field experiment, *Academy of Management Journal*, 29, 280-304.

Womack, J.P., Jones, D.T. and Roos, R., 1990, *The Machine That Changed The World*, Macmillan.

Wood, S., 1993, Are human resource practices in Japanese transplants truly different?, paper to the British Universities Industrial Relations Association Annual Conference, York University, 2-4 June 1993.

Wray, D.E., 1949, Marginal men of industry: The foremen, *American Journal of Sociology*, 54, 298-301.

AUTHORS' BIOGRAPHY

Peter Carr is an Associate Professor in the Centre For Innovative Management's Internet- based MBA program at Athabasca University, Canada, and specializes in Operations Management. He has a Ph.D. from Cranfield University, U.K., in manufacturing systems and has published papers focusing on implementation of group technology-based lean production. He has consulted widely in this field, particularly in automotive assembly and component sectors, and in aluminum and print industries. While undertaking his Ph.D., Peter was an internal consultant for Alcan Aluminum on the introduction of new production systems. His research work is continuing in this field, with a particular emphasis on development of methods to enhance relationship between academia and manufacturing sector in implementation of change through the use of innovative communications technologies, like the Internet.

Gwyn Groves is a Senior Lecturer in the School of Industrial and Manufacturing Science at Cranfield University, U.K She teaches on post-graduate courses and industrial short courses in manufacturing management, and was Course Director of their Masters Programme in Engineering and Management of Manufacturing Systems for a number of years. She was recently part of a Cranfield consultancy team introducing cellular manufacturing into a traditional U.K. engineering company. Before joining the Cranfield staff in 1987, she was an industrial market research consultant for a number of years, investigating a wide range of industries, products and markets. Current research interests include performance measurement, benchmarking, supply chain management and innovation activity in companies.

11

Group Technology-Based Improvements In Danish Industry

J. O. Riis and H. Mikkelsen

1. INTRODUCTION

Group Technology (GT) has played an important role in realizing significant improvements in Danish industrial enterprises for more than 25 years. The main focus of GT implementation has been on the formation of production groups as a means for simplifying production flows and management tasks. Marked improvements have been achieved in terms of reduction of throughput times, often to less than one-fifth, productivity improvements, often by a factor of two, better quality, and improved worker motivation and participation.

At the outset, to illustrate the scope and nature of these efforts, we introduce a case example about a development project in an industrial enterprise carried out in the late 1980s.

1.1. Case A

A medium-sized industrial enterprise producing farming equipment had a long tradition for employing modern production technology. Automated Guided Vehicles (AGVs) had been introduced in assembly, and robots were used for selected processes, such as grinding and deburring. In the late 1980s a major development project was undertaken mainly aimed at introducing modern machining centers and welding robots. In addition, the production manager was interested in involving workers more directly in daily operations as a means for creating a more dedicated and enthusiastic workforce. But he did not really know how to achieve this.

During the course of the project two external consultants proposed to combine the introduction of production technology with the formation of production groups. At a meeting with the production management group they pointed out that, in fact, the parts to be manufactured required different production processes and run in different volumes. At first the group did not believe that this would be possible taking into account that the plant produced more than 20,000 different parts. But during the meeting a first sketch of a structuring of the plant into about 10 production groups was accepted as a basis for further studies.

A detailed production flow analysis was carried out drawing on data from the Production Management System (PMS). Among other things it revealed that many parts were processed interchangeably at two plants located five miles apart. A part would be processed at one plant, then moved to the other to be processed there, and then back again. By moving a few machine tools it was possible to complete the production of most parts at just one plant.

A planned extension of one of the plants, budgeted to several million dollars, was suspended when the first analyses projected a significant reduction in needed production space, mainly due to the projected reduction in work-in-progress.

When looking for distinct production tasks, it became obvious that the production of spare parts was quite different from the remaining production tasks, because of the high volume and similarity in the sequence of production processes. Therefore, a pilot study was initiated for spare parts, and a virtual production group was formed with processing equipment physically located throughout the plant.

The operators along the production flow were identified and told about their belonging to the new production group. A continuous flow manufacturing principle was introduced. First the PMS was programmed to skip the planned stock in between processes, and the internal transportation operators were asked to move parts directly from one operator to the next person in the flow chain. Then, without asking for permission the batch size was cut to one half, and the transportation boxes were painted red for easy identification. As a result, during the first month of operation the number of boxes was reduced from 140 to about 40 as an indication of a significant reduction of work-in-progress and throughput time. Also the productivity increased markedly. The pilot study indicated a new way of managing production towards more decentralization and worker involvement.

At the strategic level the production management group was able to place the new production technology in a context directly related to the fulfillment of customer demands, and thereby ensuring an appropriate utilization of the equipment. In addition, by drawing a map of the proposed production groups along the production flow to form what was called a production concept the production management adjusted their strategy for future process and systems development.

The proposed production groups were implemented and new production planning and control systems introduced which resulted in a highly flexible production system.

1.2. Group Technology Approaches

The main use of GT has been based on classification of production processes and flow, following a Burbidge tradition (Burbidge 1989). Early studies in the '70s of the Opitz classification principles involved the development of systems for retrieving drawings and process specifications of parts in several case companies (Luef 1980). However, the main thrust of applying Group Technology has been centered around grouping the production processes and forming production groups, for which reason production flow analysis has played an important role.

This chapter will first describe the initiatives taken in the 1980s, the context in which Group Technology was used, the methods employed, the results achieved, and organizational and management issues related to implementation. This will be compared with the situation in the 1990s in terms of market situation, scope and focus, as well as barriers and facilitators. The paper will conclude with an outlook for the future role of Group Technology in industrial enterprises.

2. PRODUCTION GROUPS IN THE 1980s

The application of Group Technology in the 1980s was dominated by an effort to initiate innovative changes in industrial enterprises marked by a breaking away from traditional mode of operation.

Much research and development in the 1970s helped prepare for this situation. A study of different approaches to GT starting with the Opitz classification scheme led to opinion that a classification of production processes into similar sequences held great potential for the production area, cf. (Sant 1977). Also, in the 1970s a theoretical foundation was developed for applying an engineering design approach to production management emphasizing the need to focus on the production task (Riis 1978).

Inspired by similar attempts in Japan and Sweden, a major R&D project on the development of production systems (UPS) was initiated in the early years (Rode & Sant 1984) and followed by an R&D project on production management (ViPS) in the mid0ss (Riis 1990). The projects were supported by numerous case studies which substantiated the method developed and supported the practical impact they had on Danish industry. The two R&D projects were characterized by the following elements of the proposed development process:

- Identification of strategic challenges. This was centered around the traditional Strengths-Weaknesses-Opportunities-Threats (SWOT) analysis.
- Defining production tasks, which included identification of market requirements in terms of customer demand and the competitive situation and internal constraints, defined by product structure, production processes, etc.; and formulation of objectives and goals, i.e. management's selection of key objectives. In practice it was possible very often to identify two or more distinct production tasks with different objectives, volume, process requirements, etc. Thus, the identification of a number of pertinent production tasks for an industrial enterprise served as a valuable input to a discussion in the next phase of how to compose an appropriate production system. Originating from an engineering design approach, the notion of production tasks could naturally draw on the results of Skinner's proposed Focused Factory (Skinner 1978).
- Developing an idealized system, inspired by the traditional "dreaming" step in creative problem solving along the lines of Nadler (1970). The idea was to develop a solution which would be void of any constraints imposed on current

operations, such as existing plant facilities, machine tools, skills of the workforce. We also called this solution a concept or a vision, because it was meant to capture the overall conceptual elements of a new solution in an integrated fashion.

- Designing the subsystems. The main subsystems considered were production technology, plant layout, production planning & control, organizational structure, and wage system. An effort was made to align the product structure, sequences of production processes and production planning & control structure. In the early '80s the main emphasis was on structuring the plant layout into production groups using GT principles and methods. In the mid '80s the focus turned to the production planning & control system to simplify this task by building on production groups and to combine traditional push principle (MRP) and pull principle (Just-in-Time, Kanban).

The situation in Danish industrial enterprises in the early 1980s was ripe for drastic improvements. It was characterized by long throughput times (with processing times of only 2%), modest quality levels (the number of defects typically was around 10%) and little focus on setup time reduction. MRP principles were prevailing and applied through EDP systems which suffered from poor data discipline and long response times. In fact the production manager, planners and foremen had only limited support from PMS to manage daily operations, let alone any effort to identify trends and future challenges. A part to be manufactured would typically experience more than 5-10 shifts of responsibility as it was handed over from one production unit to the next.

2.1. The Typical Solution

In companies participating in the first development project (UPS) the typical "ideal solution" included a partitioning of the production plant into a number of production groups (typically around 10) arranged so as to support a "one way" flow of parts and components. Often four to six different flows were identified corresponding to main types of parts and components representing different production tasks identified. In this way the number of responsibility shifts for a part could be reduced to about three. The idealized solution often represented a mental jump and opened the eyes of many production managers to a different way of producing. It was transformed into a realistic plant layout most often by accepting the current building and production facilities.

The pivotal change was in the plant layout, but it automatically directed the attention to production technology, planning and control and the organization and management of the production groups.

A typical consequence of grouping similar parts to be processed at the same machines was an increase in the annual volume. This spurred analysis of setup times and possibility to design fixtures to be used for similar parts. Thus, the formation of production groups directed attention to utilization of the most expensive production facilities and in this way paved the way for the later focus on bottlenecks (OPT) and the reduction of setup times (SMED).

The introduction of production groups represented a drastic change to the operators: (i) Group working - preferably with a group bonus incentive wage scheme; (ii) Capability to master several operations at different machine tools in the production group; (iii) Involvement in the planning and scheduling of the group's work. The role of the foreman was basically kept unchanged, although new roles were initiated eventually to leave part of the scheduling to the group.

In the first development project (UPS) very few changes were made in production planning and control, not because they were not warranted, but mostly because of the inability to relate the new potentials for decentralized planning to the planning people who were operating their MRP-systems often without really understanding what the system did. In the second development project (VIPS) this issue was addressed, and methods were developed to provide a deeper insight in the actual planning and control behavior of the persons and departments involved. The development of a conceptual system provided industrial enterprises with an overall picture of the way planning and control was to be carried out. This helped them use their MRP-system in a more appropriate way and often led to the introduction of Just-in-Time (kanban) methods for selected parts and components. Very seldom did an enterprise develop a completely new production planning and control system, partly because no such system was available in the market. It was felt by many enterprises that because most of the commonly-used PMSs were based on the MRP thinking of the '70s, they were constrained in introducing more Just-in-Time methods supporting a pull mode of producing.

Thus, the formation of production groups served as a means for integrating different disciplines of production systems, such as plant layout, production processes, planning and control, and organization aspects.

2.2. Methods Of Analysis

A production flow analysis was the main method of analysis providing a good basis for the formation of production groups. The analysis was often carried out in two steps; first an intuitive classification of parts was made, either on the basis of a look at the product program and the components and parts manufactured, or by laying out on the floor a sample of all the parts. This visual mode of analysis was aimed at identifying similar parts with respect to the production processes needed and served as an effective start of the grouping process.

The second stage of analysis was based on data of the production processes and the sequences. As proposed by Burbidge (1989), the machine tool required, processing times, and the sequence were used as input. Due regard was taken to the capacity of key machine tools, whereas the capacity of minor machines could be neglected as long as it exceeded the need. Because the main criterion for forming production groups was to complete the processing of a part to be ready for sub-assembly, the number of workers in a group would vary. It was considered a good rule of thumb to hold the number below 15 members. In this range a formal structure and formal procedures for the group would have to be developed. Preferably the group size would be around 8 - 10. It was important to assess whether informal social relationships and team spirit would have a chance to develop; e.g. considering the complexity of the tasks, the degree of visibility in the group, the social interaction taken place during work.

In most cases it was possible by use of a mixture of data manipulation and intuitive thinking to identify a number of production groups without having to buy new expensive machine tools. However, it often was necessary to duplicate cheaper machines.

A critical constraint was experienced when an expensive facility was placed in the middle of the production sequence, e.g. surface treatment, soldering of printed circuit boards. If several production groups would feed into this facility and the capacity was limited, unplanned delays were likely to appear. A duplication of the equipment would not be justified; so either the capacity was increased to well above the demand, or management had to focus attention on handling planning conflicts which would arise at the bottleneck.

2.3. Typical Results Obtained

Some of the potential for drastically reducing the throughput times was realized in most cases, typically to one third or one fifth in the first place. At a later stage when also the PMS system became tuned to utilize the fewer planning points and bar coding led to more precision of data the throughput times were further reduced.

To the surprise of many production managers, the productivity also was improved significantly, typically by a factor of two. Although most production managers knew of some potential for increasing productivity, the marked simplification of the production flow which the introduction of production groups implied showed that many operations were unnecessary, e.g. transportation and handling. In addition it became possible to simplify planning and thereby reduce much of turbulence previously so prevalent.

Similarly, many production managers were surprised to discover that the new plant layout and the use of production groups required less space than they already had. Prior to the initiative, management in many companies felt that they needed more space. When looking at their shop floor overfilled with work-in-progress, one was tempted to agree to this solution. But as the throughput times were reduced, so was the work-in-progress, and it was possible to make visible the flow of parts and components through the plant.

Although the direct focus and objective was to reduce the throughput times, the formation of production groups also gave grounds for initiating a dialogue with production engineers, product engineering design and with sales and marketing. As already mentioned, the identification of key machine tools in production groups often led to discussions with engineers to reduce the setup times. Also engineers from product design were often invited to a production or assembly group to discuss ways of simplifying the work. And sales were consulted about the major concern of customers.

The introduction of production groups led to improved worker-employee relations. Reminiscence of the often adversary relation of the '70s in which any productivity improvement was believed to reduce job satisfaction were still active in the early '80s. But the introduction of production groups offered a possibility to reconcile the two objectives. Many components of what would constitute a good job proposed in the '70s would be accounted for in a production group, e.g. job enlargement by asking an operator to be able to perform several types of operations, and job enrichment by involving operators in the planning of the work in their group. The cycle time often was increased significantly which gave rise to much discussion among workers and production engineers; for example, with a cycle time of half a day would the operator be able to remember what he has learned. Examples similar to the assembly line at Volvo were seen also in Denmark with a group responsible for the final assembly of complex equipment.

The introduction of production groups implied changes of working traditions and habits, as well as attitude. For instance, a marked change of attitude and self-esteem was seen when operators were asked to sign-off their performed tasks as a result of quality inspection. Often operators were hesitant to accept a group bonus incentive scheme, having been used to an individual incentive system. But by an large, the workers and their unions could see the potentials of production groups to improve their working life conditions, and the introduction of production groups was conducive to a marked improvement in the worker-employee relations in the '80s.

2.4. Implementation

The process proposed by the two development projects to define the strategic challenges and to develop an idealized system represented to most industrial enterprises a completely new way of going about improvement. Most production managers had been used to make incremental changes; but this called for a shift of paradigm. Not surprisingly this gave rise to different approaches.

In several cases production managers became so excited about the idealized system and its potential that they adopted a Big Leap approach and tried to implement production groups overnight. A few managed to obtained significant results within a short period of time, while most production managers had to realize that development of the detailed system, e.g. plant layout, routings, etc., took much longer time than first perceived (Andersen & Mikkelsen, 1988). The expectations of top management were often unrealistically high with the consequence that the production manager was asked to leave when the promised results failed to materialize.

Other production managers adopted a Phased approach in which a number of phases and milestones were defined for a 3-4 year implementation process. In one company a cascading process was introduced leading to a gradual involvement of foremen and workers in the development process.

As briefly mentioned, operators were challenged in many ways when they were to become member of a production group. They should be trained to master several production processes, instead of typically just one; they were asked to perform a quality inspection for their own work and be responsible for the result; their active participation in scheduling the group's work was encouraged; and they were supposed to support a group incentive scheme.

In many ways, issues of managing organizational change had to be dealt with, e.g. to develop a comprehensive understanding of the new mode of operating, including the content of their own role, and to achieve an acceptance of the new organization and a will to go along. The capability to illustrate and visualize the new mode of operation has played a key role. In view of these challenges a role-playing game was developed, the Ruler Game, in which the operations of a production group was simulated (Andersen et al., 1979). The group was to produce wooden rulers of different types on 5 - 7 machines which were built around simple hobby electric drilling machines. The planning system was taken from a real company, and a group incentive scheme was used. During the two days of playing the game, the participants would experience different incidents associated with the day-to-day life of a production group. Although initially meant for a specific industrial company, we used the Ruler Game in a large number of companies either in the phase of considering introducing production groups or in the process of implementing them. During the '80s the game was adopted as part of a training program for workers all over the country and has played a significant role in facilitating the implementation of production groups in Danish industrial enterprises.

In other instances we developed company-specific simulation games to demonstrate a new way of carrying out production planning and control based on production groups. In this way new principles such as Continuous Flow Manufacturing, Just-in-Time Production, and Period Batch Control were demonstrated for the products and production processes of the specific company (Mitens, Mikkelsen & Riis 1991).

2.5. Opposing Approaches

The '80s was in many respects a turbulent period for manufacturing. The awareness of the persuading Japanese results and the appearance of new radically different solutions, as discussed above, initiated a search for new means and approaches. Several production paradigms were discussed, such as the Just-in-Time, Total Quality Management, Kaizen. In Denmark, and probably also in other countries, two opposing approaches were challenging industrial enterprises. One approach followed the lines described above in which the degree of complexity was attempted to be reduced primarily by introducing production groups which were considered as small factories within a factory. This was in many respects a Simplify approach and represented only minor investment in modern technology. The other approach could be named an Automate approach according to which the potential of new production technology and information technology gave promise to significant improvements. The concept of Computer Integrated Manufacturing (CIM) caught much attention. Several industrial companies successfully implemented parts of CIM paying attention to the required organizational changes. But other enterprises were let to believe that CIM could be introduced overnight and failed to realize any significant improvement, despite the heavy investment. Or they applied modern facilities and systems to a traditional concept of manufacturing and production management and obtained only minor improvements.

The battle between the two opposing approaches looked towards the end of the '80s very much to favor the Automate approach. This led us to wonder why an industrial enterprise, with the prospect of being able to reduce throughput times and increase productivity drastically by introducing production groups with little investment required, instead would prefer to invest in modern IT and manufacturing technology (Burbidge, Falster & Riis, 1990). It appears that the development in the '90s has led to an appreciation that the two opposing approaches are both needed and, in fact, may support one another. As will be discussed in the next section, this was also in due time, because of new challenges to industrial enterprises.

3. CONTINUOUS IMPROVEMENT IN THE 1990s

In the early '90s a new wave of improvement effort gained momentum inspired heavily by the philosophy and practical experience of Kaizen. It may be seen as a contrast to the top-down approach of UPS and to the technology-oriented CIM approach. The central Danish initiative was named Production Development through Employee Involvement (MAPU, 1993). As indicated by its title, it focused on a bottom-up approach to engage employees actively not only in the daily operations, but also in the improvement of operations, as a means for increasing productivity. As in Kaizen, reduction of waste of any kind was one of the main vehicles for the process. Management also saw this effort as a means for achieving a higher degree of flexibility required by the explosion of product variants and reduced delivery times. Management was interested in a more direct mode of operating making it possible to cut many administrative and managerial activities away, thus also supporting the concept of Lean Production (Womack & Jones, 1990 and 1996).

3.1. Case B

A medium-sized company manufactures tools for both the professional market (carpenters, fitters, electricians, etc.) and do-it-yourself customers. In the mid '80s it undertook a development project in production. Based on production flow analyses and identification of a number of different production tasks, the plant was divided into production groups. As a result, the throughput times were drastically reduced and productivity increased. Furthermore, the project gave rise to a strategic plan for investment in production technology and a search for an MPS system which was supportive of the Just-in-Time principle.

Five years later, management wanted to revive the spirit of the development project. But this time it should be based on a bottom-up approach. The MAPU initiative was chosen as an appropriate vehicle. Following the process and methods of Kaizen, almost all workers were involved in Continuous Improvement groups. Suggestions were encouraged and acted upon. Workers were gradually being involved in scheduling of production orders in their production groups. As a result, the productivity went up by 50% over a three-year period, at the same time as work-in-progress was cut by more than 60 %. This made it possible for the company to become more adaptive to customer request. The bottom-up approach created a positive attitude towards change which was used at a later stage when the production groups were aligned along flow lines.

3.2. Situation In 1990s

The results achieved in Case B were typical of the enterprises which adopted a MAPU approach. The single production unit was the primary object of the improvement effort. It is our impression that enterprises which had based their production units on a well-thought out production flow analysis benefited the most, mainly because the production units were formed around rather well-defined production tasks implying some degrees of freedom for the production group.

As to the organization of production units a high degree of experimentation took place centered around nominating one of the operators to serve as a coordinator for the group. The main responsibility of the coordinator has been to prepare a work plan for the operators, often including several shifts in the light of the dynamic incoming flow of customer orders.

The '90s was also marked by a continued pressure from competition to increase productivity and quality, to decrease delivery times and to ensure precision in delivery. An increased appreciation could be observed of the need to combine and integrate several means, e.g. process technology, production planning and control, production configuration, and organizational issues.

Past the middle of the '90s the situation may be characterized in the following way:

- The functioning of production planning and control systems has been greatly improved mainly through the introduction of bar coding systems for data collection and more frequent computer runs of the MRP net demand. Some enterprises have added scheduling systems to their MRP-based systems. However, the computer systems are still based on the traditional MRP planning principles and have difficulty with supporting the effort to adopt Just-in-Time modes of operating and to meet customers' request for being quoted a precise delivery date. This deficiency was pointed out early and repeatedly by Burbidge.

- A better mastering of production technology and the IT-based integrated systems created a much more reliable and deterministic basis for planning and scheduling. This led to the definition of realistic objectives and commitments (e.g. due dates).
- The very short delivery times realized in many enterprises has left very little discretion for the single production unit. This imposes a severe constraint on the effort at the local level to engage employees actively into scheduling. As observed in several enterprises, the Continuous Improvement process which has taken the single production unit as the primary object suffered from loosing momentum. Several initiatives have been taken to overcome this embedded conflict, one in the direction of identifying which control tasks are most suited to be located decentralized and which tasks should be dealt with by a central, coordinating unit, recognizing that the issue is a both-and issue instead of the often discussed either-or issue. Furthermore, the analysis has pointed to the need for employees in a single production unit to be able to see their own role along the total production flow, calling for an overall vision of production made available to individual employees.
- The groups have primarily been involved in capacity planning, and marked improvements have been achieved. Instead of treating capacity planning as a formal exercise, the coordinator of the group have been able to also include "soft issues", such as to persuade an operator to work overtime and to accept the request of an operator for taking half a day off. The involvement of operators in planning their own work has increased the flexibility of the production plant, a great challenge to modern manufacturing with the short delivery times.
- Barriers to successful implementation of groups are for example: Resistance from middle management to adopt a new, supportive and coaching role; Failure to introduce a wage system, which stimulates team work and the attainment of specified goals; Lack of top management support.
- Facilitators to successful implementation of groups are for example: Timely and relevant feed-back to the groups of its performance; The general cooperative spirit between workers and management, supported by the unions and employers' associations; The willingness and competence of workers to take on greater responsibility. The capability to visualize the overall planning in a transparent way.
- We have observed that any enterprise devoted to improving their performance will have many different improvement initiatives in progress at any point in time, such as production process improvement, quality management projects, adjustments of the MPS system, product improvements, Continuous Improvement activities at the local level, adjustments of the wage system, etc. Very often they are not coordinated, because each organizational unit is carrying out their own projects. This calls for an effort to develop an overall conceptual

image (vision) of how the coming production system will look and an ability to orchestrate parallel improvement projects (Riis & Knopp, 1996).

Compared with the implicit role GT played in the early '90s, we envisage that GT-based on production flow analysis will experience a renaissance in the near future. A recent study of four industrial enterprises supports the impression of the industrial interest in combining a Kaizen approach to involve workers in an improvement process and the use of GT to structure the production plant into appropriate flow groups (Foged & Jessen 1996). The effort supports the flow aspect advocated in Business Process Reengineering and Lean Production, and at the same time it attempts to identify the nature and degree of discretion available in a production unit. However, it is necessary to combine production groups into flow lines.

3.3. Case C

A small electronic enterprise was in the middle of the '90s confronted with a profound crisis. The price was too high, the quality not satisfactory, and the delivery time was long and unreliable. With the increased motivation to change that standing with the back against the wall had created throughout the company, the introduction of production groups served as a vehicle for turning the company completely around within three quarters of a year.

The background was a functional plant layout, unclear process flows and production initiated on the basis of uncertain forecasts. Thus, a production system overdue for a drastic change.

Two production groups were defined with an even distribution of the 30 workers. The first group included the first operations connected with mounting printed circuit boards, and the second group took care of the proceeding operations of mechanical assembly, calibration, testing and packing.

The reason for this definition was the different nature of customization. Whereas the operations in the first group were the same for all products, the second group had to adjust and calibrate the products to customers' specification. Thus, the two groups called for different production management methods to be applied. In fact, a Kanban principle was adopted for the interaction between the two groups.

All walls in the old plant were removed which allowed for a smooth and visible flow within production groups. Each operator was asked to test his/her own work, and the group was to plan its own work supported by one foreman for both groups. The appointment of a coordinator for each group facilitated direct interaction with sales, engineering design and purchase.

The precision of internal deliveries and throughput times were measured in production, and sales recorded precision in delivery to customers. In addition, the suppliers' quality and time of delivery were measured. The workers continued to be paid on the basis of a fixed hourly rate.

Within less than three quarters of a year the throughput time was cut from 10-12 weeks to 10 days at the same time as precision of delivery and productivity were increased significantly.

Initially, great effort was needed to have workers understand and accept the basic ideas and potential of introducing production groups, and to realize the changes of the plant layout without disturbing the daily operations. However, the company nevertheless succeeded in creating a large commitment and enthusiasm among workers facilitating the change in working habit and attitude. This may be explained by the felt need to introduce drastic change and the small size of the company.

4. FUTURE CHALLENGES AND OPPORTUNITIES

As the customers may be expected to become more demanding all over the world, and as the strive for perfection in manufacturing to offer value to the customers continue, industrial enterprises will be confronted with new challenges. In this section we shall outline some of the trends we see relating to the further use of GT in manufacturing.

The challenge to reconcile the increased demand for shorter delivery times with the implied quest for flexibility and adaptability and the desire to involve employees more actively will most likely continue to exist and even become more acute. One direction would be to let employees become more engage in improvement initiatives, as for example already known from Kaizen and Total Productive Maintenance. Another direction would be to develop decision support systems for scheduling to help employees schedule the work in their production group in view of the incoming flow of customer orders and available capacity.

A third direction is seen in the two recent concepts, Fractal Company and Holonic Manufacturing, cf. Warnecke (1993) and Van Brussel (1994). Although they draw on chaos theory and cybernetic ideas of self-organizing systems, the idea appears to be the same as in the early '80s to form small factories within the factory which would offer room for self-organizing principles. However, the competitive environment is different now, and there is a need to explore the possibility of establishing a higher degree of self-organization at the lower level without the risk of sub-optimization.

Developments in recent years have demonstrated the need to see production in a broader context. Two directions are dominating: along supply and distribution chains, and product development process. Following the emerging management principle of involving those whose are directly affected, we will expect that production and assembly groups would be involved in the introduction of new products, especially the start-up of production. Similarly, we will expect that more direct links will be made between production groups and their external suppliers, in particular the call-off of deliveries. Also, the use of GT may point to new roles of suppliers in terms of completing the processes of a component or part.

TABLE 1. Production Groups: A Check List Of Factors To Consider.

Factors	Examples
Strategic advantages Which strategic impact is expected? Why introduce production groups? What urges the company to consider this action?	Shorter delivery times; Increased productivity; Worker involvement and empowerment; Improved quality.
The formation of groups Which criteria are relevant? How are the groups defined? Do they cut across functions and disciplines? Do they include both blue and white collar employees? In which parts of the production plant are there not formed production groups, Why not?	Criteria: Similar production processes; Similar parts, belonging to the same product family; Similar production management task (volume, required quality and delivery times, etc.); Development of team spirit and collaborative working modes.
The tasks of the groups In addition to the well-defined production tasks which tasks are given to the production groups?	Scheduling; Quality inspection and assurance; Change of equipment; Maintenance; Training of operators; Continuous Improvement.
Consequences for interaction with other units Which changes in the tasks of other units have been necessary? How will the interplay between the production group and the other units be affected? How will this affect the management systems? Where are the degrees of freedom for the production group?	Management systems: Fewer planning areas and points with more aggregate measures; New role for the system: To support decentralized decision making. Areas of discretion: Capacity planning, scheduling, customer orders, call-off deliveries from suppliers, joint planning with other production groups.
Performance measurement Which performance measures are used? How often? How is the wage system? If it includes any incentive scheme, how is it constructed?	Possible measures of performance: Delivery precision; Quality level; Worker productivity; Wage system: Fixed hourly pay with variation according to qualifications and commitment. Bonus (Approx. 15 - 20% of total pay) dependent on delivery, productivity and quality
Experiences gained What has been successful and which results have been achieved? What has been difficult? Why? What would be done differently, if the company had a chance to start all over again?	Keep record of improvements; Hold experience sharing meetings

TABLE 2. Group Technology: Some Trends in Danish Industry.

	1980s	1990s	Coming years
The situation	Long lead and throughput times. Poor data discipline	Short throughput times. Higher precision in production, due to bar coding and production technology	Customization
Main focus	Reduction of delivery and throughput times. Delivery of promised quality	Increased productivity and employee involvement. Reliable deliveries.	Value to the customers. Perfection in production.
Strategy	Simplify or automate	Simplify and automate	Orchestrate a multi-focused effort
Assumptions about workers	Emerging co-operative spirit. Idealistic and intrinsic values of job enlargement and enrichment.	A belief that workers have the capabilities of to work together in an intelligent way.	A belief that built-in flexibility of groups can be capitalized. A belief that workers can adopt a customer oriented view.
Role of GT	To enable the formation of groups of similar production processes and sequences.	To identify local areas of discretion, and to assist in forming production flow lines.	To identify areas where plant within plant is possible. To support formation of administrative groups working on similar tasks.
Challenges	To introduce pull principles without the support of MPSs. To change attitudes and behavior of workers towards collaborative group thinking	To combine continuous improvement and innovative, top-down initiatives. To reconcile the need for adaptability and employee involvement.	To integrate production and administrative processes along flows. To increase flexibility with shorter delivery times and increased productivity.

Project management has successfully been used to bring people together from different disciplines to solve complex development tasks. We find that this idea may also be used for solving complex tasks in the daily operations bringing people together from different departments and even different companies. Several examples are emerging. However, the issue of defining appropriate sets of tasks to be addressed by project groups appears to be essential for the possibility to maintain a reasonable amount of discretion in the group. We believe that the principles of GT may be useful when applied to administrative tasks. Perhaps new models for describing and grouping administrative tasks into similar sets may have to be developed. But there is a great need for improving the administrative processes surrounding production, and GT holds in our opinion great potentials for contributing.

5. SUMMARY AND CONCLUSIONS

Many factors influence the formation and management of production groups. Looking back on the many examples we have seen and been involved in we have summarized the most relevant factors to consider in a check list shown in Table 1. This may help understand the key role of GT, but also to appreciate that other factors are important for the successful implementation of production groups.

In the chapter we have shown how GT played an important role in the '80s and a more implicit role in the '90s. Yet, the current situation indicates that, in our view, GT again may come to play a significant role in the future if further developed for administrative and managerial tasks. Being aware of the risk to oversimplify in Table 2 we have summarized the main development trends in Danish industry.

The paper has also clearly demonstrated that the context and the situation of industrial enterprises have changed dramatically in the past 20 years calling for new approaches and refinement of existing methods.

REFERENCES

Andersen, O. S., and Mikkelsen, H., 1988, *UPS - A short-cut to innovation?* (in Danish), Promet, Copenhagen.

Andersen, S., Iversen, J., Riis, J. O., and Sant, K. L., 1979, A production game as a miniature production cell, *Proceedings of the International Conference on Production Research*, Amsterdam.

Burbidge, J. L., 1989, *Production Flow Analysis*, Clarendon Press, Oxford.

Burbidge, J. L., Falster, P., and Riis, J. O., 1991, Why is it difficult to sell GT and JIT to industry?, *Production Planning & Control*, 2, 2 , 160 - 166.

Foged, S., and Jessen, N., 1996, *The future workplace: An empirical study of four companies' work with production groups* (in Danish), Unpublished report, Dept. of Production, Aalborg University.

Luef, H., 1980, *Industrial retrieval systems* (in Danish), Ph.D. thesis, Dept. of Industrial Engineering and Production Management, Technical University of Denmark.

MAPU - Production Development through Employee Involvement (in Danish), 1993, Confederation of Danish Industries.

Mitens, L., Mikkelsen, H., and Riis, J. O., 1990, *Development of Company-specific Games for Production Management* (in Danish), A ViPS-report, Dept. of Production, Aalborg University.

Nadler, G., 1970, *Work Design: A Systems Concept*, R.D.Irwin, Illinois.

Riis, J. O., 1978, *Design of Management Systems - An Analytical Approach*, Akademisk Forlag, Copenhagen.
Riis, J. O., 1990, The use of production management concept in the design of production management systems. *Production Planning And Control*, 1, 45-52.
Riis, J. O. and Knopp, J., 1996, Implementing new production management modes: orchestration of simultaneous improvement activities, in *Advances in Production Management Systems*, (Eds. N. Okino, H. Tamura and S. Fujii), IFIP, Laxenburg, Austria, 331-337.
Rode, J. and Sant, K. L., 1984, *Development of Production Systems* (in Danish), UPS-report.
Sant, K. L., 1977, *Group Technology in Industrial Enterprises* (in Danish), Polyteknisk Forlag, Copenhagen.
Skinner, W., 1978, *Manufacturing in the Corporate Strategy*, Wiley, New York.
Van Brussel, H., 1994, Holonic manufacturing systems: The vision matching the problem, *Proceedings of The First European Conf. on Holonic Manufacturing Systems*, Hannover.
Womack, J. P., Jones, D. T., and Roos, D., 1990, *The Machine that Changed the World*, Macmillan, New York.
Womack, J. P. and Jones, D. T., 1996, *Lean Thinking*, Simon and Schuster, New York.
Warnecke, H-J., 1993, *The Fractal Company*, Springer-Verlag, Berlin.

AUTHORS' BIOGRAPHY

Jens O. Riis is Professor of Industrial Management Systems at Department of Production, Aalborg University, Denmark. He holds an M.Sc. in Mechanical Engineering from Technical University of Denmark and a Ph.D. in Operations Research from University of Pennsylvania, USA. His main teaching and research areas are design of production management systems, technology management, project management, and integrated production systems. He worked with John L. Burbidge in the late '80s on integration in manufacturing and exchanged on many occasions ideas and experiences about forming production groups. Prof. Riis is currently heading two research programs in integrated production systems and technology management, and is a member of the IFIP Working Group 5.7 on Computer Aided Production Management Systems and of the international editorial board of several international journals.

Hans Mikkelsen is a Senior Consultant at Sant + Bendix, a member of Coopers & Lybrand. He holds an M.Sc. in Mechanical Engineering from the Technical University of Denmark and has been involved in consultant work and development projects in the areas of design of production system, production management systems, project management and product development both from a system's perspective and from an organizational and management point of view.

12

Benefits From PFA In Two Make-To-Order Manufacturing Firms in Finland

S. Karvonen, J. Holmström and E. Eloranta

1. INTRODUCTION

This chapter demonstrates the benefits of applying Production Flow Analysis (PFA) to redesign factory organization and reduce setups in two make-to-order manufacturing environments in Finland. In the first, one-of-a-kind manufacturing operation, the primary benefits were improved change management and increased capacity. In the second, subcontractor case, the main benefit was to launch a design for manufacturability process with key customers. In the first case Group Analysis and Tooling Analysis was used together. In the second case only Tooling Analysis was applied.

Factory layout can be based either on process or product organization. Product organization, in most cases, means group technology (GT) (Burbidge 1989). PFA is a progressive technique which uses Company Flow, Factory Flow, and Group Analysis to find the division into groups and families. Line Analysis is used to study the flow of materials within a group and Tooling Analysis is used to find tooling families of parts that can be made at the same set-up on a machine.

In make-to-order environments functional process organization is the norm rather than the exception. The rationale for this is that frequent changes in product design makes it difficult to organize the production process around products. Consequently it is difficult to find make/design to order manufacturing companies where PFA techniques have been applied.

This chapter presents two cases of applying PFA techniques in make-to-order manufacturing environments. The first case is from a one-of-a-kind production environment. The second case is from a subcontractor. The objective is to show the benefit of PFA techniques in these most typical job shop environments.

In the one-of-a-kind production environment two benefits of PFA techniques will be demonstrated. The introduction of GT results in a much simplified Engineering change management process. A systematic approach to tooling, i.e. Tooling Analysis, in bottleneck work centers enables moving parts to non-bottlenecks and substantially increasing the capacity of the plant.

In the subcontracting case Tooling Analysis is introduced as the basis for a Design-for-manufacturability process. The Tooling Analysis technique is used by the design staff of customer and subcontractor to standardize non-critical dimensions.

2. CASE 1: APPLYING PFA IN ONE-OF-A-KIND PRODUCTION

The first case company is engaged in one-of-a-kind production. The firm employs 350 people of which 90 are employed in manufacturing. The product is a wood processing system for pulp factories. The market is highly competitive and dependent on the investment cycle in the volatile pulp and paper industry. There are several basic models, but each produced unit is unique from a manufacturing perspective. The dimensions and exact configuration of the basic models are varied according to customer specifications. Production is controlled by the installation of the wood processing system on the customer site. This means that manufacturing must quickly deal with engineering changes.

PFA was applied in the case company to introduce GT. The purpose of the PFA techniques is to reorganize production into groups of machines and work centers that can perform all operations required to produce a range of parts or products from start to finish. This method of organizing the manufacturing process in batch production forms a basic of GT.

The objective was an improved material flow system, but also, to increase engineering change management capabilities (Hameri & Karvonen 1994). From the production control point of view, the initial process organization of the company resulted in the material flow system seen in Figure 1. This material flow system is complex and throughput times are exceedingly long due to the long transfer time between operations. With several project deliveries simultaneously in process the situation was frequently on the verge of the chaotic. Each product design change had to be communicated to most work centers, causing confusion and production delays.

PFA consists of several sequential stages. The first stage is Company Flow Analysis (CFA), which is useful for analyzing the operations of large corporations or complex supplier networks. The next step is Factory Flow Analysis (FFA). The objective of FFA is to divide a large factory into focused factories. Group Analysis (GA) is used to form groups of machines that can manufacture parts or larger components from start to finish. The material flow within the group is optimized using the Line Analysis (LA). The last step, Tooling Analysis (TA), is used to analyze the individual machines and work centers of a Group.

2.1. Group Analysis

The first step in the formation of groups in the case company was the Group Analysis. The machines were classified and the production groups identified based on the main components (Table 1). Table 2 shows the situation after reallocation of operations and combining two initial groups. Reallocation means that exceptional operations in the production groups are reallocated to related machine types inside the same group.

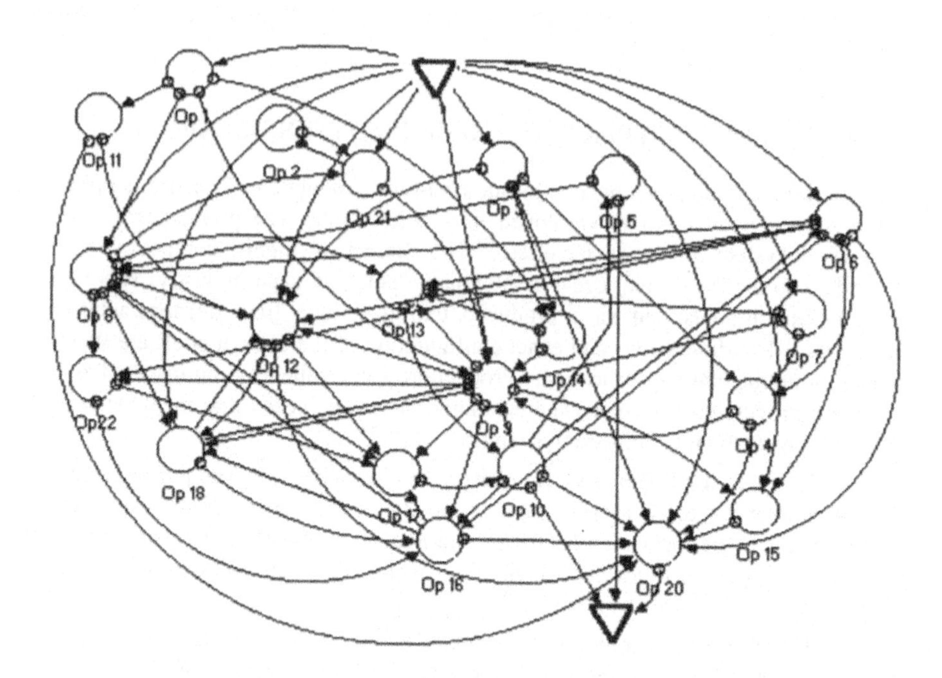

FIGURE 1. Material flow with process organization (Hameri & Karvonen 1994).

The key to eliminating the exceptions by reallocation is a detailed analysis of part specifications and processing methods. In the case company the broaching drills were particularly difficult to place in a group. Initially the large broaching drill was placed in group G, where it was used for machining six parts. The smaller broaching drills were placed in groups B and G. By examining the parts and machines in detail it was realized that the wheel support of the debarking drum could also be manufactured with the small drill, and not only with the large broaching drill. The large drill could therefore be moved to group B, where it can be used to process 64 parts. Now the small broaching drill can be moved to group D, where it can process all parts of the radial drilling and milling machine. For the large broaching drill this now leaves exceptions in only groups C and G. For the small broaching drill there is exceptions in groups C and E+F.

The implementation process of GT in the plant was started by forming the Group B. This decision was made after the Group Analysis. The difficulty to allocate the broaching drill capacity gave strong indications that the production of the parts in Group B was in fact the bottleneck of the operation.

TABLE 1. Machine Classification (Hameri & Karvonen 1994).

Name	Cd	N	F	Distribution of machines/operations between groups						
				A	B	C	D	E	F	G
Faceplate lathe	S	1	3	1/2	1					
Keywaying m/c	S	1	4			1/2			2	
Broaching drill (L)	S	1	11		4	1				1/6
Broaching drill (S)	C	2	121		1/61	8	7	6		1/39
Radial drilling m/c	C	5	758	13	2/338	1/151	1/73	64	35	1/84
Milling m/c	C	3	179		1/52	1/40	33	1/38	16	
Stampel pl. clipper	C	1	14				6		1/8	
Pillar drilling m/c	C	1	33						1/33	
Welding table	I	2	86	2/86						
Centering lathe (L)	I	3	98		1/12	9	1/9	8		1/60
Centering lathe (S)	C	4	307	16	1/96	28	5	1/45	1/85	1/32
CNC - lathe	C	1	8						1/8	
Submerged arc	C	3	5	3/5						

Legend: S - special, C- common, I - intermediate; N: total number of machines of the same type; F: total number of different parts with operations on a machine type; Underlined numbers: number of parts with no manufacturing resources in the production group.

TABLE 2. After Reallocation of Operations and Combining Groups (Hameri & Karvonen 1994).

Name	Cd	N	F	Distribution of machines/operations between groups					
				A	B	C	D	E+F	G
Faceplate lathe	S	1	3	1/2	1				
Keywaying m/c	S	1	4			1/2		2	
Broaching drill (L)	S	1	67		1/64	1			2
Broaching drill (S)	C	2	140			8	1/83	6	1/43
Radial drilling m/c	C	5	643	13	2/296	1/151		1/99	1/84
Milling m/c	C	3	146		1/52	1/40		1/54	
Stampel pl. clipper	C	1	56		1/42		6	8	
Pillar drilling m/c	C	1	33						
Welding table	I	2	86	2/86					
Centering lathe (L)	I	3	126		1/12		1/14	8	1/92
Centering lathe (S)	C	4	194	1/16	1/96	1/37		1/45	
CNC - lathe	C	1	93					1/93	
Submerged arc	C	3	5	3/5					

2.2. Line Analysis

The next step of the PFA is to focus on the flow of materials in a group. The Line Analysis step gives information on how to organize the layout of the group. The analysis is essentially to determine all the possible operation route numbers needed to process a part. The routes are individually named and the frequency of each route is calculated. In the case company a line analysis was conducted and served as the starting point for the layout designs. However, in an old facility with heavy equipment it was not possible to realize optimal material flows in all the groups.

Heavy machine tools, such as the broaching drills and the face plate drill, must lay on a foundation of concrete. The construction of these concrete foundations are the reason for the very high moving costs per machine. Moving lighter machine tools, such as radial drilling machines, is much simpler because they are fixed to the floor simply by nuts and bolts.

2.3. Management of Design Changes After Group Implementation

Figure 2 shows the material flow system of the plant with a product organization. It must be taken into consideration that three groups out of six have been implemented and the rest three groups are implemented nowadays. The primary effect of GT is that it radically reduces the lead-time for introducing a design change. The business benefit of reduced engineering change lead-time is less waste and a reduction of risk. This is because the notification about a design change can be immediately communicated to the right people. After the introduction of GT the notification of the team leader of one specific group is enough. Before it was necessary to inform several foremen and production planners about each design change. Previously some people who should have been notified was frequently overlooked - resulting in paying out significant penalties to the customer or the installation contractors. Today the change notification can be communicated in 5 minutes to the affected group. Before it took up to 24 hours to notify all affected work centers.

2.4. Tooling Analysis Of Bottleneck

The final step in PFA is the Tooling Analysis. The main objective is to find families of parts that can be produced with the same set-up and to plan the loading sequence for minimum set-up time. The result is reduced batch sizes and more capacity. This is particularly valuable in bottleneck work centers.

To perform a tooling analysis it is necessary to know for each part:

- Material type (e.g. steel, aluminum)
- Material shape (e.g. bar, tube, casting)
- Material size

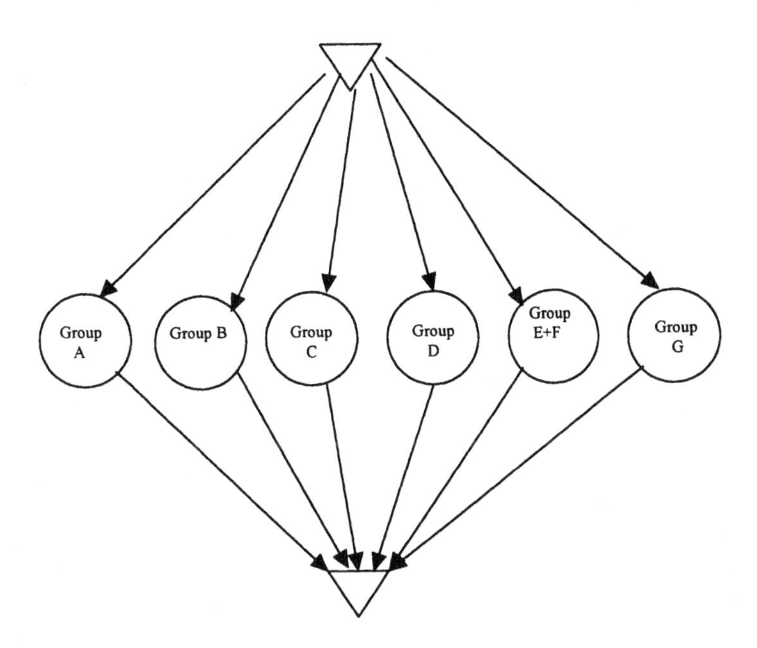

FIGURE 2. Production flow between groups after GT. Three groups out of six implemented and remaining three are being implemented.

It is also necessary to know the tools and jigs used for different material types, shapes and sizes. Burbidge (1989) describes how the Tooling Analysis is performed in detail. First the parts and operations at each work center is listed. The parts are then divided into subsets by type of material, form and size. Next, the parts are divided into subsets by material holding tooling. For each category identified in the previous step a tooling matrix is produced. With the help of the tooling matrix tooling families are found. Then the loading sequence is studied to minimize set-up time. Finally, the possibilities to reduce variety is studied.

The tooling analysis was conducted in the case company once the groups had been formed. The most important result of this step was that it revealed more parts that could be moved away from the bottleneck broaching drill. This is in group B, which had become the bottleneck of the whole factory.

The parts processed in the work center are different types of sub assemblies. Therefore, no analysis based on form is made. Figure 3 shows the tooling families of the broaching drill after the first stage of the analysis.

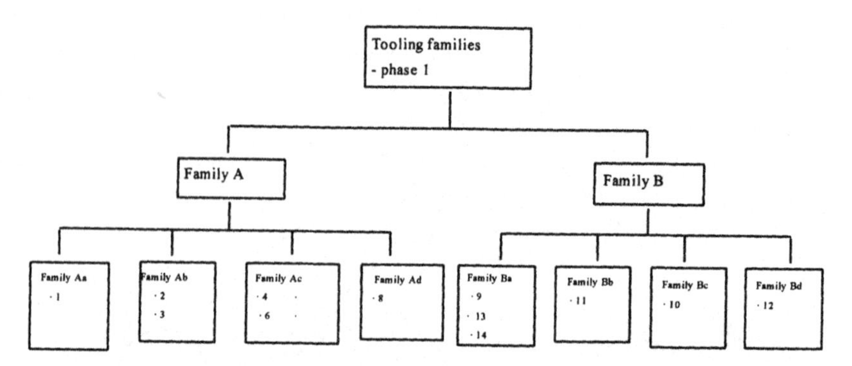

FIGURE 3. The tooling families of the broaching drill after the first stage of analysis. (Karvonen & Holmström 1996)

The parts belonging to the same family were found from the analysis of holding tools and set-ups of the parts machined on the broaching drill shown in table 3. The slowest holding setups are the angle element, a small v-notch piece and v-notch piece (normal). Therefore, the tooling family 'A' is formed from parts that do not require these set-ups. The tooling family 'A' consists of part numbers 1 to 8. The tooling family 'B' is formed of the parts that uses the angle element, the small v-notch piece and the v-notch piece holding tools, i.e. part numbers 9 to 14.

TABLE 3. Holding Tools and Number of Setups of Parts Machined on Broaching Drill.

		Parts														
		1	2	3	4	5	6	7	8	9	10	11	12	13	14	Σ
	number of set-ups	2	6	6	9	2	2	2	4	2	3	1	5	10	8	62
1	Fastener A	•														1
2	Fastener B								•							1
3	Fastener C		•	•	•					•	•		•	•	•	8
4	Fastener D					•	•	•								3
5	The bolts of the fastener	•	•	•	•	•	•	•	•	•	•		•	•	•	13
6	Raising pieces		•	•	•		•	•					•			6
7	Shock		•	•	•	•	•	•	•				•	•	•	10
8	Angle element A									•				•	•	3
9	Angle element B												•			1
10	Small v-notch piece										•					1
11	Clamps												•	•	•	3
12	Support pieces of a baffle plate												•			1
13	Angle jig													•		1
14	V-notch piece (normal)											•				1
15	Special fastener											•				1
	Holding tools/part	2	4	4	4	3	4	4	3	3	3	2	7	6	5	-

The next step of the Tooling Analysis is to explore the possibilities to use the milling and radial drilling work centers of the group to perform some of the operations of the broaching drill. It was found that the work load of the broaching drill can be reduced by moving parts number 10,13, and 14 to the milling work center. The move of parts 13 and 14 requires that a special jig is constructed for holding these two parts. The operations for part number 9 can also be moved to the radial drilling and milling work centers. The milling work center machines the surfaces and the drill the holes. All the operations for part number 5 can also be moved to the radial drilling and milling work centers. The move requires the construction of a new jig.

After moving these parts to alternative work centers the load of the broaching drill is reduced by 110 hours per sub-system. Figure 4 shows the tooling families after the move of parts to alternative work centers. The operations of part number 5 of the tooling family 'A' is transferred to non-bottleneck machines. From the tooling family 'B' the sub-families 'B.b' and 'B.c' are transferred to other work centers.

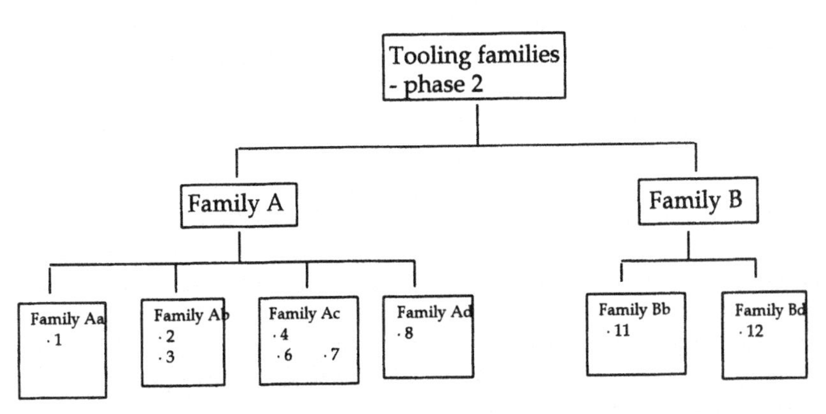

FIGURE 4. Tooling families of broaching drill after second phase of analysis.

The primary advantage of moving to GT in the case company was improved change management. In the make-to-order project business changes occur frequently for both internal and external reasons. However, with a product organization the project engineer who makes a drawing change can communicate directly with the group. Previously, with a process organization, it was necessary to use the production planner as an intermediary to communicate and manage the execution of engineering changes. The benefit of change management was realized when forming the groups based on the Group Analysis.

The full benefit of PFA was achieved with the Tooling Analysis on the bottleneck work center of the bottleneck group. In the analysis several parts which were not previously identified as parts which could be produced on alternative machines were found. The transfer only required the construction of two new jigs. With the transfer the number of set-ups on the bottleneck machine was reduced from

62 to 37. Seventeen tools were also transferred to other machines with the parts. With these actions the throughput time on the bottleneck was reduced by 110 hours. The throughput time of the group was cut by ten days for a complete subsystem. Effectively the throughput time of the group as a whole was reduced by almost 30%.

3. CASE 2: DFM LAUNCHED BY TOOLING ANALYSIS

The case describes how Tooling Analysis was used to launch a Design for Manufacturability process between KSH-Productor Oy and its customers. KSH-Productor Oy is a subcontractor for parts requiring precision machining in the medical instruments and telecommunications industries. The company was formed after a Management Buy Out of the machining operations of a multinational engineering company. It has at the moment 60 employees and a customer base of 40 companies from the medical instruments and electronics industries. The plant is situated in Espoo, Finland, 20 km from Helsinki. The production volumes of the subcontracted parts are from 10 up to 1000 units. The unit value of the manufactured parts varies from $400-$60,000. Even though the subcontractor business is very competitive and vulnerable to recession KSH-Productor Oy has been very successful, achieving a 20 per cent growth per annum.

There are four bottleneck machines in the operation. One of these, a CNC , was chosen as the pilot for introducing Tooling Analysis. The original objective of the analysis was to reduce set-ups by forming tooling families. Reduced set-ups on a bottleneck work center increases capacity and allows for reduced lot sizes. Investment in new plant can be delayed, which is important in a fast growing operation. Smaller lot sizes improves delivery service and reduces subcontractor's Work-in-Process inventory and customer's component stocks.

With the Tooling Analysis technique the reduction of set-up time is realized by forming tooling families. A tooling family is composed of parts that can be machined with the same set-up. Based on the tooling families it is possible to determine the optimal machining sequence, i.e. the sequence that minimizes fixture and tooling changes.

However, in this case it was discovered that the greatest improvement potential was in a standardization of product designs. The Tooling Analysis was used to launch a Design for Manufacturability process between the case company and its customers.

The parts machined in the work center chosen for the TA pilot are either aluminum or plastic. Since aluminum and plastic can be machined with the same tools it was not necessary to perform a grouping of parts based on the material. Nor was a grouping based on the geometry of the parts necessary, since the parts were asymmetrical almost without exception. To perform the Tooling Analysis it was therefore necessary to have only the information on the tools and fixtures of the parts. In the subcontractor case the first step of the Tooling Analysis was to analyze fixtures and tooling.

Fixtures for the parts are mostly clamps, chucks, and part specific jigs. The consequence of this is that tooling families are difficult to form based on the currently used fixtures. Additionally, it is not possible to machine parts using the same fixtures without set-up. The parts have to be fixed and released individually when clamps and chucks are used.

In the case company there is a considerable improvement potential in better fixture design. The objective for fixture design should be to enable several parts to be fixed by the same jig. This way these parts can be machined in sequence without set-up.

On the pilot work center 79 different tools were initially used to produce a total of 39 different parts. The operations performed are drilling and milling . The number of tools needed to produce the parts is disproportionately high. The tools needed to produce the parts are presented in table 4. The tools are sorted according to type and size. In the tooling analysis five tooling families were found. The tooling families are designated A, B, C, D and E. The parts in a tooling family are produced with the same tools.

On the work center only 10 out of 39 parts could be allocated to a tooling family. With this result it is obvious that an alternative venue for improvement must be found. The next step was to look closer at the design of the parts with a designer from the design department of both the subcontractor and a major customer. The objective was to find the non-critical dimensions of the design. For example the more unusual non-critical hole dimensions are eliminated by using either a bigger or smaller drill. This substitution is designated with an arrow in table 4.

Based on the analysis of the parts design the subcontractor and customer designers were able to eliminate the following 8 tools. The elimination of tools is based on the principle of standardization. The most suitable tools to use as standard tools are according to the analysis:

- Ø 9 mm center bit (needed to produce 32 parts, see table 4)
- Ø 10 mm cutter (27 parts)
- Ø 20 mm cutter (19 parts)
- Ø 50 mm face milling cutter (17 parts)
- Ø 3,3 mm halftwist bit (14 parts)
- Ø 4,2 mm halftwist bit (13 parts)
- Ø 6,0 mm cutter (12 parts)

Table 5 shows that after transferring machining operations to standard tools three new tooling families emerge. These are designated F, G, and H. Additionally, tooling families B and D were merged and part 887995 was added to tooling family C. In table 4 there are also tools that are not used for any part. These tools have been used in the past and are included for future requirements.

TABLE 4. Standardizing Non-critical Dimensions of Design.

	Part number																																						
	Tooling families																																							
	A		B		C		D		E																															
Tool (mm)	885346	885336	801086	755-B	801281	801262	801074	801037	801089	747-B	889622	894090	893901	884598	884630	830992	831436	831329	800153	800164	894123	894088	831236	887995	801076	801087	831276	801137	801002	800691	800479	800207	800097	801091	893897	894250	744-B	801347	801334	Parts/tool
Cutter 2			1	1																		1			1									1			1			6
2,5											1											1																		2
2,8																						1																		1
3			1	1			1	1										1							1					1							1		1	9
3,5											1				1																									2
4														1								1									1									3
5			1	1			1	1																	1															5
6			1	1					1	1					1	1									1	1	1						1	1			1			12
6 special					1	1																								1										3
8												1					○↕					1																	1	3
8,95/2°																																								
9,6	1	1																																				1		3
10			1	1	1	1	1	1			1	1	1	1		1		1	1	1	1		1	1	1	1	1				1		1		1		1	1	1	26
10 long																					1									1										2
16														1									1													1				3
20					1	1						1	1	1	1	1		1	1	1			1				1		1	1		1	1	1				1	1	19
20 long																																						1		1
25			1	1			1	1																	1												1			6
Halftwist bit 1														1																										1
1,5 special																																								
1,7 long																																								
1,9 special																																							1	1
2,1			1	1			1	1																									1							5
2,5													1		1						1				1	1				1				1		1			1	9
2,6																																					1			1
2,8																		1											1											2
3																																					1			1
3,1																																								
3,3											1		1	1	1								1						1	1		1			1	1		1	1	12
3,5									1	1							○↑																	○↑			1			5
3,8														1																					1					2
3,9													1																											1
4																														○↓						○↓				2
4,1																○↓											○↓													2
4,2												1	1		1				1				1									1			1				1	10
4,5											1			○↑																1										3
5					1	1						1											1				1									1			1	7
5,2													○↓																											1
5,5																																1								1
5,6											○↑																													1
5,8																																								
6											○↓																													1
6/90°																																		1			1			2
6,2																																								
6,8																			1				1									1				1				4
8													○↓	○↓																										2
8,3																						○↓																		1
8,5	1	1																																						2
9																																1								1
9,2																													1											1
10																												1												1
Screw tap M3																									1									1			1			3
M4													1	1									1						1			1				1		1	1	8
M5													1						1													1								3
M6																							1																	1
M8																			1				1									1				1				4
Centre bit 5													1				1													1										3
9	1	1	1	1	1	1	1	1	1	1	1	1	1	1	1	1		1	1		1	1	1		1	1	1	1	1			1	1		1	1		1	1	32
Scraping tool 3 H7																		1											1											2
4 H7													1	1																										2
6 H7																																								
8 H7																																								
20 H7														1																										1
Face milling cutter 50			1	1					1	1									1	1			1		1	1		1	1				1	1	1	1	1		1	17
80												1			1																							1		3
80 (special)																																								
Ball milling cutter 5												1															1													2
18																																								
Crusher 16												1																1												2
Cutter R 1																																								
R 2 curved									1	1																														2
R 3 (special)																									1	1												1		3
R 4 curved																																								
R 5 curved																○																								1
Bevelling bit 25/30°																																								
26/45°																															1		1						1	3
Special bit 15	1	1																																						2
22																							1																	1
Toothed cutt												1																												1
Angle cutter														1																										1
Tools/parts	4	4	9	9	5	5	6	6	5	5	8	10	13	14	8	6	3	6	8	3	4	7	13	1	11	6	7	4	8	10	3	10	7	8	6	11	11	9	13	

TABLE 5. Operations: Unique and Common to Only 2 Parts.

Tool (mm)	801066	750	801074	801077	801281	801282	802390	801083	747	805346	805326	832901	831438	800162	800194	801076	801087	804630	802882	804538	831329	831409	[illegible]	831236	805227	804090	801127	801062	831276	806831	804473	800287	800087	801081	832897	804260	744-B	801347	801334	Parts/tool
Tooling families	B	B	B	B	C	C	C	E	E	A	A	F	F	G	G	H	H																							
Cutter 2	1	1														1							1											1			1			6
2,5																							(1)		(1)															2
2,8																							(1)																	1
3	1	1	1	1												1					1									1							1		1	9
3,5																		(1)							(1)															2
4																				1			1								1									3
5	1	1	1	1												1																								5
6	1	1						1	1							1	1	1	1										1				1	1			1			12
6,0 special					1	1																								1										3
8																							1			1													1	3
9,4										1	1																											1		3
10	1	1	1	1	1	1	1					1	1	1	1	1	1		1	1	1	1		1	1	1			1		1		1		1		1	1	1	27
10 long																						(1)								(1)										2
16																				1				1												1				3
20					1	1						1		1	1			1	1	1	1			1		1		1	1	1		1	1	1				1	1	19
20 long																																						(1)		1
25	1	1	1	1												1																					1			6
Halftwist bit 1																				(1)																				1
1,9 special																																							(1)	1
2,1	1	1	1	1																													1							5
2,5												1				1	1	1				1								1				1		1			1	9
2,6																																					(1)			1
2,8																					(1)							(1)												2
3																																					(1)			1
3,3												1	1					1		1				1	1			1		1		1		1	1	1		1	1	14
3,5								1	1																												1			3
3,8																				(1)															(1)					2
3,9												(1)																												1
4,2												1		1				1	1	1				1		1			1	1		1			1	1			1	13
4,5																									(1)					(1)										2
5					1	1																		1		1			1							1			1	7
5,5												1													1							1								3
6/90°																									1									1			1			3
6,8														1										1								1				1				4
8,5										1	1	1								1			1																	5
9																																(1)								1
9,2																												(1)												1
10																											(1)													1
Screw tap M3																1																		1			1			3
M4												1								1				1				1				1				1		1	1	8
M5												1		1																		1								3
M6																								(1)																1
M8														1										1								1				1				4
Centre bit 5												1	1																	1										3
9	1	1	1	1	1	1		1	1	1	1	1		1		1	1	1	1	1	1	1	1	1	1	1	1	1	1			1	1		1	1		1	1	32
Scraping tool 3 H7																					(1)							(1)												2
4 H7												(1)								(1)																				2
20 H7																				(1)																				1
Face milling cutter 50	1	1						1	1					1	1	1	1							1			1	1					1	1	1	1	1		1	17
80																		1								1												1		3
Ball milling cutter 5																										(1)			(1)											2
Crusher 16																										(1)	(1)													2
Cutter R2 curv								(1)	(1)																															2
R3 (special)																1	1																					1		3
Bevelling bit 26/45°																															1		1						1	3
Special bit 15										(1)	(1)																													2
22																								(1)																1
Toothed cutt.																										(1)														1
Angle cutt.																				(1)																				1
Tools/part	9	9	6	6	5	5	1	5	5	4	4	13	3	8	3	11	6	8	5	14	6	4	7	13	8	10	4	8	7	9	3	10	7	8	6	10	11	9	13	

In Table 5 the operations that are only performed for one or two parts are circled. When one of these unique operations is eliminated by a redesign of the part one tool is also eliminated. This observation was taken as the basis for a Design for Manufacturability co-operation between the subcontractor and its customers. The objective is to eliminate the unique operations and this way improve capacity and service.

Table 6 shows potential improvement by a redesign of parts. If all unique operations can be eliminated by a redesign of the parts, then the parts could be produced with only 44 tools. If also the operations common only to two parts, then all the parts could be produced with only 32 tools. This is a reduction of 59 per cent.

TABLE 6. Improvement Potential.

	Part number																																						
	Tooling families																																						
	A				C			D		E		F		G		H																							
Tool (mm)	801086	795-32	801074	801077	801281	801262	887559	801089	747-32	885345	885336	893901	831438	800163	800164	801076	801087	894630	830392	884098	831329	894173	894088	831236	889677	889450	801137	801002	837276	800691	800479	800207	800057	801051	833897	894250	744-32	801347	801334	Parts/tool
Cutter 2	1	1														1							1											1			1			6
3	1	1	1	1												1					1									1							1		1	9
4																				1			1								1									3
5	1	1	1	1												1																								5
6	1	1						1	1							1	1	1	1										1				1	1			1			12
6,0 special					1	1																								1										3
8																							1			1													1	3
9,6										1	1																											1		3
10	1	1	1	1	1	1	1					1	1	1	1	1	1		1	1	1	1		1	1	1			1		1		1		1		1	1	1	27
16																				1				1												1				3
20					1	1						1		1	1			1	1	1	1			1		1		1	1	1		1	1	1				1	1	19
25	1	1	1	1												1																					1			6
Halftwist bit 2,1	1	1	1	1																													1							5
2,5												1				1	1	1				1								1				1		1			1	9
3,3												1	1					1		1				1	1		1			1		1		1	1	1		1	1	14
3,5								1	1																												1			3
4,2												1		1				1	1	1				1		1			1	1		1			1	1			1	13
5					1	1																		1		1			1							1			1	7
5,5												1													1							1								3
6/90°																									1									1			1			3
6,8														1										1								1				1				4
8,5										1	1	1								1			1																	5
Screw tap M3																1																		1			1			3
M4												1								1				1				1				1				1		1	1	8
M5												1		1																		1								3
M8														1										1								1				1				4
Centre bit 5												1	1																	1										3
9	1	1	1	1	1	1		1	1	1	1	1		1		1	1	1	1	1	1	1	1	1	1	1	1	1	1			1	1		1	1		1	1	32
Face milling cutter 50	1	1						1	1					1	1	1	1							1			1	1					1	1	1	1	1		1	17
80																		1								1												1		3
Cutter R3 (special)																1	1																					1		3
Bevelling bit 26/45°																															1		1						1	3
Yhl. Työkaluja	9	9	6	6	5	5	1	4	4	4	3	11	3	8	3	11	6	7	5	9	4	3	5	12	5	8	2	5	6	7	3	9	7	8	5	10	9	8	12	

The process for standardizing the parts on the bottleneck work center can be described in three steps: 1) Tooling Analysis to identify set-ups for parts; 2) Subcontractor and customer design engineers identify parts with unique operations and agree to eliminate these; and, 3) The case company and the customer review with each new order the design of the parts with unique operations.

This way the parts on the bottleneck are standardized one by one based on the Tooling Analysis technique. The Tooling Analysis is the common language needed to set goals and agree on procedures to achieve the goals. Figure 5 shows the improvement process. The starting point was 79 tools for 39 parts. The first meeting between subcontractor and customer eliminated 8 tools. After two months 15 more tools have been eliminated. Currently parts are redesigned with every new order. *The goal 32 tools will have been reached in six months with current order frequency.* To reach the improvement target a total of 24 parts have to be redesigned.

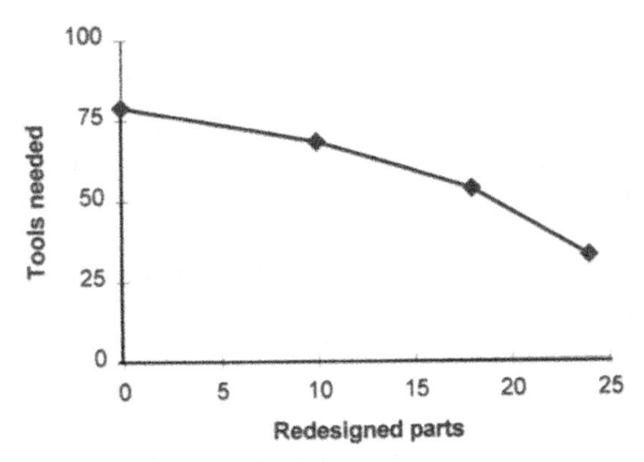

FIGURE 5. The results of the DFM process.

The incremental approach is an advantage when dealing with bottleneck operations. Improvements can be introduced gradually, without disrupting operations. In the case company the valuable bottleneck resource was used continuously, while involving production ,design and customers in the parts redesign and set-up reduction. The basis for the whole process was the Tooling Analysis technique. The benefit of the redesign is currently a 15 per cent reduction in set-up and after the goal (table 6) will have been reached the benefit is estimated as a 30 per cent reduction in set-up.

4. CONCLUSIONS

The case studies demonstrate clear benefits from the application of PFA techniques for companies operating in a make-to-order manufacturing environment. The introduction of GT resulted in a much simplified Engineering change management process in the one-of-a-kind case. The Tooling Analysis in bottleneck work centers enabled moving parts to non-bottleneck work centers and increased the potential capacity of the one-of-a-kind operation by 40 per cent. There was no more need for new capacity or extra shifts on the bottleneck work center.

The Tooling Analysis can also be used as the basis for driving a Design-for-manufacturability process with customers. The Tooling Analysis was used in the subcontractor case by the design staff of subcontractor and customers to standardize non-critical dimensions. The process immediately yielded a 15 per cent reduction in set-up now and potentially the benefit is a 30 per cent reduction in set-up on the bottleneck work center.

The benefits of PFA is particularly clear in the case company operating in a project oriented manufacturing environment. Here, the situation is as follows:

- the company has a basic design of its product. This is a good indicator that a product organization of the manufacturing process is feasible.
- each produced unit is unique. The exact design of the product is determined by the process and site where the product will be installed.

In this situation with a basic product, but where each produced unit is unique, an effective engineering change process is critical for the success of the business. The direct benefit from better engineering change management is that the penalty fees for delays and errors can be substantially reduced. The indirect benefit is that the capacity is more efficiently used, which allows for more business and better profitability.

The key insight from the application of PFA in the project oriented operation is that the reorganization of the material flow system needs to be followed by a concentrated effort to reduce set-ups in bottleneck work centers. In other words Group Analysis needs the support of Tooling Analysis in bottleneck work centers. Due to the high cost of moving heavy machine tools the optimization of material flows within groups was not pursued, i.e. Line Analysis was not systematically applied in the case company.

The change of the material flow system also requires that the work organization is changed. Previously the case plant had operated with a traditional job shop work organization. However, GT made it necessary to introduce a more team based organization with shared responsibilities. The tasks of production scheduling , engineering change management and quality control was much simplified by GT. The personnel resource freed from these administrative tasks could be put to much more productive use on the shop floor and in contacts with customers and installation crews.

In the subcontractor case Group Analysis was not applied. In this case it was evident from the start that a product organization is difficult to realize in a situation with tens of customers and a wide range of very different products. However, in a business where you compete with your ability to machine parts and components for the high tech products of your customers it is important to have a design for manufacturability process in place. Only this way can you over time improve your competitive position vis a vis other subcontractors and in-house production.

The first step in the implementation sequence for launching the design for manufacturability process was to identify the full range of tools needed to produce the current designs. For this task Tooling Analysis was used. The second step was to standardize designs in co-operation with the customer with each new order.

The main organizational change required was to establish a direct co-operation between the customer's design engineers and the subcontractor manufacturing engineer. Here, the key enabler was the clear development goals that could be set based on the Tooling Analysis. Both design and manufacturing engineers were challenged by the development potential indicated in the Tooling Analysis. The work procedures to standardize designs could be immediately established in the first meeting between customer and subcontractor design engineers.

REFERENCES

Burbidge, J. L., 1989, *Production flow analysis - For planning group technology*, Clarendon Press, Oxford.

Hameri, A. P., and Karvonen, S., 1994, Using production flow analysis in project oriented industry, *International Journal of Manufacturing System Design*, 1, 2, 111-117.

Karvonen, S., and Holmström, J., 1996, How to use tooling analysis to reduce setup and improve throughput, *Production and Inventory Management Journal*, 37, 3, 75-80.

AUTHORS' BIOGRAPHY

Sauli Karvonen, M.Sc.(Eng), is a management consultant and managing director of SKA-Research Oy. His main interests are manufacturing system design, group technology and mass customization. Previously he worked as a researcher at the Helsinki University of Technology in IIA-Research Center. He received his MSc in industrial economics from Helsinki University of Technology in 1991. He is a graduate student at Helsinki University of Technology. He is also member of APICS and he has published in *Production and Inventory Management Journal* and *International Journal of Manufacturing System Design.*

Jan Holmström, PhD, is an Honorary Research Fellow (Docent) at Helsinki University of Technology. His main research interests are supply chain management, information systems applications in logistics and manufacturing system design. He recieved his PhD in industrial economics from Helsinki University of Technology in 1995. He has published in the *Production and Inventory Management Journal, International Journal of Production Economics, Production Planning & Control, Business Process Re-engineering & Management Journal, Supply Chain Management: An International Journal, and the Journal of Integrated Manufacturing Systems.*

Dr. Eero Eloranta has been Professor of Industrial Management at Helsinki University of Technology since 1991. Currently Prof. Eloranta is on sabbatical leave at Nokia Corporation as logistics chain integrator for global logistics of Nokia

Mobile Phones. He received his M.Sc. in information systems, industrial engineering and management science in 1976, and Ph.D. in 1981, both at Helsinki University of Technology. He has also worked at Eindhoven University of Technology in the Netherlands and else where. Prof. Eloranta is chairman of IFIP (International Federation of Information Processing) working group 5.7: Computer Aided Production Management, and also president-elect of EEMA (European Engineering and Management Association). Prof. Eloranta and his research teams have strongly contributed and participated in international research programs such as ESPRIT, RACE and IMS. He has published numerous reviewed journal papers, books and research reports. His hobbies are primarily related to long distance physical excercise (jogging, cross-country skiing, swimming, cycling, canoeing), sailing, other outdoor life and some music.

13

Cellular Manufacturing at Zanussi-Electrolux Plant, Susegana, Italy

R. Panizzolo

1. INTRODUCTION

This chapter describes a case study of a firm in Italy which has completely transformed its production processes (without, incidentally, stopping production even for a single day) and which has undergone a radical organizational transformation and dramatic changes in attitudes, beliefs and working methods.

This case study attempts to document the complex process of transition from conventional manufacturing to cellular manufacturing systems at the Zanussi-Electrolux plant at Susegana in northern Italy. This was a project undertaken by a group of men and women who, for more than three years, struggled together and shared the joys, difficulties, regrets, delusions and fatigue involved in such a complex endeavor. It also attempts to detail several noteworthy aspects relating to product and process technologies, including machine tools, robots, and advanced technology, all of which were effectively integrated in the end towards the production of quality products and the improvement of the people's quality of life.

This chapter is organized as follows. Sections 2 and 3 describe, respectively, the scenario and decision making processes involved in the adoption of cellular manufacturing (CM); Section 4 outlines critical factors in the implementation process; Section 5 illustrates the general structure of the cellular layout of Susegana plant; Section 6 reports on the changes required in product structure when implementing CM; Sections 7, 8 and 9 describe, respectively, the characteristics of the cellular production system, the management information system and how production planning and control are carried out.

2. IMPLEMENTATION SCENARIO

In 1983 a seemingly insurmountable financial crisis hit Zanussi and many rescue plans were put forward, at a national level, in an attempt to save a firm that employed 28,000 people. These plans included mass redundancy schemes, none of which, thanks partly to Trade Union action, went ahead. In 1984, it was suggested to Mr. Wallenberg, then president of the Swedish holding company that owned Electrolux, that he consider saving a famous Italian firm which, in the 1970s, had dominated most of the important European markets with its products.

Experts at Electrolux were doubtful; though attracted by the potential synergy in production, research and development and marketing, they were wary of both the somewhat shaky political and trade union climate, and of the disastrous economic situation of the firm. Despite these doubts, negotiations did go ahead and, in December 1985, it was decided to increase of Zanussi's capital, 49% of which was underwritten by Electrolux. Towards the end of 1986 Electrolux's quota was increased to 94.8%.

In those years the Electrolux group's mission was to become the market leader. The firm's strategy was based on making a large number of acquisitions in strategically important markets and making equally large investments in production technology which was to become one of the strong points of the group. When the Electrolux group took control of Zanussi, many programs for improving the group's profitability were already under way in various plants all over the world. Interventions in the production processes were aimed at: adapting production capacity to meet the needs of the market (in terms of both quantity and breadth of mix); reducing production costs and amount of capital tied up in stocks, and, improving production organization by restructuring activities on the production line. Lastly, great efforts were being made to improve the group's quality image. In brief, in order to keep profits high, Electrolux consistently followed a strategy wherein investments in new products and new production technologies played a central role.

In this scenario, the investment program began, to the tune of $250 million, of which about $90 million went towards redesigning the plant in Susegana, Italy. The plan was to convert the Susegana plant so that it would specialize in the production of refrigerators for domestic use, while production of other products was moved to other Zanussi plants in Italy. Electrolux managers intended that the Susegana plant be restructured into a state-of-the-art refrigerator production plant for the group and a laboratory for experimenting with new technologies.

3. ADOPTION AND JUSTIFICATION OF CM

Accordingly, the Susegana plant was completely restructured and its production process was redesigned according to the philosophy of CM with the adoption of integrated and computerized production technologies.

The adoption of a production system based on cells was considered to be a strategically vital, a winning ploy able to offer fundamental competitive advantages over competitors, in addition to profits and costs. Such restructuring was seen to be of paramount importance because through innovation, the plant would be able to win a position of technological leadership, thus forcing the competitors to follow suit. Thus, the strategic ramifications of implementing CM outweighed the operational or tactical consequences. The focus was long-range, emphasizing strategic advantages such as improved flexibility, ability to respond to customer demand, decreased time-to-market, and improved product quality. One manager summarized thus: "We were faced with business pressure and foreign competition able to deliver not the same product but an equivalent product at a significantly lower cost with a significantly reduced schedule. It was felt that there was no way we could compete without implementing cellular manufacturing".

Thus, it was clear that the decision to adopt CM had involved not only financial investment analysis techniques, but also evaluations of a strategic or competitive nature, with techniques based on risk analysis, strategic positioning analysis, and managerial and organizational impact analysis. All these techniques allowed the decision makers to better understand the real costs, benefits and risks of the choices they were making.

Three levels of participants were involved in the justification process: technical personnel, middle management, and top management. The decision making process took a long time: almost one and a half years from idea initiation to final approval of the project. It took so long because it was not easy for anyone to fully understand all aspects of the project. As one manager stated: "We are talking about a $90 million system.. we had never spent anything like that for equipment before... no history for it.. no real solid justification.. no numbers for the financial guys to get their teeth into.. So, you can imagine the wailing and gnashing of teeth".

Many difficulties were encountered. In the first place, invariably, some factors had been overlooked and were responsible for additional costs during project implementation. Moreover, given the complexity of the project, it was difficult to identify and quantify all benefits, both tangible and intangible. Some participants said: "It was a new technology that we had never used before. We did not know what it was going to do for us. We didn't know if it was even going to work. When you do it (the system) you change the way you do business. It has an impact on so many things that you couldn't possibly do it correctly".

Not only were the costs and benefits somewhat vague, but so too were the perceived risks associated with the system's implementation. Some of these risks were of a financial nature: if the new system did not work at all or was only partially functional, then the ability to ship products was threatened. In addition to the financial risks, there were considerable risks for the organization: individuals were particularly sensitive to "career risks". Resistance to change by middle management was particular critical. One manager stated: "The problem we have in our company is that some of our people hate advanced technology. They absolutely hate it. They are basically old mechanical engineers, or they are business guys, who have been trained in different technologies. They are skeptical. Part of this is just learning... part of it is fear of the unknown".

In conclusion, one could say that the "interpersonal component" played an important role in winning approval for the project. A combination of credibility and commitment by top management and its hand-picked team, was used to convince everyone to go ahead.

4. CRITICAL FACTORS IN IMPLEMENTATION PROCESS

Given the complexity and strategic importance of the project, the decision was mad to divide development and implementation into a series of temporal and logic phases which were then entrusted to different work groups, each coordinated by project leader. This project leader was a senior manager who had to answer directl to Electrolux management and who was appointed to lead the implementatio process. The project leader devoted 100% of his time to system implementation.

In order to ensure the effectiveness of implementation, a formal projec implementation team was set up with members drawn from each of the department and areas affected by the changes (for example, production control, operations industrial engineering, information systems, cost accounting, and so forth).

The firms developed an implementation plan to guide the organizatio through the implementation process. This was a very detailed, carefully prepare plan which also sought to anticipate potential problem areas. This plan predicted tha about two years would be required to set up the new system and reach an optimun level of operations. During the implementation phase, the project schedule wa carefully monitored. Costs were tracked in order to identify any deviation from th project budget allocated. Classical project management techniques were used tc monitor the time and cost variables.

However, various difficulties and problems arose which delayed completion of the project for almost one year. Some of these difficulties were of a technical nature, and arose from the problems of interfacing the new process and new machines with the old. But it was managerial problems rather than technical problems that often presented greater challenges to the organization. The implementation of the system had required changes in virtually all areas of the organization: changes in management methods, human resource allocation, organization structure and design, and manufacturing processes, to name a few.

In particular, transforming the production process into a cell system changed the firm's management philosophy. From a centralized decision making process, the firm moved to one where responsibility was decentralized. Effecting this type of change, managing the inevitable conflicts of competencies, and the inevitable resistance and defense of positions of power acquired over the years, which were difficult to settle and reshape, proved to be a long and difficult task.

The most important impact on the organization was changes in the nature of jobs and job responsibilities. Direct labor employees were given more responsibility and their skills were upgraded. In general, the workers' jobs were described as being cleaner, less demanding physically, and more demanding mentally. In addition, employees were expected to analyze and solve problems experienced in their areas.

Implementation of CM resulted in significant changes in organizational structure. Entire departments were eliminated, or drastically reduced in number. For example, the implementation of the automated warehouse and material handling system resulted in major cutbacks in the number of handlers of materials. As a result of the reduction in the number of employees, the manager in charge lost power within the organization.

Organizational culture was another area where the firm reported major changes. Many participants perceived that a dramatic shift had taken place in the organization. Indeed, there were shifts in attitude, beliefs and customs that emphasized and encouraged automation. A gradual but definite change in culture was experienced by the firm.

Education and training were stressed as the most critical factors for overcoming managerial problems. A formal department responsible for on-the-job training and education provided a comprehensive program, covering the interfaces and linkages between systems and processes. The company also relied on external resources such as consultants and suppliers to provide training. Typically, the consultants' services were in the area of conceptual education and were focused on middle and top management.

External resources also played other roles in the implementation phase. For example, consultants performed functions such as providing technical assistance on system conversions or programming support. Suppliers provided useful information on new product and competitors, technical expertise, product enhancement and technical support. However, some negative impact was felt. Some of the most serious problems were: delays in equipment deliveries and software development, marketing "hype" and exaggerating product characteristics, inadequate technical support, unreliability of some suppliers, and unrealistic or incongruous expectations.

5. THE CELLULAR MANUFACTURING SYSTEM AT SUSEGANA

As stated earlier, it was more demanding market needs and rising international competitiveness that primarily led Electrolux to restructure the Zanussi plant. A production process characterized by a high level of responsiveness to market needs, besides customary efficiency of manufacturing activities, was sought.

According to Electrolux managers, overcoming the traditional productivity-flexibility dilemma was indissolubly linked to achieving "operational similitudes" in the areas of design, production processes and production planning (these similitudes refer, for example, to component standardization in the area of design, to routing standardization in the area of production and to leveled production in the area of production planning). It would be easier to obtain these similitudes with a cell production system, set up according to Group Technology (GT) principles. Furthermore, they recognized that the use of new manufacturing technologies was a mandate for a CM approach. This 'mandate' for CM derives from the application of robotics and other forms of mechanized/automated material handling systems, as well as from a desire to build closely-coupled manufacturing systems with low throughput times. Efficient use of these philosophies and technologies, in essence, requires a cell-structure approach to manufacturing.

The design of the cell production system at Susegana has required careful designing of the product structure and of both system architecture and operationa procedures. As a first step, a careful analysis of the characteristics of th components/products manufactured was carried out to investigate the possibility o being able to group them in homogeneous and sufficiently large production families Traditionally, this approach leads to the setting up of cells organized for machinin families of parts. Each cell carries out, in ideal situations, all the operations in th production routing for the parts belonging to a particular family, and each cel operates independently.

In the case of the Zanussi-Electrolux plant, the search for efficient, an sufficiently flexible production systems led to the development of a unique cellula manufacturing system. It was first decided to dedicate the entire transformatio process to one product family (refrigerators) and then to structure it according t areas, or technological cells, designed to carry out all the operations relative to a particular stage in the transformation process.

In this system, the cells, equipped with highly specialized machinery, were arranged according to more or less complex combinations with one main branch, that could be subdivided into two or more secondary branches, and other auxiliary branches which converged on the main branch but which were able to bypass entire cells. The term "sequence cells" was used to identify this type of production system which is composed of technological cells - each of which carries out one stage of the production process - that are arranged in such a way as to reflect the main direction of the production flow so as to facilitate the flow of materials. For this reason, a sequence cell system presents aspects similar to those of line flow production systems. At the same time, flexibility is maintained at a relatively high level, because complex stages of different product routings are carried out inside the cells.

6. CHANGES REQUIRED IN PRODUCT STRUCTURE DUE TO CM

Before describing the characteristics of the cellular plant of Susegana further, it is important to understand that setting up the cellular layout required a number of interventions, not only in the production process, but also in the product structure. Indeed, one could say that the transition to CM was only made possible through integrated product/process design without which the project would have been doomed to failure. Had the product not been redesigned to take into account the characteristics of the new production process, it is unlikely that the benefits usually associated with the adoption of a cellular system would have been felt. The main modifications made to product structure are described below.

Before the transition to a cellular layout, the refrigerator was made up of a "C"-shaped sheet, shaped so as to form the sides and the top, while struts to support the compressor, and to fix the sides, were in the base. However, this refrigerator structure was not flexible enough to satisfy the demands of the market. For example, it was difficult to alter either height or width once cabinets had been put into warehouses.

This problem encouraged a re-evaluation of the product structure, which led to the idea of producing a refrigerator in "panels". In this case, the cabinet, instead of being a "C"-shaped sheet, became a group made up of draw sides, press-forged ties and plastic back. In particular, the cabinet was made up of two side panels with another panel for the top. This structure made it much easier to alter the measurements of the refrigerator; also, only less valuable coils, instead of sub-assemblies had to be held in warehouses. This also required much less storage space. Hence, not only were there economic benefits but also improvements in the firm's capability to respond to customer demands.

The panel structure also allowed the aesthetic aspects of the refrigerator to be improved as different materials in different colors could be used for the sides, top and door which made the refrigerator more adaptable to different situations (greater customization). Lastly, a structure based on panels meant that the sheet metal cutting process could be eliminated (which was not a flexible phase).

Other changes made in the product structure included:

- the door of the refrigerator, in painted sheet metal, which now contains the door liner and the insulation;
- the inner liner, which is now directly linked to the cabinet by foaming.

To conclude, the products were redesigned in terms of a modular architecture: the product remained substantially the same in functional terms, but underwent deep revisions in its design. It is now structured in sub-assemblies: the sides, the door, the inner liner, the cabinet, the electrical components, the compressor bar, etc. In this way, the refrigerator is made of parts and components which can be produced and assembled with a good degree of flexibility.

7. THE NEW PRODUCTION SYSTEM

As stated earlier, redesign of the Zanussi-Electrolux plant was carried out with the aim of specializing in production of refrigerators for domestic use. The Electrolux managers also intended that there should be a high production volume of products, with high levels of efficiency and a fairly wide mix.

After the transition to cellular manufacturing systems, the plant was characterized by the following main features: (i) high production volumes, with more than 1,100,000 parts a year, produced at a rate of about 4,200 a day, (ii) wide range: about 40 basic models, produced with a total of 1,000 variants, (iii) ability to produce the entire range of products in reduced times (i.e. one week), (iv) small lot sizes, equal to 16 units.

A number of interventions in the production process were necessary to obtain a high degree of efficiency and flexibility. Basically, the production process was subdivided into separate production stages (i.e. technological cells designed to carry out all the operations relative to that particular stage). Among other tasks, it involved:

- elimination of the die operation;
- simplification and de-specialization of painting operations;
- introduction of the foaming operation, which is an assembly stage;
- introduction of stages for production of main sub-assemblies;
- revision of material picking with the introduction of the kit concept. Forming specific assembly kits, according to a plan, facilitates distribution of components for assembly to workplaces and ensures that all components are present before an assembly operations begin.

The cellular layout is organized into 14 cells arranged in sequence, with two parallel branches in the middle stages that are dedicated to fabricating cabinets and to forming kits of components before final assembly, as shown in Figure 1.

This production system provides high efficiency and mix flexibility. Efficiency is achieved through production in automatic cells, while flexibility is obtained at two levels: at a global level, thanks to a production philosophy which carries out refrigerator production through final assembly of a cabinet and a kit of components; at a local level, thanks to the cell's independence of the mix. This production philosophy, at a global level, justifies the presence of two parallel branches in the middle production stages, dedicated respectively to the fabrication of the cabinet and to kit formation respectively.

The factory is, therefore, a production system which is made up of a group of autonomous production areas with a reduced number of intra-operational warehouses and a high level of automation and integration. The arrangement of production cells has been defined in such a way as to permit continuous flow through the factory. The cells have been designed to concentrate either automatic or manual operations in one cell, so that operating problems are homogeneous and management is simplified.

Automation has also been introduced into the basic operations: cabinet and door automatic assembly/foaming, compressor assembly, soldering, packing and automatic material handling. Each cell includes one or more input warehouses and some machines which enable a part to be fabricated. Single cells are connected by material and information flows. As for material handling, the movement is by standard quantities, or multiples of standard, to reduce and simplify appointments within one cell or between different cells.

From a logistics-flow point of view, the cells can be classified into three categories (numbers refer to Figure 1).

***head cells**:* processing only purchased materials: cutting coils (1), evaporator production (5), vacuum forming (8) and pre-assembly (11);

appointment cells: processing semi-worked parts coming from upstream cells: handling (4), inner liner pre-assembly (6), cabinet automatic assembly and foaming (7), door pre-assembly and foaming (9), kitting (10), match point kit-cabinet (12);

flow cells: processing semi-worked parts coming from one upstream cell: sheet metal processing (2), painting (3), final assembly (13) and packing (14).

The cells can also be classified, according to their degree of flexibility, into three types:

limited flexibility pre-production cells: amongst which are cutting coils (1) and evaporator production (5); these are cells with very little automation and with production machines guided by an operator, for the production of semi-machined parts;

limited flexibility main production cells: amongst which are sheet metal processing (2) and vacuum forming (8); these use the semi-machined parts from pre-production cells and other components to feed downstream cells; in particular, the cell with large dedicated machines, which carries out vacuum forming of inner liners, is critical for the set-up times which condition downstream production;

high flexibility main production cells: located downstream from the previous cells, which are unaffected by the mix: painting (3), handling (4), inner liner pre-assembly (6), cabinet automatic assembly and foaming (7), door pre-assembly and foaming (9), kitting (10), pre-assembly (11), match point kit-cabinet (12), final assembly (13) and packing (14).

8. MANAGEMENT INFORMATION SYSTEMS

To satisfy the needs of an automated, efficient and flexible production system, an information system was designed to support various production management activities. The system had to provide effective interfaces among the diverse components of the production system and also with other, external, components.

To ensure the efficacy of such an information system, a high degree of reliability and integration of the constituent elements must be guaranteed. As regards reliability, the choices made at Susegana have followed a policy of standardization at all levels, particularly in the data communications system, which should be common to the majority of computer and Programmable Logic Controller (PLC) producers, thus avoiding the problem of having to use too many protocols.

The information system of the Susegana plant is part of a wider information system structured at more than one level. The levels are (see Figure 2): Level 1: machine control system; Level 2: area control system; Level 3: factory operating system; Level 4: division information system; and, Level 5: company information system. The information system of the Susegana plant includes levels 1, 2 and 3 of this classification.

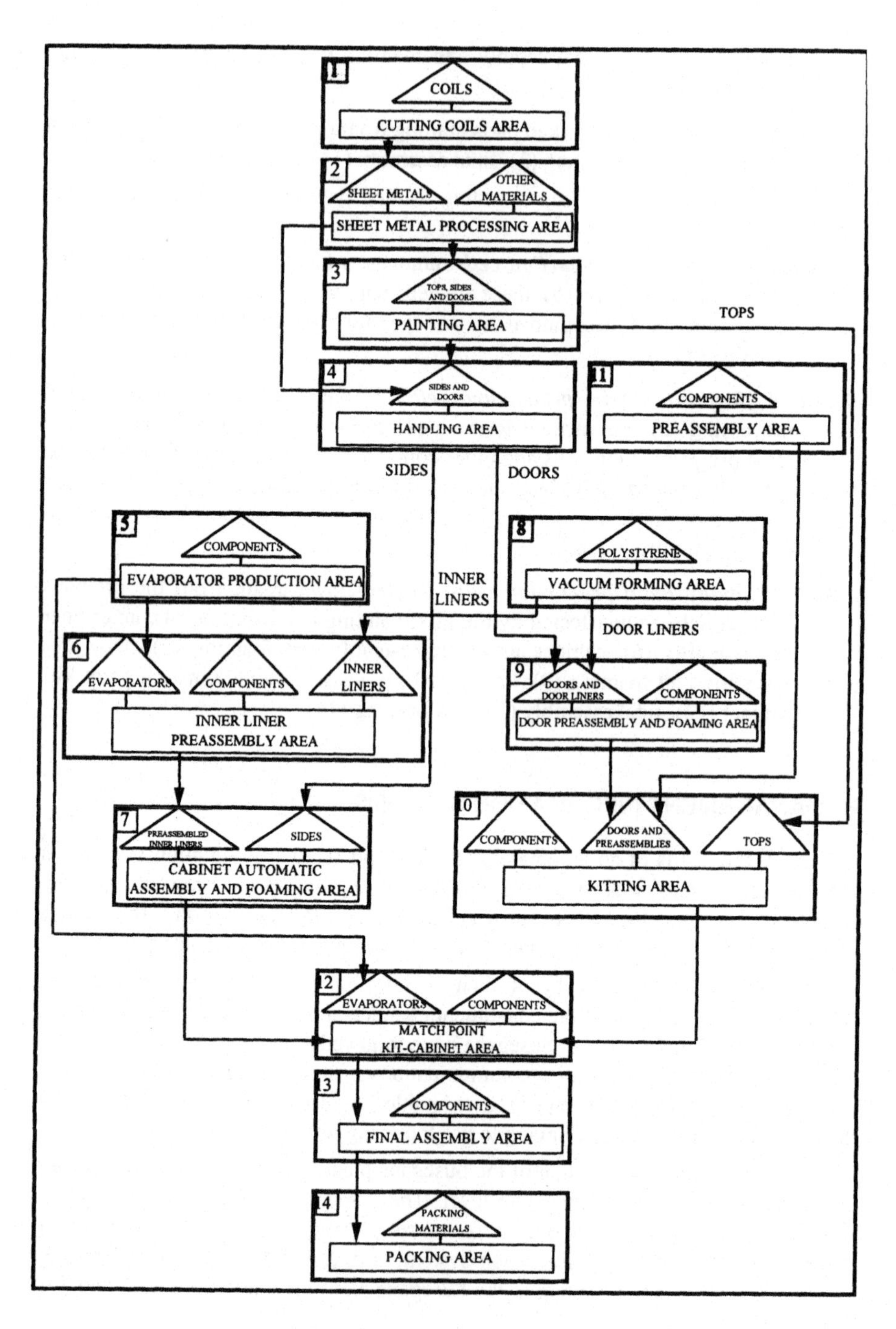

FIGURE 1. The cellular layout.

Considering the top level first, Level 5 is involved with sales and order processing, and formulation of the requirements for finished products which are then passed down to level 4. This level involves the whole company.

Level 4 manages the technical data related to the products (bill of materials and routings), it determines purchased material requirements, sends orders to suppliers and monitors finished-product warehouses. Each day, level 4 informs level 3 about new production requirements and variations in those that were previously communicated. Level 4 also notifies level 3, on a daily basis, of the latest supplies according to the purchasing plan. This level is dedicated to a specific product line.

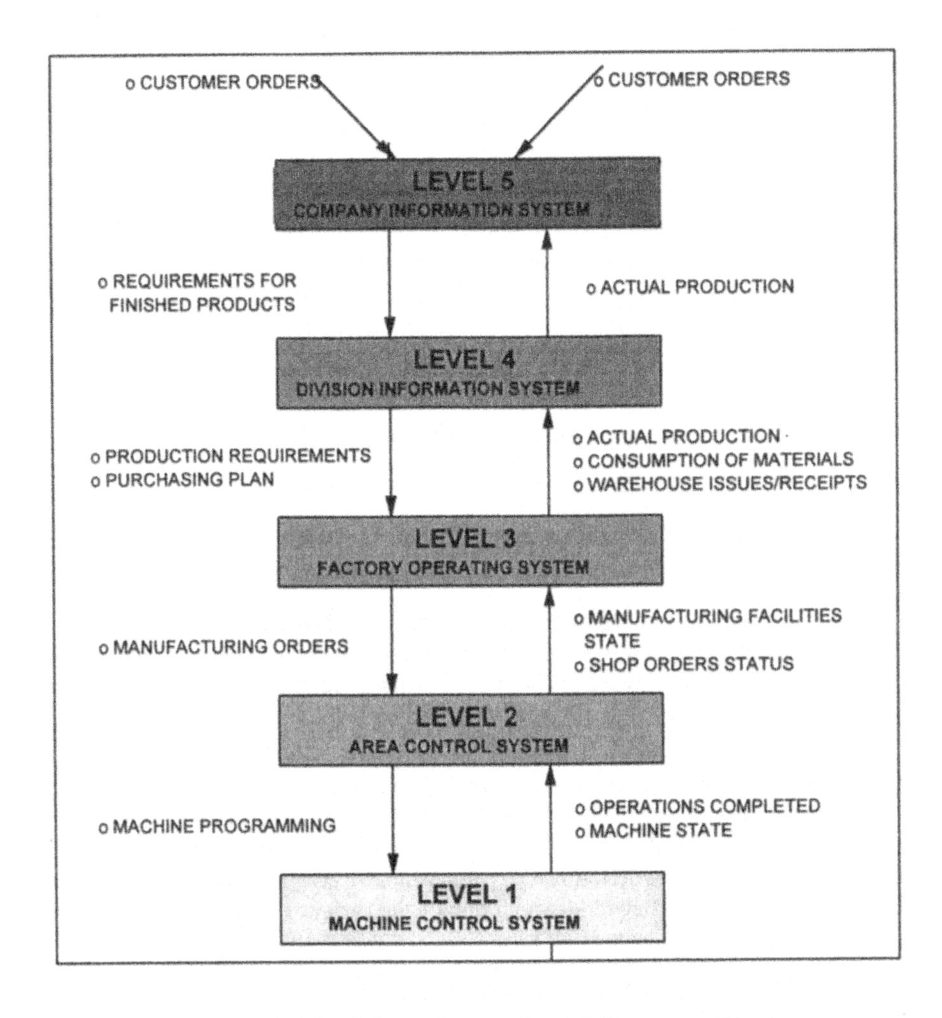

FIGURE 2. The information system of Susegana Plant.

Level 3 is dedicated to manufacturing planning and control and i coordinates the activities of the various production areas. It monitors the productio process, recording and updating data which comes from level 2. Level 3 sends leve 4 a daily report of the quantities actually produced. This report, which outlines th state of production, can be used to let customers know the status of their orders Each day, level 3 informs level 4 about consumption of materials and all warehous issues/receipts. Level 4 needs this information to draw up an accurate purchase material plan. But level 3 feeds the computers of level 2 in real time, transmittin shop orders data and other detailed information required for the production process.

Level 2 is made up of a group of software modules that are usually specifi for each area or cell. Its function is to optimize activities carried out in each cell i accordance with overall production plan. It also monitors cell warehouse and interna movements. The main flow of information passing from levels 2 to 3 concern updating status of shop orders and cell state (level of work in process, queues machine utilization, etc.). This information allows level 3 to synchronize operation among cells in real time. Lastly, level 2 sends level 1 all the information the PLC require to make each machine tool in the cell function correctly.

Level 1 oversees the management of a particular machine tool, and ensure compliance with cycle times and supplies diagnostic information regarding the stat of the machine. Level 1 sends level 2 information regarding the real progress o operations planned and the state of the machines.

The different levels are integrated by means of 4 communication networks:

- "company network", which links the various divisions;
- "division network", which links the various factories;
- "factory network", which links the various production areas;
- "area network", which links the various machines within an area.

The first two networks are characterized by a low number of connections low rate of interchange and both cover a very wide physical area. The opposite is true for the other two networks. These have a large number of connections, very high rate of interchange and cover restricted physical areas.

The factory network is an ETHERNET (CSMA/CD) type which corresponds to the standard IEEE 802.3. This choice was made on the grounds of the variety of products the standard IEEE 802.3 can deal with, and of the large number of applications it has in the area of factory automation. For reasons of homogeneity, and due to favorable cost/performance relationship, SIEMENS (SINEC L1 and H1) were chosen for the area networks which link up the various PLCs within each area.

9. PRODUCTION PLANNING AND CONTROL SYSTEMS

Among the various levels, the factory operating system (level 3) is in a central position in relation to the broader company information system and uses specific hardware and software dedicated to factory management. Level 3 software allows for subdivision into many modules, the most important of which are: 1) Production Planning, and, 2) Operation Control in Real Time.

9.1. Production Planning

This module is designed to create an optimum production plan on the basis of production requirements, as stipulated by level 4, and of the state of the production process (machine, equipment, labor, stock levels) in the period being examined. Here, production planning is carried out by defining three production programs which are updated daily and used for short-term scheduling:

o the "operating program": this is the factory production program which defines the quantity of finished products with respect to the production capacity and to present and expected component and raw material availability;

o the "executive program": is the program which, on the basis of the operating program and of production capacity, gives each production cell its detailed program and machining sequences;

o the "production program": is the program which, for limited flexibility cells only, is extracted from the executive program which refers to the preceding day, and constitutes an anticipated production program for the downstream warehouses; this program is necessary to overcome the limited flexibility of these cells which is the result of the longer set-up times they require.

Raw materials and purchased components must be available for the head cells to carry out the three programs defined above. Whereas, for the appointment and flow cells, not only purchased components, but also the semi-machined parts which come from upstream cells, must be available.

9.2. Operation Control In Real Time

All activities within the production process must be monitored in real time if production operations are to be carried out efficiently. The aim of this module is to keep track of the status of each machine and of shop orders and, also, to synchronize materials flows to ensure correct production order release.

Two different methods were used for production order release at Susegana plant, as illustrated in Figure 3. In the converging upstream cells, where the semi-machined parts are the input for a downstream appointment cell, production order release is either on the basis of a "material issue" or a "semi-machined part receipt".

Figure 3a shows the regulation of production order release on the basis of material issue. After a production order release (time T1) to cell C2, the issue of the materials from the internal warehouse is communicated externally (time T2). This information is used to regulate production order release to cell C1 (time T3). The production order release for cell C3 (time T4) takes place after semi-machined parts from both C1 and C2 have been received in the internal warehouses of cell C3.

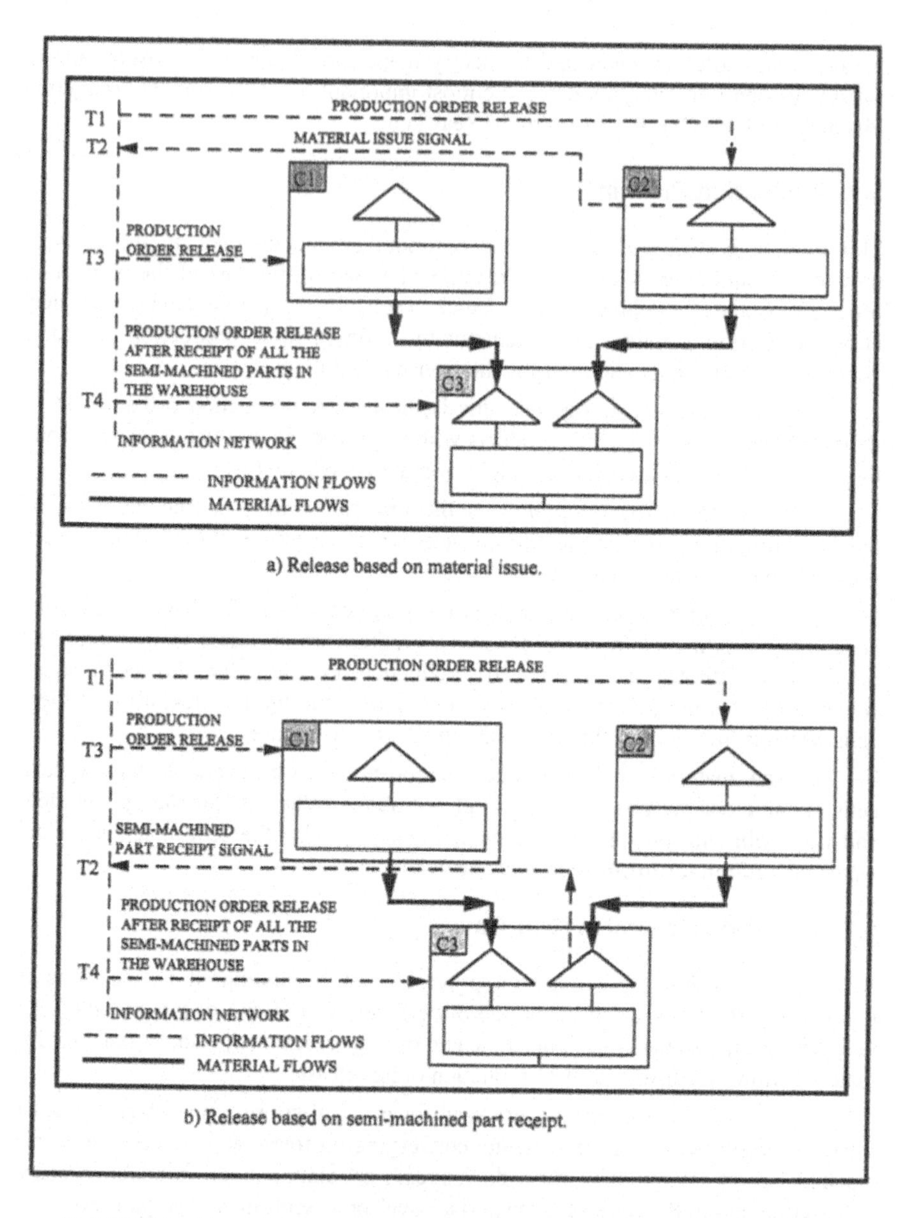

FIGURE 3. Two different methods used for production order release.

Figure 3b shows the way in which production order release is regulated on the basis of semi-machined part receipt. Starting with a production order release to cell C2 (time T1), but only after the receipt of semi-machined parts into the internal warehouse of cell C3, a signal is sent to communicate the receipt (time T2). This information is used to regulate the production order release to cell C1 (time T3). Release to cell C3 (time T4), occurs only after the semi-machined parts from cell C1 have reached the internal warehouse of C3.

Given high variety of semi-worked parts and components, the control system should synchronize appointments at various stages. Throughput times at different cells are not constant. If appointments are not managed correctly, some semi-machined parts may arrive long before others, resulting in long waits and imbalances. To avoid this, the "conditioned production order release", which is based on an "if .. then" logic, is used. This technique, which can be applied to both production order release criteria in Figures 3a and 3b, is shown in Figure 4.

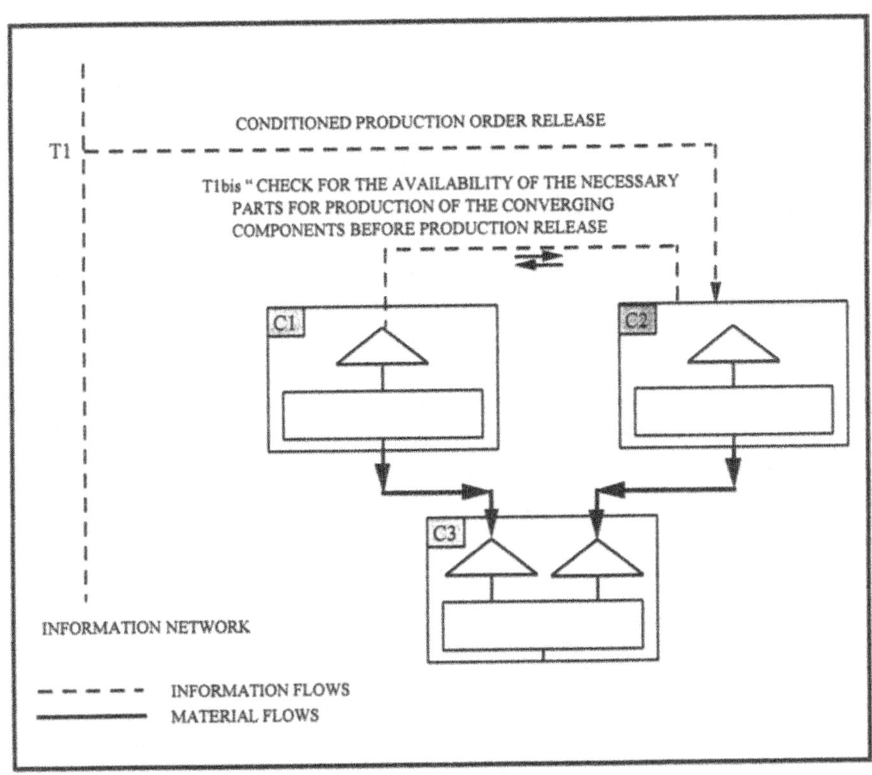

FIGURE 4. Conditioned production order release.

After a production order release to cell C2 (time T1), in order to avoid a long wait for components in the C3 input warehouse, the production release to cell C2 is checked (time T1bis), in order to discover the availability, in the cell C1 warehouse, of the parts required to produce the converging components at the same appointment in C3. If the reply is affirmative, the production order release to C2 is carried out. After this, the production order releases to cells C1 and C3 are, in their turn, issued, using one of the two methods described in Figures 3a and 3b.

10. CONCLUSIONS

In this paper, the production system with a sequence-cell layout at the Zanussi-Electrolux plant, at Susegana, Italy was described. Some distinctive characteristics of this cellular manufacturing system are the capacity to manufacture a wide mix of parts and the ability to produce the entire range of products in greatly reduced times. These characteristics derive from the identification of product families and portions of the production process which constitute independent units within the factory. Structuring the plant in separate production units, each one totally dedicated to a precise stage of the total production process, permits simplicity of flows, quick decisions and less managerial and organizational problems.

The implementation of CM in Zanussi has resulted in substantial reduction in work-in-process inventory (70% reduction) and materials movement, improved quality (quality inspection is currently a responsibility of the workers in the cells), and an overall increase in productivity. This was achieved through correct balancing of the cells, creating an almost continuous flow-type production. Production planning and control have also been improved by the implementation of CM. Due to continuous flow-type operations, many intermediate component numbers have been eliminated from production planning and control software. Furthermore, capacity is more flexible and responsive as a result of improved capability of knowing delivery timing and requirements. These improvements have enabled the plant to meet demand more successfully.

Finally, it should be stressed that the management of Zanussi-Electrolux plant have always viewed the transition to CM not simply as the rearrangement of machines on the factory floor; the strategic ramifications of implementing CM outweighed the operational or tactical consequences.

AUTHOR'S BIOGRAPHY

Roberto Panizzolo is Assistant professor at Department of Industrial Innovation and Management, University of Padua, Italy. He holds an M.Sc. degree in Electronic Engineering and a Ph.D. in Industrial Engineering from the University of Padua. He is a member of European Operations Management Association (EurOMA) and American Production and Inventory Control Society (APICS). His current research interests and teaching are in Technology Management and Operations Management areas, with particular emphasis in production systems design, manufacturing planning & control systems, cellular manufacturing, and management of advanced manufacturing technology. He has published many articles in international journals in these areas, and has participated in a number of international conferences. In addition to his academic duties, Dr. Panizzolo is also actively involved in several national and international research projects.

14

Microsoft Ireland: Realigning Plant, Sales, Key Suppliers By Customer Family

R. J. Schonberger

1. INTRODUCTION

Microsoft grew exponentially in the 1980s to become the world's leading software company. Its market value of $21 billion exceeded even that of General Motors. Founded by Chairman Bill Gates in 1975, Microsoft emerged in the 1990s, according to *Business Week*, "as clearly the most important single force in the entire computer industry."

The company introduced 48 new products in 1992, including well over 100 international versions. Among those was a phenomenal success: Microsoft Windows 3.1, which soon was running on more than 12 million systems worldwide. Windows became the fastest selling graphical user interface ever. In the two years since its introduction its sales represented a new customer every 10 seconds. Other leading software products included Microsoft Word, Excel, Powerpoint and Project. The company established manufacturing facilities in Bothell, Washington, serving North America and other parts of the world, Nunacao in Puerto Rico serving North and South America, and Dublin, Ireland, serving Europe.

2. MICROSOFT IRELAND

Microsoft Ireland, located in Dublin since 1985, is the European manufacturing base for Microsoft Corporation. The 80,000 square-foot facility employs 350 people. From there, the company supplies software packages to all major European markets, with Britain, Germany, and France accounting for over 60 per cent of all sales.

Providing logistics service support to the European marketplace from a detached island location has freight and time-to-market disadvantages. Crowley (1992) documents the problems encountered by exporters located in peripheral countries such as Ireland. They include the following:

- Ireland has a small open economy that is further split into two sub-economies. Port and shipping terminal investments must therefore be highly concentrated to gain economies of scale sufficient to be competitive.

- Ireland is an island and, with the opening of the England-France channel tunne the only European Union member without a land-link to Europe.
- Ireland is remote from the center of gravity of the European market whe compared to the other member states. Hence the role of logistics proportionately higher for Irish industry.
- Ireland is impeded geographically from the European mainland by Britain, s single-mode straight-line routings cannot be followed to central Europea markets.
- Ireland lacks an abundance of raw materials, which "doubles" its logistica demands; that is, industry must both import raw materials and export finishe goods.

These five handicaps lead to an estimated transportation costs for Irish exporters to the European mainland of over nine percent of export sales value, o twice those incurred by other European member states. Additional logistics overheads add to the competitive disadvantages faced by Irish exporters.

3. MANUFACTURING OPERATIONS (1985-88)

When initially established, the Dublin plant had direct responsibility for manufacturing and shipping to the U.K. and European destinations. Marketing, customer service, and technical support was provided by each national sales subsidiary, of which there were thirteen in Europe. Manufacturing at Microsoft Ireland essentially is a two-stage process. The first stage is duplication of software packages from master disks. The second is assembly of the finished software package. Assembly is labor intensive, consisting of placing the duplicated disks, labels, manuals, license agreement, packing materials in the appropriate carton; and then shrink-wrapping the carton and awaiting shipment.

Initially, Microsoft operated like most other manufacturers: long production runs, large inventories, lengthy setup times, quality control problems, and multiple suppliers. From a total product range of 280 products, high volume lines such as Word would be produced in batches of 10,000 units once a week, with lower volume lines being assembled just once a month. The primary objective was minimizing costs associated with long setup times. This called for bulk deliveries of raw materials from suppliers and required a warehouse of 40,000 square feet capable of housing eight weeks of inventory with associated storage costs. At the end of a production run, the finished goods were moved back to the finished goods warehouse where they awaited shipment. Delivery to customers occurred at the end of the month. This approach resulted in a three-week order cycle and lent itself to stockouts, inasmuch as production capacity was locked into a given line for considerable periods of time.

The structure of the distribution channel at this time was typical of the industry. Microsoft Ireland would ship large batches intermittently to the warehouses of the thirteen sales and support subsidiaries around Europe who were responsible for onward logistics. For example, Microsoft Ireland shipped product directly to the U.K. subsidiary's warehouse. From there, Microsoft UK would ship to a mix of about 200 distributors and dealers using contract delivery for large distributors and couriers for small orders to dealers. Backorder rates were typically in the order of fifteen percent of total sales.

4. JOURNEY TO LEAN PRODUCTION (1988-90)

In 1988, Microsoft Ireland decided to confront these problems of working capital tied up in inventory, quality, and product availability. They employed a consulting firm, World Class International (WCI). WCI's studies showed that on average, Microsoft's composite lead time was 151 days: 60 days in raw material, one day in work-in-process (WIP), and 90 days in finished goods. Moreover, the product received value for only four minutes (the time it took the package to be assembled) during a normal production run. Faced with a response ratio of 151 days to four minutes, the company's reaction was immediate: Emphasizing cycle time, the new policy would be to produce smaller lots more frequently. The object was to receive supplies daily and build (assemble) daily. The consultants and Microsoft identified four critical dimensions in the implementation process.

4.1. Supplier Reduction

The company's supplier base included indigenous printing companies (manuals), packaging manufacturers, disk manufacturers, and freight forwarders. Microsoft decided to initiate a process of selecting a reduced number of strategic partners. In return for providing its suppliers with a long-term commitment, standardization of product design, and rolling sales forecasts, cost reductions would be shared 50-50 between Microsoft and the supplier. Microsoft received assurances with regard to mutual cost reduction and daily deliveries. These commitments were based not on legally binding contractual agreements but rather on "gentlemen's agreements" coupled with quarterly reviews. The strategic partnership effort reduced the supplier base by 70 percent, which significantly lowered transaction and communication costs. Raw material inventories were also reduced 70 percent.

4.2. Cutting Production Batch Sizes In Half

To facilitate shorter production runs, lower inventories, and assembly of all products on a just-in-time (JIT) basis, setup times had to be dramatically reduced. In assembly, setup was taking 35 minutes. This involved clearing out the previous job, after checking (counting) it; then ensuring that all the disks, manuals, and packaging were available at the appropriate work stations on the 30-foot long assembly line.

The first improvement occurred while the plant was still under-utilized. While some assemblers were completing one job, others pre-staged the next job on a second idle assembly line located parallel to the first. But as the plant got busier, the second assembly line was needed for production. A different way of achieving quick setup was needed. The company came up with an imaginative and novel approach in which its 30 foot-long assembly lines would be replaced.

The classic JIT configuration—a U-shaped line in which all assemblers would be in close proximity—was attractive. However, Jim Gleason, production manager, thought that for this process a completely round carousel configuration would work even better. Plant manager Brian Reynolds liked the idea but suggested making it a dual-level carousel. With two levels, while operatives assemble from one level of the carousel, teammates who have completed their tasks for that run set up the other carousel level for the next production run. Their creation is shown in Figure 1. The simple tools used in the setups (scissors, tape dispenser, etc.) were placed on easy-to-see shadow boards.

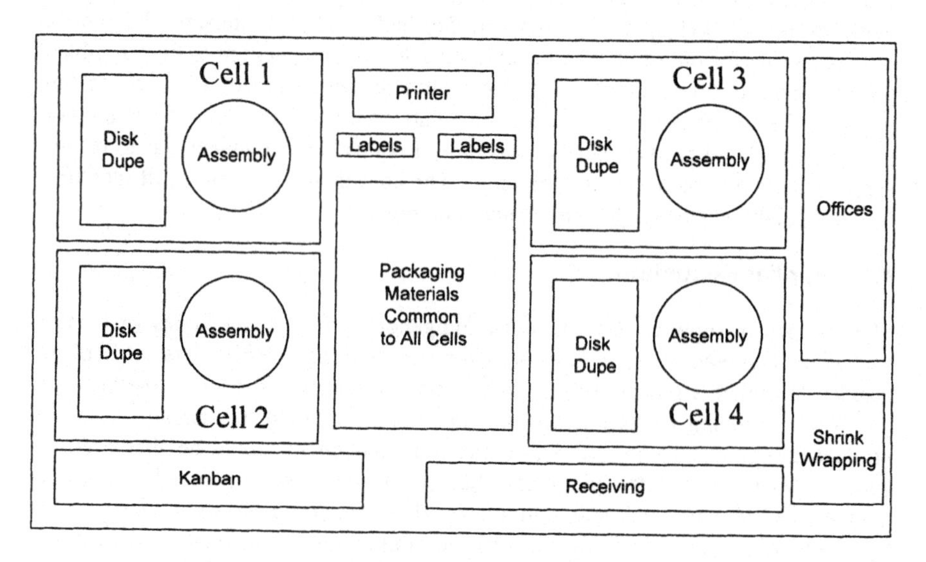

FIGURE 1. Dual-level carousel assembly cells.

In addition to eliminating setup-time, quality control improved: In a production batch of ten units, it became immediately obvious at the end of the process if any disks or manuals were not included in the carton, as they would remain highly visible on the carousel. By producing in smaller batches, quality problems were immediately and clearly identified.

A major piece of equipment in each cell is a Kallfass packaging machine, which shrink-wraps packages of disks. The operator of the Kallfass conducted round after round of detailed studies of time on the equipment. His analyses resulted in changeover time being progressively reduced from 35 minutes to one minute. Each improvement was plotted on a large trend chart near the Kallfass.

4.3. Employee Involvement

Overcoming resistance to change on the shop floor required radical changes in the way people were managed. Managers identified employees who they felt would be suitable facilitators in training operatives in JIT and total quality management (TQM) techniques. Considerable resources were devoted to company training.

For example, management recognized the Kallfass operator's talents and bestowed upon him the extra assignment of running training classes in quick setup for other plant personnel. During a plant visit, one of the case writers saw a wall notice announcing the next class and listing who would be trained; plant manager Reynolds' name was included on the list.

At first the company did not offer more pay for additional skills. Within a year, however, it became clear that the expansion of employee skills was paying big dividends. As a result, a revised compensation system would pay based on number of new skills employees acquire. In the new system, employees certified as "world-class operators" would receive five percent more pay, first choice of jobs, and opportunities to work on engineering and other projects.

The training was also geared toward development of quality-focused teams that used brainstorming to improve manufacturing processes. Emphasis on continuous improvement yielded 3,700 employee suggestions in the first two years of operation, an annual rate of over 5 per employee.

4.4. Focused Factories

Plant layout and organization was still another target for improvement. Microsoft Ireland would employ the focused-factory concept, pioneered by Skinner (1974) in the 1970s. Skinner maintained that a plant would perform better if it limited itself to a focused number of tasks, processes, or products. Initially, Microsoft established focused cells around common product families, such as Windows or Excel. As business expanded, the number of such cells grew from one to ten.

While the results were good, managers later decided on a different object of focus: customers. Since the geographic destinations of the software packages were language-related, four focused "factories within the factory," were introduced: one (Euro) for English-language customers, a second for Germany, a third for France, and the fourth for the rest of Europe (several languages). Referring again to Figure 1, we see a typical layout for one of the four focused units.

The focused factories deal with the customers of specific geographic markets. Each had its own independent manufacturing cells, complete with production equipment (duplicating machines and carousels) and work teams (see Figure 2). Each of the four cells in the French factory includes a Kallfass shrink-wrap machine. The other focused factories do not include shrink-wrap, inasmuch as customers in the other markets have accepted reduced packaging.

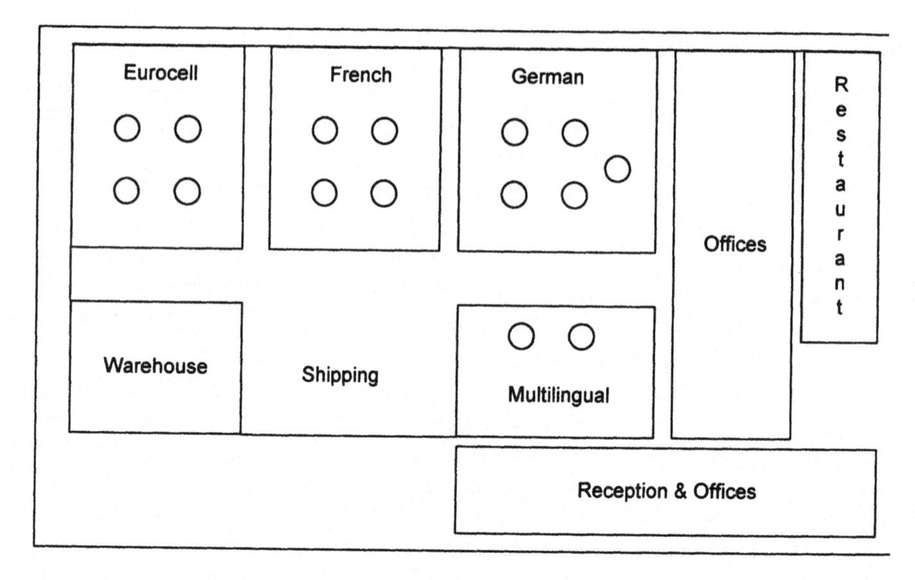

FIGURE 2. Four customer-focused cells.

Prior to 1988, Microsoft-Ireland had 42 major suppliers. A supplier reduction campaign reduced that to 11 suppliers, four of which were printers. It was not possible to cut to just one printer, because no printer in Ireland had enough capacity to handle all of Microsoft's demand. Also, with the work split among four printers, each focused factory could then have its own printer, specializing in the language of that factory. The printer for the French focused factory is good at running larger lots. In contrast, the printer for the multilingual factory has learned how to run small lots (e.g., 300 units). The following contributed to making small lots feasible for this printer:

- The printer for the multilingual factory placed a folding machine right next to each printing machine—a mini-cell—instead of all folding machines in a separate department.
- Quick-drying inks eliminate the need to allow time for ink to dry before folding (which overcomes a reason why printers often claim not to be able to offer quick response).
- Earlier, Microsoft had standardized page sizes, which eliminates the need for the printer to reset machines for different sizes.
- Microsoft had the printer equip its machines with blank-page detectors, which eliminates a critical quality control problem—and time-consuming inspections and rework.

The national flags of the destination markets were in evidence at each focused factory, making it easy to see where to deliver. With this kind of simplification and with each cell maintaining direct contact with and deliveries from its own suppliers, paperwork and administrative costs plunged. Within months, cost of goods sold had been reduced by 25 percent, while inventory levels in the plant had been cut by 70 percent, and customer lead times reduced to one day.

5. CHANGES IN CHANNEL STRUCTURE IN THE U.K.

As the radical changes in operations in the Dublin plant were beginning to pay off in improved in manufacturing performance, other developments and changes began to emerge in the channels of distribution.

5.1. Relocation Of Microsoft UK's Headquarters

In the 1980s, Microsoft operated a single distribution center in the UK, plus five more distribution centers in five other European locations. The distribution centers, in turn, served many large and small distributors and dealers.

In 1990 Microsoft UK (the sales and service subsidiary in Britain) was in the process of relocating to a new site at Winnersh in the southeast of England. As there was a lack of suitable warehousing space at this new headquarters, the parent company felt it was now opportune to evaluate the future warehousing requirements for the United Kingdom. There was some concern that the phenomenal growth rates in the industry could lead to customer-service and cost problems. As a result, the company decided to evaluate separating the warehousing function from the marketing and technical support functions currently performed by Microsoft UK.

5.2. Emergence Of A Low-Cost Channel Of Distribution

Approximately nine major distributors accounted for 80 percent of Microsoft's business in the United Kingdom. The remaining 20 percent included some 200 smaller dealers, educational establishments, and original equipment manufacturers. A notable feature of the industry is that channel structure has been historically based on computer hardware. In the mid-1980s, distributors concentrated on the sale of PCs, which offered high profit margins. However, this situation changed considerably in the 1990s as low-cost equipment began to appear on the market. The mystery surrounding the use of PCs began to evaporate with the spread of interface software such as Windows. Hardware manufacturers now began to use more direct forms of distribution to retailers as lower-cost distribution channels became the norm. Direct marketers such as Dell and Gateway developed mail and toll-free telephone ordering. Ever narrowing margins on PCs became less attractive to major distributors, and their primary focus shifted to sale of software products and ancillary hardware such as printers.

6. FROM LEAN PRODUCTION TO LEAN LOGISTICS (1990-92)

In light of these developments, Microsoft Europe managers felt it was timely t examine logistics strategy. Main issues were warehousing and structure of th channels of distribution. Following a comprehensive analysis of options, th decision was to relocate the UK distribution hub to the manufacturing facility i Dublin. While this might appear to have been a very radical decision, given th longer distribution distances, it considered the broader issue of logistics and channe strategy, rather than simply warehouse location. Rationale for selecting th Microsoft-Ireland location included the following:

- Rate-based manufacturing raised significant implications for logistics: Th ability to manufacture each product daily offered the opportunity to delive principle products daily.
- Locating the UK warehouse in Dublin facilitated "one-touch inventory"; that is once finished goods were shipped from the production line in Dublin (one touch), there were no further intermediate stocking (touching) points a Microsoft UK.
- A more direct, low-cost channel structure emerged. The company's distributior network in the UK was reduced from two hundred geographically dispersec dealers down to just nine major, concentrated distributors.
- Implementation of lean production with just-in-time supplier deliveries releasec 40,000 square feet of space formerly used for warehousing purchased materials. This space at the Dublin site could now be used as the distribution hub.

7. MANAGING SERVICE LOGISTICS FROM THE EDGE

Managers at Microsoft Ireland were aware that relocation of the UK distribution warehouse to Dublin presented a number of inter-dependent challenges, which included:

- *Control of inventory and customer service levels.* A major factor to be considered was the perceived loss of control by Microsoft UK over the entire distribution function. Concern had been expressed about separating the sales and warehousing functions and the effects on customer service levels in Britain.
- *Loss of visibility.* Other concerns revolved around Microsoft UK's loss of visibility of inventory levels for all stock keeping units. Not being able to actually see the inventory creates feelings of discomfort. The need was for some means of counterbalancing this fear factor with a comfort factor.

These concerns led to development of a phased implementation plan: Microsoft UK would initially define customer service measures and levels.

Providing a comfort factor, daily E-mail updates would flow between Ireland and the UK. Finally, any manufacturing-marketing interface difficulties relating to customer service would be addressed at weekly problem clinics at the UK headquarters.

8. IMPACT OF LEAN LOGISTICS

Within a year, the impact of lean logistics (relocation, direct shipment, and one-touch inventory) on performance was as significant as had been achieved earlier though world-class manufacturing (Schonberger, 1986). In a marketplace where rapid product introductions and revisions mirrored the importance of time-based competitive strategies, the company was able to record the following improvements in logistic performance:

- Delivery lead times were cut to one day for principal products. For slow sellers, customers were each given a certain day of the week that they would receive delivery.
- Inventory savings of over $4 million were achieved in the first year of operation.
- Backorder levels fell from 15 percent to just five percent of total orders.

Furthermore, from Microsoft UK's perspective, more time and resources were now available to devote to their core competency: marketing and technical support.

9. CONCLUSIONS

This case study illustrates the synergistic effects of world-class manufacturing, notably, a cellular structure focused on customer families, and distribution. While logistics decisions have often been a source of conflict between marketing and manufacturing, Microsoft's experience shows that mutually beneficial results are possible. A key to success is trust: The sales and service subsidiary gained confidence that logistics could not only effectively be managed by manufacturing but effectively managed from a peripheral location.

The case also illustrates the interaction between distribution channel developments and the management of logistics. A wide range of distribution channels must be serviced by Microsoft, each requiring different approaches. These channels are far from being static, so flexible strategies are necessary. Indeed, Stern *et al.* (1993) argue that, ultimately, organizations like Microsoft "instead of merely watching their channels develop organically or playing the reactionary game of catch-up, must think creatively about how they can deliver superior value to their customers." Increased product customization will drive the concept of "tailored channels," which in turn will drive logistical differentiation. And it is logistics, La

Londe and Power (1993) contend, that will ultimately integrate the product-service chain in the next century.

Acknowledgment. This case study is adapted from Fynes and Ennis (1994), with permission from Elsevier Science Ltd., Pergamon Imprint, The Boulevard, Langford Lane, Kidlington 0X5 1GB, U.K. An earlier version of this case also appears in Schonberger and Knod (1997; pages 465-471).

REFERENCES

Crowley, J.A., 1992, 1992 and the transport sector, *EUROPEAN Bureau of the Department of the Taoseach*, Dublin.

Fynes, F. and Ennis, S., 1994, From lean production to lean logistics: The case of Microsoft Ireland, *European Management Journal*, 12, 3, September, 322-330.

La Londe, B.J. and Powers, R.F., Disintegration and Re-integration: Logistics of the twenty-first century, *International Journal of Logistics Management*, 4, 2, 1-12.

Schonberger, Richard J., 1986, *World Class Manufacturing: The Lessons of Simplicity Applied*, Free Press, New York, N.Y.

Schonberger, R.J. and Knod, E.M., 1997, *Operations Management: Customer Focused Principles*, Richard D. Irwin, Burr Ridge, Ill., 6th Edition.

Skinner, W., 1974, The focused factory, *Harvard Business Review*, May-June, 113-121.

Stern, L.W., Sturdivant, F.D. and Getz, G.A., 1993, Accomplishing marketing channel change: Paths, and pitfalls, *European Management Journal*, 11, 1, 1-8.

AUTHOR'S BIOGRAPHY

Dr. Richard J. Schonberger is President, Schonberger & Associates, Inc., providing seminars and advisory services, and affiliate professor at the University of Washington, Seattle, USA. Richard is author of over 100 articles and papers and several books, including: *World Class Manufacturing—The Next Decade: Building Power, Strength, and Value* (1996) in five languages; *Building a Chain of Customers* (1990) in six languages; *World Class Manufacturing Casebook* (1987); *World Class Manufacturing: The Lessons of Simplicity Applied* (1986) in eight languages; *Japanese Manufacturing Techniques: Nine Hidden Lessons in Simplicity* (1982) in nine languages; and a textbook, *Operations Management: Customer Focused Principles*, 6th ed. (1997, with Edward M. Knod, Jr.). The Schonberger video program, "World Class Manufacturing," acclaimed by *Quality Digest*, is available as a 12-tape set. Dr. Schonberger's was inducted into the 1995 Academy of the Shingo Prize for Excellence in Manufacturing; was awarded the British Institution of Production Engineers' 1990 International Award, and received the Institute of Industrial Engineers' 1988 Production and Inventory Control Award.

15

Group Technology at John Deere: Production Planning and Control Issues

M. S. Spencer

1. INTRODUCTION

Production Planning and Control (PPC) continues to be a fascinating area of study. With the introduction and implementation of Group Technology (GT), Just-in-Time (JIT) methods and the wide-spread use of Material Requirements Planning (MRP), there is a wide variety of potential interactions available to study from master production scheduling to the shop floor execution.

Group technology is a broad philosophy of production that encompasses engineering aspects, physical layout issues, changes to production planning and control systems, and principles that affect several other functional areas (Burbidge, 1975; New, 1977; Suresh, 1979, Spencer, 1980; Hyer & Wemmerlöv, 1982; and Spencer, 1988). This article discusses the implementation of GT at John Deere from its inception, when the major emphasis was on product design aspects of GT, to latter day concerns regarding its use in facilitating just-in-time production, and day-to-day methods of planning production. The evolution of GT at John Deere serves to validate the belief that the success of GT does not rest solely upon the classification and coding systems, but also on its use as an effective management tool.

This chapter will present a brief history of GT at Deere, the problems that arose, and the use of GT and JIT in a case study that overcame the problems. Implementation of JIT programs sometimes begins in an assembly department with the use of kanban (cards) used as a trigger replenishment mechanism for assembly line inventory. When the JIT implementation program begins in the fabrication and machining areas supplying the assembly department the most serious problems with JIT arise. The farther back along the supply chain that JIT is implemented, the more difficult management problems become. This is especially true in the management of outside suppliers such as foundries where the economies of scale still require the batching of jobs. Faced with supplying a tractor assembly line, Department 500 at the John Deere Waterloo Works, a producer of heavy cast iron transmission cases using transfer lines and MRP, implemented GT and kanban not only to supply the assembly area but also back to their own supplying foundries. The results were positive and suggest a need for creativity in JIT implementation rather than a strict adherence to a cookbook approach.

2. BACKGROUND

The agricultural implement industry has undergone significant changes over the past 25 years, soaring to its peak in the late 1970s then plunging into a severe world-wide recession in the early 1980s. Bankruptcies and consolidation plagued the industry with company names that had been standard-bearers for over 100 years, such as International Harvester, White Farm Equipment, and J.I. Case forced to either merge or exit the industry. Since the mid-1980s the surviving implement companies, lead by John Deere and Company, have been at the forefront of production planning and control systems innovation in order to reduce cost and maintain profitability. Group Technology and Just-in-Time production methods have been widely implemented and significant out sourcing of components have spread throughout the industry.

2.1. The Industry And Company

The John Deere Company is the world's leading manufacturer of heavy-duty farm equipment. Heavy-duty farm equipment consists of two-wheel row crop tractors, larger four-wheel drive tractors, combines, and self-propelled forage equipment. The company also produces a variety of construction and industrial equipment and consumer lawn-care products. The company's primary heavy equipment manufacturing facilities are in Waterloo, Iowa (for tractors) and Moline, Illinois (for combines) in the United States, and in Mannhiem, Germany and in several locations in France. Overall the company had sales of approximately \$6B (US) in 1996. Tractor sales account for about 25% of the total.

Prior to the industry consolidation in the 1980s, Deere prided itself in producing the vast majority of its components. Since the company has been a major producer of farm equipment since the 1920s, the production facilities developed over a long period of time and are located in multiple-story buildings spread over a large area. Factories were dedicated to specific types of vehicles and as much fabrication and assembly was conducted within a single complex. The company's largest single complex is the Waterloo Works which once produced all of the company's row crop and four-wheel drive tractors. As a result of JIT implementation and cost pressures the company now out sources a substantial portion of the fabrication operations to supplier companies. Only the most critical components, such as diesel engines and transmissions are still manufactured within the company. A state-of-the-art tractor assembly factory went into production in Waterloo, IA in 1982 located about 5 miles from the original manufacturing complex. The down-sizing process has reduced the overall manufacturing floor space used at the original Waterloo Works site by at least 50%.

2.2. The Waterloo Works

The John Deere Waterloo Works was the original site of farm tractor production beginning in the 1920s. The word "works" is used by Deere to describe a production complex consisting of many separate factories or facilities. At one time, the

Waterloo Works consisted of two separate foundries, six major multiple-story manufacturing buildings and over 50 smaller production buildings. The John Deere Waterloo Works as recently as 1979 employed over 15,000 salaried and wage employees and produced over 165 tractors per day. With the relocation of diesel engine production and tractor assembly to new sites, and the out-sourcing program, current employment is about 5000 while supporting production of approximately 120 tractors per day.

The Waterloo Works is located next to the John Deere Electric Foundry. The Waterloo Works conducts itself as a separate business so that all components are sourced by a bidding process. Even though the Electric Foundry is located adjacent to the Waterloo Works, there is no guarantee that a component will be sourced there. In fact, considerable out-sourcing of foundry castings has occurred as a result of the bidding process.

Likewise, the John Deere Tractor Assembly Division utilizes the same bidding process to source its components. There is no guarantee that a component currently produced at the Waterloo Works will continue to be purchased by the Tractor Division, unless successful in the bidding process. Component price, quality and delivery reliability are three major criteria in the bidding process.

2.3. Department 500

This chapter focuses on the production planning and control methods used by department 500 at the Waterloo Works as an example of the integration of GT methods and practices within the company. The department manufacturers cast iron cases called "casings" for the tractor transmissions. The transmissions are assembled in another department, in another building within the Waterloo Works. The transmission casings measure approximately three feet wide, three feet high and very from two feet to four feet long. Casings weigh between 100 to 150 pounds. Department 500 manufacturers 23 different casings.

In the 1970s the department was created by rearranging approximately 30 machines (a few new machines were purchased) into seven manufacturing cells from what had been a functional layout spread throughout several floors in the factory (see figure one for a layout diagram). Machine cells were arranged by grouping similar products using group technology methods. For a period of time the department was viewed by management as the showcase for the company's group technology efforts.

The average routing for a transmission case is six machine centers, although many operations are performed at a single machine center. Average set-ups between major part families range from two to four hours. There are currently 24 wage employees assigned to department 500. The castings for the transmission cases are sourced from either the John Deere Electric Foundry, and from an outside independent foundry.

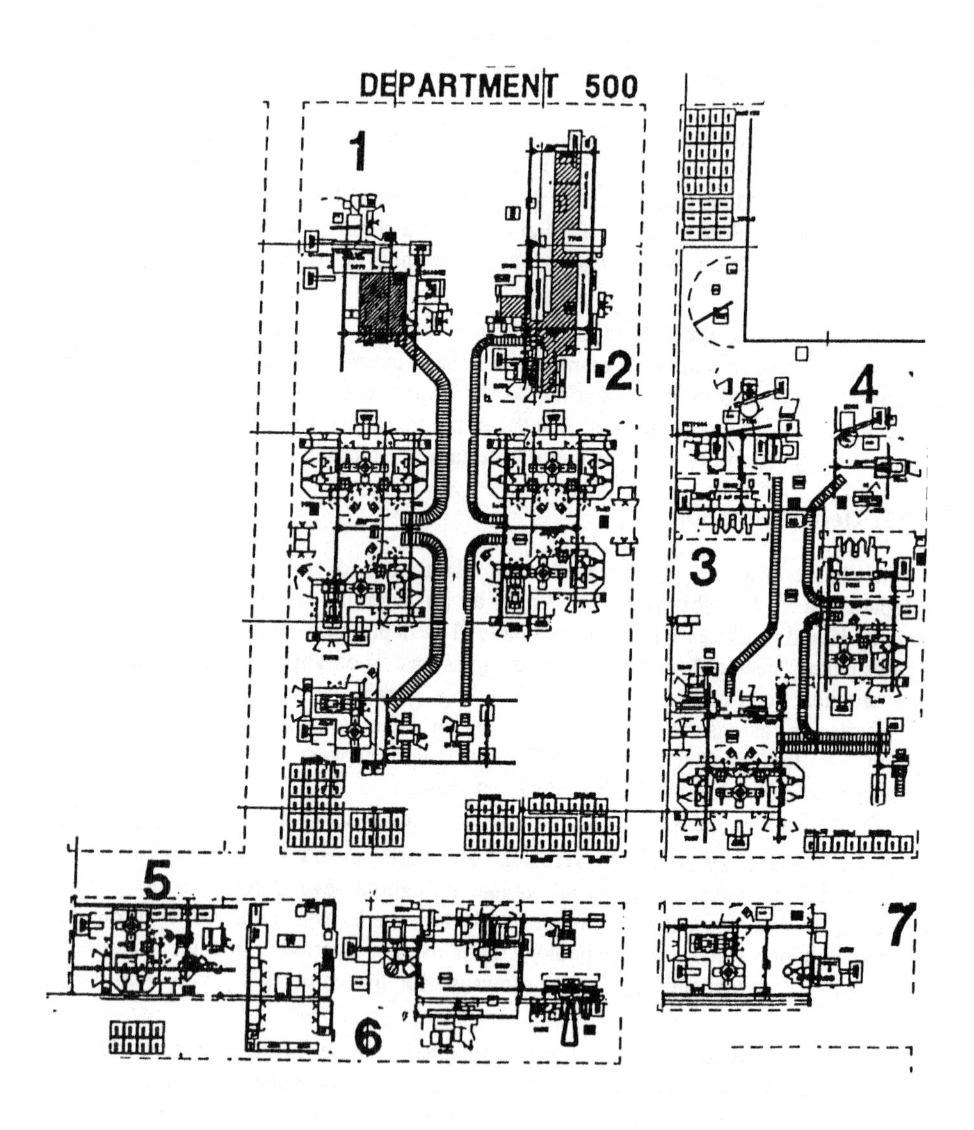

FIGURE 1. Department 500.

3. GROUP TECHNOLOGY AT JOHN DEERE

In the mid-1970s, when plans were first being discussed for the building of the new tractor assemble facility, the decision was made at Waterloo to search for a compatible Group Technology system to facilitate layout decisions. After several commercially available systems were evaluated, MICLASS was chosen and installed in 1976. A Dutch research organization, TNO had developed the idea of computerizing the classification and coding process itself, as well as the analysis programs used for forming families. Deere was an early implementer of the TNO software.

Testing at the Waterloo factory lead to the recognition of the limitation place by the 12 digit code used in MICLASS and a Group Technology Task Force was formed in 1978 with members from manufacturing engineering, industrial engineering, and technical services from Deere Headquarters in Moline, IL, and personnel from the Industrial Factory in Dubuque, IA, Combine Factory in East Moline, IL, as well as the Tractor factory in Waterloo, IA.

The result of this task force was the development of an in-house 35 digit code classification system. Once the proprietary system was developed, Waterloo Works engineers began to code their parts. It took about 7 people about six months to complete the coding for the current 100,000 parts. There are three major categories that emerged from the coding effort: rotational parts having circular features (shafts, gears and hubs); non-rotational parts typically made from steel or cast iron (housings, pump bodies, and major weldments); and sheet steel (tubing, hoods and fenders).

Over 200 Waterloo engineers, production supervisors, process planners and schedulers were then trained on the use of the computer GT system. The GT system itself use managed from Deere headquarters by the John Deere Group Technology System Group under the Computer-Aided Manufacturing Services Division. Indicating the importance place on GT by Deere Managers, that division reported directly to the Vice-President for Engineering along with the Director of Industrial Engineering and the Director of Manufacturing Engineering.

The heart of the JDGTS is an interactive classification program which generates data that references the part's geometric features (see figure two). The system also processes production data such as routings, planned requirements, machine load data, and cost details. The analysis component includes the Data Module used for coding, the Analysis Module used to analyze machine loading calculations, the Modify Module used to perform "what-if" scenarios, and the File Module that interfaced with statistical analysis programs. Additionally, the was a Help Module used to pilot users through any rough spots. The system was used to a considerable degree in the development and introduction of the 1982 Row Crop Tractor line, the new generation of diesel engines, and the four wheel drive line of tractors all introduced in the early 1980's. The focus was primarily on the JDGTS design component.

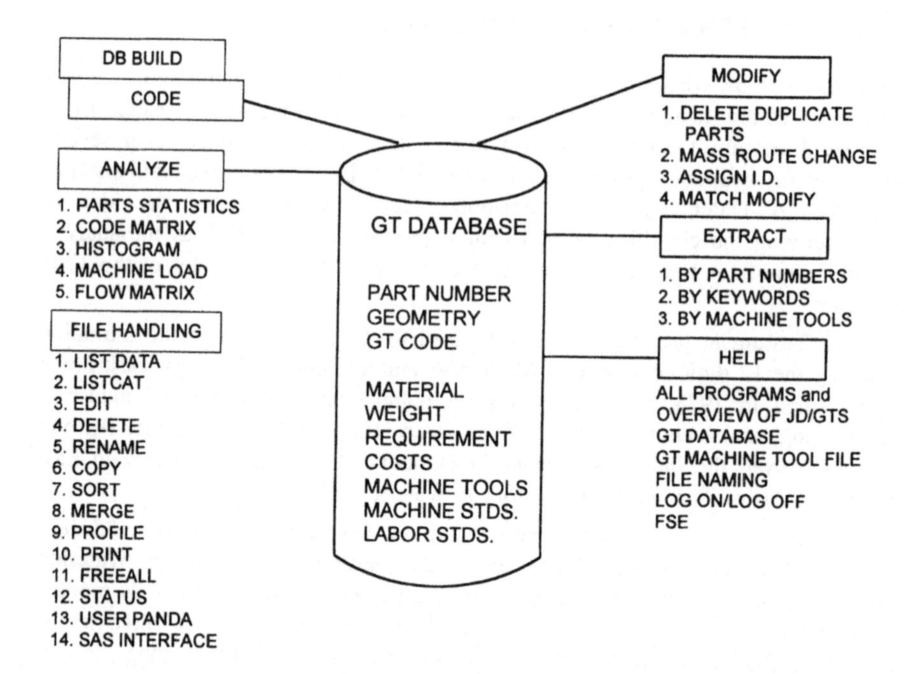

FIGURE 2. The John Deere/GT System (JD/GTS) software package.

Throughout the mid to late 1980s the JDGTS the focus shifted as the system was primarily used to identify part families and reorganize various factories along the cellular approaches defined using the GT system. The focus of management's attention was on the layout grouping component of GT apart from the design and production planning components. Major implementations included the John Deere Harvester Works, the Hydraulic cylinder department in the East Moline factory, the Ottumwa Forage Equipment factory, and the factory in Monterrey, Mexico, among others (Dutton, 1986, 1987; Farnum, 1988, Pesch, Jarvis & Troyer, 1993; and Gamboa, 1996). At the Waterloo Works GT was used to reconfigure, among others, the following major departments: Gears and gear finishing, Screw machining, Transmission and clutch shafts, PTO's, and Transmission Casings. The later being the department used for this case study.

As mentioned previously, the agricultural economy entered into a prolonged depression in 1982 which lasted until 1988. During that time no Deere factories were closed, unlike many competitors, but a new strategy emerged that would have profound implications. Deere embrace the Just-in-Time production methodology lead initially by the John Deere Engine Division and facilitated by a joint venture with the Yanmar Tractor Company in Japan. GT concepts embraced JIT to such an extent that more than one Deere manager concluded the GT is the best technique available to achieve JIT (Welke & Overbeeke, 1988; and HBS, 1988).

The focus of the JDGTS shifted to its use to group components into part families for out sourcing decisions. Deere made excellent use of the system to shed non-core production to suppliers and thus developed a highly integrated value-chain approach to production.

Among the successes there were concerns that arose in the implementation of GT. The chief areas were production scheduling concerns, human resource management concerns, and cost accounting concerns. Each will be discussed but the scheduling concerns will be examined in some depth as its examination involves the marriage of GT/JIT into the traditional MRP system.

4. CHARACTERISTICS OF THE PRODUCTION SYSTEM

The production planning and control system used at department 500 began with the implementation an in-housed developed Material Requirements Planning (MRP) system in 1978. The MRP system was modified as considerable company resources were devoted to the development of a state-of-the-art computer integrated manufacturing system designed to link the various factories worldwide. The severe industry recession that began in 1982 precipitated a joint venture with the Yanmar Tractor Company of Japan as a strategic alliance. This joint venture allowed the company to examine, in depth, the use of JIT at Yanmar. Beginning in 1982 and for the following ten years the company implemented JIT throughout its worldwide manufacturing facilities as the primary tool to maintain its competitiveness in the marketplace. Because of its position as the major profit center for the company, the Waterloo Works was chosen to lead in the implementation of both systems.

4.1. Material Requirements Planning II

A common worldwide inter-factory system is the communications link among all Deere factories. The inter-factory system operates on-line with real time transmission of computer generated requirements for all components and vehicles. Each factory operates its own MRP system that was developed jointly in-house among all factories. As such, the electronic exchange of data became a key competitive advantage for the company. Each factory also operates a commonly developed master scheduling system (MPS) that feeds requirements into the MRP system.

Net requirements from the MRP system feed the commonly-developed purchasing management system if components are purchased from outside sources. If components are manufactured at another Deere factory, the net requirements are transmitted through the inter-factory system and feed directly into the MRP system through the supplying factory's MPS system (see figure three for a diagram of the system flow).

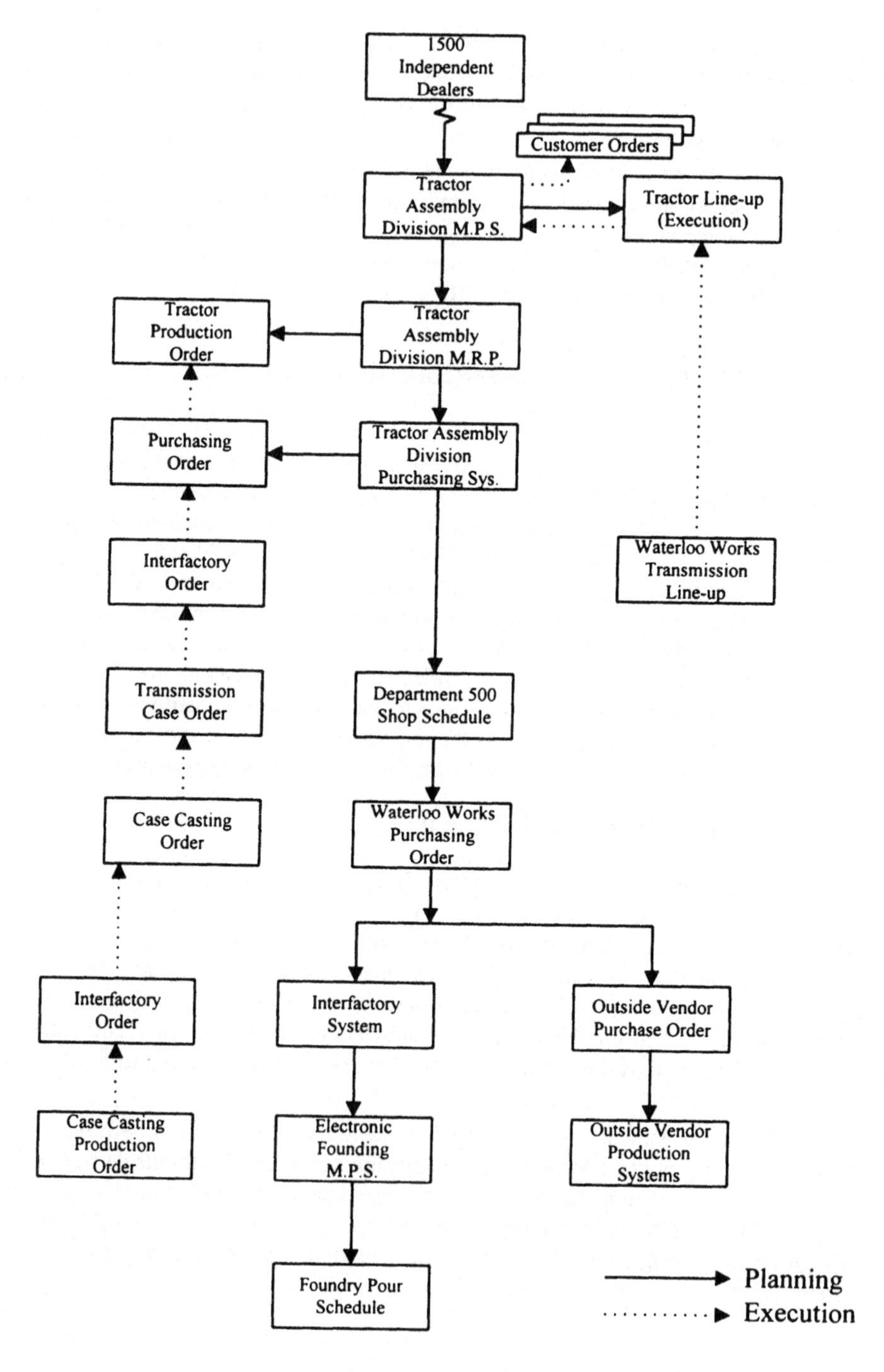

FIGURE 3. Systems Flow.

As orders for farm tractors are written for farmers by the John Deere dealer network, the order, consisting of a due date and the various tractor options, is feed into the MPS at the Tractor Assembly Division. The MPS is feed into the MRP system where net requirements for components are calculated. For example, one component would be for a particular transmission. Each transmission has a unique serial number that is tied to a specific tractor. The requirements are then sent into the inter-factory system and transmitted to the supplying factory. In the case of a transmission, the requirement is sent to the Waterloo Works. The Waterloo Works' MPS is then updated with the new requirement with a corresponding due date to the tractor assembly build date. The MPS is then transmitted daily on third shift to the Waterloo Works' MRP system.

The Waterloo Works MRP system explodes the transmission order through the bill-of-material and calculates the net requirement for the transmission casing manufactured within department 500. It is here that the scheduling problems arise as there is no logic in the MRP system that allows grouping of part families by GT classification. Instead, the MRP system drives its net requirements via the product's bill of material which may not have any relationship with the grouping accomplished by the GT system.

Additionally, the transmission casing requires a casting to be procured. The Waterloo Works MRP system then transmits the net requirement for the transmission casing casting into the purchasing system.

Since the transmission casings are procured from both the Electric Foundry and an outside vendor, the purchasing system selects, based on predetermined constraints, the amount of casting required from the outside vender and the amount needed from the Electric Foundry as well as the corresponding due dates. Since the Electric Foundry is a Deere factory, those requirements are automatically entered into the inter-factory system and transmitted into the Electric Foundry's MPS. The MPS feeds its requirements into the foundry's MRP where production schedules are altered as required to meet the calculated production due dates.

Execution of the production plan is problematic. Changes can occur throughout the production planning cycle. For example, customers may change the configuration of an ordered tractor, in addition to the normal changes seen in an MRP system. The execution of the production requirements requires manual intervention. Within department 500 the changeover times vary from two to four hours depending on the family of parts being manufactured. Within a family of parts the changeovers are relatively small. However, going from one part family to another requires a major changeover tending to the longer times. In addition, there are requirements for preventive maintenance, as well as experimental production and unplanned machine breakdowns that occur. A production scheduler was assigned to manually review all the requirements from the MRP system and develop supporting production schedules. It was these manual schedules that are executed. This scheduling problem, although not unique to GT applications, is the most often cited as the key problem with the use of the JDGTS by both production personnel and Deere managers and in the literature (Burbidge, 1975; New, 1977; Suresh, 1979; and Spencer, 1980).

A similar series of activities are required at the foundry. The economics of foundry operations require that considerable batching a casting requirements occur. Foundry pours are each unique and are somewhat unpredictable. Sometimes the pour of iron into molds goes smoothly while at other times there is considerable difficulty. Additionally, the chemistry of each pour must be determined as early as possible in order to maintain quality throughout the pour. As such, a production scheduler manually determines the foundry pour schedule using the net requirements form the MRP system. Towards the end of this case the use of GT at the foundry to resolve some of the problems will be discussed.

4.2. Just-in-Time Production Program

A relatively complete JIT system including the use of Total Quality Management and the kanban trigger mechanism was implemented at both the Tractor Assembly Division and the Waterloo Works Transmission area during the late 1980s. Since JIT is a continuous improvement system, the factory's management does not say that JIT has been implemented, however by all usual measures, the JIT program is quite advanced. Specific discussion of the JIT program is beyond the scope of this article, however, the kanban trigger method will be discussed. The kanban trigger system adds more linkage problems among components routed across GT cells as kanban does not provide any grouping logic either, although other components of JIT, such as minimum set-ups, improved preventive maintenance, and total quality management, are supportive of GT methods.

The date from the original sales order triggers a computerized final assembly schedule called "the line-up" to be generated 15 days before the due date at the Tractor Assembly Division. The line-up schedule is transmitted to the components factories such as the transmission assembly at the Waterloo Works, and the Diesel Engine Factory also located in Waterloo, IA. The components factories then use the Tractor Assembly Division's line-up to modify their own computer generated line-up if required. Normally, only small adjustments in dates are required to balance the final assembly line. The line-up for the transmissions is also transmitted to department 500 as is used to match against the open MRP production orders to ensure the supply of casings.

Kanban triggers are used two days prior to the tractor assembly build date to pull material into the final assembly line matched in sequence to the line-up schedule. Recall that the major tractor components including the transmission and engine have serial numbers that are assigned to a specific custom tractor. The kanbans move from the consumption point on the tractor assembly line back into the staging areas. The staging areas are point of use areas immediately behind and along the assembly areas. Once triggered by a kanban from the staging areas, the empty location triggers a replenishment from the receiving dock. The transmissions are unloaded from a truck into the receiving area dock as trucks arrive from the Waterloo Works. The sequence is maintained by matching the kanban from the staging area to the serial number at the receiving dock.

At the Waterloo Works the transmissions are assembled along an assembly line which is similar in operation to the tractor line described above. Kanbans trigger replenishment from the staging areas at the Waterloo Works. Kanbans trigger replenishment into the staging areas from department 500. Because of the relatively long changeovers and the need to batch in department 500's cells, kanbans are not used to move material within the department but are used to trigger replenishment of raw casting from the Electric Foundry and from the outside supplier foundry. Neither foundries use kanban within their facility except to trigger shipments of raw castings to department 500.

5. REVISIONS TO PRODUCTION PLANNING AND CONTROL

As a result of the implementation and use of GT and JIT within the Waterloo Works, inventory levels a of transmission casings decreased from 15 days on hand to 5.5 days on hand. Transmission casings are valued at approximately $400 to $600 (US) each. Quality increased as measured by the defect rate and rework costs declined approximately 25%. However, continuous improvement, a deeply embedded philosophy at the John Deere Company, seemed to stop at that level, however. Engineering studies were conducted using the Analysis and Modify Modules to reduce changeover durations as that appeared to be the factor limiting further improvements. The results of these studies were disappointing as they indicated further changeover reductions were not able to be cost justified using the traditional return on investment calculations. Simply, the changes to the machine tools were more expensive than the calculated reductions in inventories. Recall also that the prime motivator for the Waterloo Works is to maintain profitability and reduce cost as the competitive bidding process continues to be used to measure in-house production versus out-sourcing by independent suppliers.

As a result of the apparent impasse, management took the step of reorganizing department 500 into a Quality Circle and used a team approach to seek continuing improvements. The original team was composed of the shop foreman as leader, the production scheduler and production control supervisor, the materials inventory analyst, the production scheduler form the transmission assembly line (department 500's in-house customer), the master production scheduler for the Waterloo Works, and two department 500 operators.

In addition to the original teams, several staff support personnel were used throughout the project to add varies expertise as required. For example, several industrial engineers, product design engineers, and mechanical engineers as well as computer analysts were assigned to the team at various times. The department 500 foreman, the production scheduler, and the two production workers kept the remaining workers informed of team activities throughout the process both informally and at formal weekly meetings.

The result of the team's activities after four months was the design of a "leapfrog" kanban system used to trigger replenishment of castings from the foundries based on consumption at the tractor assembly line. The team was unable to uncover any flaws in the various engineering studies that could have been used to reduce the changeovers. The team also found that management could not support undocumented changes simply because of the feeling that intangible improvements would result. The team did essentially look at the entire process from the final product customer back to the supplier of raw casting to devise a revision that was successful in moving the continuous improvements to a higher level. As a result the scheduling problems posed by the MRP system could be leapfrogged to a degree, by allowing an early alert to the foundry and to Department 500 of a replenishment trigger. The additional time allowed for some grouping by families to occur.

The revision required significant computer programming revisions but used the existing inter-factory system, MPS and MRP systems to establish an electronic kanban trigger that leaped over some of the departments to collapse replenishment time while still retaining the ability to batch production at department 500 and at the foundries.

Under the "leapfrog" kanban method, a bar code is scanned at the Tractor Assembly Division's assembly line. The bar code is an electronic kanban that triggers actions at both department 500 and the supplying foundries. The bar code originates from the tractor line-up and is located on each transmission's move document. By scanning the bar code at the tractor assembly line a replenishment is triggered at department 500. Kanban triggers continue to be employed as previously described to orchestrate material delivery from Department 500 to the transmission assembly line.

At the same moment that the bar code reading triggers a replenishment order at department 500 for a new casing, a replenishment order is triggered at the supplying foundry from a new casting to be delivered into department 500. In this manner both the foundries and department 500 know of the consumption of a transmission at the tractor assembly line as it occurs. This establishes a sequence of production through the entire production system. Once established, the sequence is maintained by the triggering system without manual intervention (see figure four).

In fact, the department 500 production scheduler was reassigned and not replaced, his duties being accomplished automatically by the leapfrog kanban system. The lot sizing and family grouping that was being done manually by the scheduler is now able to be maintained, once established, by the replenishment triggers. The lot sizes and groupings are able to remain constant because of the replenishment by the foundry. This foundry replenishment is made possible by the early alert of a transmission being used at the tractor final assembly line so that enough time exist for the foundries to batch their pour.

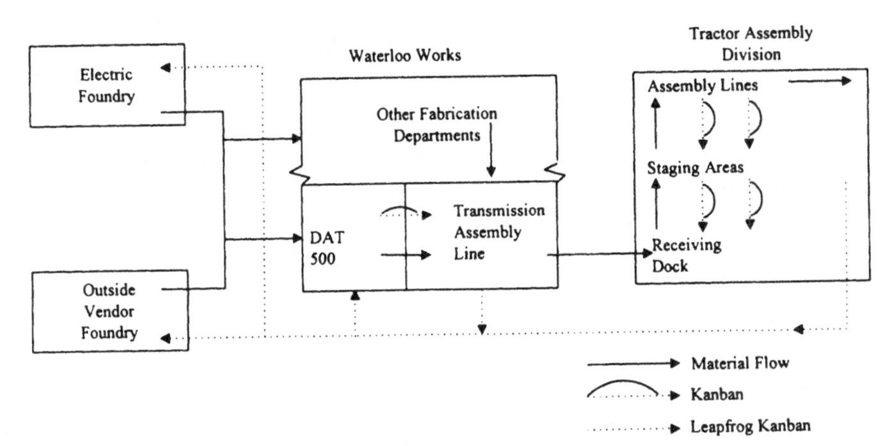

FIGURE 4. Materials flow.

6. RESULTS AT DEPARTMENT 500

The implementation of the leapfrog kanban method has allowed an overall reduction of transmission casing inventory from $300,000 (US) in 1989 to less than $100,000 (US) in 1995, during a period of constant production levels. Delivery of raw casting form the foundries now occurs daily into department 500. Average work-in-process inventory is now 1.5 days. Quality as measured by the costs of rework has declined over the same time period by 25%. Further, department 500 now delivers finished transmission casings into the transmission assembly line multiple times per shift rather than the once daily deliveries previously used. Periodic inventory auditing reports that the department 500 inventory accuracy has greatly improved.

The results achieved from department 500 now serve as a model for continuous improvements using teams throughout the Waterloo Works and Tractor Assembly Division. Several other major components are now being scheduled using the leapfrog kanban trigger approach. The team continues to operate and now consists of the production supervisor and all production operators. New operators assigned to department 500 now receive two weeks of JIT training including group problem solving methods and team building before ever receiving training on the operation of a specific machine tool. Additional operator team-based revisions have also made improvements in quality of raw castings and new product introductions.

Management reports that its major concern for department 500 is the level of kanbans within the system and the need for periodic review of lot sizes and part families given new product introductions that are currently planned. Both concerns will be given to the production team of their resolution.

7. HUMAN RESOURCE MANAGEMENT CONCERNS

One rather unexpected area of concern arose in the human relations aspect of group technology at the Waterloo Complex. Wage employees in Waterloo are represented by the United Auto Workers (UAW). Most employees are paid through an incentive plan where compensation is determined by the total production over a predetermined minimum. Set-ups are compensated at the hourly rate only, usually about 33% less than on-incentive earnings. Set-ups, then are something to be avoided in favor of long runs of the same component.

The GT program restricted, as did MRP scheduling, the long runs as grouping caused, in some cases more frequent set-ups than under previous circumstances. Further, JIT and GT both encourage teamwork within a manufacturing cell rather then the individual maximizing that was the pervious case. The success of the cell depends on the team rather than an individual maximizing his or her pay.

These concerns where sufficiently widespread throughout Deere that the company engaged the United Auto Workers in collective bargaining to remove some of the features of the incentive program and replace them with a gainsharing plan called the continuous improvement pay plan (CIPP). After negotiation the UAW agreed with the new pay plan. The work force is now rewarded on a 50% split of any savings obtained over the predetermined standard. several GT core cell teams have received large semi-annual payouts for the savings generated under the cellular concept.

8. ACCOUNTING ISSUES

A third area of concern was concentrated in the accounting field. Accountants, as was previously described in the Department 500 case, found that traditional cost accounting could stifle management initiative to use GT for continuous improvements. It was more difficult to capture an accurate cost in a multi-functional manufacturing cell then in a single function machining department. It was necessary to revise budget reports and replace elements of the traditional cost gathering system to capture the true costs of production within the cells. Deere was actively implementing the Activity Based Accounting (ABC) process in the early 1990s which allowed for most of these inaccuracies to be remedied. An open question is the effect the traditional cost accounting procedures would have had if ABC had not been implemented.

9. GT INVESTIGATION AT DEERE FOUNDRY AND EDI WITH GT

The current direction of the GT philosophy at the Waterloo Complex follows two paths. First there is active investigation of GT implementation at the Deere Electric

Foundry. The second path is the use of GT codes through Electronic Data Interchange (EDI) to support continued out sourcing of non-core production.

The foundry began exploration of GT in 1992 in the core room, Department 787. The core room processes the cylinder block core assemblies used in the production of 4-, 6-, and 8-cylinder block castings for diesel engines. Department 787 consists of one CB100 machines, five CB300 machines, and two 356VP machines arranged in a cell using GT techniques. The department operates on a single shift with nine wage employees and a supervisor. While the implementation is still underway, results are encouraging. Foundry management is active in the on-going review of the process.

Deere has been using the Bulletin Board System (BBS) to transmit engineering drawings and specifications over the Internet to suppliers for quotes sing 1993. Currently there are over 600 potential suppliers who can download the part's geometry and use the data to prepare a quote. The BBS provides a toll free phone number that enable potential bidders to talk directly to the design engineers to clarify a specification or to make suggestions to improve the design and/or manufacturability.

10. CONCLUSIONS

While the results from a single case cannot be generalized, observations from leading companies may always be useful for research. First, there appears to be a synergistic combination of using elements from GT, MRP and JIT to enhance production results rather than an earlier belief that JIT would replace MRP. Second, MRP appears to best serve as a platform for the transmission and collection of production information with GT methods serving as a motivator for continuous improvements and JIT as a means of execution of a production plan. Third, operational improvements are possible in a single department using JIT, but far greater improvements are possible if the whole process throughout the supply chain is considered even when independent companies exist. Finally, the use of kanban, normally viewed as a manual pull system, is able to be exploited across wide geographic distances and functional boundaries through computer data interchange.

Calculations done at Deere Headquarters reported an average saving of over $10 million per year and an annual saving of over 96,000 work-hours (Vogt, 1988). Cited it that report was the results form the Hydraulic Cylinder Division in Moline where the part count dropped 81% as the result of out sourcing, set-up time reduction of 75% lead time reduction of 42%, and scrap reduction of 80%.

In 1993 with the development of the new 8000 series tractor line (replacing the 1982 series discussed previously) group technology, again, played a major role. Cast machining cells were integrated into the assembly line for the clutch drum, manifold and rear cover. The cells feed directly into the assembly line. As outsourcing activities were completed at the Waterloo complex, and production consolidated to its current configuration new coding activities ceased at the Waterloo

Production Engineering center, although the JDGTS continues to receive support from the Technical Services Division at Headquarters. The shift away from the design component of GT and the layout component of GT indicates the emergence once again of the production element. The success of GT at the John Deere Waterloo Works can best be indicated by its acceptance throughout the factory as "just the way production is done" rather than being identified as a special engineering technique.

Acknowledgment. Special thanks go to the 20 or so John Deere Waterloo employees whose work and interviews went into the preparation of this case.

REFERENCES

Burbidge, J. L., 1975, *The Introduction of Group Technology*, Heineman, London.

Dutton, B., 1986, GT harvests parts reduction and quality, *Manufacturing Systems*, 4, 7, 36.

Dutton, B., 1987, Deere: An FMS update, *Manufacturing Systems*, 5, 11, 39-40.

Farnum, G. T., 1988, The X factor, *Manufacturing Systems*, 6, 3, 87-88.

Gamboa, F. V., 1996, Advanced scheduling at Deere, *Manufacturing Systems*, 14, 3, 22A-24A.

Harvard Business School Case 9-687-053, 1988, Deere & Company (A) The computer-aided manufacturing services division, HBS, Boston.

Hyer, N. L. and Wemmerlov, U., 1982, MRP/GT: A framework for production planning and control of cellular manufacturing, *Decision Sciences*, 13, 4, 681-689.

Pesch, M. J., Jarvis, L. L. and Troyer, L., 1993, Turning around a rust-belt factory: The $1.98 solution, *Production and Inventory Management Journal*, 34, 2, 57-62.

New, C. C., 1977, MRP and GT: A new strategy for component production, *Production and Inventory Management*, 18, 3, 51-62.

Spencer, M. S., 1980, Scheduling components for group technology lines, *Production and Inventory Management*, 21, 4, 43-49.

Spencer, M. S., 1988, Developing finite schedules for cellular manufacturing, *Production and Inventory Management Journal*, 29, 1, 74-79.

Suresh, N. C., 1979, Optimizing intermittent production systems through group technology and an MRP system, *Production and Inventory Management*, 20, 4, 77-84.

Vogt, C. F., 1988, Group technology: CIM's sleeping giant, *Design News*, 44, 16, 22-23.

Welke, H. and Overbeeke, J., 1988, Cellular Manufacturing: A good technique for implementing just-in-time and total quality control, *Industrial Engineering*, 20, 11, 36.

AUTHOR'S BIOGRAPHY

Dr. Michael S. Spencer is an Associate Professor of Production and Operations Management at the University of Northern Iowa. He received his Ph.D. in Operations Management from the University of Georgia in 1992. Dr. Spencer previously held various materials management positions at the John Deere Engine Division where he implemented both MRP and JIT systems. Dr. Spencer is certified as a fellow of American Production and Inventory Control Society (APICS) and has also served on their board of directors. He is currently Vice-President of APICS Educational and Research Foundation.

16

Design and Reengineering of Production Systems: Yugoslavian (IISE) Approaches

D. Zelenović, I. Cosić and R. Maksimović

1. HOW WE WERE LED TO GROUP TECHNOLOGY IDEAS

Satisfying customer needs with respect to product quality, price and delivery terms forms a basic goal of manufacturing. Analysis of work processes in real systems, conducted over a long period of time (1965-1978) at Institute for Industrial Systems Engineering (IISE) in Novi Sad, Yugoslavia consistently showed a lack of effectiveness with respect to quality, lead time and costs, in a time of rapid technology development, fast growth of competition, unstable markets, economic uncertainty and increasing social demands.

Our analyses of real systems consistently highlighted the need for reengineering material flows first, followed by other functions in companies. Our basic conclusions were that most of the inadequacies resulted from a complex, ***process*** type of flow (Figure 1). Such process-types of flows were found to result in:

- Long lead times for processes ($T_{cp)}$, implying poor relationship with objectively-established operation times (Σtii);
- Large amount of set up and passive times, in the entire time to market;
- Long waiting lines between operations, large amount of unfinished manufacturing and low turnover ratios;
- Inefficient information flows and procedures, due to inadequate methods, producing down times and extended manufacturing periods;
- Unresolved problems in finding the best sequence of job orders for production;
- Difficulties in realization of synergistic effects in system;
- Problems in motivating all participants in production for keeping production processes within planned tolerances.

The research team at IISE was led to conclude that there was considerable need for work process reengineering, and developing more effective approaches to production systems planning and design.

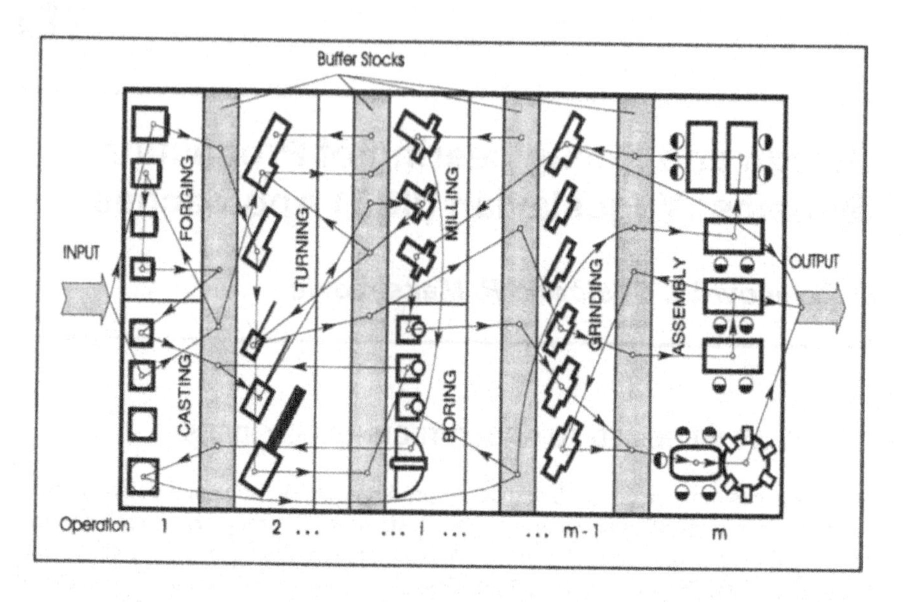

FIGURE 1. Process type of flow.

2. BASIS OF IISE-APPROACH FOR MATERIAL FLOW DESIGN

Applying a ***group approach*** in design of material flows, and a ***product approach*** in systems organization are essentially the basis of IISE-approach for redesign and revitalization of industrial systems.

The *group approach* was developed on the basis of *part similarity* in manufacturing programs, a restricted number of part shapes in real conditions and unification of all parts with *similar characteristics* into an *operational group*. The operational group is the basic unit in development of group flows and it is determined on the basis of classification systems. This approach enables conditions for increasing number of parts (q_j) on the relation input-output of working process.

The term *characteristics similarity* of parts means similarity of those characteristics which determine production possibilities on the same machine or group of machines with same characteristics. That way we design material flows of operational groups instead of single flows, which significantly "increases" amount of parts in flows and, with many other effects, raises quality of flows.

The proportion of basic values referred to processes of flow design for the case of group flows is given like load/capacity ratio as follows:

$$k^*_{ser} = \frac{\text{load}}{\text{capacity}} = \frac{\sum_{i=1}^{i=m}\left(\sum_{j=1}^{j=k} q_j \cdot t_{ii}^{(j)}\right)}{K_e} \gtrless 1,$$

where q_j = [units/time period]; Σt_{ii} = [time units/product, a degree of product technology complexity]; and K_e = [units/time period, effective capacity].

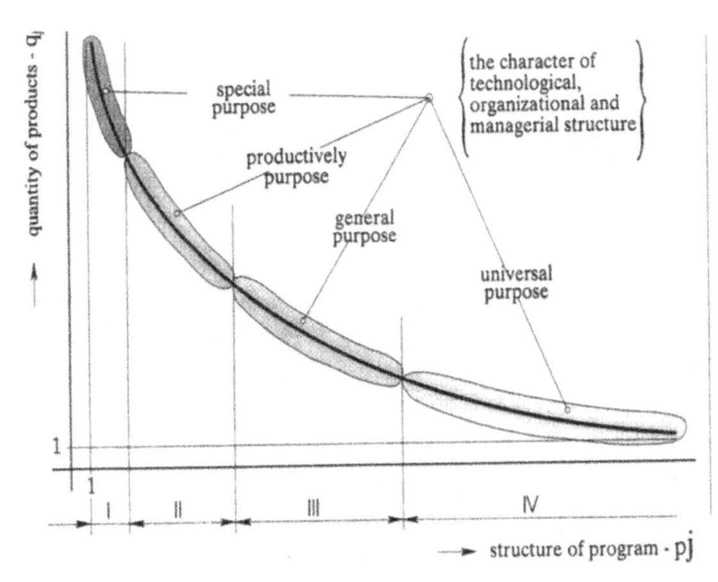

FIGURE 2

Detailed study of load/capacity ratios and product proportions (p_j - q_j) enabled the development of ***a general flow model*** (Figure 2). This model divides material flows in industry into four types (types 1.1, 1.2, 2.1 and 2.2). Their essential characteristics, conditions and criteria for design are shown in Figure 3.

Our experience at IISE, based on more than 25 revitalization projects, showed us that the group approach removes the constraints which single approach generates, to a large extent. The group approach enables:

- restriction of object characteristics (shape, materials, etc.) to a minimum;
- objective criteria for determining technology coefficients; reduction in cost of processing and technology, because of designing technology for group of similar parts, instead of every one part;
- reduced setup times with only one setup time for a group of simiular parts;
- significant simplification of flows in system, making possible comparisons of different variations of job order sequences to minimize manufacturing time, unfinished production level, and to meeting deadlines.

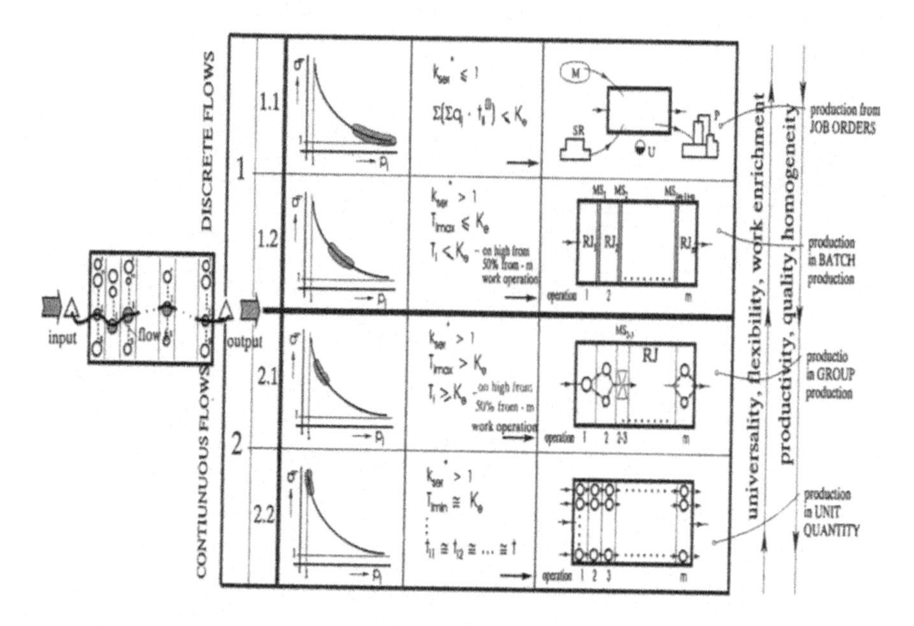

FIGURE 3

The ***product approach*** is based on arranging manufacturing structures on principle of *working units* which, as shown on Figure 4, consists of machines and workers necessary for making a group of similar parts. The working units are independent structures capable of completely manufacturing groups of similar parts.

Product approach, especially in combination with group approach, ensures:

- conditions for effective production planning;
- simplification of flows, and convenient control of work processes;
- increased responsibility for results, quality and dead lines;
- possibility of implementing new approaches to organizational technology of industrial systems;

and represents a structure of work process characterized by:

- maximal flexibility in labor-intensive processes,
- high flexibility and high productivity in capital-intensive processes, and relatively high flexibility and maximal productivity in knowledge-intensive processes.

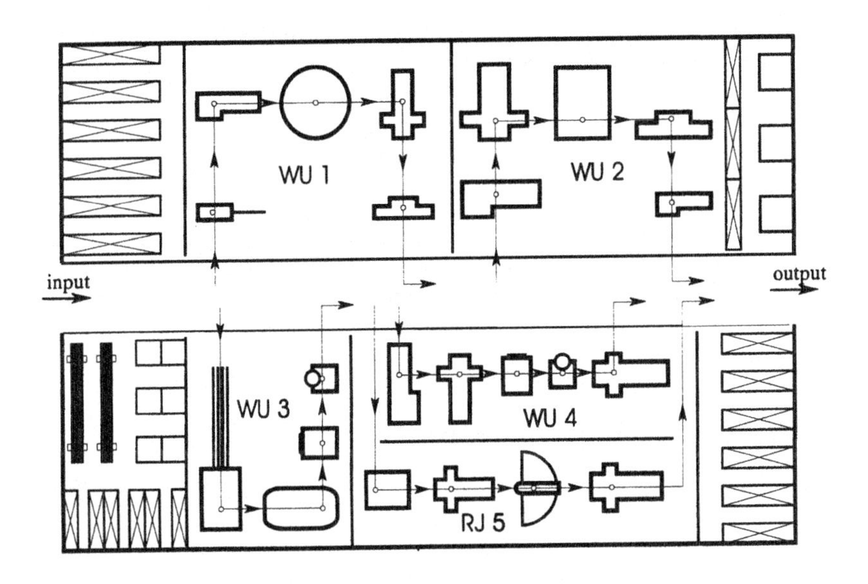

FIGURE 4

3. GT AS A BASIS FOR IISE APPROACHES

The IISE approach for development of effective production systems is based on a consideration of various external and internal factors, and it is based on:

a classification of parts with similar characteristics;

development of parts on the basis of ability to exist independently, and

applying ***methods for rational integration*** of part structures in totality, using connections with necessary strength and direction, for ensuring synergies.

This, somewhat organic approach, relies on the following principles:

- *Principle of Similarity*: part characteristics and manufacturing methods which allow *simplicity* and minimum complexity in system flows;
- *Principle of Independent Durability*: based on a natural law of separating (like living organisms) and creating full individuality and conditions for autonomous operation. The term autonomous operation requires sufficient connections with environment and supporting systems, because we are talking about open systems;
- *Principle of Managing Toward Goals*: based on the need for keeping process parameters within design limits in real time and given environmental constraints; the basic challenge is the capability for resisting environmental impacts and disorders in working processes;

- *Principle of Ability For Effective Work*: The basic characteristic of high quality systems, based on the *principle of independent durability*, is the ability for accomplishing programmed effects/investment rate. It is possible by development of flexible structures with minimum complexity, development of real-time control procedures, for turning system from breakdown to operating stage in planned time;
- *Principle of Effective Motivation* and conditions for work assurance: based on a thorough investigation of relations between various participants, hierarchy of positions, work place integrity, and accomplishing a pleasant and productive atmosphere.

The steps involved in the IISE-approach are as follows.

3.1. Step 1: Operational Groups Design

Operational group is, as it is said above, basic element in development of group flows and represents group of parts with similar characteristics which determine possibility of production of these parts on the same machine or on the group of machines with same characteristics. Procedure of operational groups design is consisted of the following phases:

3.1.1. Classification Of Work Pieces.

The procedure for operational groups modeling consists of categorization of work pieces on the basis of their similarity - using a *classification system*. The system includes classification criteria which allow, on the basis of similar classification character, grouping of work pieces suitable for operation on the same machine(s).

The basic structure for classification system of industrial systems elements, is shown like classification system KS-IISE-08 (Figure 5). Figure 6 represents part of the classification system for work pieces - parts made by removing of material. It is shown that classification criteria are represented by *characteristics* (which cover all attributes relevant for possibility of manufacturing on the particular machine) and by tolerance areas in the ranges of classification system which correspond to machine working areas. Implementing classification system on all work pieces, we get *classification codes*.

3.1.2. Design Of Modules.

A group of work pieces with *same classification code* (***module*** in Figure 7) is the group with highest homogeneity, whose characteristics are within the tolerances determined by classification system criteria.

3.1.3. Design Of Operational Groups.

Procedure for design of operational groups is based on the machine characteristics analysis and determining its limits by all characteristics - fields of classification system or, in other words, by determining ***classification code profile*** (Figure 7).

3.2. Step 2. Elaboration Of Group Technology

3.2.1. Selection / Design Of Complex Work Piece.

The operational group is an essential unit in group flow design. This procedure means that we select *real* or design *imaginary* complex work piece (part or assembly) for which we create technology. Complex part must includes relevant characteristics of all parts from operational group.

3.2.2. Elaboration Of Group Technology.

For the selected *real* or designed *imaginary* complex part, a group technology is created. Group technology must be of corresponding quality, for selecting the most adequate working procedure, order of operations and theirs elements (Figure 8). In accordance with the principle of similarity (characteristics similarity of parts in group), the technology for complex part is applied on each part from operational group. Certain modifications, due to particular characteristics of part - measures, surface shapes, sort of material, tolerances and other ones, are unavoidable. That way, higher quality of technology is obtained because of possibility for concentration on group characteristics, assurance of approach conformity and reliable elements for determining times for each operation (Figure 9).

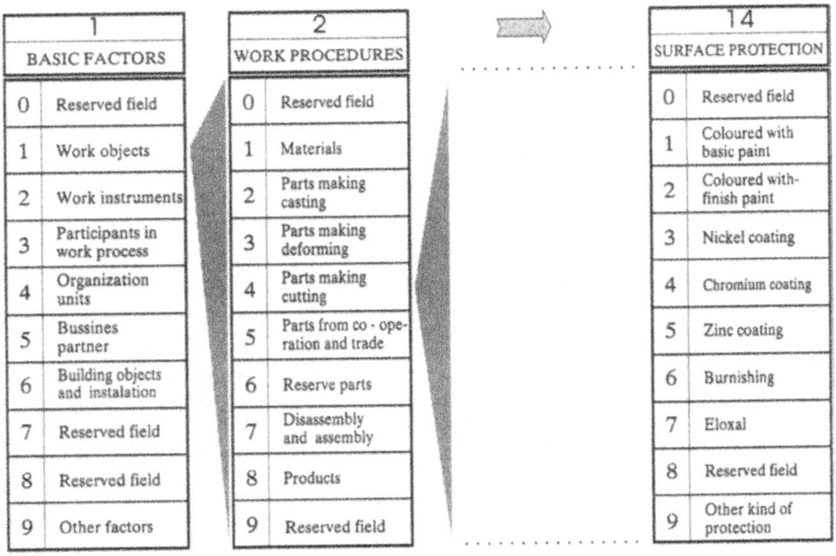

1	
BASIC FACTORS	
0	Reserved field
1	Work objects
2	Work instruments
3	Participants in work process
4	Organization units
5	Bussines partner
6	Building objects and instalation
7	Reserved field
8	Reserved field
9	Other factors

2	
WORK PROCEDURES	
0	Reserved field
1	Materials
2	Parts making casting
3	Parts making deforming
4	Parts making cutting
5	Parts from co - operation and trade
6	Reserve parts
7	Disassembly and assembly
8	Products
9	Reserved field

14	
SURFACE PROTECTION	
0	Reserved field
1	Coloured with basic paint
2	Coloured with-finish paint
3	Nickel coating
4	Chromium coating
5	Zinc coating
6	Burnishing
7	Eloxal
8	Reserved field
9	Other kind of protection

FIGURE 5

field \ character	BASIC GROUP OF FACTORS	PATH OF GETING FACTORS	BASIC DIVISION	BASIC DIMENSION RATIO	BASIC DIMENSION	FORM OF BASIC OUTER SURFACE	FORM OF BASIC INTERNAL SURFACE	SPECIAL OUTER SURFACE	SPECIAL INTERNAL SURFACE	KIND OF MATERIAL	SHAPE OF HALFPRODUCTS	KIND OF HEAT TREATMENT	SURFACE QUALITY	SURFACE PROTECTION
	1	2	3	4	5	6	7	8	9	10	11	12	13	14
0			TURNING, SYMETRIC WITHOUT GEARING	L/D< 0,1	D< 6	Smoothly	Without	Without	Without	Steels with [illegible] composition	Round bars	Without heat treatment	N12 50 µm	Without protection
1	WORK PIECE			0,1< L/D< 0,5	6< D< 10	With step from one side	Center neat	Milling surface	Axial holes or opens without division	Carbon steels	Profile bars	Hardening and annealing	N111 25 µm	Painting with basic colours
2				0,5< L/D< 1	10< D< 16	With step from both side	Smoothly or with step from one side	Impressing surface	Axial holes or opens with division	Alloy steels	Pipes	Carburising	N10 12,5 µm	Japan
3				1< L/D< 2	16< D< 25	1 + taper	With step from both side	Knurling surface	Radial holes or opens	Steels for automate	Profile I,T,L,U	Induction hardening	N9 6,3 µm	Nickel coating
4		PARTS MAKING CUTTING		2< L/D< 5	25< D< 50	2 + taper	1 + taper	Coil rolling	combination 1, 2 i 3	Easy metals	Tins, tapes, plates	Heating	N8 3,15 µm	Chromium coating
5				5< L/D< 10	50> D> 100	1,2,3,4 + profile surface	2 + taper	Gear rolling	Stew holes or opens	Coloured metals	Parts from casting	Normalizacija	N7 1,6 µm	Zinc coating
6				L/D> 10	100< D< 150	0,1,3 + coil	1,2,3,4 or 5 + coil	1 + 2	Holes or opens with coil	Plastic materials	Parts from [illegible] and stamping	Nitriranje	N6 0,8 µm	Burnishing
7					D> 150	2,4,5 + coil	1-6 + vertical flutes	3 + 4	Polygon opens	Bakelite	Arcing parts	Alitiranje	N5 0,4 µm	Eluxel
8						0-6 + small, grinding	Holes or open with ratio l/d > 5	3 + 5		Tekstolite	Assembly		N4 0,2 µm	Lubrication
9						0-8 + small, rolling				Other materials	Other		N3 0,1 µm	5 + 2

FIGURE 6

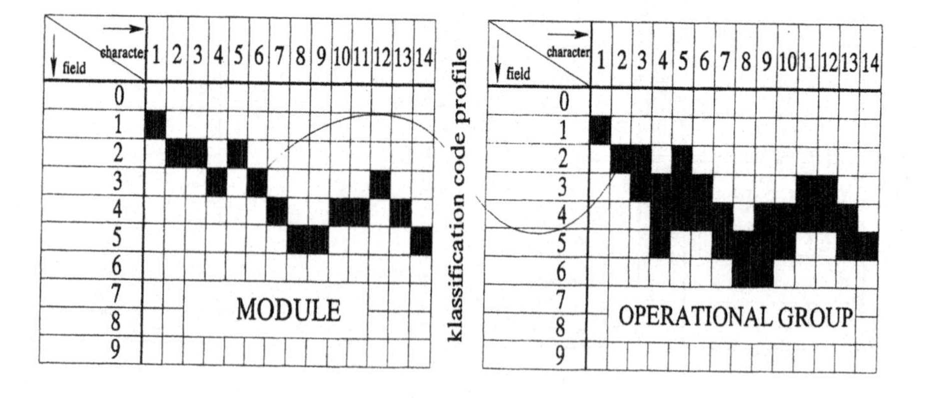

FIGURE 7

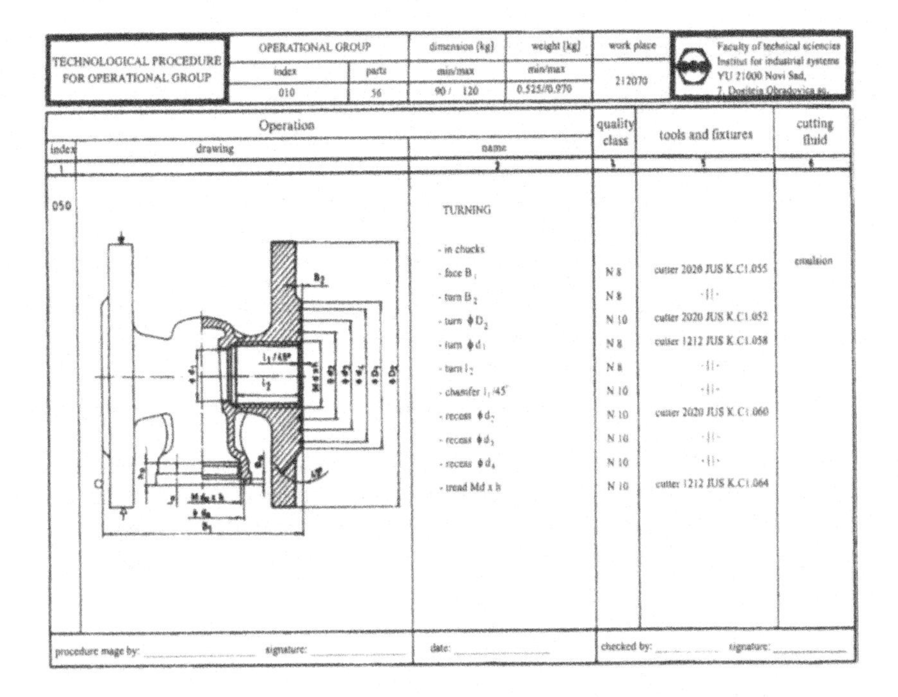

TECHNOLOGICAL PROCEDURE FOR OPERATIONAL GROUP	OPERATIONAL GROUP		dimension (kg)	weight [kg]	work place	Faculty of technical sciencies Institut for industrial systems YU 21000 Novi Sad, 7. Dositeja Obradovica sq.
	index	parts	min/max	min/max	212070	
	010	56	90 / 120	0.525/0.970		

Operation index	drawing	name	quality class	tools and fixtures	cutting fluid
1		2	3	5	6
050		TURNING			
		- in chucks			emulsion
		- face B_1	N 8	cutter 2020 JUS K.C1.055	
		- turn B_2	N 8	-''-	
		- turn ϕD_2	N 10	cutter 2020 JUS K.C1.052	
		- turn ϕd_1	N 8	cutter 1212 JUS K.C1.058	
		- turn l_2	N 8	-''-	
		- chamfer l_1 /45°	N 10	-''-	
		- recess ϕd_2	N 10	cutter 2020 JUS K.C1.060	
		- recess ϕd_3	N 10	-''-	
		- recess ϕd_4	N 10	-''-	
		- tread Md x h	N 10	cutter 1212 JUS K.C1.064	

procedure mage by: ______ signature: ______ date: ______ checked by: ______ signature: ______

FIGURE 8. Example of technology for complex part.

TECHNOLOGY PROCEDURE FOR GROUP PART	OPERATIONAL GROUP index. 010 name: PERTS, ROTATING, SHORT	page/pages: 3/12	Faculty of technical sciencies Institut for industrial systems YU 21000 Novi Sad, 7. Dositeja Obradovica sq.

part	operation index	operation work place	pass name	pass dimension	tools and fixtures	cutting condition i / d	v / n	s / s_z	a / i	time structure t_{pz}: t_{pzo}	t_{pzd}	t_{ii}: t_{io}	t_{ip}	t_{id}
1	2	3	4	5	6	7	8	9	10	11	12	13	14	15
605157	050	212070	TURNING											
			- in chucks										0,50	
			- face	ϕ40 x 65	cutter 2020 JUS K.C1.055	25	750	0.25	1			0.13	0.15	
			- face	ϕ75 x 3		15	500	0.25	1			0.12	0.05	
			- turn	ϕ75 x 10	cutter 2020 JUS K.C1.052	12	500	0.40	2			0.12	0.05	
			- turn	ϕ20 x 25		30	1000	0.30	2			0.20	0.10	
				ϕ20 x 3	cutter 1212 JUS K.C1.058	5	500	0.10	1			0.10	0.05	
			- recess		cutter 2020 JUS K.C1.060							0.87	0.50	0.23
										25		1.60		
	060	216410	DRILLING											

	date:	time structure by:	date:	verified by:	date:	page/pages:

FIGURE 9. Example of technology for part of operational group.

3.3. Step 3: Design Of Work Units

The following load/capacity ratio enables determination of the flow type in accordance with the material flows model shown in Figure 3:

$$k^{*}_{ser} = \frac{\text{load}}{\text{capacity}} = \frac{\sum_{i=1}^{i=m}\left(\sum_{j=1}^{j=k} q_j \cdot t_{ii}^{(j)}\right)}{K_e} \gtrless 1,$$

Succeeding steps in group flow design involve analyzing possibilities for making group large enough for which proportion load/capacity:

$$T_i \gtrless K_e$$ (T_i [time units/period of time] - load of subsystem in observation),

is on a level that allows creation of an independent part of system structures. The case of $T_i < K_e$ for more sections of flow show a need for unification of more similar flows in independent entities until we have $T_i \geq z \cdot K_e$ (z = 1, 2, ... n - real number), for larger number of flows section in observation, which is shown on Figure 10.

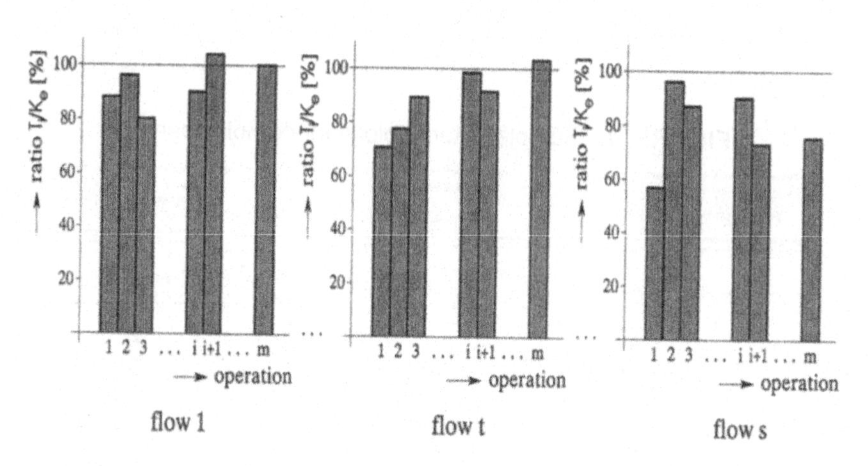

FIGURE 10

Unification of flow characteristics is ensured by connecting operational groups with similar classification code profile, in which work pieces, based on similarity, have similar production processes and manufactured on the same work centers. Groups of work pieces and flows it generates, satisfying necessary conditions, separate into groups in which all projected operations for all work pieces are started and finished.

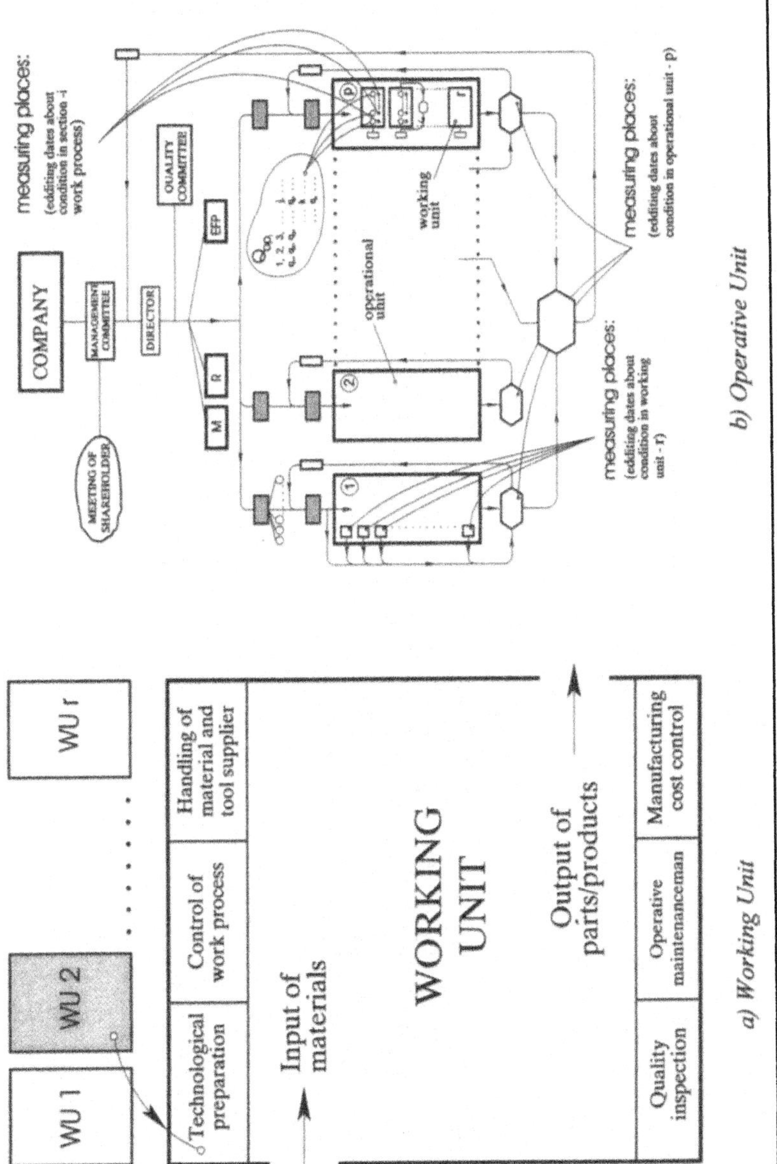

a) *Working Unit*

b) *Operative Unit*

FIGURE 11

Necessary conditions for ensuring responsibility of effect, quality and dead lines - than complete manufacture of work pieces in the part of manufacture program, is getting by transfer to the determine working process (instructionally, directionally, handling the material, preparing tools, quality inspection, operational maintenance and cost control), as shown on Figure 11. On the level of working unit.

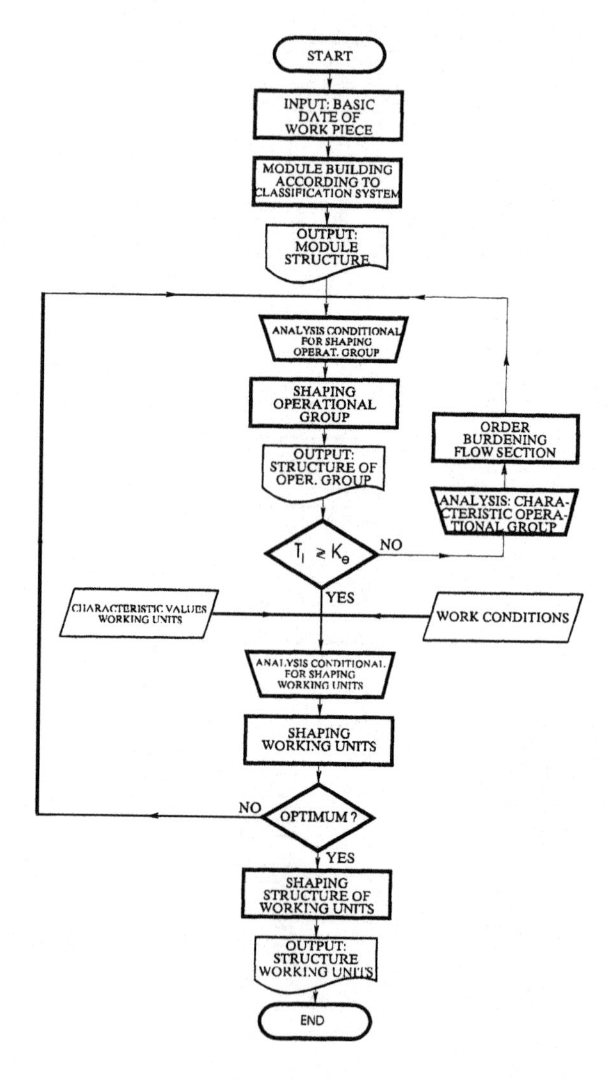

FIGURE 12

The group approach enabled the development of a new approach for designing and managing industrial systems based on two types of organizational units (Figure 11). The *working unit* is a part of the structure capable of taking responsibility for complete manufacturing of groups of similar parts; a part technologically equipped with needed and sufficient process participants. Group of working units make up an *operative unit* that is capable of complete performance of a certain program of the company. A group of operating units make up the company. Experience with real programs, with need for real time solutions, led to development and use of a software tool, referred to as **APOPS-08** at **IISE,** and shown as a flow chart in Figure 12.

4. AN APPLICATION EXAMPLE AND RESULTS

The manufacturing program considered here contains 350 different electric equipment products for the car industry. We briefly illustrate here the procedure and results of a reengineering effort in accordance with IISE approach.

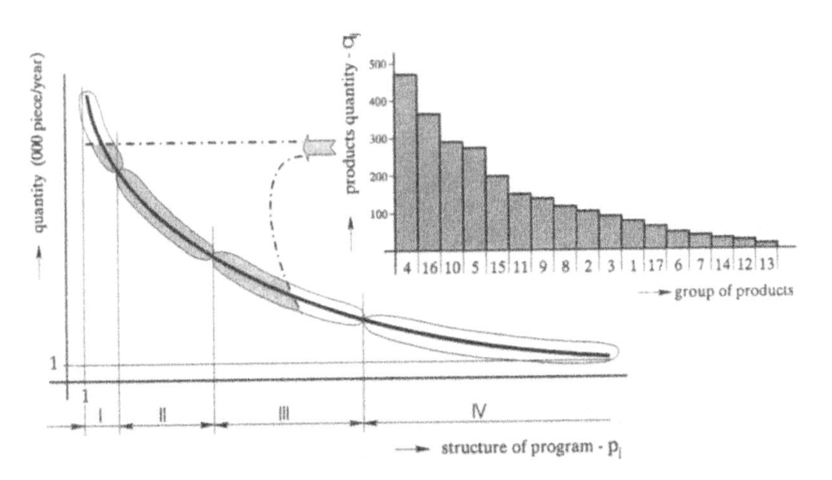

FIGURE 13

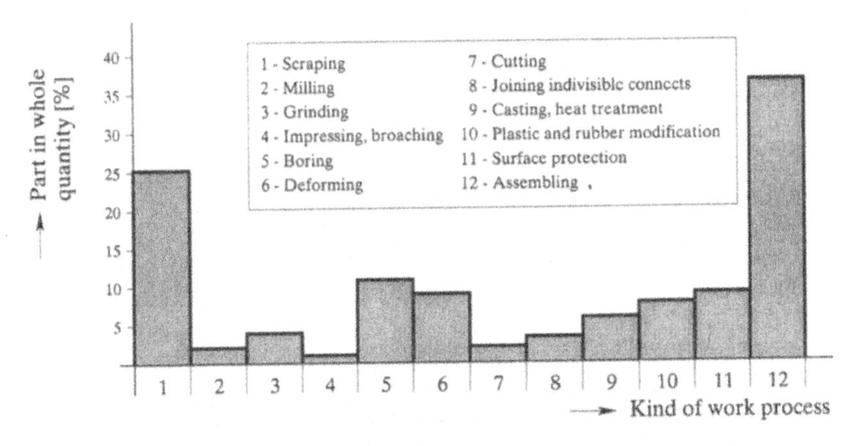

FIGURE 14

First, the relation between structures and quantity. It occurs aound fields II and III and a smaller part of field I, as shown in Figure 13. The occurrence of various types of operations is shown in Figure 14.

A detailed analysis of characteristics of parts in production was conducted as shown in Table 1.

TABLE 1.

OPERATION		Quantity %
Cutting	Rotational Parts: no gear operation	90
	Rotational Parts: with gear operation	4
	Housing parts	5
Analysis of Rotational Parts	By diameter	
	D < 6 mm	40
	6 < D < 10 mm	30
	10 < D < 60 mm	20
	By length	
	1 < L/D < 2	25
	2 < L/D < 5	35
	5 < L/D < 10	20
Heat treatment	Parts without heat treatment	85
	Hardening and annealing parts	5
	Induction hardening parts	3
	Carbonizing and hardening parts	3
...		
Surface protection	Parts without surface protection	75
	Parts protection with zinc	20
	Parts protection with nickel	2

Analysis Of ($T_{cp} / \Sigma t_{ii}$) Relationships.
This ratio (where T_{cp} = real lead time in system, and Σt_{ii} = sum of operational times) was computed for a representative sample of finished job orders, as shown in Ttable 2.

Analysis Of Materials Flows.
Analysis of materials flows as shown here were complex since the system has been designed according to *processes approach* in shaping given structures and *individual*

approach in shaping flows. The material flows and other analyses clearly show a need for reengineering to a *group approach* and suitable organization structures.

TABLE 2.

ASSEMBLY PRODUCT	Order	Start Date (S)	End Date (E)	T_{cp}= (E-S) Hrs	Σt_{ii}	$T_{cp} / \Sigma t_{ii}$
ST -085.2	3210	07.04.'9 3	09.07.'93	10.968	122,00	8.990
MB - 40	8521	16.04.'9 3	18.05.'93	9.528	74,11	12.856
RPS - 16	3142	10.01.'9 2	13.01.'93	8.832	79,36	11.073
:	:	:	:	:	:	:
BN - 16	8667	16.09.'9 2	16.02.'93	3.600	36,91	9.753
Machining						
Motor	3429	06.05.'92	12.05.'93	8.904	13,72	64.879
Shaft	4745	31.03.'93	30.07.'93	2.856	7,05	40.510
Cap	8350	20.03.'92	19.08.'92	3.600	7,10	50.704
Cover	8351	11.11.'92	23.06.'93	5.328	4,50	118.400

In the next step, based on profiles of classification index, **306** operational groups in processing and **120** in assembling were designed. At the same time, based on load/capacity ratios and the capability of individual existence, based on equipment and participants needed to make a viable group in claimed period and quality, and other conditions which provide individual existence, finally 7 ***working units*** were designed, which were capable of independent existence. The basic structure before and after reengineering are shown in Figure 16.

Other effects due to reengineering were decrease in set up & inter-operation times by factors of 3 to 12 in total lead times.

It also led to simplified material and information flows, which ensured better management and release of job order in shorter, equal time intervals:

	Prodn. cycle T_{CP}		RATIOS $T_{CP} / \sum t_{ii}$		Capacity T_{CP}		Input variety for N orders N!	
	Before	After	Before	After	Before	After	Before	After
Factory 1	112	14	75.020	25	2,46	8,54	250!	5!
Factory 2	520	22	110	12	4,52	15,12	210!	7!
Factory 3	235	18	9.300	45	0,51	2,65	1.350!	30!
:	:	:	:	:	:	:	:	:
Factory 25	95	4	155.25	250	2,47	12,05	450!	8!

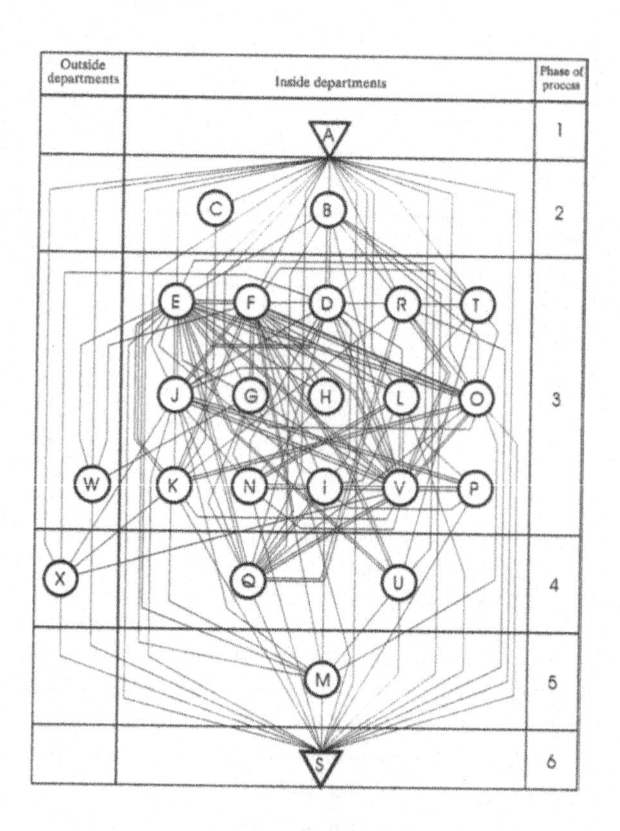

FIGURE 15

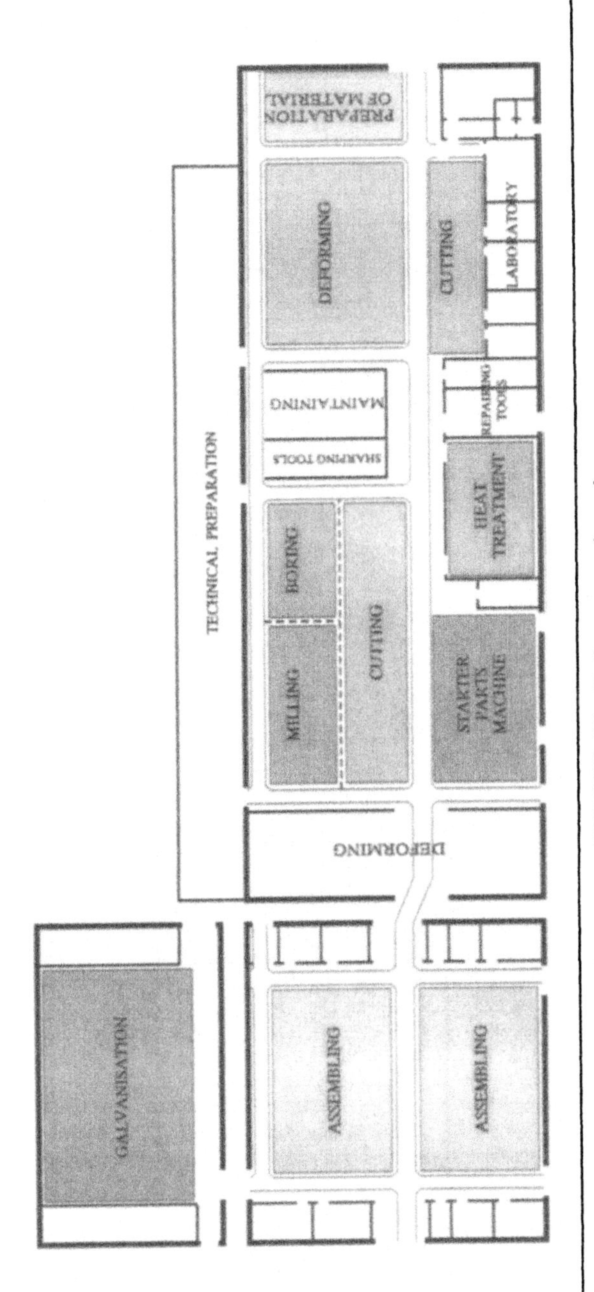

FIGURE 15. Before reengineering.

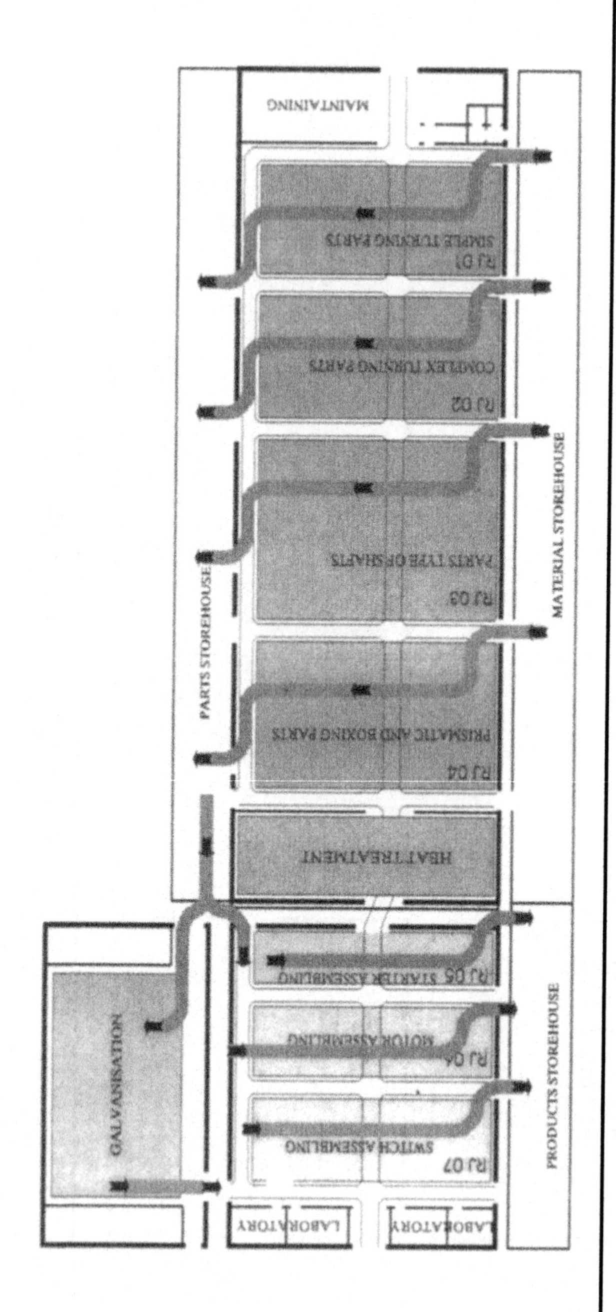

FIGURE 16. After reengineering.

The effects also included:

- Decrease in unfinished manufacture, by factors of 4 -15 times;
- Increase in turnover ratios, 10 to 12 times;
- Reduction of manufacturing lead times, by 8 -25 times.
- Providing good conditions for team work, higher responsibility for group efforts, primarily for quality, better motivation level within teams, etc.;
- Better operating plan execution, upto (95-100)% levels;
- Higher suitability for automation towards flexible automated cells;

So far, the rengineering appraoch developed at ***IISE*** has been found to be very effective, and it should be possible to apply this approach in manufacturing programs elsewhere. Finally, it is necessary to emphasize that this approach presents the basis for reengineering of other company functions (top management, marketing, purchasing, etc.) for developing effective, total production structures.

REFERENCES

Sokolovski, A. P., 1949, *Kurs Technologii Masinostroenia*, Masigz, Leningrad.

Mitrofanov, S. P., 1965, *Naucnie Osnovi Technologiceskoj Podgotovki Grupovogo Proizvodstva*, Masinostroenie, Moscow.

Opitz, H., 1965, *Moderne Productionstechnik*, W. Giradet, Essen.

Opitz, H., 1965, *Werstücksystematik Und Teilefamilien - Fertigung*, W. Giradet, Essen.

Burbidge J. L., 1978, *The Principles of Production Control*, Macdonald and Evans Ltd., Plymouth.

Burbidge J. L., 1978, *The Introduction Of Group Technology*, Heineman, London.

Burbidge J. L., Falster, P., Riigs, J., 1989, Integration audit, *Computer Integrated Production Systems*, 148-161.

Burbidge, J. L. and Zelenović, D., 1983, Using PFA. to plan for a new factory, *Material Flow*, 1, 129-140.

Zelenović, D., 1992, *Toward Fully Integrated Functions Of Production Systems (new philosophy of production - personal aproach)* (in Serbian), Academy of Sciences and Arts of Vojvodina, Novi Sad.

Zelenović, D., 1987, *Design Of Production System* (in Serbian), Naucna knjiga, Beograd.

Zelenović, D., 1984, *Production Control* (in Serbian), Naucna knjiga, Beograd.

Zelenović, D., 1996, *Technology Of Organization Of Industrial Systems - Enterprises* (in Serbian), Naucna knjiga, Beograd.

Zelenović, D., Cosić, I., Maksimović, R., 1991, Ein Beitrag Zur Erforchung Der Bedingungen Für Die Herabsetzung Des Komplexitätsgrads Von Produktionssystemstrukturen, wt *Werkstattstechnik*, 11/91, 656-660, Springer - Verlag, Berlin.

Zelenović, D., Burbidge, L. J., Cosić. I., Maksimović, R., 1995, The Division of Large Complex Production Systems into Indeependent Autonomous Units, *Proceedings of*

Cosić, I., 1983, Contribution To Development of Production Structures with Higher Level of Flexibility, Ph. D. thesis (in Serbian), Novi Sad.

Maksimović, R., 1989, Investigation of Working Characteristics of Effective Production Structures, M. Sc. Thesis (in Serbian), Novi Sad, 1989.

Cvijić, M., 1994, Contribution to Deveelopment of Effective Production Systems On the Principles of Ability for Independent Existence, M. Sc. Thesis (in Serbian), Novi Sad.

AUTHORS' BIOGRAPHY

Dragutin Zelenović is Professor and Corresponding member of Serbian Academy of Sciences and Arts. Since 1992 he has been Professor of Industrial Systems, Effective Management and Organization of Production Systems at University for Peace established by United Nations. He has a Ph.D. degree from Faculty of Mechanical Engineering, Novi Sad. From 1957 to 1966 he worked at the Cutting Tool Factory on engineering jobs in product design and production control. In 1966, he moved to University of Novi Sad, Faculty of Mechanical Engineering. From 1975 to 1989 he served as Dean of the Faculty of Technical Sciences and Rector of the University of Novi Sad. Besides involvement in more than 25 major research projects, he has published 155 papers in Yugoslavia, 25 reviewed papers abroad and 12 books, all of which have been cited in the papers of numerous authors.

Ilija Cosić is Professor at Faculty of Technical Sciences University of Novi Sad covering the areas of assembly systems and design of production systems. He has a Ph.D. from the same institution, where, from 1983 to 1988 he was senior lecturer. Professor Cosić is also a managing director of Institute of Industrial Systems which belongs to Faculty of Technical Sciences in Novi Sad. He is project manager of one scientific research project "Investigation and the Development of Computer Integrated Robotised Flexible Production Systems". Ilija Cosić is the author of 12 books and he has published over 150 research papers.

Rado Maksimović is research assistant at Faculty of Technical Sciences in Novi Sad for Production Systems, Production Control and Organization. He has an MS degree from the University of Novi Sad. He has been involved in more than 15 research projects in design of production systems, group technology and flexible automation, and he has published about 100 papers in Yugoslavia and abroad. He was the designer of a flexible multifingered robot hand developed at Institute of Industrial Systems in Novi Sad in cooperation with USC, Los Angeles. For this design he received two gold medals at: 1st International robot Olympics, Glasgow '90, and EUREKA'90, Brussels.

APPENDIX

Bibliography: Works of John L. Burbidge

BOOKS

The following pages list the books, papers, reports and unpublished reports stored at Cranfield University Library.

Burbidge, J.L., 1960, *Standard Batch Control*, Macdonald and Evans, Plymouth.

Burbidge, J.L., 1962, *The Principles of Production Control*, (4th edition), Macdonald and Evans, Plymouth.

Burbidge, J.L., 1968, *Group Technology*, Turin International Centre, Turin.

Burbidge, J.L. (ed.), 1970, *Group Technology: Proceedings of an International Seminar held at the Turin International Centre from 8th to 13th September 1969*, Turin International Centre, Turin.

Burbidge, J.L., 1971, *Production Planning*, Heinemann, London.

Burbidge, J.L., 1975, *The Introduction of Group Technology*, Heinemann, London.

Burbidge, J.L., 1975, *Final Report on the Study of the Effects of Group Technology Production Methods on the Humanisation of Work*, Turin International Centre, Turin.

Burbidge, J.L. (ed.), 1976, *Effects of group production methods on the humanisation of work: proceedings of an international seminar held at the Turin International Centre in July 1975*, Turin International Centre, Turin.

Burbidge, J.L., 1976, *Group Production Methods and Humanisation of Work: The Evidence in Industrialised Countries*, Research series 10, ILO, Geneva.

Burbidge, J.L., 1979, *Group Technology in the Engineering Industry: A Report on Research, financed by the Science Research Council by Birmingham, Bradford and Salford Universities, and by the London School of Business Studies*, Mechanical Engineering Publications, London.

Yoshikawa, H. and Burbidge, J.L. (eds.), 1987, *IFIP TC 5/WG 5-7 Working Conference on New Technologies for Production and Management Systems, Tokyo*, Elsevier, Amsterdam.

Burbidge, J.L., 1987, *IFIP Glossary of Terms Used in Production Control*, Elsevier, Amsterdam.

Burbidge, J.L., 1989, *Production Flow Analysis for Planning Group Technology*, Clarendon Press, Oxford.

Companys R., Falster, P. and Burbidge, J.L., (eds.), 1990, *Databases for Production Management: Proceedings of the IFIP Working Group 5-7 on Design, Implementation and Operation of Databases for Production Management*, Barcelona, Spain, 10-12 May 1990, Elsevier, Amsterdam.

Burbidge, J.L., 1995, *Period Batch Control*, Clarendon Press, Oxford.

JOURNAL ARTICLES & CONFERENCE PAPERS

Burbidge, J.L., 1958, A new approach to production control, *I.Prod.E. Journal*, 7, 5, 2-16.

Burbidge, J.L., 1959, A new approach to the batch quantity decision, *Productivity Measurement Review*, 17, 27-40.

Burbidge, J.L., 1959, Integrated control, *The manager*, November, 1-6.

Burbidge, J.L., 1960, Are our stocks really necessary?, *The Manager*, December.

Burbidge, J.L., 1961, The economic batch quantity - is it economic?, *The Cost Accountant*, 39, 10, 350-352.

Burbidge, J.L., 1961, The new approach to production, *The Production Engineer*, December, 3-19.

Burbidge, J.L., 1963, Production flow analysis, *The Production Engineer*, December, 2-11.

Burbidge, J.L., 1964, The case against the economic batch quantity, *The Manager*, January.

Burbidge, J.L., 1964, A new classification and coding of management, *Scientific Business*, November.

Burbidge, J.L., 1969, Production flow analysis, *The Production Engineer*, April/May.

Burbidge, J.L., 1971, Is conventional stock control outmoded?' *Industrial Purchasing News*, September, 34-36.

Burbidge, J.L., 1972, Group technology - the state of the art, *CME*, February 1973, 74-76.

Burbidge, J.L., 1972, Production flow analysis, *Work Study*, 21, 8.

Burbidge, J.L., 1973, Aida and group technology, *International Journal of Production Research*, 11, 4, 315-324.

Burbidge, J.L., 1973, Problems when running a factory in a developing country, in *Proceedings 3rd convegno sul organizzazione aziendale*, 4-5 October, Turin, Italy.

Burbidge, J.L., 1977, A manual method of production flow analysis, *The Production Engineer*, October.

Burbidge, J.L., 1978, Whatever happened to GT?, *Management Today*, September, 87-90.

Burbidge, J.L., 1978, What is wrong with Britain's big companies?, *The Director*, July, 52-55.

Burbidge, J.L., 1978, The case against computerised operation scheduling, *Production Management and Control*, 6, 3, 2-6.

Burbidge, J.L., 1979, Garbage in, Garbage out, *Production Management and Control,* 1-3.

Burbidge, J.L., 1980, What is wrong with materials requirement planning, *The Production Engineer*, October.

Burbidge J.L., 1981, The simplification of material flow systems, in *6th International Conference on Production Research*, 24-26 August, Novi Sad and *International Journal of production Research*, 20, 3, 339-347.

Burbidge, J.L., 1981 New methods of organisation to improve production efficiency, in Davies, B. J. (ed.), *22nd International Machine Tool Design and Research Conference* (MATADOR), 16-18 September 1981.

Burbidge, J.L., 1981, Britain's counter-productive plants, *Management Today*, November, 86-89, 166.

Burbidge, J.L., 1981, Cutting stock for survival, *Production Engineer*, April, 29-31.

Burbidge, J.L., 1982, The Japanese system, *Production Management and Control*, 10, 1, 1-5.

Burbidge, J.L., 1982, Comparison of Kanban and other production control systems, *Production management and control*, 10, 3, 15-23.

Burbidge, J.L., 1983, Production flow analysis, *Institute of Production Control Journal*, May/June, 1-10.

Burbidge, J.L., and Zelenovic, D., 1983, Using PFA to plan GT for a new factory, *Material Flow*, 1, 129-140.

Burbidge, J.L., 1983, Five golden rules to avoid bankruptcy, *The Production Engineer*, October, 13-14.

Burbidge, J.L., 1984, Lessons from Mondragon's success, *Initiatives*, 8, February, 26-28.

Burbidge, J.L., 1984, A classification of production system variables, in Hubner (ed.), *Production management systems: strategies and tools for design: proceedings of the IFIP WG 5-7 working conference on strategies for design and economic analysis of computer-supported production management systems*, 28-30 September 1983, Vienna.

Burbidge, J.L., 1984, Production control for flexible production systems, in Doumeingts, Guy and Carter, William A. (eds.), *Advances in production management systems: production management systems in the eighties: IFIP WG 5-7 Working Conference on Advances in Production Management Systems, APMS '82*, 24-27 August 1982, Bordeaux.

Burbidge, J.L. and Dale, B.G., 1984, Planning the introduction and predicting the benefits of GT, *Engineering costs and production economics*, 8, 117-128.

Dale, B.G., Burbidge, J.L., Cottam, M.J., 1984, Planning the introduction of group technology, *International Journal of Operations and Production Management*, 4, 1, 34-48.

Burbidge, J.L., 1985, The design of production systems, in Ahmed, S. I. (ed.), *Production research as a means of improving productivity: proceedings of the 7th International Conference on Production Research*, 1993, Ontario.

Burbidge, J.L., 1984, *Production flow analysis (PFA) for Kongsberg*, Cranfield University.

Burbidge, J.L., 1985, Automated production control with a simulation capability, in Falster, P. (ed.), *Modelling Production Management Systems: IFIP WG 5-7 working conference*, 29-31 August 1984, Copenhagen.

Burbidge, J.L., 1985, Production planning and control - a personal philosophy, in Augustin, S. et al (eds.), *Decentralized production management systems: Proceedings of the IFIP WG 5-7 Working Conference on Decentralized Production Management Systems*, 28-29 March 1985, Munich.

Burbidge, J.L., 1985, Production flow analysis, in Bullinger, H. J. and Warnecke, H. J. (eds.), *Towards the factory of the future: proceedings of the 8th International Conference on Production Research*, 20-22 August 1985, Stuttgart.

Burbidge, J.L., 1986, Period batch control, in *2nd National Conference on Production Research*, September 1986, Edinburgh.

Burbidge, J.L., 1986, A future for the engineering industry, in Leech, D.J. et al. (eds.), *International Conference on Engineering Management: theory and applications*, 15-19 September 1986, Swansea.

Burbidge, J.L., 1986, Production flow analysis and the design of FMSs, in Szelke, E. and Browne, J. (eds.), *Advances in production management systems '85: proceedings of the 2nd IFIP Working Conference on Advances in Production Management Systems, APMS '85*, 27-30 August 1985, Budapest.

Burbidge, J.L., 1986, Economic advantages of group assembly, in McGoldrick, P.F. (ed.), *Advances in Manufacturing Technology: Proceedings of the 1st National Conference on Production Research*, September 1985, University of Nottingham.

Burbidge, J.L., 1987, JIT for batch production using PBC, in Hundy, B.B. (ed.), *Proceedings of the 4th European Conference on Automated Manufacturing*, 12-14 May 1987, Birmingham, UK.

Burbidge, J.L., 1987, Connectance in production management systems, in Bo, K. et al. (eds.), *Computer applications in production and engineering - CAPE '86: proceedings of the 2nd International IFIP Conference on Computer Applications in Production and Engineering*, 20-23 May 1986, Copenhagen.

Burbidge, J.L., 1987, Group technology for better production control, in *2nd World Congress of Production and Inventory Control*, 7-9 April 1987, Geneva.

Burbidge, J.L., 1987, The design of integrated production systems, in Kusiak, Andrew (ed.), *Modern production management systems: proceedings of the IFIP TC/WG 5-7 Working Conference on Advances in Production Management Systems - APMS '87*, 11-14 August 1987, Winnipeg.

Burbidge, J.L., et al., 1987, Integration in manufacturing, *Computers in Industry*, 9, 297-305.

Burbidge, J.L., 1987, Low stock manufacturing, in McGoldrick, P.F. (ed.), *Advances in manufacturing technology II: proceedings of the 3rd National Conference on Production Research*, September 1987, Nottingham.

Burbidge, J.L., 1987, Production control systems for assembly industries, Discussion paper, written for *Production Engineer*.

Burbidge, J.L., 1987, Period batch control, in *Proceedings of the APICS World Congress*, Geneva.

Burbidge, J.L., 1987, Group technology in Yugoslavia, *The Production Engineer*, December, 11-13.

Burbidge, J.L., 1987, Throughput time and profitability, in Yoshikawa, H. and Burbidge, J.L. (eds.), *New technologies for production management systems: proceedings of the IFIP WG 5-7 Working Conference*, 1-3 October 1986, Tokyo.

Burbidge, J.L., 1988, Group Technology - the state of the art, in Worthington, B. (ed.), *Advances in manufacturing technology III: proceedings of the 4th National Conference on Production Research*, September 1988, Sheffield City Polytechnic.

Burbidge, J.L., 1988, Operation scheduling with GT and PBC, *International Journal of Production Research*, 26, 3, 429-442.

Burbidge, J.L., 1988, IM before CIM, in Davies, B.J. (ed.), *Proceedings of the 21st International MATADOR Conference*, 20-21 April, UMIST, Manchester.

Burbidge, J.L., 1988, Period batch control, in Rolstadas, A. (ed.), *Computer-aided Production Management*, Springer-Verlag, Berlin. Chapter 5, 71-77.

Burbidge, J.L., 1989, AI and capacity planning with GT and PBC, in Browne, Jim (ed.), *Knowledge based production management systems: proceedings of the IFIP WG 5-7 Working Conference on Knowledge Based Production Management Systems*, 23-25 August 1988, Galway.

Burbidge, J.L., 1989, Group Technology, in Hashni, M.S.J. (ed.), *Advanced manufacturing technology: proceedings of the 6th Irish Manufacturing Committee*, 31 August - 1 September 1989, Dublin.

Burbidge, J.L., 1989, Production control: the future choice, in Chandler, Jeff (ed.), *Advances in manufacturing technology IV: proceedings of the 5th National Conference on Production Research*, September 1989, Huddersfield Polytechnic.

Burbidge, J.L., 1989, Group technology and period batch control, in McGoldrick, P.F. (ed.), *Extended summaries of papers presented to the 10th International Conference on Production Research*, August 1989, University of Nottingham.

Burbidge, J.L., Falster, P. and Riis, J.O., 1989, Integration audit, *Computer Integrated Manufacturing Systems*, 2, 3, 148-162.

Burbidge, J.L., 1989, Group technology, in Wild, Ray, (ed.), *International Handbook of Production and Operations Management*, Cassell Education Ltd, London.

Burbidge, J.L., 1990, Production planning - a universal conceptual framework, *Production Planning and Control*, 1, 1, 3-16.

Burbidge, J.L., Partridge, J.T. and Aitchison, K.J., 1990, Planning GT for Davy Morris using PFA, *Production Planning and Control*, 2, 1, 59-72.

Burbidge, J.L., 1990, Plant layout for GT and PBC, in Eloranta, Eero (ed.), *Advances in production management systems: proceedings of the 4th IFIP TC5/WG 5-7 International Conference on Advances in Production Management Systems - APMS '90*, 20-22 August, Espoo, Finland.

Burbidge, J.L., 1990, How to introduce GT plus JIT, in Monaghan, J. and Lyons, C.G. (eds.), *Advanced manufacturing technology and systems; proceedings of the Seventh Conference of the Irish Manufacturing Committee*, 29-31 August, Trinity College Dublin.

Burbidge, J.L., 1990, Production control for GT, in Carrie, A.S. and Simpson, I. (eds.), *Advances in manufacturing technology V: proceedings of the 6th National Conference on Production Research*, September 1990, Strathclyde University.

Burbidge, J.L., 1990, A database for the control of stocks - when using GT and single cycle ordering, in Companys, Roman et al. (eds.), *Databases for production management: proceedings of the IFIP WG 5.7 Working Conference on Design, Implementation and Operations of Databases for Production Management*, 10-12 May 1990, Barcelona.

Burbidge, J.L., Falster, P and Riis, J.O., 1991, Why is it difficult to sell GT and JIT to industry? *Production Planning and Control*, 2, 2, 160-166.

Burbidge, J.L., 1991, Period batch control with GT - the way forward from MRP, in *Planning and control through turbulent times: proceedings of the British Production and Inventory Control Society 26th annual conference*, 14-16 November, Birmingham.

Burbidge, J.L., 1992, Change to GT. Process organisation is obsolete, *International Journal of Production Research*, 30, 5, 1209-1219.

Burbidge, J.L., 1992, Production flow analysis (PFA) for planning GT, *Journal of Operations Management: Special Issue on Group Technology and Cellular Manufacturing*, 10, 1, 5-27.

Burbidge, J.L., 1992, Launch sequence scheduling, in Browne, D.J. (ed.), *Technology in manufacturing for Europe 1992: proceedings of the 9th Conference of the Irish Manufacturing Committee, IMC 9*, 2-4 September 1992, University College Dublin.

Burbidge, J.L., 1993, The small company and growth with GT, in Bramley, A. and Mileham, T. (eds.), *Advances in manufacturing technology VII: proceedings of the 9th National Conference on Manufacturing Research*, 6-9 September 1993, University of Bath.

Burbidge, J.L., 1993, The design of lean manufacturing systems, in 26th *International Symposium on Automotive Technology and Automation: dedicated conference on lean manufacturing in the automotive industries*, 13-17 September 1993, Aachen.

Burbidge, J.L., 1993, GT. Where do we go from here? in Pappas, I. A. and Tatsiopoulos, I.P. (eds.), *Advances in production management systems: proceedings of the IFIP TC5/WG5.7 5th International Conference on Advances in Production Management Systems - APMS'93*, 28-30 September 1993, Athens.

Burbidge, J.L., 1993, Reducing delivery times for OKP products, *Production Planning and Control*, 4 ,1, 77-83.

Burbidge, J.L., 1993, Introducing GT, in Orpana, V. and Lukka, A. (eds.), *Production research 1993: proceedings of the 12th International Conference on Production Research*, 16-20 August 1993, Lappeenranta, Finland.

Burbidge, J.L., 1993, Manufacturing and operations management, in Koshal, D. (ed.), *Manufacturing Engineer's Reference Book*. Butterworth-Heinemann. Chapter 15.

Burbidge, J.L., 1994, The use of PBC in the implosive industry, *Production planning and control*, 5, 1, 97-102.

Burbidge, J.L., 1994, The strategy for the introduction of GT, *International Journal of Manufacturing System Design*, 1, 1, 19-29.

Burbidge, J.L., 1994, Group technology and growth at Shalibane, *Production Planning and Control*, 5, 2.

Burbidge, J.L., 1994, The material conversion classification, in *Proceedings of the IFIP WG 5.7 Working Conference*, 16-18 June 1994, Trondheim.

Burbidge, J.L., 1994, Group technology and cellular production, in Case, K. and Newman, S.T. (eds.), *Advances in manufacturing technology VIII: proceedings of the 10th*

National Conference on Manufacturing Research, 5-7 September, Loughborough University of Technology.

Burbidge, J.L., 1994, What must we do to get lean? in Storrar, A.M. (ed.), *Lean production: from concept to product: proceedings of the 11th Conference of the Irish Manufacturing Committee*, 31 August - 2 September 1994, Queens University Belfast

Burbidge, J.L., 1995, Back to production management, *Manufacturing Engineer*, 74, 2, 66-71.

REPORTS

Burbidge, J.L., 1964, Production flow analysis: a technique for finding the families and groups for Group Technology by analysis of the information contained in the component route cards, International Centre for Advanced Technical and Vocational Training, Turin.

Burbidge, J.L., 1970, Turin material flow simulation programme: the perfect flow project, International Centre for Advanced Technical and Vocational Training, Turin.

Burbidge, J.L., 1970, The case against stock control, International Centre for Advanced Technical and Vocational Training, Turin. First presented to the 5th European Conference of the British Production and Inventory Control Society in Edinburgh, October 1970.

Burbidge, J.L., 1970, Period batch control, International Centre for Advanced Technical and Vocational Training, Turin.

Burbidge, J.L., 1971, The Olivetti OFF 50 project - an application of PFA, A case study, International Centre for Advanced Technical and Vocational Training, Turin.

Burbidge, J.L., 1972, Group technology: a study of the effects of group production methods on the humanisation of work, Background paper no. 8, International Centre for Advanced Technical and Vocational Training, Turin.

Burbidge, J.L., 1972, Using AIDA to plan a new production system, A case study, International Centre for Advanced Technical and Vocational Training, Turin.

Burbidge, J.L., 1973, Research report: Production Flow Analysis at Black and Decker - Part III, factory flow analysis, A case study, International Centre for Advanced Technical and Vocational Training, Turin.

Burbidge, J.L., 1973, Spare parts management, International Centre for Advanced Technical and Vocational Training, Turin.

Burbidge, J.L., 1973, Production flow analysis on the computer, International Centre for Advanced Technical and Vocational Training, Turin. Paper presented at the Third Annual Conference of the G.T. Division of the Institute of Production Engineers at Hallam Tower Hotel, Sheffield, 20/21 November 1973.

Burbidge, J.L., 1974, Why has management been slow to adopt GT? Discussion paper, International Centre for Advanced Technical and Vocational Training, Turin.

Burbidge, J.L., 1974, The effect of group production methods on workers participation in decisions, International Centre for Advanced Technical and Vocational Training, Turin. Presented at Proceedings of the ILO Symposium on Workers Participation, 20-30 August, 1974, Oslo.

Burbidge, J.L., 1975, The stability of socio-technological systems, Discussion paper, International Centre for Advanced Technical and Vocational Training, Turin.

UNPUBLISHED PAPERS

Burbidge, J.L., 1969, Group technology bibliography, Papers from 1958 - 1969 Turin International Centre for Advanced Technical and Vocational Training, Turin.
Burbidge, J.L., 1970, The design of production systems, Draft chapter for new book.
Burbidge, J.L., 1970, The perfect flow project, Research report, International Centre for Advanced Technical and Vocational Training, Turin.
Burbidge, J.L., 1973, The case against Adam Smith, Discussion paper, International Centre for Advanced Technical and Vocational Training, Turin.
Burbidge, J.L., 1976, The management of production, Discussion paper.
Burbidge, J.L., 1980, The pseudo-economic batch quantity, Discussion paper.
Burbidge, J.L., 1982, The British business schools - where next? Discussion paper.
Burbidge, J.L., Dale, B.G. and Cottam, M.J., 1983, Planning the introduction of group technology, A case study.
Burbidge, J.L., 1984, Why production flow analysis? Discussion paper.
Burbidge, J.L., 1985, The laws of production system design, Discussion paper.
Burbidge, J.L., 1986, Stock reduction for survival and a future, Discussion paper.
Burbidge, J.L., 1987, The design of integrated production systems, Discussion paper.
Burbidge, J.L., 1988, Organisational aspects of integrated manufacturing, Discussion paper.
Burbidge, J.L., 1988, Production management in the university, Discussion paper.
Burbidge, J.L., 1989, Group technology, Booklet written for DTI.
Burbidge, J.L., 1990, Group technology, Discussion paper.
Burbidge, J.L., 1990, GT - the state of the art, Discussion paper.
Burbidge, J.L., 1991, GT can always be substituted for process organisation, Discussion paper.
Burbidge, J.L., 1991, GT is a universally applicable concept, Discussion paper.
Burbidge, J.L., 1993, The advantages of GT, Discussion paper.
Burbidge, J.L., 1993, The introduction of GT, Discussion paper.
Burbidge, J.L., 1993, An engineer looks at the NHS.
Burbidge, J.L., 1993, Finding GT groups with PFA, Discussion paper.
Burbidge, J.L., Organisation for automation, Discussion paper.
Burbidge, J.L., Material flow systems and production control, Discussion paper.
Burbidge, J.L., The production system variable connectance model, Discussion paper.
Burbidge, J.L., Stock reduction for survival, Discussion paper.

INDEX

abduction 124, 128
absenteeism 158, 242, 250, 259, 261, 262, 263, 266, 267
accountability 6, 405, 413
adaptive resonance theory (ART) 88, 89, 173, 174, 175, 180-182
alternative process plan 104-106, 108, 109-111, 252
analytical model 225, 273, 276, 277, 363, 365, 375, 377, 386, 397
apparel industry 11, 255-272
artificial intelligence (AI), 5, 55, 93, 99, 101, 108-110, 119, 145, 422
assembly 9-12, 21-25, 27, 29, 30, 32, 40-43, 46-50, 52-59, 66, 91, 102, 120, 165-166, 232, 234, 246, 256, 262, 272, 291, 293, 295, 297, 298, 306, 310, 313, 314, 322, 344, 364, 381, 382, 393, 403, 407-409, 413, 440, 441, 446, 447, 452, 453, 482, 483, 492-494, 501, 502, 509-512, 515, 523, 536
assembly cell 11, 12, 122, 255-272, 293, 364, 413, 494
automatic feature recognition 80
automation 2, 11, 14, 15, 110, 121, 153, 221, 255, 338, 339, 344, 381-386, 390-397, 479, 482, 483, 486, 534, 536
backpropagation 86, 92, 182
Baldwin effect 195, 204
bill of material (BOM) 71, 297, 485, 509
bond energy analysis (BEA), 142, 146-150, 155, 160, 162-165, 186
bottleneck 142, 144, 153-155, 241, 261, 312, 339, 446, 458, 460-466, 471-473
breadth of skill 7
Brisch classification system 3, 17, 34, 392
business process reengineering (also see: reengineering) 7, 237, 238, 284, 452
business process 7-10, 12, 152, 221-227, 236-238, 284, 452, 474
capacity 6, 108-109, 115-116, 143-144, 153, 155, 199, 202- 203, 209, 212-213, 219, 224-225, 229, 233, 237, 241-246, 248-252, 275, 278, 286, 293-299, 311-316, 343, 345, 350, 358, 364, 370, 373, 385, 387, 389, 394, 397, 400, 404, 416, 445-446, 451, 453, 458, 460, 462, 466, 471-473, 476, 487, 490, 492, 496, 518, 519, 526, 534
cell design 2, 17, 58, 90, 93, 95, 106, 116, 131, 133, 148, 150, 183, 186-187, 190, 194, 200-207, 210, 217, 258-259, 268, 270, 286, 316, 358, 367, 413
cell formation 58, 76, 82, 83, 93, 106, 131-135, 138-153, 155, 158, 159, 164, 167-168, 175, 181-193, 196-197, 199-200, 202-204, 217-219, 252
cellular layout 1, 2, 6, 8, 9, 11, 12, 14, 16, 37, 112, 114-115, 129, 131, 148, 168, 171, 173, 198, 273-287, 340, 352, 395, 475, 480, 482, 484
cellular manufacturing 15, 16, 21, 33, 56, 58, 76, 79, 91, 92, 93, 131, 135, 141, 144, 149-152, 153, 165-168, 181-185, 199, 202-203, 205, 217-219, 223, 239, 246, 250-251, 253-254, 271, 273, 274, 276, 280, 286-289, 295-296, 299, 302, 304-309,

312, 320-322, 324-325, 334, 336-338, 347, 350, 359, 360-362, 381-382, 391, 393-396, 399, 401, 420-423, 425, 434, 440, 475, 477, 479- 481, 490, 515
chromosome 133, 135, 187-190, 193, 196, 200, 206, 210-211, 217-218
classification and coding 3, 9, 10, 12, 13, 16-17, 21, 35, 37, 55-61, 63, 67-68, 71-75, 77-79, 82, 86, 92, 96-98, 182, 392, 501, 504
cluster analysis 58, 92, 133, 135-138, 141-146, 169, 180, 183, 202
CNC 11, 382, 383
code dictionary 52, 53
CODE 3, 392
coding 3, 4, 39, 40, 47, 48, 53-54, 60, 75, 96, 114, 116, 135, 158, 321, 514, 522
company flow analysis 155, 392, 459
composite part 7
computer aided design (CAD) 3, 10, 13, 17, 36, 40-42, 48, 54, 56, 76-77, 79, 81-82, 91, 93, 100, 103-105, 107-108, 111, 115, 119, 124, 128, 183, 224, 272, 383, 395
computer aided process planning (CAPP) 2, 5, 10, 12, 15-17, 23, 93, 98-99, 101, 103, 107, 110, 112, 114-115, 117-121, 123, 126, 128
computer integrated manufacturing (CIM) 2, 15, 110, 222, 398, 449, 506, 515
computer numerical control (CNC) 11, 45, 85, 114, 120, 166, 199, 382, 383, 393, 461, 466
concurrent engineering 9, 10, 12, 16, 18, 19, 21, 33, 36, 91, 94, 109, 111, 122, 168, 224, 284, 422
cost accounting 1, 7, 478, 506, 513
data base 3-5, 10, 24, 27, 29, 37, 39-43, 48, 54-57, 59, 65, 68-77, 91, 96-99, 101, 104-105, 108, 119-125, 126, 128, 150, 159, 202, 256, 260-263, 265-266, 268, 270, 272, 351, 355-356
database management system 10, 59, 74- 75, 98
decision support system (DSS) 149, 182, 255-259, 263, 268, 271, 336, 453
deduction 126, 127
design cycle, 8, 18-21, 49
design for assembly (DFA) 21-22, 24-27, 30, 32, 40, 43, 47, 50, 54, 56, 91
design for manufacturability (DFM) 9, 12, 15, 18, 21, 23, 34-35, 114, 458, 466, 471, 473
design for manufacturability and assembly (DFMA) 9, 12, 15-16, 19-23, 29, 34
design for service (DFS) 16, 21, 26-27, 29, 34-35
design retrieval 17, 38-42, 79, 90, 114, 184
direct numerical control (DNC) 383
disassembly 22, 27, 29, 30-36
disjoint subclasses 61, 62
economic order quantity (EOQ or EBQ), 7, 301-305, 307, 313
empowerment 410
factory flow analysis 156, 165, 392, 459
fitness value 201
fixture 45-46, 57, 71, 383, 387, 466, 467
flexibility 11, 79, 108, 141, 147, 166, 167, 173, 186-187, 193, 195, 200, 221, 224, 237, 252, 255, 274, 279, 282-283, 287, 295, 299, 303, 309, 311, 321-322, 336, 341, 360, 365, 378, 381, 383, 385-386, 388-391, 394, 396-399, 402, 405, 421, 425,

434, 437, 451, 453, 455, 476, 479, 480-482, 483, 487, 520
flexible automation 2, 11, 15, 121, 381- 386, 391-397, 536
flexible manufacturing 11, 57, 150, 168, 181-184, 203, 222, 338, 385, 386-387, 390, 392-393, 396-400
flexible manufacturing cell (FMC) 184
flow control 7, 294, 297, 311, 317
FMS 11, 57, 150, 168, 181-182, 203, 222, 293, 338, 385-387, 392-394, 396-400, 515
functional layout 6, 151, 165-167, 202, 240, 248-250, 252, 273-278, 282-287, 393, 402, 503
fuzzy logic methods 145, 169, 174, 177, 179-181, 356
generative method (CAPP) 5, 10, 90, 93, 99, 100, 101, 103, 104, 106-109, 118, 122-123
genetic algorithm 12, 133, 145, 181, 185- 190, 192-206, 210, 211, 213, 217-219, 356, 361, 459
graph theory 142, 144
group analysis 78, 133, 155, 156, 158, 160, 164-168, 392, 458-460, 465, 473
group layout 6, 7, 311, 317
human factors 11, 271
human resource management 7, 11, 401, 403, 406, 414, 416, 420, 506
hybrid code 60, 75
induction 124, 530
inventory control 1, 71
inventory level 74, 145, 241, 243, 302, 304, 340, 435, 487, 497, 498, 511
Jaccard similarity coefficient 136-141, 144, 146, 301
just-in-time (JIT) system 2, 7, 222, 255, 256, 294, 298, 303, 305, 315, 325, 409, 425, 435, 456, 493, 494, 495, 501-502, 506, 510-511, 513-515
knowledge-based system (KBS) 356-357
labor 6, 9, 12, 21, 25, 30, 131, 132, 202, 240, 248, 251, 276-283, 285, 303, 327, 339-341, 350, 361, 364-365, 368, 378, 408, 410, 416, 421, 425, 431, 478, 487, 492, 520
labor assignment 278, 283, 364-365, 368
labor flexibility 365, 378
Lamarckian evolution 195-197, 199, 200, 204
lead time, 1, 4, 5, 6, 8, 21, 145, 199, 221, 224, 226, 229-231, 234, 236, 239-241, 243-246, 248-253, 255, 273, 277, 280, 285, 287, 294-297, 309-314, 317-319, 340, 363, 423, 435, 493, 497, 499, 514, 517, 530, 534
lot sizes 114, 121, 199, 241, 242, 248, 257, 277-278, 280, 282, 293, 296, 301, 303, 309, 311, 313-316, 327, 362, 366, 442, 462, 466, 481, 512, 513
machine learning 10, 57, 77, 78, 91-92, 202
machining center 7, 322, 382, 441
maintainable design 4
master production schedule (MPS) 294, 296-297, 309-310, 314-318, 343, 351, 354-358, 450-451, 507, 509, 511
material handling 6, 186, 250, 275, 367, 369, 383, 391-392, 396, 478, 479, 482
materials requirements planning (MRP) 7, 119, 158, 202, 290-299, 305-307, 309-315, 318-319, 320, 343, 351, 355, 357, 360, 444- 445, 450, 501, 506-507, 509-515

mathematical programming 77, 133, 142, 153, 169, 177, 179-180, 185, 226, 347
MICLASS 3, 13, 17, 35, 96, 98, 114-115, 118, 392, 504
MIPLAN 98
module 68, 71, 72, 99, 100, 159-160, 161-163, 264-265, 269, 351, 428-431, 433, 436, 487, 522
MRP II 296, 297, 305
multiple families 82
neural networks 10, 12, 92, 133, 145, 148, 169, 170-171, 175, 177, 181-182, 185, 202, 356
new product development 8, 221, 223, 227-229, 237
object-oriented modeling 59, 61, 63, 64, 66-67, 70-71, 75
operating policy 11, 284, 300, 362-363, 365-369, 372, 375, 377-379
operators 61, 68, 99, 162, 164-165, 168, 240, 246, 250, 255, 257, 259-268, 270-271, 290-292, 347-379, 382-383, 390-391, 405, 413-414, 417-418, 423, 433, 442, 445, 447-452, 454, 483, 494-495, 511, 513
Opitz 3, 13, 17, 35-36, 44, 56, 96, 110, 114, 442-443, 535
optimized production technology / timetable (OPT) 294, 298, 302, 445
overlapped operation 6, 62, 284
part geometry 47, 60, 97, 100, 118, 391
part volume 141, 146
part-family scheduling, 11, 285, 303
partitioning 39, 144, 149, 217, 218, 248, 249, 276-281, 374, 444
part-machine grouping 2, 78, 131-152, 144, 150, 169, 171, 173-182
pattern recognition 2, 10, 12, 38, 55, 77, 92, 145, 152, 180, 183
period batch control (PBC) 7, 11, 295 297, 298, 305, 309, 310, 311 312-319, 344, 363
pooling 253, 276-279, 282, 285, 288, 307, 338
process planning 10, 93, 95, 99-100, 103-105, 107-109, 123, 272
production control 1, 13, 34, 95, 158, 165, 167, 288-291, 295, 302, 304, 307-308, 424, 459, 478, 536
production flow analysis (PFA) 3, 10, 12, 13, 15, 56, 58, 76-78, 82, 86, 106, 109, 112, 114-115, 117, 132-135, 150, 153, 155, 165, 182, 218, 253, 392, 442, 445, 450, 452, 456, 458, 459, 461, 462, 465, 472-474, 535
production planning and control 2, 11, 250, 281, 289-292, 294-299, 302-306, 363, 442, 445, 448, 450, 475, 490
productivity 4, 38, 42, 79, 112, 115, 119, 152, 250, 259, 268, 270, 281, 301, 403, 404, 406, 408, 418, 423, 426, 432, 435, 442, 446-447, 449-450, 453-455, 479, 490, 520
proliferation of parts / processes 2, 4, 10, 112, 114-118, 121, 174, 175
quality control 2, 7, 157, 246, 393, 405, 409, 424, 473, 492, 494, 496
quality function deployment (QFD) 16, 19, 36, 224
queuing networks 10, 226, 228, 233, 237
rank order clustering (ROC) 77, 132, 133, 141-142, 147, 179, 182, 218
rationalization 4
recycling 29, 30, 31, 32, 34, 56
reengineering (also see: business process reengineering) 7, 13,

226, 237, 238, 284, 452, 517, 529, 530, 533, 534
relational data base 10, 71, 99, 104, 119, 120, 125
reorder point 7, 291, 297, 311, 314
retrieving 5, 38, 42, 45, 71, 121, 442
routings 1, 7, 58, 97, 114, 115, 122-128, 133, 140, 142, 157-158, 193-194, 199, 202-203, 242, 246-247, 251-252, 294, 297, 317, 325, 447, 480, 485, 492
run quantity 302
scheduling rule 12, 299-300, 322, 324, 326, 334, 363, 364, 379, 398
scheduling 11-12, 57, 94-95, 104, 108, 152, 204, 218-229, 233-235, 237-238, 246, 250, 255, 268, 271-274, 284-287, 289-290, 294, 296, 299-300, 303-308, 311-330, 334-364, 375, 377-379, 382, 388, 392, 396-400, 412, 417, 445, 448, 450-454, 473, 487
sequence-dependent scheduling 193, 330, 368
sequencing 103-104, 111, 141, 211, 233, 275, 287, 312-314, 317, 322-323, 337, 361, 364-365, 368, 379, 395
setup 6, 114, 121, 132, 142, 144, 145, 156, 193, 234, 242, 243, 248, 250, 275, 276-286, 289, 295, 296, 300-303, 309, 311-315, 317-330, 333-334, 337, 363, 367-368, 374, 388, 444, 446, 474, 492-495, 519
setup reduction 114, 121, 275-276, 279-285, 314
setup time 6, 132, 142, 144, 145, 193, 234, 242, 248, 250, 275, 280, 284, 286, 296, 301, 311-313, 315, 317-318, 321, 324-327, 329-330, 333-334, 363, 444, 446, 492, 493, 519
simplified 15, 22, 97, 157, 295, 299, 330, 391, 393, 404, 410, 411, 458, 472, 473, 482
simplified work flow 6
simulated annealing 185, 204, 218, 307, 326, 338
simulation 2, 10, 11, 12, 14, 27, 29, 44, 57, 146, 234, 237, 241, 256, 261, 263, 266, 268, 270, 271, 272, 273, 274, 275, 276, 277, 278, 281, 282, 285, 286, 287, 288, 300, 304, 314, 324, 325, 327, 328, 336, 338, 363, 364, 365, 366, 372, 373, 377, 378, 379, 448
simultaneous engineering 94
small lot 241, 311, 313, 315, 481, 496
SMED 445
solid models 80, 92, 124, 183
specialization 6, 7, 165, 166, 167, 223, 425, 482
standardization 4, 5, 15, 17, 105, 121, 431, 466, 467, 479, 483, 493
stock control 7, 297
supply chain management 8, 221, 223, 224, 232, 233, 234, 237, 238, 379, 440, 474
synchronization 233, 234, 298, 343, 344, 345, 346, 348, 349, 350, 351, 361, 425
team leader 246, 418, 423, 424, 426, 428, 429, 430, 431, 432, 433, 434, 435, 436, 437, 438, 462
TNO 17, 31, 114, 128
tooling 5, 9, 23, 26, 42, 46, 47, 56, 58, 63, 64, 115, 121, 132, 141, 143, 156, 251, 281, 289, 392, 458, 459, 462, 463, 464, 465, 466, 467, 471, 472, 473, 474
tooling family 464, 465, 466, 467
total quality management (TQM) 222, 255, 409, 410, 495, 7
transfer batches 298, 316, 327, 337, 362, 365, 368, 370, 372, 373, 374

transfer quantity 302
variability 224, 226, 230, 231, 244, 248, 249, 250, 277, 280, 281, 296, 300, 303, 316, 328, 360, 365, 372, 373, 374, 413
variant method (CAPP) 5, 95, 96, 98, 108, 118
variety reduction 4
visual and intuitive approaches 3
workers 21, 38, 132, 156-158, 162, 164, 166-167, 248, 251-252, 277-278, 282, 298, 347, 362, 364-374, 390, 394, 404-408, 413, 416, 423, 441, 447-448, 450-453, 455, 478, 490
work-in-process (WIP) inventory 6, 241, 243-244, 273-274, 277-282, 285, 365, 367, 369-370, 372, 375, 493